DIGITAL SIGNAL PROCESSING APPLICATIONS USING THE ADSP-2100 FAMILY

ANALOG DEVICES TECHNICAL REFERENCE BOOKS

Published by Prentice Hall

- Analog-Digital Conversion Handbook
- Digital Signal Processing in VLSI
- Digital Signal Processing Applications Using the ADSP-2100 Family

Published by Analog Devices

- Nonlinear Circuits Handbook
- Transducer Interfacing Handbook
- Synchro & Resolver Conversion

DIGITAL SIGNAL PROCESSING APPLICATIONS USING THE ADSP-2100 FAMILY

VOLUME 1

by

The Applications Engineering Staff of

Analog Devices, DSP Division

Edited by Amy Mar

PRENTICE HALL, Englewood Cliffs, NJ 07632

Published by Prentice-Hall, Inc.
A division of Simon & Schuster
Englewood Cliffs, New Jersey 07632

The publisher offers discounts on this book when ordered in bulk quantities. For more information, write:

Special Sales/College Marketing
Prentice-Hall, Inc.
College Technical and Reference Division
Englewood Cliffs, New Jersey 07632

Printed in the United States of America

10 9 8 7 6 5 4 3 2

ISBN 0-13-219726-X

Prentice Hall International (UK) Limited, *London*
Prentice Hall of Australia Pty. Limited, *Sydney*
Prentice Hall Canada Inc., *Toronto*
Prentice Hall Hispanoamericana, S.A., *Mexico*
Prentice Hall of India Private Limited, *New Delhi*
Prentice Hall of Japan, Inc., *Tokyo*
Simon & Schuster Asia Pte. Ltd., *Singapore*
Editora Prentice-Hall do Brasil, Ltda., *Rio de Janeiro*

Table of Contents

Page

15. Sonar Beamforming

16. Memory Interface

17. Multiprocessing

Preface

This book is about bridging the gap between digital signal-processing (DSP) algorithms and their real-world implementations on state-of-the-art digital signal processors. Each chapter tackles a specific application topic, briefly describing the algorithm and discussing its implementation on the ADSP-2100 family of DSP chips.

Anyone who wants to understand how a processor optimized for digital signal processing, such as the ADSP-2100, is used to solve a particular problem will find this book informative. The areas addressed include–but are not limited to–traditional signal processing, since graphics and numerical applications also benefit from the features of a DSP processor.

We do not attempt to explain the signal processing theory of any application in full detail. Our readers are assumed to already understand the theory and practice applying to their own areas of interest. *Digital Signal Processing in VLSI,** a companion book in the Analog Devices technical reference set, provides much of the necessary basics. The references listed at the end of each chapter provide a wealth of additional information.

This volume spans topics ranging from the very simple to the moderately complex. Here is a brief summary of each section's contents:

- *Fixed-point arithmetic operations*

  How basic fixed-point arithmetic operations are mapped onto the hardware of the ADSP-2100.

- *Floating-point arithmetic operations*

  How to convert from fixed-point to floating-point representation–and vice versa–and how to perform basic floating-point arithmetic operations using the ADSP-2100. *Block floating-point* operations are discussed in the chapter on fast Fourier transforms.

- *Function approximations*

  How to perform numerical approximations of some useful functions.

*Higgins, Richard J., *Digital Signal Processing in VLSI*. Englewood Cliffs, NJ: Prentice Hall, 1990.

- *Digital filters*

  Implementations of several finite impulse-response (FIR) and infinite impulse-response (IIR) filters that have fixed coefficients. Also described are multirate filters, which change the sampling rate of digitally represented signals. This section also discusses adaptive filters (with time-varying coefficients).

- *One-dimensional fast Fourier transforms*

  Implementations of several one-dimensional fast Fourier transform (FFT) algorithms and the related operations of bit reversal, digit reversal, block floating-point scaling, and windowing. How to optimize the FFT programs for speed.

- *Two-dimensional fast Fourier transforms*

  An implementation of an FFT in two dimensions.

- *Image processing*

  Implementations of several algorithms used in processing digitized images.

- *Graphics*

  A graphics subsystem based on the ADSP-2100, complete with all software routines and support circuitry.

- *Linear predictive speech coding*

  Techniques used to analyze, encode, and synthesize speech signals.

- *Pulse-code modulation*

  An ADSP-2100 implementation of the CCITT standard pulse-code modulation (PCM) algorithm. Encoding and decoding are shown, employing both μ-law and A-law companding methods.

- *Adaptive differential pulse-code modulation (ADPCM)*

  An ADSP-2100 implementation of the CCITT standard ADPCM algorithm. A nonstandard program that is suitable for some applications is also described.

- *Modem algorithms*

  Several algorithms used in implementing high-speed modems.

- *Dual-tone multifrequency coding (DTMF)*

  How to generate and detect the CCITT standard DTMF signals.

- *Sonar beamforming*

  Both software and hardware for a digital beamforming system for passive sonar.

- *Memory interface*

  A design example that shows considerations for implementing an interface between the ADSP-2100 and various types of memory and I/O.

- *Multiprocessing*

  An interface between two ADSP-2100s operating in parallel. Dual-port memory and software issues are addressed.

- *Host interface*

  How to use the ADSP-2100 as a coprocessor to a host CPU, using the Motorola 68000 as an example.

The text provides comprehensive source-code listings, complete with comments and accompanied by explanatory text. A supplementary diskette—furnished with the book—contains the program listings.

ACKNOWLEDGEMENTS

The substance of this book was contributed by the applications engineers of the Analog Devices DSP Division. They designed, developed and tested the software and the hardware systems presented here, drafted the accompanying documentation and reviewed the final publication. Over time, and with feedback from many customers who put these applications to use, the applications group has also refined much of this information. Besides Bob Fine, who heads the group, contributors include: Dan Ash, Chris Cavigioli, Ron Coughlin, Steve Cox, Jeff Cuthbert, Fares Eidi, Cole Erskine, Hayley Greenberg, Matt Johnson, Kapriel Karagozyan, Gerald McGuire, Gordon Sterling and Bruce Wolfeld.

Jim McQuaid provided significant editorial feedback on all chapters; Adele Hastings produced virtually all drawings and layout; Sandra Perry and other Marketing engineers gave comments and input.

Amy Mar

Norwood, Massachusetts

Introduction 1

1.1 OVERVIEW

This book presents a compilation of routines for a variety of common digital signal processing applications based on the ADSP 2100 DSP microprocessor family. These routines may be used as is or they may serve as a jumping-off point for the development of routines tailored to your particular application. Each routine is prefaced by a discussion of the algorithm or data formats underlying the code.

Besides showing the specific applications, the set of routines demonstrates a variety of programming tactics for getting the most performance out of the ADSP-2100 family processors, for example, the proper way to segment loops to utilize the ADSP-2100's cache memory. We believe that readers will benefit from reading every chapter, even if their present application interests concern only a single topic.

The material in this book was originally published as Volumes 1, 2 and 3 of the *ADSP-2100 Family Applications Handbook*. The information in those volumes has been updated and integrated into this book; it supersedes those earlier publications.

1.2 ADSP-2100 FAMILY OF PROCESSORS

This section briefly describes the ADSP-2100 family of processors. For more complete information, refer to the *ADSP-2100 User's Manual*, *ADSP-2101 User's Manual*, and *ADSP-2111 User's Manual*, available from Analog Devices or the *ADSP-2100 Family User's Manual*, available from Prentice Hall and Analog Devices. For the applications in this book, "ADSP-2100" refers to *any* processor in the ADSP-2100 family unless otherwise noted.

The ADSP-2100 is a programmable single-chip *microprocessor* optimized for digital signal processing (DSP) and other high-speed numeric processing applications. The ADSP-2100 chip contains an ALU, a multiplier/accumulator, a barrel shifter, two data address generators and a program sequencer; data and program memories are external. The ADSP-2100A is a pin- and code-compatible version of the original ADSP-2100 fabricated, in 1.0-µm CMOS. It can operate at a faster clock rate than the ADSP-2100.

1 Introduction

The ADSP-2101 is a programmable single-chip *microcomputer* based on the ADSP-2100. Like the ADSP-2100, the ADSP-2101 contains computational units, as well as a program sequencer and dual address generators. Additionally, there are 1K words of data memory and 2K words of program memory on chip, two serial ports, a timer, boot circuitry (for loading on-chip program memory at reset), and enhanced interrupt capabilities. Because the ADSP-2101 is code-compatible with the ADSP-2100, the programs in this book can be executed on these chips as well (some modifications for interrupt vectors may be necessary), although not all programs are designed to make use of the extra features and functions of the ADSP-2101.

1.2.1 ADSP-2100 Architecture

This section gives a broad overview of the ADSP-2100 internal architecture, using Figure 1.1 to show the architecture of the ADSP-2100 processor.

The ADSP-2100 processor contains three full-function and independent computational units: an arithmetic/logic unit, a multiplier/accumulator and a barrel shifter. The computational units process 16-bit data directly and provide for multiprecision computation.

Two dedicated data address generators and a complete program sequencer supply addresses. The sequencer supports single-cycle conditional branching and executes program loops with zero overhead. Dual address generators allow the processor to output simultaneous addresses for dual operand fetches. Together the sequencer and data address generators allow computational operations to execute with maximum efficiency. The ADSP-2100 family uses a modified Harvard architecture in which data memory stores data, and program memory stores both instructions and data. Able to store data in both program and data memory, ADSP-2100 processors are capable of fetching two operands on the same instruction cycle.

The internal components are supported by five internal buses.

- Program Memory Address (PMA) bus
- Program Memory Data (PMD) bus
- Data Memory Address (DMA) bus
- Data Memory Data (DMD) bus
- Result (R) bus (which interconnects the computational units)

Introduction 1

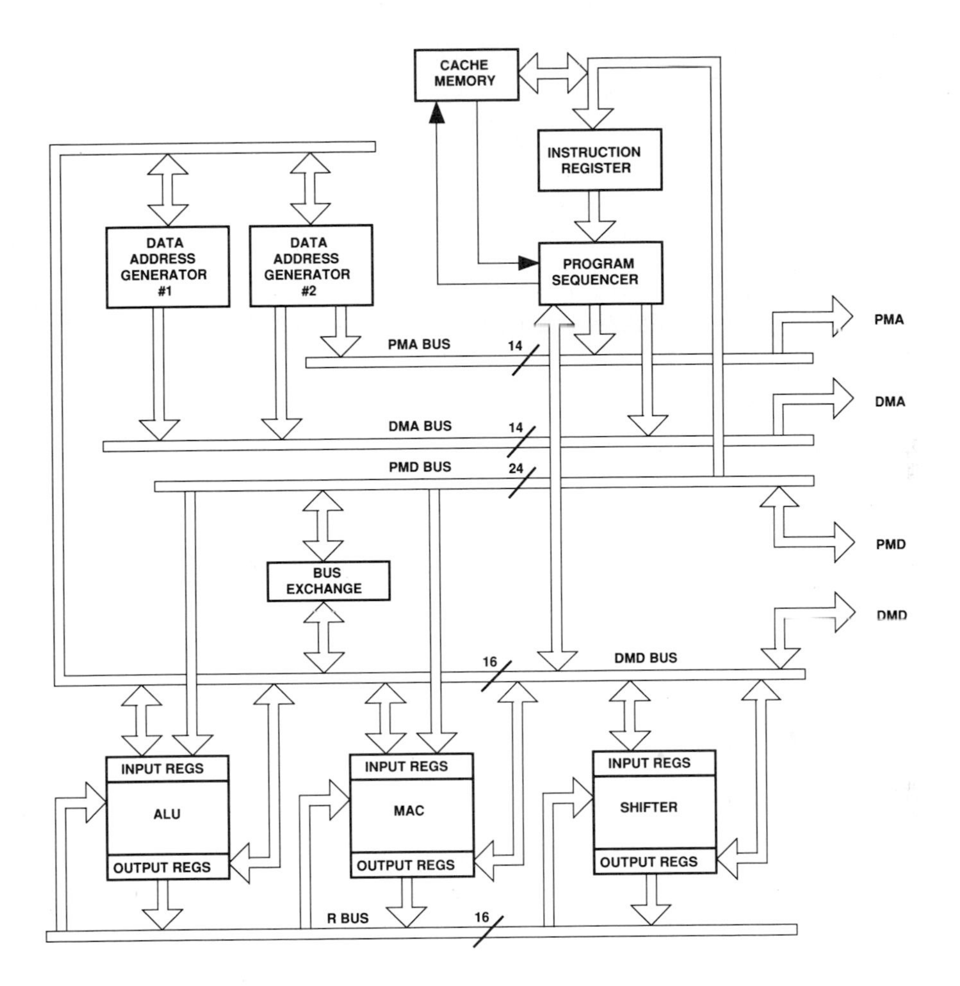

Figure 1.1 ADSP-2100 Internal Architecture

On the ADSP-2100, the four memory buses are extended off-chip for direct connection to external memories.

The program memory data (PMD) bus serves primarily to transfer instructions from off-chip memory to the internal instruction register. Instructions are fetched and loaded into the instruction register during one processor cycle; they execute during the following cycle while the next instruction is being fetched. The instruction register introduces a single level of pipelining in the program flow. Instructions loaded into the instruction register are also written into the cache memory, to be described below.

The next instruction address is generated by the program sequencer depending on the current instruction and internal processor status. This address is placed on the program memory address (PMA) bus. The program sequencer uses features such as conditional branching, loop counters and zero-overhead looping to minimize program flow overhead. The program memory address (PMA) bus is 14 bits wide, allowing direct access to up to 16K words of instruction code and 16K words of data. The state of the PMDA pin distinguishes between code and data access of program memory. The program memory data (PMD) bus, like the processor's instruction words, is 24 bits wide.

The data memory address (DMA) bus is 14 bits wide allowing direct access of up to 16K words of data. The data memory data (DMD) bus is 16 bits wide. The data memory data (DMD) bus provides a path for the contents of any register in the processor to be transferred to any other register, or to any external data memory location, *in a single cycle*. The data memory address can come from two sources: an absolute value specified in the instruction code (direct addressing) or the output of a data address generator (indirect addressing). Only indirect addressing is supported for data fetches via the program memory bus.

The program memory data (PMD) bus can also be used to transfer data to and from the computational units through direct paths or via the PMD-DMD bus exchange unit. The PMD-DMD bus exchange unit permits data to be passed from one bus to the other. It contains hardware to overcome the 8-bit width discrepancy between the two buses when necessary.

Each computational unit contains a set of dedicated input and output registers. Computational operations generally take their operands from input registers and load the result into an output register. The registers act as a stopover point for data between the external memory and the computational circuitry, effectively introducing one pipeline level on input and one level on output. The computational units are arranged side by side rather than in cascade. To avoid excessive pipeline delays when a series of different operations are performed, the internal result (R) bus allows any of the output registers to be used directly (without delay) as the input to another computation.

For a wide variety of calculations, it is desirable to fetch two operands at the same time—one from data memory and one from program memory. Fetching data from program memory, however, makes it impossible to fetch the next instruction from program memory on the same cycle; an additional cycle would be required. To avoid this overhead, the ADSP-2100 incorporates an instruction cache which holds sixteen words. The benefit of the cache architecture is most apparent when executing a program loop that can be totally contained in the cache memory. In this situation, the ADSP-2100 works like a three-bus system with an instruction fetch and two operand fetches taking place at the same time. Many algorithms are readily coded in loops of sixteen instructions or less because of the parallelism and high-level syntax of the ADSP-2100 assembly language.

Here's how the cache functions: Every instruction loaded into the instruction register is also written into cache memory. As additional instructions are fetched, they overwrite the current contents of cache in a circular fashion. When the current instruction does a program memory data access, the cache automatically sources the instruction register if its contents are valid. Operation of the cache is completely transparent to user.

There are two independent data address generators (DAGs). As a pair, they allow the simultaneous fetch of data stored in program and in data memory for executing dual-operand instructions in a single cycle. One data address generator (DAG1) can supply addresses to the data memory only; the other (DAG2) can supply addresses to either the data memory or the program memory. Each DAG can handle linear addressing as well as modulo addressing for circular buffers.

With its multiple bus structure, the ADSP-2100 supports a high degree of operational parallelism. In a single cycle, the ADSP-2100 can fetch an instruction, compute the next instruction address, perform one or two data transfers, update one or two data address pointers and perform a computation. Every instruction executes in a single cycle.

Figure 1.2, on the next page, is a simplified representation of the ADSP-2100 in a system context. The figure shows the two external memories used by the processor. Program memory stores instructions and is also used to store data. Data memory stores only data. The data memory address space may be shared with memory-mapped peripherals, if desired. Both memories may be accessed by external devices, such as a system host, if desired. Figure 1.2 also shows the processor control interface signals, (RESET, HALT and TRAP) the four interrupt request lines, the bus request and bus grant lines (BR and BG) and the clock input (CLKIN) and output (CLKOUT).

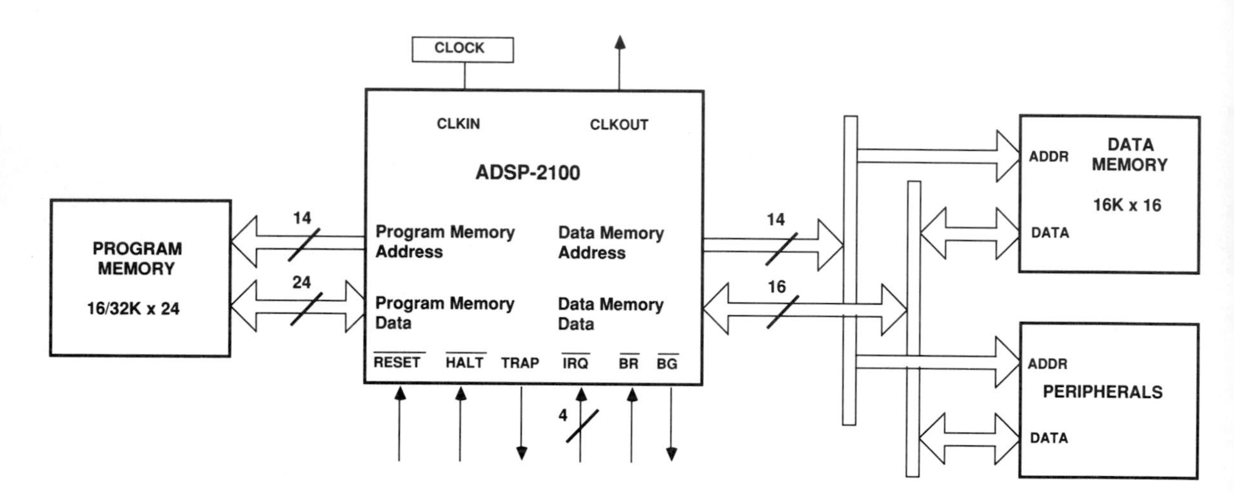

Figure 1.2 ADSP-2100 System

1.2.2 ADSP-2101 Architecture

Figure 1.3 shows the architecture of the ADSP-2101 processor. Like the ADSP-2100, the ADSP-2101 contains an arithmetic/logic unit, a multiplier/accumulator, and a barrel shifter—plus two data address generators and a program sequencer.

The ADSP-2101 has 1K words of 16-bit data memory on-chip and 2K words of 24-bit program memory on-chip. The processor can fetch an operand from on-chip data memory, an operand from on-chip program memory and the next instruction from on-chip program memory in a single cycle. (The speed of on-board memory access makes this possible and eliminates the need for cache memory as on the ADSP-2100.)

This scheme is extended off-chip via a single external memory address bus and data bus which may be used for either program or data memory access and for booting. Consequently, the processor can access external memory once in any cycle.

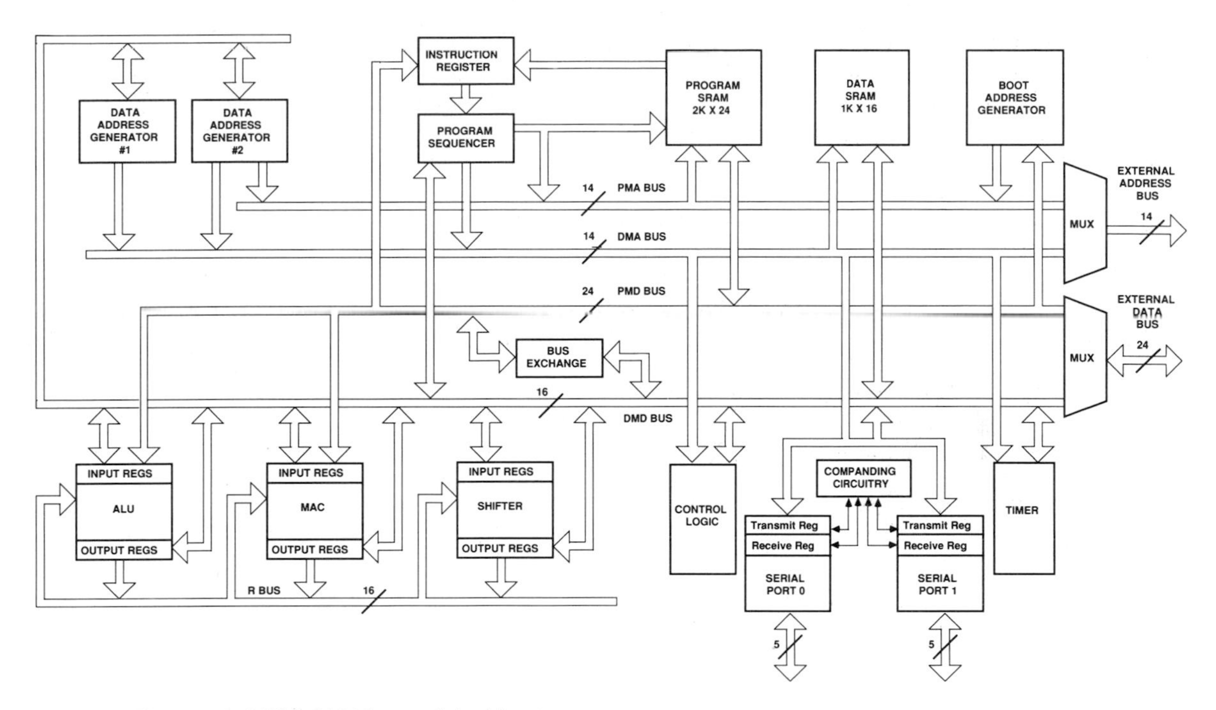

Figure 1.3 ADSP-2101 Internal Architecture

Boot circuitry provides for loading on-chip program memory automatically after reset. Wait states are generated automatically for interfacing to a single low-cost EPROM. Multiple programs can be selected and loaded from the EPROM with no additional hardware.

The memory interface supports memory-mapped peripherals with programmable wait-state generation. External devices can gain control of buses with bus request/grant signals ($\overline{BR}$ and $\overline{BG}$). An optional execution mode allows the ADSP-2101 to continue running while the buses are granted to another master as long as an external memory operation is not required.

The ADSP-2101 can respond to six user interrupts. There can be up to three external interrupts, configured as edge- or level-sensitive. Internal interrupts can be generated from the timer and the serial ports. There is also a master $\overline{RESET}$ signal.

The two serial ports ("SPORTs") provide a complete serial interface; they interface easily and directly to a wide variety of popular serial devices. They have inherent hardware companding (data compression and expansion) with both μ-law and A-law available. Each port can generate an internal programmable clock or accept an external clock.

The SPORTs are synchronous and use framing signals to control data flow. Each SPORT can generate its serial clock internally or use an external clock. The framing synchronization signals may be generated internally or by an external device. Word lengths may vary from three to sixteen bits. One SPORT (SPORT0) has a multichannel capability which allows the receiving or transmitting of arbitrary data words from a 24-word or 32-word bitstream. The SPORT1 pins have alternate functions and can be configured as two additional external interrupt pins and the Flag Out (FO) and Flag In (FI) pins.

The programmable interval timer provides periodic interrupt generation. An 8-bit prescaler register allows the timer to decrement a 16-bit count register over a range from each cycle to every 256 cycles. An interrupt is generated when this count register reaches zero. The count register is automatically reloaded from a 16-bit period register, and the count resumes immediately.

1.3 ASSEMBLY LANGUAGE OVERVIEW

The ADSP-2100 family's assembly language uses an algebraic syntax for ease of coding and readability. The sources and destinations of computations and data movements are written explicitly in each assembly statement, eliminating cryptic assembler mnemonics. Each assembly statement, however, corresponds to a single 24-bit instruction, executable in one cycle. Register mnemonics, listed below, are concise and easy to remember.

Mnemonic	*Definition*
AX0, AX1, AY0, AY1	ALU inputs
AR	ALU result
AF	ALU feedback
MX0, MX1, MY0, MY1	Multiplier inputs
MR0, MR1, MR2	Multiplier result (3 parts)
MF	Multiplier feedback
SI	Shifter input
SE	Shifter exponent
SR0, SR1	Shifter result (2 parts)
SB	Shifter block (for block floating-point format)
PX	PMD-DMD bus exchange
I0 - I7	DAG index registers
M0 - M7	DAG modify registers
L0 - L7	DAG length registers (for circular buffers)
PC	Program counter
CNTR	Counter for loops
ASTAT	Arithmetic status
MSTAT	Mode status
SSTAT	Stack status
IMASK	Interrupt mask
ICNTL	Interrupt control modes
RX0, RX1	Receive data registers (not on ADSP-2100)
TX0, TX1	Transmit data registers (not on ADSP-2100)

The ADSP-2101 instruction set is an upward-compatible superset of the ADSP-2100 instruction set; thus, programs written for the ADSP-2100 can be executed on the ADSP-2101 with minimal changes.

Here are some examples of the ADSP-2100 assembly language. The statement

```
MR = MR + MX1*MY1;
```

performs a multiply/accumulate operation. It multiplies the input values in registers MX1 and MY1, adds that product to the current value of the MR register (the result of the previous multiplication) and then writes the new result to MR.

The statement

```
DM(buffer1) = AX0;
```

writes the value of register AX0 to data memory at the location which is the value of the variable *buffer1*.

1.4 DEVELOPMENT SYSTEM

The ADSP-2100 family is supported with a complete set of software and hardware development tools. The ADSP-2100 Family Development System consists of Development Software to aid in software design and in-circuit emulators to facilitate the debug cycle. EZ-ICE™ evaluation boards are available for evaluating the processors. Additional development tool capabilities continue to be added as new members of the processor family are introduced.

The Development Software includes:

- System Builder

This module allows the designer to specify the amount of RAM and ROM available, the allocation of program and data memory and any memory-mapped I/O ports for the target hardware environment. It uses high-level constructs to simplify this task. This specification is used by the linker, simulators, and emulators.

- Assembler

This module assembles a user's source code and data modules. It supports the high-level syntax of the instruction set. To support modular code development, the Assembler provides flexible macro processing and "include" files. It provides a full range of diagnostics.

- Linker

The Linker links separately assembled modules. It maps the linked code and data output to the target system hardware, as specified by the System Builder output.

- Simulator

This module performs an instruction-level simulated execution of ADSP-2100 family assembly code. The interactive user interface supports full symbolic assembly and disassembly of simulated instructions. The Simulator fully simulates the hardware configuration described by the System Builder module. It flags illegal operations and provides several displays of the internal operations of the processor.

- PROM Splitter

This module reads the Linker output and generates PROM-programmer-compatible files.

- C Compiler

The C Compiler reads ANSI C source and outputs source code ready to be assembled. It also supports inline assembler code.

In-circuit emulators provide stand-alone, real-time, in-circuit emulation. The emulators provide program execution with little or no degradation in processor performance. The emulators duplicate the simulators' interactive and symbolic user interface.

Complete information on development tools is available from Analog Devices.

1.5 CONVENTIONS OF NOTATION

The following conventions are used throughout this book:

- All listings begin with a comment block that summarizes the calling parameters, the return values, the registers that are altered, and the computation time of the routine (in terms of the routine's parameters, in some cases).
- In listings, all keywords are uppercase; user-defined names (such as labels, variables, and data buffers) are lowercase. In text, keywords are uppercase and user-defined names are lowercase italics. Note that this convention is for readability only.
- In comments, register values are indicated by "=" if the register contains the value or by "—>" if the register points to the value in memory.
- All numbers are decimal unless otherwise specified. In listings, constant values are specified in binary, octal, decimal, or hexadecimal by the prefixes B#, O#, D#, and H#, respectively.

1.6 PROGRAMS ON DISK

This book includes an IBM PC 5¼ inch high-density diskette containing the routines that appear in this book. As with the printed routines, we cannot guarantee suitability for your application. The diskette also contains a demonstration version of the ADSP-2101 Simulator. This demonstration is self-running and documented on-line.

1.7 FOR FURTHER SUPPORT

You can reach Analog Devices in the following ways:

- By contacting your local Analog Devices Sales Representative
- For information on DSP product features, availability, and pricing, call (617) 461-3881 in Norwood, Massachusetts, USA
- For Applications Engineering information, call either the DSP applications group at (617) 461-3672 in Norwood, Massachusetts, USA, or the Linear applications group at (617) 461-2628 in Wilmington, Massachusetts, USA
- The DSP Norwood office Fax number is (617) 461-3010
- The factory may also be reached by
  Telex: 924491
  TWX: 710/394-6577
  Cables: ANALOGNORWOODMASS
- Through the DSP Group's Bulletin Board Service; it can be reached at 300, 1200, or 2400 baud, no parity, 8 bits data, 1 stop bit by dialing:
  (617) 461-4258
- By writing to:
  Analog Devices
  SPD-DSP
  One Technology Way
  P.O. Box 9106
  Norwood, MA 02062-9106
  USA

Fixed-Point Arithmetic ■ 2

2.1 OVERVIEW

Binary number representations usually include a sign and a radix point, as well as a magnitude. The sign shows whether the number is positive or negative. The radix point separates the integer and fractional parts of the number.

The sign of a binary number can be represented with one bit. In most representations, a zero indicates positive and a one indicates negative. The sign bit is usually in the leftmost location (most significant bit).

There are several formats for representing negative numbers, including signed-magnitude, ones complement, and twos complement. The most common method, and the one used by the ADSP-2100, is twos complement. The advantage of twos-complement format is that it provides a unique representation for zero, whereas the other formats have both a positive and a negative zero. In twos-complement format, zero is considered positive; therefore, the magnitude of the largest negative number that can be represented with a given number of bits is one greater than the magnitude of the largest positive number. A twos-complement number of k+1 bits (one bit indicates the sign and k bits indicate the magnitude) can represent the range of numbers from $2^k - 1$ to -2^k.

The twos complement of a binary number can be calculated in one of two ways: 1) invert all the bits and add one to the least significant bit, or 2) invert all bits to the left of the least significant 1. For example:

```
Binary +72        0100 1000
Invert bits       1011 0111
Add 1           + 0000 0001
                -----------
Binary -72        1011 1000
```

or

```
Binary +72                  0100 1000
Invert all bits left of       |
least significant 1         invert
Binary -72                  1011 1000
```

2 Fixed-Point Arithmetic

A radix point is placed between two bits in a number. The bits to the left of the radix point represent the integer part of the number; the bits to the right of the radix point represent the fractional part of the number. There are two ways to specify the location of the radix point:

- *Fixed-point format* places the radix point at a single, predetermined location. Often this location is to the left of all bits (all bits are fractional) or to the right of all bits (all bits are integer). Because the location of the radix point is assumed by software, it does not need to be represented explicitly. Arithmetic operations (such as multiplication) can change the radix-point position so that shifting may be necessary to keep the number in the same fixed-point format.

- *Floating-point format* uses two numbers to represent a value: a mantissa and an exponent. The exponent indicates the location of the radix point. The exponent may be stored along with the mantissa or in a separate register.

The ADSP-2100 represents numbers in a fixed-point format. In this publication, the location of the radix point is given by the format designation I.Q, in which I is the number of bits to the left of the radix point and Q is the number of bits to the right. For example, the 1.15 format indicates signed full fractional numbers; one integer bit indicates the sign, and 15 fractional bits indicate the fractional magnitude. Full integer number representation is 16.0 format. For most signal processing applications, fractional numbers (1.15 format) are assumed. The multiplier and divider of the ADSP-2100 are optimized for use with this format.

ADSP-2100 addition, subtraction, and multiplication primitives operate directly on single-precision (16-bit) numbers. In this chapter, we show an example of how to program the ADSP-2100 to perform single-precision, fixed-point division. We also include explanations of extended-precision, fixed-point arithmetic. Example implementations of addition, subtraction, and multiplication are shown in both double precision (32-bit operands) and triple precision (48-bit operands). Division is shown using a 64-bit dividend and a 32-bit divisor.

Double-precision operations are most common, so we show examples that can be implemented directly. Triple-precision arithmetic is also shown to demonstrate how to handle the middle words of extended-precision numbers. Repeating the middle-word operations allows extension to any precision.

2.2 SINGLE-PRECISION FIXED-POINT DIVISION

The ADSP-2100 instruction set includes two divide primitives, DIVS and DIVQ, to compute fixed-point division. The DIVS instruction calculates the sign bit of the quotient, and the DIVQ calculates a single bit of the quotient. These instructions can be used to implement a nonrestoring (remainder is invalid) add/subtract division algorithm that works with signed or unsigned operands. The operands must be either both signed or both unsigned. Because each instruction produces one bit of the quotient, dividing a 16-bit divisor into a 32-bit dividend to produce a 16-bit quotient requires 16 instructions, and therefore 16 cycles. Block diagrams of the DIVS and DIVQ operations are shown in Figures 2.1 and 2.2, on the following page.

The division algorithm performs either an addition or subtraction based on the signs of the divisor and the partial remainder. Mano, 1982, gives an excellent explanation of a similar algorithm.

In the ADSP-2100 implementation of the division algorithm for signed operands, the divisor can be stored in AX0, AX1, or any register on the R bus. The MSW of the dividend can be loaded into AY1 or AF, and the dividend's LSW is loaded into AY0. To calculate the quotient, the ADSP-2100 first executes a DIVS instruction to compute the sign of the quotient, followed by 15 DIVQ instructions to compute 15 quotient bits. A signed fixed-point division routine is shown in Listing 2.1. This routine takes the divisor from AX0, the dividend's MSW from AF, and the dividend's LSW from AY0. The quotient is returned in AY0.

In unsigned division, the dividend's MSW must be loaded into AF, and the ASTAT register must be cleared to set the AQ bit to zero. The ADSP-2100 executes 16 DIVQ instructions. Listing 2.2 shows a subroutine to perform an unsigned division. The registers must be preloaded with the same values as for the signed division routine: the divisor in AX0, the dividend's MSW in AF, and the dividend's LSW in AY0.

The format of the quotient is determined by the format of the two operands. If the dividend is in P.Q format, and the divisor is in M.N format, the quotient will be in (P–M+1).(Q–N–1) format. Some format manipulation may be necessary to guarantee the validity of the quotient. For example, if both operands are signed and fully fractional (dividend in 1.31 format and divisor in 1.15 format) then the result is fully fractional (in 1.15 format), and therefore the dividend must be smaller than the divisor for a valid quotient.

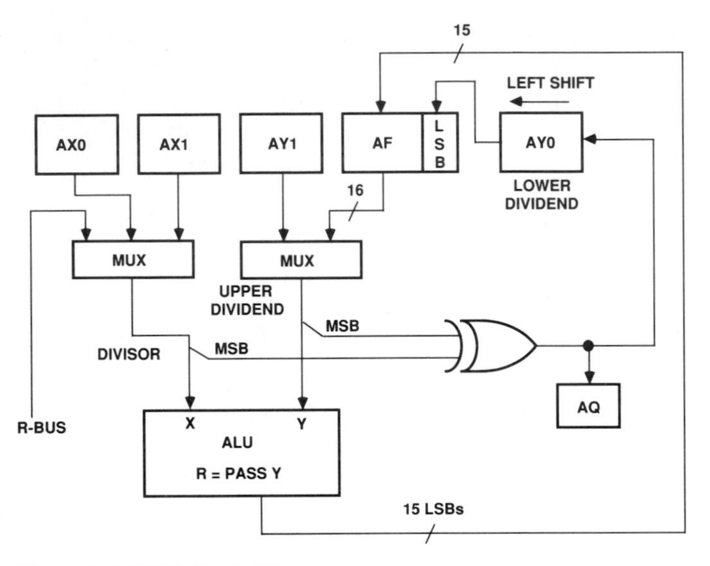

Figure 2.1 DIVS Block Diagram

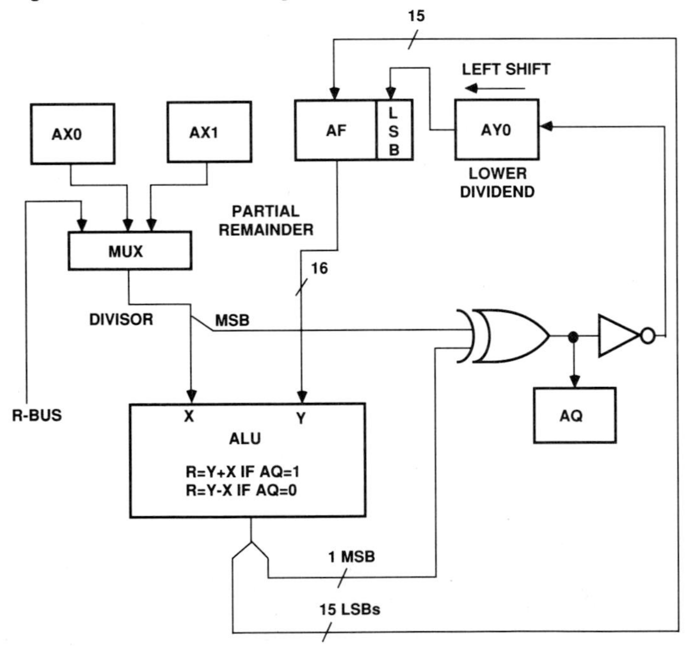

Figure 2.2 DIVQ Block Diagram

To divide two integers (dividend in 32.0 format and divisor in 16.0 format) and produce an integer quotient (in 16.0 format), you must shift the dividend one bit to the left (into 31.1 format) before dividing.

```
.MODULE Signed_SP_Divide;

{
   Signed Single-Precision Divide

   Calling Parameters
        AF = MSW of dividend
        AY0 = LSW of dividend
        AX0 = 16-bit divisor

   Return Values
        AY0 = 16-bit result

   Altered Registers
        AY0, AF

   Computation Time
        17 cycles

}
.ENTRY  sdivs;

sdivs:  DIVS AF,AX0;                            {Compute sign bit}
        DIVQ AX0; DIVQ AX0; DIVQ AX0;           {Compute 15 quotient bits}
        DIVQ AX0; DIVQ AX0; DIVQ AX0;
        DIVQ AX0; DIVQ AX0; DIVQ AX0;
        DIVQ AX0; DIVQ AX0; DIVQ AX0;
        DIVQ AX0; DIVQ AX0; DIVQ AX0;
        RTS;
.ENDMOD;
```

Listing 2.1 Single-Precision Divide, Signed

```
.MODULE Unsigned_SP_Divide;

{
   Unsigned Single-Precision Divide

   Calling Parameters
        AF = MSW of dividend
        AY0 = LSW of dividend
        AX0 = 16-bit divisor

   Return Values
        AY0 = 16-bit result

   Altered Registers
        AY0, AF

   Computation Time
        18 cycles

}

.ENTRY   sdivq;

sdivq:   ASTAT=0;                          {Clear AQ bit of ASTAT}
         DIVQ AX0;                         {Compute 16 quotient bits}
         DIVQ AX0; DIVQ AX0; DIVQ AX0;
         DIVQ AX0; DIVQ AX0; DIVQ AX0;
         DIVQ AX0; DIVQ AX0; DIVQ AX0;
         DIVQ AX0; DIVQ AX0; DIVQ AX0;
         DIVQ AX0; DIVQ AX0; DIVQ AX0;
         RTS;
.ENDMOD;
```

Listing 2.2 Single-Precision Divide, Unsigned

2.3 MULTIPRECISION FIXED-POINT ADDITION

The following algorithm adds two multiprecision operands together:

1. Add the two LSWs to produce the LSW of the result and a carry bit.
2. Add the next word of each operand plus the carry from the previous word to produce the next word of the result and a carry bit.
3. Repeat step 2 until every word of the result has been computed. After the MSW has been computed, the status flags of the ALU will be valid for the multiprecision sum.

To produce a valid number, the radix points must be in the same location in both operands. The result will be in the same format as the operands.

Listing 2.3 shows a subroutine that implements the addition algorithm for two double-precision numbers. The LSW of the augend is stored in AX0, and its MSW is stored in AX1. The LSW of the addend is stored in AY0, and its MSW is stored in AY1. The LSW of the result is returned in SR0, and its MSW is returned in SR1.

Listing 2.4 shows a subroutine that performs triple-precision addition. This routine retrieves the augend from data memory, LSW first, starting at DM(I0). The addend is read from data memory, LSW first, starting at DM(I1). The result is stored in data memory, LSW first, starting at DM(I2). Each data memory access modifies the address by adding the value of M0 to the I register used in the access. Before executing the routine, you must ensure that the augend and addend are loaded at the correct data memory locations, and you must initialize I0, I1, I2, and M0. You must also set the buffer length registers L0, L1, and L2 to zero to disable circular buffers and allow the ADSP-2100 to read the full numbers. Note that most of the subroutine instructions are concerned with either reading operands or writing the result.

```
.MODULE Double_Precision_Add;

{  Double-Precision Addition
        Z = X + Y

   Calling Parameters
        AX0 = LSW of X
        AX1 = MSW of X
        AY0 = LSW of Y
        AY1 = MSW of Y

   Return Values
        SR0 = LSW of Z
        SR1 = MSW of Z

   Altered Registers
        AR,SR

   Computation Time
        4 cycles
}
```

(listing continues on next page)

```
.ENTRY   dpa;

dpa:     AR=AX0+AY0;                       {Add LSWs}
         SR0=AR, AR=AX1+AY1+C;             {Add MSWs}
         SR1=AR;
         RTS;
.ENDMOD;
```

Listing 2.3 Double-Precision Addition

```
.MODULE Triple_Precision_Add;

{
   Triple-Precision Addition
        Z = X + Y

   Calling Parameters
        I0 —> Buffer for X                 L0 = 0
        I1 —> Buffer for Y                 L1 = 0
        I2 —> Storage Buffer for Z         L2 = 0
        M0 = 1

   Return Values
        Z Buffer is filled

   Altered Registers
        I0, I1, I2, AR, AX0, AY0

   Computation Time
        10 cycles
}

.ENTRY   tpa;

tpa:     AX0=DM(I0,M0);                    {Fetch LSWs}
         AY0=DM(I1,M0);
         AX0=DM(I0,M0), AR=AX0+AY0;        {Add LSWs}
         AY0=DM(I1,M0);
         DM(I2,M0)=AR, AR=AX0+AY0+C;       {Save LSW, add middle words}
         AX0=DM(I0,M0);                    {Fetch MSWs}
         AY0=DM(I1,M0);
         DM(I2,M0)=AR, AR=AX0+AY0+C;       {Add MSWs}
         DM(I2,M0)=AR;                     {Save MSW}
         RTS;
.ENDMOD;
```

Listing 2.4 Triple-Precision Addition

2.4 MULTIPRECISION FIXED-POINT SUBTRACTION

The subtraction algorithm is very similar to the addition algorithm. In fact, subtraction can be accomplished by adding the twos complement of the subtrahend to the minuend. Multiprecision subtraction is performed by the following steps:

1. Subtract the LSW of the subtrahend from the LSW of the minuend to produce the LSW of the result and a borrow ("carry–1") bit.
2. Subtract the next word of the subtrahend from the next word of the minuend and add to it the "carry – 1," producing the next word of the result and a carry bit. The "carry – 1" value effectively implements a borrow signal from the previous word.
3. Repeating step 2 until every word of the result has been computed. After the MSW has been computed, the status flags of the ALU will be valid.

Listing 2.5 shows a double-precision subtraction routine. This routine assumes that the minuend's LSW is stored in AX0, the minuend's MSW is in AX1, the subtrahend's LSW is in AY0, and the subtrahend's MSW is in AY1. The result is returned in the SR registers (LSW in SR0, MSW in SR1).

Listing 2.6 shows the triple-precision subtraction routine. Most of its instructions perform data I/O. The minuend is read, LSW first, starting at DM(I0); the subtrahend is read, LSW first, starting at DM(I1). The result is written, LSW first, starting at DM(I2). Before calling this routine, you must initialize I0, I1, I2 to the correct data memory locations, M0 to one (the memory spacing value) and L0, L1, and L2 to zero (to disable circular buffers).

```
.MODULE Double_Precision_Subtract;

{  Double-Precision Subtraction
        Z = X - Y

   Calling Parameters
        AX0 = LSW of X
        AX1 = MSW of X
        AY0 = LSW of Y
        AY1 = MSW of Y

   Return Values
        SR0 = LSW of Z
        SR1 = MSW of Z
```

(listing continues on next page)

```
   Altered Registers
        AR, SR

   Computation Time
        4 cycles
}

.ENTRY  dps;

dps:    AR=AX0-AY0;                  {Subtract LSWs}
        SR0=AR, AR=AX1-AY1+C-1;      {Subtract MSWs}
        SR1=AR;
        RTS;
.ENDMOD;
```

Listing 2.5 Double-Precision Subtraction

```
.MODULE Triple_Precision_Subtract;

{  Triple-Precision Subtraction
        Z = X - Y

   Calling Parameters
        I0 -> X Buffer               L0 = 0
        I1 -> Y Buffer               L1 = 0
        I2 -> Z Buffer               L2 = 0
        M0 = 1

   Return Values
        Z Buffer is filled

   Altered Registers
        I0, I1, I2, AR, AX0, AY0

   Computation Time
        10 cycles
}

.ENTRY  tps;

tps:    AX0=DM(I0,M0);                     {Fetch LSWs}
        AY0=DM(I1,M0);
        AX0=DM(I0,M0), AR=AX0-AY0;         {Subtract LSWs}
        AY0=DM(I1,M0);
        DM(I2,M0)=AR, AR=AX0-AY0+C-1;      {Store LSW, fetch middle}
                                           {words}
        AX0=DM(I0,M0);                     {Fetch MSWs}
        AY0=DM(I1,M0);
        DM(I2,M0)=AR, AR=AX0-AY0+C-1;        {Subtract MSWs}
        DM(I2,M0)=AR;                        {Store MSW}
        RTS;
.ENDMOD;
```

Listing 2.6 Triple-Precision Subtraction

2.5 MULTIPRECISION FIXED-POINT MULTIPLICATION

Multiplication is more complicated than either addition or subtraction. An important task is placing the radix point in the product. In addition and subtraction, the radix point location in the result is the same as for both operands. In multiplication, the radix point may be in a different location in the two operands, and its location in the product depends on the locations in the two operands. The result of multiplying a number in P.Q format by a number in M.N format produces a result in (P+M–1).(Q+N+1) format. The ADSP-2100 multiplier automatically shifts the product one bit to the left to eliminate the redundant sign bit of the result. Therefore, multiplication of two numbers in a full fractional format (1.15 format, the most common format) returns the result in a full fractional format (1.31). If two numbers in full integer format (16.0) are multiplied, the result is in 31.1 format. To produce an integer result (32.0 format), the product must be shifted to the right one bit before storing it in memory.

Multiplication is performed according to the following procedure, which is illustrated in Figure 2.3, on the next page.

1. Multiply each word of the multiplicand by the LSW of the multiplier and an appropriate power of two to shift it left to its correct position.
2. Multiply each word of the multiplicand by the next word of the multiplier and the appropriate power of two.
3. Repeat step 2 until each word of the multiplicand has been multiplied by every word of the multiplier.
4. Add all the partial products together.

In computing the partial products, the signed/unsigned switch of the multiplier determines whether the multiplication is signed, mixed-mode, or unsigned. Multiplication of the MSWs should be signed, and multiplication of a less significant word by either MSW should be mixed-mode. All other multiplication should be unsigned.

U and N are 16-bit words

$$
\begin{array}{r}
N_n\,N_{n-1}\,\cdots\,N_2\,N_1\,N_0 \\
\times \qquad U_n\,U_{n-1}\,\cdots\,U_2\,U_1\,U_0 \\
\hline
2^{16n}U_0N_n + 2^{16(n-1)}U_0N_{n-1} + \ldots 2^{32}U_0N_2 + 2^{16}U_0N_1 + U_0N_0 \\
2^{16(n+1)}U_1N_n + 2^{16n}U_1N_{n-1} + \ldots \quad 2^{48}U_1N_2 + 2^{32}U_1N_1 + 2^{16}U_1N_0 \\
\vdots \\
+\, 2^{16(2n)}U_nN_n + \ldots 2^{16(n+1)}U_nN_1 + 2^{16n}U_nN_0
\end{array}
$$

Figure 2.3 Multiprecision Multiplication

Listing 2.7 shows a double-precision multiplication routine for fractional operands. The routine assumes that the multiplicand's LSW is stored in MX0 and its MSW is stored in MX1, and that the multiplier's LSW is stored in MY0 and its MSW is stored in MY1. This routine produces a 64-bit product that is stored in data memory, LSW first, starting at DM(I0). M0 and L0 should be set to one and zero, respectively, before the routine is executed.

The product of the LSWs of the operands is computed first, and the LSW of the result is written to DM(I0). The MR register is then shifted 16 bits to the right. The inner products are computed and added to the shifted value in MR; MR0 is written to the next data memory location. The MR register is again shifted 16 bits to the right, and the product of the MSWs is computed and added to MR. MR0 is written to data memory, followed by MR1, to complete the 64-bit product. Note how the signed/unsigned switch indicates the type of multiplication for each partial product.

The triple-precision multiplication routine is shown in Listing 2.8. In this routine, the multiplicand and the multiplier are both stored in data memory, LSW first, the multiplicand starting at DM(I0) and the multiplier starting at DM(I1). The routine produces a 96-bit result stored LSW first, starting at DM(I2). The X and Y buffers must be declared (and located in memory) as circular buffers; L0 and L1 are set to three because the routine circles back to refetch the first words of the operands. Before executing the routine, you should set M0 to one and L2 to zero.

Listings 2.9 and 2.10 show the double-precision routine and the triple-precision routine, respectively, for integer multiplication. These routines differ from the multiplication routines already described only in that they shift the result one bit to the right before writing it to memory, in order to generate a full integer product.

```
.MODULE Double_Precision_Multiply;

{  Double-Precision Multiplication
        Z = X × Y

   Calling Parameters
        I0 -> Address of Z Buffer   L0 = 0
        M0 = 1
        MX0 = LSW of X
        MX1 = MSW of X
        MY0 = LSW of Y
        MY1 = MSW of Y

   Return Values
        Z Buffer Filled

   Altered Registers
        MR, I0

   Computation Time
        13 cycles

}

.ENTRY   dpm;

dpm:    MR=MX0*MY0(UU);          {Compute LSW}
        DM(I0,M0)=MR0;           {Save LSW}
        MR0=MR1;                 {Shift right 16 bits}
        MR1=MR2;
        MR=MR+MX1*MY0(SU);       {Compute inner product}
        MR=MR+MX0*MY1(US);
        DM(I0,M0)=MR0;           {Shift right 16 bits}
        MR0=MR1;
        MR1=MR2;
        MR=MR+MX1*MY1(SS);       {Compute MSW}
        DM(I0,M0)=MR0;
        DM(I0,M0)=MR1;           {Store MSW}
        RTS;
.ENDMOD;
```

Listing 2.7 Double-Precision Multiplication

```
.MODULE Triple_Precision_Multiply;

{  Triple-Precision Multiplication
        Z = X × Y

   Calling Parameters
        I0 -> X Buffer                    L0 = 3
        I1 -> Y Buffer                    L1 = 3
        I2 -> Z Buffer                    L2 = 0
        M0 = 1

   Return Values
        Z Buffer Filled

   Altered Registers
        MX1,MX0,MY1,MY0,MR,I0,I1,I2

   Computation Time
        26 cycles
}

.ENTRY  tpm;

tpm:    MY0=DM(I0,M0);
        MX0=DM(I1,M0);
        MX1=DM(I1,M0), MR=MX0*MY0(UU);     {Compute LSW}
        DM(I2,M0)=MR0;                     {Save LSW}
        MR0=MR1;                           {Shift right 16 bits
        MR1=MR2;
        MY1=DM(I0,M0), MR=MR+MX1*MY0(UU);
        MR=MR+MX0*MY1(UU);
        DM(I2,M0)=MR0;
        MR0=MR1;                           {Shift right 16 bits}
        MR1=MR2;
        MY1=DM(I0,M0), MR=MR+MX1*MY1(UU);
        MX0=DM(I1,M0), MR=MR+MX0*MY1(US);
        MY0=DM(I0,M0), MR=MR+MX0*MY0(SU); {Skip 1st word, LSW}
        DM(I2,M0)=MR0;
        MR0=MR1;                           {Shift right 16 bits}
        MR1=MR2;
        MY0=DM(I0,M0), MR=MR+MX1*MY1(US);
        MR=MR+MX0*MY0(SU);
        DM(I2,M0)=MR0;
        MR0=MR1;
        MR1=MR2;
        MR=MR+MX0*MY1(SS);
        DM(I2,M0)=MR0;
        DM(I2,M0)=MR1;                     {Save MSW}
        RTS;
.ENDMOD;
```

Listing 2.8 Triple-Precision Multiplication

```
.MODULE Integer_DPM;

{
   Integer Double-Precision Multiplication
        Z = X × Y

   Calling Parameters
        I0 -> Z Buffer                  L0 = 0
        M0 = 1
        MX0 = LSW of X
        MX1 = MSW of X
        MY0 = LSW of Y
        MY1 = MSW of Y
        SE = -1

   Return Values
        Z Buffer Filled

   Altered Registers
        I0, MR, SR

   Computation Time
        14 cycles

}

.ENTRY  idpm;

idpm:   MR=MX0*MY0(UU);                     {Compute LSW}
        MR0=MR1, SR=LSHIFT MR0(LO);         {Shift LSW right 1 bit}
        MR1=MR2, SR=SR OR LSHIFT MR1(HI);   {before saving}
        DM(I0,M0)=SR0, MR=MR+MX1*MY0(SU);
        MR=MR+MX0*MY1(US);
        MR0=MR1, SR=LSHIFT MR0(LO);
        MR1=MR2, SR=SR OR LSHIFT MR1(HI);
        DM(I0,M0)=SR0, MR=MR+MX1*MY1(SS);
        SR=LSHIFT MR0(LO);
        SR=SR OR LSHIFT MR1(HI);
        SR=SR OR LSHIFT MR2 BY 15(HI);
        DM(I0,M0)=SR0;
        DM(I0,M0)=SR1;                  {Save MSW after shifting}
        RTS;
.ENDMOD;
```

Listing 2.9 Integer Double-Precision Multiplication

```
.MODULE Integer_TPM;

{   Integer Triple-Precision Multiplication
        Z = X × Y

    Calling Parameters
        I0 -> X Buffer                      L0 = 3
        I1 -> Y Buffer                      L1 = 3
        I2 -> Storage for Z                 L2 = 0
        M0 = 1
        SE = -1

    Return Values
        Z Buffer Filled

    Altered Registers
        MX0,MX1,MY0,MY1,MR,I0,I1,I2,SR

    Computation Time
        28 cycles
}

.ENTRY  itpm;

itpm:   MY0=DM(I0,M0);                      {Fetch LSWs}
        MX0=DM(I1,M0);
        MY1=DM(I0,M0), MR=MX0*MY0(UU);      {Compute LSW}
        MX1=DM(I1,M0);
        MR0=MR1, SR=LSHIFT MR0(LO);         {Shift LSW}
        MR1=MR2, SR=SR OR LSHIFT MR1(HI);
        DM(I2,M0)=SR0, MR=MR+MX0*MY1(UU); {Store LSW}
        MR=MR+MX1*MY0(UU);
        MR0=MR1, SR=LSHIFT MR0(LO);
        MR1=MR2, SR=SR OR LSHIFT MR1(HI);
        DM(I2,M0)=SR0, MR=MR+MX1*MY1(UU);
        MY1=DM(I0,M0);
        MR=MR+MX0*MY1(US), MX0=DM(I1,M0);
        MY0=DM(I0,M0), MR=MR+MX0*MY0(SU); {Skip 1st word}
        MR0=MR1, SR=LSHIFT MR0(LO);
        MR1=MR2, SR=SR OR LSHIFT MR1(HI);
        DM(I2,M0)=SR0;
        MY0=DM(I0,M0), MR=MR+MX1*MY1(US);
        MR=MR+MX0*MY0(SU);
        MR0=MR1, SR=LSHIFT MR0(LO);
        MR1=MR2, SR=SR OR LSHIFT MR1(HI);
        DM(I2,M0)=SR0, MR=MR+MX0*MY1(SS);
        SR=LSHIFT MR0(LO);                  {Shift MSW}
        SR=SR OR LSHIFT MR1(HI);
        SR=SR OR LSHIFT MR2 BY 15(HI);
        DM(I2,M0)=SR0;
        DM(I2,M0)=SR1;                      {Save MSW}
        RTS;
.ENDMOD;
```

Listing 2.10 Integer Triple-Precision Multiplication

2.6 MULTIPRECISION FIXED-POINT DIVISION

The routine shown in Listing 2.11 provides a convenient method for doing double-precision division (64-bit dividend, 32-bit divisor). It is a double-precision software implementation of the algorithm used by the hardware instructions DIVS and DIVQ (see section 2.2). Mano, 1982, gives an excellent explanation of a similar algorithm. The division algorithm generates one bit of the quotient by comparing the divisor and the partial remainder. If the divisor is less than or equal to the partial remainder, the quotient bit is a one; otherwise, the quotient bit is a zero. The divisor is subtracted from the partial remainder if it is less than the partial remainder, and then the divisor is shifted right one bit and compared to the partial remainder to generate the next bit. This routine shifts the dividend to the left rather than the divisor to the right, accomplishing the same comparison.

Before calling the routine, you must load SE with –15. You must also load the dividend into the SR and MR registers. Its MSW should be loaded into SR1, the next word in SR0, the next in MR1, and its LSW in MR0. The divisor should be loaded into the AY registers, LSW in AY0 and MSW in AY1. The result is returned in the MR registers, LSW in MR0 and MSW in MR1.

The division subroutine has two entry points; *ddivs* should be called to execute a signed division, and *ddivq* should be called to execute an unsigned division. The section of the routine starting at the *ddivs* label calculates the sign bit by comparing the MSW of the divisor with the MSW of the dividend and shifts the dividend one bit left. The CNTR register is set to 31, the number of bits that remain to be calculated. The section of the routine starting at the *ddivq* label sets the CNTR register to 32 and AX1, which is used to compare the divisor and the partial remainder, to zero. The portion of the routine common to both signed and unsigned division is the *ddivu* loop. In this loop, the divisor is subtracted from the partial remainder if both have the same sign; otherwise, the divisor is added to the partial remainder. The quotient bit is determined by comparing the MSW of the divisor (in AX0) with the partial remainder (in AF). The dividend is shifted one bit left, and the loop is repeated until all 32 bits of the quotient have been computed.

```
.MODULE      Double_Precision_Divide;

{
       Double-Precision Division
              Z = X ÷ Y

       Calling Parameters
              AY0 = LSW of Y
              AY1 = MSW of Y
              SR1 = MSW of X
              SR0 = Next Significant Word of X
              MR1 = Next Significant Word of X
              MR0 = LSW of X
              SE = -15

       Return Values
              MR1 = MSW of Z
              MR0 = LSW of Z

       Altered Registers
              AF,AR,AX1,AX0,SI,SR,MR

       Computation Time
              485 cycles (maximum)
}

.ENTRY       ddivs;
.ENTRY       ddivq;

ddivs:       AF=PASS SR1;
             SI=SR0, AR=SR1 XOR AY1;    {Exclusive OR sign bits}
             AX1=AR;
             SR=LSHIFT MR0 BY 1(LO);    {shift dividend up 1 bit}
             SR=SR OR LSHIFT MR1 BY 1(HI);
             SR=SR OR LSHIFT AR(LO);    {shift in quotient-sign bit}
             AR=PASS AF, MR0=SR0;
             MR1=SR1, SR=LSHIFT MR1(LO);
             SR=SR OR LSHIFT SI BY 1(LO);
             SR=SR OR LSHIFT AR BY 1(HI);
             CNTR=31;
             JUMP ddiv;
```

```
ddivq:      CNTR=32;
            AX1=0;
ddiv:       AX0=AY1;
            DO ddivu UNTIL CE;
                   AR=ABS AX1;                {is quotient bit set?}
                   IF POS JUMP aqz;           {no, -divisor from partial remainder}
aqo:               AR=SR0+AY0;                {yes, +divisor to partial remainder}
                   SI=AR, AF=SR1+AY1+C;
                   JUMP ddivi;
aqz:               AR=SR0-AY0;
                   SI=AR, AF=SR1-AY1+C-1;
ddivi:             SR=LSHIFT MR0 BY 1(LO);    {shift dividend 1 bit}
                   SR=SR OR LSHIFT MR1 BY 1(HI);
                   AR=AX0 XOR AF;             {compute quotient bit}
                   AX1=AR;                    {save quotient bit}
                   AR=NOT AX1;
                   SR=SR OR LSHIFT AR(LO);    {shift in new bit}
                   MR0=SR0, AR=PASS AF;
                   MR1=SR1, SR=LSHIFT MR1(LO);
                   SR=SR OR LSHIFT SI BY 1(LO);
ddivu:             SR=SR OR LSHIFT AR BY 1(HI);
                   RTS;
.ENDMOD;
```

Listing 2.11 Double-Precision Division

2.7 REFERENCES

Mano, M. Morris. 1982. *Computer System Architecture, Second Edition.* Englewood Cliffs, N.J.: Prentice-Hall, Inc.

Floating-Point Arithmetic ■ 3

3.1 OVERVIEW

In fixed-point number representation, the radix point is always at the same location. While this convention simplifies numeric operations and conserves memory, it places a limit on the magnitude and the precision of the number representation. In situations that require a large range of numbers or high resolution, a relocatable radix point is desirable. Very large and very small numbers can be represented in floating-point format.

Floating-point format is scientific notation; a floating-point number consists of a mantissa and an exponent. Each part of the floating-point number is stored in a fixed-point format. The mantissa is usually in a full fractional format, and the exponent is in a full integer format (see Chapter 2 for a discussion of fixed-point formats). In some cases, a constant (excess code or bias) is added to the exponent so that it is always positive.

A floating-point number is "normalized" if it contains no redundant sign bits; all bits are significant. Normalization provides the highest precision for the number of bits available. It also simplifies the comparison of magnitudes, because the number with the greater exponent has the greater magnitude; only if the exponents are equal is it necessary to compare the fractions. Most routines (and all the routines presented in this chapter) assume normalized input and produce normalized results.

Floating-point numbers are inherently inexact because each number has multiple representations that differ only in precision. This fact introduces error into floating-point calculations (relative to the exact result). Floating-point multiplication and division do not magnify this error much, but addition or subtraction can cause significant increases in the error. Therefore, the associative law does not always hold for floating-point calculations. For an excellent discussion of floating-point accuracy see Knuth, 1969.

The routines in this chapter demonstrate ways of performing standard mathematical operations on floating-point numbers using the ADSP-2100. Because floating-point numbers can be stored in a variety of formats, each example assumes that the input numbers are converted to a standard format before the routine is called. The standard format, called "two-word" format, provides one word (16 bits) for the exponent and one word for the fraction. Signed twos-complement notation is assumed for both the fraction and the exponent. The exponent has the option of an excess code; if the excess code is not needed, it should be set to zero for all the routines that use it.

If additional precision is required, the fraction can be expanded. If additional range is needed, the exponent can be expanded. Expanding either the fraction or the exponent can be accomplished by substituting a multiprecision fixed-point arithmetic operation (see Chapter 2) for the basic operation used in the floating-point routine.

The two-word format is tailored for the ADSP-2100. It takes advantage of ADSP-2100 instructions to make calculations or conversions fast and simple. However, certain applications may require the use of IEEE 754 standard floating-point format. Details of the IEEE format can be found in the IEEE-STD-754 document, 1985. The major ways in which IEEE format differs from two-word format are:

- The number consists of a 32-bit doubleword divided into fields for (from left to right) sign bit, exponent, and fraction (mantissa)
- The exponent field is 8-bits wide (unsigned) and biased by +127
- The fraction field is 23-bits wide (unsigned)
- A "hidden bit" with a value 1 is assumed to be to the left of the fraction.

You can choose the numeric format (fixed-point or floating-point) that is better for a particular situation. In this chapter, we present routines that convert numbers from fixed-point format into floating-point format (both IEEE 754 and two-word) and vice versa. These routines are followed by examples of routines for performing basic arithmetic operations on numbers in two-word floating-point format.

3.2 FIXED-POINT TO FLOATING-POINT CONVERSION

Conversion of numbers from 1.15 fixed-point format into IEEE 754 and two-word floating-point format is discussed in this section. The corresponding floating-point to fixed-point conversions are described in the next section.

Two ADSP-2100 instructions used in fixed-point to floating-point conversion are EXP, which derives an exponent, and NORM, which normalizes (eliminates nonsignificant digits from) the mantissa.

3.2.1 Fixed-Point (1.15) to IEEE Floating-Point

Because all numbers that can be represented in 1.15 fixed-point format can also be represented in IEEE 754 floating-point format, the conversion from the fixed-point format to the floating-point format is exact. Conversion from floating-point back to fixed-point format cannot always be exact and therefore generates errors, as discussed in section 3.3.1.

The subroutine in Listing 3.1 converts a number in 1.15 format stored in AX0 into IEEE 754 floating-point format, stored in SR1 (MSW) and SR0 (LSW). The routine first checks for two special-case integer input values, –1 and 0. If the input value is either –1 or 0, the routine outputs the IEEE 754-format values explicitly. All other 1.15 numbers are fractions that can be converted to IEEE 754 format by the section that begins at *cvt*. At this point, the input has been made positive by the absolute value function, and the original sign bit has been preserved in a flag. An exponent is derived, and the number is normalized. The exponent is biased by 126 (the IEEE 754 bias of 127 minus one, because of the hidden bit). The 32-bit result is put together, using the shifter to place the output values into the correct fields.

```
.MODULE cvt_fixed_to_ieee_floating;

{  Converts 1.15 fixed-point to 32-bit IEEE 754 floating-point

   Calling Parameters
        AX0 = 1.15 fixed-point number

   Return Values
        SR1 = MSW of IEEE 754 floating-point number
        SR0 = LSW of IEEE 754 floating-point number

   Altered Registers
        AX1,AY0,AY1,AF,AR,SR,SE,SI

   Computation Time
        20 cycles (maximum)
}
.ENTRY  ieeeflt;

ieeeflt:    AY0=H#8000;                 {neg. one = H#8000}
            AY1=126;                    {-127 bias + 1 for hidden bit shift
            AR=AX0-AY0;                 {AY0 = -1? (H#8000)}
            IF NE JUMP numok;           {no, do conversion}
            AR=H#BF80;                  {if neg one, do float right now}
            SR=LSHIFT AR BY 0 (HI);     {this is IEEE float for -1}
            RTS;                        {skip conversion, output right now}
```

(listing continues on next page)

```
numok:      AR=PASS AX0;                  {1.15 number to convert is in AR}
            AF=ABS AR;                    {make positive}
            IF NE JUMP notzero;           {if not zero, do conversion}
            SR=LSHIFT AR BY 0 (HI);       {special case for zero}
            RTS;                          {exit if zero}
notzero:    IF NEG JUMP itisneg;
itispos:    SI=H#0000;                    {CLEAR sign bit flag if positive}
            JUMP cvt;
itisneg:    SI=H#8000;                    {SET sign bit flag if negative}
cvt:        AR=PASS AF;                   {use abs(orig_number) }
            SE=EXP AR (HI);               {derive exponent}
            SR=NORM AR (HI);              {normalize fraction}
            AX1=SE;                       {load AX1 with exponent}
            AR=AX1 + AY1;                 {add 126 to exponent}
            SR=LSHIFT SR1 BY +2 (HI);     {remove sign & hidden bit}
            SR=LSHIFT SR1 BY -9 (HI);     {open sign bit and expo field}
            SR=SR OR LSHIFT SI BY 0 (HI); {paste sign bit}
            SR=SR OR LSHIFT AR BY +7 (HI); {paste exponent}
            RTS;
.ENDMOD;
```

Listing 3.1 Fixed-Point to IEEE Floating-Point

3.2.2 Fixed-Point (1.15) to Two-Word Floating-Point

The routine shown in Listing 3.2 converts a number in 1.15 fixed-point format into two-word floating-point format. An exponent is derived, the number is normalized, and a bias is added to the exponent value. If no bias is needed, the routine should be called with the bias value set to zero.

```
.MODULE single_fixed_to_floating;

{
   Convert 1.15 fixed-point to two-word floating-point

   Calling Parameters
        AR = fixed point number             [1.15 signed twos complement]
        AX0 = exponent bias (0=unbiased)    [16.0 signed twos complement]

   Return Values
        AR = biased exponent                [16.0 signed twos complement]
        SR1 = mantissa                      [1.15 signed twos complement]
```

```
    Altered Registers
         SE,SR,AY0,AR

    Computation Time
         5 cycles
}

.ENTRY   fltone;

fltone:  SE=EXP AR (HI);                {Determine exponent}
         SR=NORM AR (HI);               {Remove redundant sign bits}
         AY0=SE;
         AR=AX0+AY0;                    {Add bias}
         RTS;
.ENDMOD;
```

Listing 3.2 Fixed-Point to Two-Word Floating-Point

3.3 FLOATING-POINT TO FIXED-POINT CONVERSION

Conversion of numbers from either IEEE 754 or two-word floating-point format into 1.15 fixed-point format is discussed in this section. The corresponding fixed-point to floating-point conversions are described in the previous section.

3.3.1 IEEE Format to Fixed-Point Format (1.15)

Not all numbers that can be represented in IEEE 754 floating-point format can be represented in 1.15 fixed-point format. Therefore, the routine that converts from IEEE 754 floating-point to 1.15 fixed-point format generates an error word. The error word indicates which error condition (positive or negative, overflow or underflow) the IEEE 754 floating-point number conversion creates and also if a loss of precision occurs due to the truncation of the 23-bit mantissa to 15 bits. Truncation forces rounding toward zero, one of four possible rounding modes defined in the IEEE 754 standard. The error word protocol is listed below.

No loss of precision (8 LSBs of IEEE 754 mantissa are all zeros):

Error Condition	*Error Word*	*Result (hexadecimal)*
none	0000	1.15 conversion result
positive overflow	F000	7FFF
positive underflow	0F00	0000
negative underflow	00F0	0000
negative overflow	000F	8001

8-bit loss of precision:

Error Condition	*Error Word*	*Result (hexadecimal)*
precision loss only	FFFF	1.15 conversion result
positive overflow	0FFF	7FFF
positive underflow	F0FF	0000
negative underflow	FF0F	0000
negative overflow	FFF0	8001

The subroutine shown in Listing 3.3 converts an IEEE 754 floating-point number to 1.15 fixed-point format. The IEEE 754 floating-point MSW is read from MR1, and the LSW is read from MR0. The 1.15 fixed-point result is returned in MX1, and the error word is returned in MX0.

The routine first checks the input for the special cases of integers –1 and 0. If the input is either of these integers, the conversion is loaded into MX1 directly and returned. If the input is fractional, the routine checks the exponent field to determine whether the input number is out of the 1.15 fixed-point format range. If it is, the input is examined further to set the appropriate error word for the error condition. Then the error word is toggled if the eight LSBs of the input number are not all zeros. The conversion ignores the eight LSBs, so precision is lost if any of these bits are nonzero.

The conversion of input that can be represented in 1.15 format is done in the section beginning at the label *convert*. First, the exponent field is extracted and unbiased, and the resulting exponent is stored in the SE register. Then, the mantissa is shifted by the amount stored in the SE register. This shift results in a valid 1.15 number in the SR1 register. Finally, if the input number is negative, the result is negated.

```
.MODULE cvt_ieee_float_to_fixed;

{
   Convert 32-bit IEEE 754 floating-point to 1.15 fixed-point

   Calling Parameters
        MR1 = MSW of IEEE 754 floating point number
        MR0 = LSW of IEEE 754 floating point number

   Return Values
        MX1 = 1.15 fixed point number
        MX0 = ERROR word

   Altered Registers
        AX0,AX1,AY0,AY1,AF,AR,MX0,MX1,SR,SE

   Computation Time
        51 cycles (maximum)
}

.ENTRY  ieeefix;

ieeefix:    AY0=H#00FF;                   {mask to extract exp, lo_bit check}
            AY1=H#0100;                   {mask for overflow sign}
            MX1=H#0000;                   {clear fixed result}
            MX0=H#0000;                   {clear ERROR status}
            AX1=H#0000;                   {clear UNDER/OVERFLOW flag}

zero:       SR=LSHIFT MR1 BY 0 (LO);      {check for flt_num = 0}
            SR=SR OR LSHIFT MR0 BY 0 (LO);
            AR=PASS SR0;
            IF EQ RTS;

negone:     AF=PASS MR0;                  {check for flt_num = -1}
            IF NE JUMP checkexpo;
            AF=PASS MR1;
            AX0=H#BF80;
            AR=AX0-AF;
            IF NE JUMP checkexpo;
            MX1=H#8000;
            MX0=H#0000;
            RTS;

checkexpo:  SR=LSHIFT MR1 BY -7 (LO);     {extract exponent}
            AF=SR0 AND AY0;               {extract exponent}
            AX0=H#007E;                   {upper valid expo - 126 decimal}
            AR=AF-AX0;                    {is expo .gt. max valid ???}
            IF GT JUMP overflow;          {if YES, then overflow}
            AX0=H#0070;                   {lower valid expo - 112 decimal}
            AR=AF-AX0;                    {is expo .lt. min valid ???}
            IF LT JUMP underflow;         {if YES, then underflow}
            JUMP eightlsb;                {go check if 8 LSBs are set}
```

(listing continues on next page)

```
overflow:   AX1=H#FFFF;                       {set UNDER/OVERFLOW FLAG true}
            AF=SR0 AND AY1;                   {extract sign bit with AND mask}
            IF NE JUMP negover;
posover:    MX0=H#F000;                       {ERROR = "positive overflow"}
            MX1=H#7FFF;                       {make fixed result max. pos. value}
            JUMP eightlsb;
negover:    MX0=H#000F;                       {ERROR = "negative overflow"}
            MX1=H#8001;                       {make fixed result max. neg. value}
            JUMP eightlsb;

underflow:  AX1=H#FFFF;                       {set UNDER/OVERFLOW FLAG true}
            AF=SR0 AND AY1;                   {extract sign bit with AND mask}
            IF NE JUMP negunder;
posunder:   MX0=H#0F00;                       {ERROR = "positive underflow"}
            JUMP eightlsb;                    {fixed result remains zero}
negunder:   MX0=H#00F0;                       {ERROR = "negative underflow"}
            JUMP eightlsb;                    {fixed result remains zero}

eightlsb:   SR=LSHIFT MR0 BY 0 (LO);          {get 16 LSBs of flt_num}
            AF=SR0 AND AY0;                   {extract lower 8 LSBs with AND mask}
            IF EQ JUMP endlsb;
            AR=MX0;                           {if any are set, toggle ERROR}
            AR=NOT AR;
            MX0=AR;                           {ERROR value stored in MX0}
endlsb:     AR=PASS AX1;                      {check for under/overflow situation}
            IF NE RTS;                        {do not convert if under/overflow}

convert:    SR=LSHIFT MR1 BY -7 (LO);         {set up exponent field to mask}
            AF=SR0 AND AY0;                   {extract exponent field by AND}
            AX0=H#007F;                       {exponent bias - 127 decimal}
            AR=AF-AX0;                        {subtract bias from exponent}
            SE=AR;                            {unbiased expo into SE register}
            SR=LSHIFT MR1 BY 8 (HI);          {paste hi mantissa word}
            SR=SR OR LSHIFT MR0 BY 8(LO);     {paste lo mantissa word}
            AY1=H#8000;                       {hidden "1" bit}
            AR=SR1 OR AY1;                    {paste hidden "1" bit}
            SR=LSHIFT AR (HI);                {denormalize mantissa}
            AF=PASS MR1;                      {set sign bit if orig. was neg.}
            AR=SR1;                           {use only 16 MSBs of result}
            IF LT AR=-SR1;                    {if neg. orig. do twos complement}
            MX1=AR;
            RTS;
.ENDMOD;
```

Listing 3.3 IEEE Floating-Point to Fixed-Point

3.3.2 Two-Word Format to Fixed-Point Format (1.15)

Converting two-word floating-point numbers to 1.15 fixed-point format is very simple using ADSP-2100 instructions. The exponent word is decremented by the exponent bias that has been passed to the conversion routine, and the result is stored in the SE register. The SE register determines the amount of shift performed by a shift instruction for which no immediate shift value is given. After the fractional part is shifted (arithmetically) the 1.15 fixed-point result is in the SR1 register. The conversion routine is shown in Listing 3.4. Note that this routine does not provide error handling for floating-point numbers that cannot be represented in fixed-point format.

```
.MODULE cvt_2word_float_to_fixed;

{
   Convert two-word floating-point to 1.15 fixed-point

   Calling Parameters
        AX0 = exponent              [16.0 signed twos complement]
        AY0 = exponent bias         [16.0 signed twos complement]
        SI = mantissa               [1.15 signed twos complement]

   Return Values
        SR1 = fixed-point number    [1.15 signed twos complement]

   Altered Registers
        AR,SE,SR

   Computation Time
        4 cycles
}

.ENTRY  fixone;

fixone: AR=AX0-AY0;                 {Compute unbiased exponent}
        SE=AR;
        SR=ASHIFT SI (HI);          {Shift fractional part}
        RTS;
.ENDMOD;
```

Listing 3.4 Two-Word Floating-Point to Fixed-Point

3.4 FLOATING-POINT ADDITION

The algorithm for adding two numbers in two-word floating-point format is as follows:

1. Determine which number has the larger exponent. Let's call this number X (= Ex, Fx) and the other number Y (= Ey, Fy).
2. Set the exponent of the result to Ex.
3. Shift Fy right by the difference between Ex and Ey, to align the radix points of Fx and Fy.
4. Add Fx and Fy to produce the fraction of the result.
5. Normalize the result.

Note that if the exponents are equal, the exponent of the result can be set to either, and no shifting of the fraction is necessary before the addition.

The ADSP-2100 version of the above algorithm is shown in Listing 3.5. The routine reads the exponents of the input operands from AX0 and AY0 and the corresponding fractions from AX1 and AY1. Upon return, AR holds the exponent of the result and SR1 holds the fraction. The routine first determines the operand with the largest exponent and shifts the fractional part of the other operand to equate the exponents. The fractions are added to form an unnormalized sum. This sum is fed to the exponent detector (in HIX mode to allow for overflow in the ALU) to determine the direction and magnitude of the shift required to normalize the number. The NORM instruction of the shifter uses the negative of the value in SE for the magnitude of the shift. The value in SE is then added to the exponent of the result to yield the normalized exponent.

```
.MODULE floating_point_add;

{
   Floating-Point Addition
        z = x + y

   Calling Parameters
        AX0 = Exponent of x
        AX1 = Fraction of x
        AY0 = Exponent of y
        AY1 = Fraction of y

   Return Values
        AR = Exponent of z
        SR1 = Fraction of z
```

```
    Altered Registers
         AX0,AY1,AY0,AF,AR,SI,SE,SR

    Computation Time
         11 cycles
}

.ENTRY   fpa;

fpa:     AF=AX0-AY0;                  {Is Ex > Ey?}
         IF GT JUMP shifty;           {Yes, shift y}
         SI=AX1, AR=PASS AF;          {No, shift x}
         SE=AR;
         SR=ASHIFT SI (HI);
         JUMP add;
shifty:  SI=AY1, AR=-AF;
         SE=AR;
         SR=ASHIFT SI (HI), AY1=AX1;
         AY0=AX0;
add:     AR=SR1+AY1;                  {Add fractional parts}
         SE=EXP AR (HIX);
         AX0=SE, SR=NORM AR (HI);     {Normalize}
         AR=AX0+AY0;                  {Compute exponent}
         RTS;
.ENDMOD;
```

Listing 3.5 Floating-Point Addition

3.5 FLOATING-POINT SUBTRACTION

The algorithm for subtracting one number from another in two-word floating-point format is as follows:

1. Determine which number has the larger exponent. Let's call this number X (= Ex, Fx) and the other number Y (= Ey, Fy).
2. Set the exponent of the result to Ex.
3. Shift Fy right by the difference between Ex and Ey, to align the radix points of Fx and Fy.
4. Subtract the fraction of the subtrahend from the fraction of the minuend to produce the fraction of the result.
5. Normalize the result.

Note that if the exponents are equal, the exponent of the result can be set to either, and no shifting of the fraction is necessary before the subtraction.

The ADSP-2100 version of the above algorithm is shown in Listing 3.6. The routine reads the exponents of the input operands from AX0 and AY0 and the corresponding fractions from AX1and AY1. Upon return, AR holds the exponent of the result and SR1 holds the fraction. The routine first determines the operand with the largest exponent and shifts the fractional part of the other operand to equate the exponents. The unnormalized difference of the fractions is then found. This difference is fed to the exponent detector (in HIX mode to allow for overflow in the ALU) to determine the direction and magnitude of the shift required to normalize the number. The NORM instruction of the shifter uses the negative of the value in SE for the magnitude of the shift. The value in SE is then added to the exponent of the result to yield the normalized exponent.

```
.MODULE floating_point_subtract;

{
   Floating-Point Subtraction
        z = x - y

   Calling Parameters
        AX0 = Exponent of x
        AX1 = Fraction of x
        AY0 = Exponent of y
        AY1 = Fraction of y

   Return Values
        AR = Exponent of z
        SR1 = Fraction of z

   Altered Registers
        AX0,AY1,AY0,AF,AR,SI,SE,SR

   Computation Time
        11 cycles
}

.ENTRY  fps;

fps:    AF=AX0-AY0;                   {Is Ex > Ey?}
        IF GT JUMP shifty;            {Yes, shift y}
        SI=AX1, AR=PASS AF;           {No, shift x}
        SE=AR;
        SR=ASHIFT SI (HI);
        AR=SR1-AY1;                   {Subtract fractions}
        JUMP subt;
```

```
shifty: SI=AY1, AR=-AF;
        SE=AR;
        SR=ASHIFT SI (HI);
        AY1=SR1;
        AY0=AX0, AR=AX1-AY1;          {Subtract fractions}
subt:   SE=EXP AR (HIX);
        AX0=SE, SR=NORM AR (HI);      {Normalize}
        AR=AX0+AY0;                   {Compute exponent}
        RTS;
.ENDMOD;
```

Listing 3.6 Floating-Point Subtraction

3.6 FLOATING-POINT MULTIPLICATION

Multiplication of two numbers in two-word floating-point format is simpler than either addition or subtraction, because there is no need to align the radix points. The algorithm to multiply two numbers *x* and *y* (Ex, Fx and Ey, Fy) whose exponents are biased by an excess code of *b* (which may be set to zero) is as follows:

1. Add Ex and Ey; subtract *b* from this sum to produce the exponent of the result.
2. Multiply Fx by Fy to produce the fraction of the result.
3. Normalize the result.

The ADSP-2100 routine shown in Listing 3.7 reads the exponents of the operands from AX0 and AY0 and the corresponding fractions from AX1 and AY1. The excess value, *b,* is read from MX0. This routine returns the exponent of the result in AR, and the fraction in SR1. After the exponent and fraction of the result are calculated, the routine checks the MV bit for overflow of the least significant 32 bits of the MR register. If MV is set, the MR register is saturated to its full scale value. Saturation is necessary because the exponent detector is unable to process overflowed numbers in the multiplier. If MR were not saturated on overflow, the routine would incorrectly compute the product of –1 and –1 as –1. The routine finishes by normalizing the product.

```
.MODULE floating_point_multiply;

{
   Floating-Point Multiply
        Z = X × Y

   Calling Parameters
        AX0 = Exponent of X
        AX1 = Fraction of X
        AY0 = Exponent of Y
        AY1 = Fraction of Y
        MX0 = Excess Code

   Return Values
        AR = Exponent of Z
        SR1 = Fraction of Z

   Altered Registers
        AF,AR,AX0,MY1,MX1,MR,SE,SR

   Computation Time
        9 cycles
}

.ENTRY  fpm;

fpm:    AF=AX0+AY0, MX1=AX1;         {Add exponents}
        MY1=AY1;
        AX0=MX0, MR=MX1*MY1 (RND);   {Multiply fractions}
        IF MV SAT MR;                {Check for overflow}
        SE=EXP MR1 (HI);
        AF=AF-AX0, AX0=SE;           {Subtract bias}
        AR=AX0+AF;                   {Compute exponent}
        SR=NORM MR1 (HI);            {Normalize}
        RTS;
.ENDMOD;
```

Listing 3.7 Floating-Point Multiplication

3.7 FLOATING-POINT DIVISION

The algorithm to divide one number X (= Ex, Fx) by another number Y (= Ey, Fy) in two-word floating-point format is as follows:

1. Subtract Ey from Ex; add the excess value (if any) to this number to form the exponent of the result.
2. Divide Fx by Fy to yield the fraction of the result.
3. Normalize the result.

The ADSP-2100 implementation of this algorithm is shown in Listing 3.8. The routine reads the exponent of X (the dividend) from AX0 and the fraction from AX1. It reads the exponent of Y (the divisor) from AY0 and the fraction from AY1. The excess code *b* is read from MX0. The routine returns the exponent of the quotient in AR, and the fraction of the quotient in SR1. Because both Fx and Fy are in 1.15 format, their division produces a 1.15 quotient. To ensure a valid (1.15 format) quotient, Fx must be less than Fy. If Fx is not less than Fy, the routine shifts Fx one bit right, and Ex is increased by one. After the shift, the division can be performed without producing an overflow. The routine finishes by normalizing the result.

```
.MODULE floating_point_divide;

{
   Floating-Point Divide
        z = x ÷ y

   Calling Parameters
        AX0 = Exponent of x
        AX1 = Fraction of x
        AY0 = Exponent of y
        AY1 = Fraction of y
        MX0 = Excess Code

   Return Values
        AR = Exponent of z
        SR1 = Fraction of z

   Altered Registers
        AF,AR,MR,SE,SI,SR,AX1,AX0,AY0

   Computation Time
        33 cycles (maximum)
}
```

(listing continues on next page)

```
.ENTRY   fpd;

fpd:     SR0=AY1, AR=ABS AX1;
         SR1=AR, AF=ABS SR0;
         SI=AX1, AR=SR1-AF;           {Is Fx > Fy?}
         IF LT JUMP divide;           {Yes, go divide}
         SR=ASHIFT SI BY -1 (LO);     {No, shift Fx right}
         AF=PASS AX0;
         AR=AF+1, AX1=SR0;            {Increase exponent}
         AX0=AR;
divide:  AX0=MX0, AF=AX0-AY0;
         MR=0;
         AR=AX0+AF, AY0=MR1;
         AF=PASS AX1, AX1=AY1;        {Add bias}
         DIVS AF, AX1;                {Divide fractions}
         DIVQ AX1; DIVQ AX1; DIVQ AX1; DIVQ AX1; DIVQ AX1;
         DIVQ AX1; DIVQ AX1; DIVQ AX1; DIVQ AX1; DIVQ AX1;
         DIVQ AX1; DIVQ AX1; DIVQ AX1; DIVQ AX1; DIVQ AX1;
         MR0=AY0, AF=PASS AR;
         SI=AY0, SE=EXP MR0 (HI);
         AX0=SE, SR=NORM SI (HI);     {Normalize}
         AR=AX0+AF;                   {Compute exponent}
         RTS;
.ENDMOD;
```

Listing 3.8 Floating-Point Division

3.8 FLOATING-POINT MULTIPLY/ACCUMULATE

The floating-point multiply/accumulate routine computes the sum of N two-operand products. This value can also be found using repeated calls to the floating-point multiplication and addition routines, but the multiply/accumulate routine functions more efficiently because it removes overhead. The multiply/accumulate algorithm is as follows:

1. Multiply the first two operands and normalize the product.
2. Multiply the next two operands and normalize the product.
3. Compare the product to the accumulated result, and shift one or the other to align the radix points.
3. Add the product to the accumulated result and normalize the sum.
4. Repeat steps 2 to 4 until all input operands are exhausted.

The routine shown in Listing 3.9 uses I0 to point to the *x* buffer, I1 to point to the *y* buffer. Each buffer should be organized with the exponent of each value first, followed by the fraction. The routine calculates the first product before entering the loop, so CNTR should store the value of the buffer length minus one. MX0 stores the excess value (which may be zero). M0 should be initialized to one. The multiply/accumulate result is returned with the exponent in AR and the fraction in SR1.

After each product is calculated, the MV bit is checked to see whether the MR register overflowed. If overflow occurs, MR is saturated to positive full scale. This saturation is necessary because the exponent detector cannot process overflowed MR register values.

```
.MODULE floating_point_multiply_accumulate;

{  Floating-Point Multiply/Accumulate
                  n
         z = Σ (x(i) × y(i))
                i=1

   Calling Parameters
         I0 -> x Buffer                  L0 = 0
         I1 -> y Buffer                  L1 = 0
         M0 = 1
         CNTR = Length of Buffer - 1
         MX0 = Excess Code

   Return Values
         AR = Exponent of z
         SR1 = Fraction of z

   Altered Registers
         AF,AR,AX0,AX1,AY0,AY1,MX1,MY1,SE,MR,SR

   Computation Time
         13 × (n-1)+16
}
.ENTRY   fpmacc;

fpmacc:  AX0=DM(I0,M0);                  {Get 1st Ex}
         AY0=DM(I1,M0);                  {Get 1st Ey}
         AF=AX0+AY0, MX1=DM(I0,M0);      {Add exp., get 1st Fx}
         AR=PASS AF, MY1=DM(I1,M0);      {Get 1st Fy}
         AX1=AR, MR=MX1*MY1(RND);        {Multiply fractions}
         IF MV SAT MR;                   {Check for overflow}
         SE=EXP MR1(HI);
         AY1=SE, SR=NORM MR1(HI);        {Normalize}
         AR=AX1+AY1, AX0=DM(I0,M0);
         AX1=AR;
         AY0=DM(I1,M0);
```

```
         DO macc UNTIL CE;
         AF=AX0+AY0, MX1=DM(I0,M0);  {Compute product exp.}
         AR=AX1-AF, MY1=DM(I1,M0);   {Sum exp. > product exp.?}
         IF GT JUMP shiftp;          {Yes, shift product}
         SE=AR, MR=MX1*MY1(RND);     {No, shift sum}
         IF MV SAT MR;
         AY1=MR1, AR=PASS AF;
         AX1=AR, SR=ASHIFT SR1(HI);
         JUMP add;
shiftp:  AF=PASS AR;
         AR=-AF;
         SE=AR, MR=MX1*MY1(RND);
         IF MV SAT MR;
         AY1=SR1, SR=ASHIFT MR1(HI);
add:     AR=SR1+AY1, AX0=DM(I0,M0);  {Accumulate}
         SE=EXP AR(HIX);
         AY1=SE, SR=NORM AR(HI);     {Normalize}
         AR= AX1+AY1, AY0=DM(I1,M0);
macc:    AX1=AR;
         SR0=MX0;                    {Get bias}
         AF=PASS SR0;
         AR=AX1-AF;                  {Subtract bias}
         RTS;
.ENDMOD;
```

Listing 3.9 Floating-Point Multiply/Accumulate

3.9 REFERENCES

Knuth, D. E. 1969. *The Art of Computer Programming: Volume 2 / Seminumerical Algorithms*. Second Edition. Reading, MA: Addison-Wesley Publishing Company.

IEEE Standard for Binary Floating-Point Arithmetic: ANSI/IEEE Std 754-1985. 1985. New York: The Institute of Electrical and Electronics Engineers, Inc.

Function Approximation 4

4.1 OVERVIEW

Transcendental functions such as sines and logarithms are often approximated by polynomial expansions. The most widely used of these expansions are the Taylor and McLaren series. They can be used to approximate almost any function whose derivative is defined over the specified input range. The ADSP-2100 routines in this chapter produce function approximations from polynomial expansions, except for random number generation, which is accomplished using the linear congruence method.

Because the ADSP-2100 performs single precision (16-bit) fixed-point operations, the accuracy of a polynomial expansion approximation decreases as the order of the polynomial increases. In order to achieve accuracy in a polynomial expansion of a limited order, we have provided optimized coefficients for the polynomials used in the function approximations in this chapter. Most of these coefficients were calculated using the statistical analysis technique of regression. The coefficients are given in the fixed-point hexadecimal format that allows the maximum precision for the necessary magnitude.

In the interests of simplicity and accuracy, some formulas used in this chapter are valid for a limited range of input. These routines employ the properties of the particular function to scale or offset the input value to a value within the valid range and thereby expand the range to accommodate virtually any input.

4.2 SINE APPROXIMATION

The following formula approximates the sine of the input variable x:

$$\sin(x) = 3.140625x + 0.02026367x^2 - 5.325196x^3 + 0.5446778x^4 + 1.800293x^5$$

The approximation is accurate for any value of x from 0° to 90° (the first quadrant). However, because $\sin(-x) = -\sin(x)$ and $\sin(x) = \sin(180° - x)$, you can infer the sine of any angle from the sine of an angle in the first quadrant.

The routine that implements this sine approximation, accurate to within two LSBs, is shown in Listing 4.1. This routine accepts input values in 1.15 format. The coefficients, which are initialized in data memory in 4.12 format, have been adjusted to reflect an input value scaled to the maximum range allowed by this format. On this scale, 180° equals the maximum positive value, H#7FFF, and –180° equals the maximum negative value, H#8000, as shown in Figure 4.1.

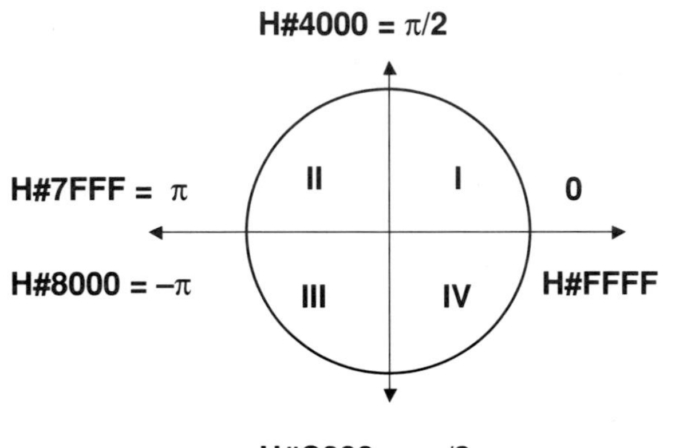

Figure 4.1 Scaled Angle Values

The routine shown in Listing 4.1 reads the scaled input angle from AX0. This angle is first modified to generate the angle in the first quadrant that will yield the same sine (or negative sine). If the input is in the second or fourth quadrant (bit 14 of the input value is a one) the input is negated to produce the twos complement, which represents an angle in the third or first quadrant, respectively. The sign bit of this angle is cleared to produce an angle in the first quadrant, and this result is stored in AR.

If the original angle is in the first quadrant, its value is unchanged. If it is in the second quadrant, negation changes it to the third quadrant, and the sign bit removal changes it to the first quadrant. If the original angle is in the third quadrant, the removal of the sign bit changes it to the first quadrant. An angle that is originally in the fourth quadrant is changed to the first quadrant by negation.

The sine of the modified angle is calculated by multiplying increasing powers of the angle by the appropriate coefficients. The square of the angle is computed and stored in MF while the first coefficient is fetched from data memory. The first term of the sine approximation is stored in the MR registers (in which the result is subsequently accumulated) in parallel with the second coefficient fetch. In the *approx* loop, the next term of the approximation is computed and added to the partial result in MR; then a multifunction instruction fetches the next coefficient and generates the next power of the angle at the same time.

Because the coefficients are in 4.12 format, a shift instruction is needed to scale the result to 1.15 format. The result is then checked for overflow. If the value in SR1 exceeds H#7FFF, the routine saturates the result at the maximum positive value, H#7FFF, which is read from AY0. Then the sign of the result is restored, if necessary. If the input angle (stored in AX0) is negative, the result must be negated.

```
.MODULE Sin_Approximation;

{
   Sine Approximation
        Y = Sin(x)

   Calling Parameters
        AX0 = x in scaled 1.15 format
        M3 = 1
        L3 = 0

   Return Values
        AR = y in 1.15 format

   Altered Registers
        AY0,AF,AR,MY1,MX1,MF,MR,SR,I3

   Computation Time
        25 cycles
}
```

(listing continues on next page)

```
.VAR/DM sin_coeff[5];

.INIT   sin_coeff : H#3240, H#0053, H#AACC, H#08B7, H#1CCE;

.ENTRY  sin;

sin:    I3=^sin_coeff;                     {Pointer to coeff. buffer}
        AY0=H#4000;
        AR=AX0, AF=AX0 AND AY0;            {Check 2nd or 4th quad.}
        IF NE AR=-AX0;                     {If yes, negate input}
        AY0=H#7FFF;
        AR=AR AND AY0;                     {Remove sign bit}
        MY1=AR;
        MF=AR*MY1 (RND), MX1=DM(I3,M3);    {MF = x2}
        MR=MX1*MY1 (SS), MX1=DM(I3,M3);    {MR = C1x}
        CNTR=3;
        DO approx UNTIL CE;
           MR=MR+MX1*MF (SS);
approx:    MF=AR*MF (RND), MX1=DM(I3,M3);
        MR=MR+MX1*MF (SS);
        SR=ASHIFT MR1 BY 3 (HI);
        SR=SR OR LSHIFT MR0 BY 3 (LO);     {Convert to 1.15 format}
        AR=PASS SR1;
        IF LT AR=PASS AY0;                 {Saturate if needed}
        AF=PASS AX0;
        IF LT AR=-AR;                      {Negate output if needed}
        RTS;
.ENDMOD;
```

Listing 4.1 Sine Approximation

4.3 ARCTANGENT APPROXIMATION

The following polynomial expansion computes the arctangent of the variable x, where $x < 1$:

$$\arctan(x) = 0.318253x + 0.003314x^2 - 0.130908x^3 + 0.068542x^4 - 0.009159x^5$$

If $x \geq 1$, the following formula can be used to derive the arctangent:

$$\arctan(x) = 0.5 - \arctan(1/x)$$

The reciprocal of x when $x \geq 1$ is a valid input for the polynomial expansion. The arctangent approximated by these equations is scaled to a range that corresponds to +90° to –90°.

The subroutine shown in Listing 4.2 computes the arctangent of a 32-bit value to within two LSBs. It reads the input in 16.16 format from MR0 (LSW) and MR1 (MSW). The absolute value of the input is calculated and written back to the MR registers. In the section beginning at the *posi* label, the fractional part of the input number (MR0) is shifted one bit to the right to put it in 1.15 format.

If the integer part of the input (in MR1) is zero, the arctangent approximation can be calculated using the input value (in AR) in the polynomial expansion directly, and execution jumps to the *noinv* label. If the MR1 value is not zero, the input is greater than one, and the reciprocal of the input value must be calculated. The value is normalized, and a one in 16.0 format is normalized with the same SE value. Dividing the normalized input value into the shifted one generates the reciprocal of the input value in 1.15 format; this value is written to the AR register.

The input value in AR is used in the polynomial expansion calculation beginning at the *noinv* label. The square of the input value is calculated while the first coefficient is fetched from data memory. The first term of the approximation is calculated while the second coefficient is fetched from data memory. In the *approx* loop, which is executed three times, the next term of the approximation is calculated and added to the partial result in MR in parallel with the fetch of the next coefficient. Then the next power of the input value is calculated and stored in MF. After the loop execution completes, one more instruction is needed to calculate the last term of the approximation and complete the result in MR.

If the input value was less than one, the calculation is complete. If the input was greater than one, (integer part greater than zero), the result must be subtracted from 0.5. The subroutine checks the integer part of the original input, which is in AY1, and if it is not zero, the result is subtracted from 0.5 (H#4000 in 1.15 format, stored in AY0). The last step determines the sign of the result; if the input was negative (determined by the sign of the integer value in AX1), the result is negative; otherwise, the result is positive.

The result in AR is in 1.15 format. It is scaled to a range in which 180° is represented by the maximum positive value (H#7FFF) and –180° is represented by the maximum negative value (H#8000). This approximation yields angles scaled to the range from 90° to –90° (represented by 0.5 to –0.5), as shown in Figure 4.1.

```
.MODULE Arctan_Approximation;

{
   Arctangent Approximation
        y = Arctan(x)

   Calling Parameters
        MR1 = Integer part of x
        MR0 = Fractional part of x
        M3 = 1
        L3 = 0

   Return Values
        AR = Arctan(x) in 1.15 scaled format
        (-0.5 for -90°, 0.5 for 90°)

   Altered Registers
        AX0,AX1,AY0,AY1,AR,AF,MY0,MY1,MX1,MF,MR,SR,SI

   Computation Time
        58 cycles (maximum)
}

.VAR/DM/RAM atn_coeff[5];

.INIT   atn_coeff : H#28BD, H#006D, H#EF3E, H#08C6, H#FED4;

.ENTRY  arctan;

arctan: I3 = ^atn_coeff;                {Point to coefficients}
        AY0=0;
        AX1=MR1;
        AR=PASS MR1;
        IF GE JUMP posi;                {Check for positive input}
        AR=-MR0;                        {Make negative number positive}
        MR0=AR;
        AR=AY0-MR1+C-1;
        MR1=AR;
posi:   SR=LSHIFT MR0 BY -1 (LO);       {Produce 1.15 value in SR0}
        AR=SR0;
        AY1=MR1;
        AF=PASS MR1;
        IF EQ JUMP noinv;               {If input < 1, no need to invert}
        SR=EXP MR1 (HI);                {Invert input}
        SR=NORM MR1 (HI);
        SR=SR OR NORM MR0 (LO);
        AX0=SR1;
        SI=H#0001;
        SR=NORM SI (HI);
        AY1=SR1;
        AY0=SR0;
```

```
        DIVS AY1,AX0;
        DIVQ AX0; DIVQ AX0; DIVQ AX0;
        DIVQ AX0; DIVQ AX0; DIVQ AX0;
        DIVQ AX0; DIVQ AX0; DIVQ AX0;
        DIVQ AX0; DIVQ AX0; DIVQ AX0;
        DIVQ AX0; DIVQ AX0; DIVQ AX0;
        AR=AY0;
noinv:  MY0=AR;
        MF=AR*MY0 (RND), MY1=DM(I3,M3);
        MR=AR*MY1 (SS), MX1=DM(I3,M3);
        CNTR=3;
        DO approx UNTIL CE;
            MR=MR+MX1*MF (SS), MX1=DM(I3,M3);
approx:     MF=AR*MF (RND);
        MR=MR+MX1*MF (SS);
        AR=MR1;
        AY0=H#4000;
        AF=PASS AY1;
        IF NE AR=AY0-MR1;
        AF=PASS AX1;
        IF LT AR=-AR;
        RTS;
.ENDMOD;
```

Listing 4.2 Arctangent Approximation

4.4 SQUARE ROOT APPROXIMATION

The following equation approximates the square root of the input value x, where $0.5 \geq x \geq 1$:

$$sqrt(x) = 1.454895x - 1.34491x^2 + 1.106812x^3 - 0.536499x^4 + 0.1121216x^5 + 0.2075806$$

To determine the square root of an input value outside the range from 0.5 to 1, you must scale the value to a number within the range. After computing the square root of the scaled number, you multiply the result by the square root of the scaling value to produce the square root of the original number. The program shown in this section performs all necessary scaling.

The exponent detector of the shifter in the ADSP-2100 can be used to calculate the necessary scaling value. It determines the amount of left shifting needed to remove redundant sign bits of the input value, if any exist, and stores a number that represents the shift amount in the SE register. Because the format of the input number is 16.16, a left shift of 15 bits (register SE = –15) indicates that the input number is already between 0.5 and 1; no scaling is needed, so the scaling value *s* is one. If the number is shifted left more than 15 bits (register SE < –15), the input number must be multiplied by a scaling value that is greater than one. If the number is shifted left fewer than 15 bits (register SE > –15), the scaling value must be less than one.

The value in SE is the negative of the power of two necessary to shift the value; therefore, the scaling value *s* is equal to 2^{SE+15}. The square root is calculated as follows:

$$X = \sqrt{Y}$$
$$Z = \sqrt{(sY)} = \sqrt{s}\,\sqrt{Y}$$
$$X = Z \div \sqrt{s}$$

The square root of *s* is found as follows:

$$s = 2^{SE+15}$$
$$\sqrt{s} = \sqrt{(2^{SE+15})} = (1 \div (1 \div (\sqrt{2}^{SE+15}))) = (1 \div \sqrt{2})^{-(SE+15)}$$

Incorporating the value of $\sqrt{s}$ into the equation for X yields:

$$X = Z \div ((1 \div \sqrt{2})^{-(SE+15)})$$

The value $(1 \div \sqrt{2})^{-(SE+15)}$, is calculated by storing the reciprocal of $\sqrt{2}$ as a 1.15 constant and multiplying this constant by itself SE+15 times, producing a result in 1.15 format. This result is the value of $\sqrt{s}$ if SE+15 is negative. If SE+15 is positive, the value of $\sqrt{s}$ is the reciprocal of the result, which is found by dividing the result into 1 (in 9.23 format) to produce a value in 9.7 format.

Listing 4.3 shows the ADSP-2100 routine to approximate the square root of *x*, where $0 \leq x < 32768$. The first part of the routine scales the input number and computes the square root of the scaled number. The constant term of the polynomial is stored as the constant *base*, which is loaded into MR to be added to the other terms as they are computed in the *approx* loop.

If the scaling value $\sqrt{s}$ is one (SE+15 = 0), the result is shifted into 8.8 *unsigned* format and returned; otherwise, $\sqrt{s}$ must be computed, beginning at the label *scale*.

The constant *sqrt2*, which is equal to 1 ÷ √2, is loaded into MR1 and MY1. The AR register is loaded with the absolute value of SE+15. If this value is one, then MR contains the value of √s or its reciprocal; if not, the *compute* loop computes the correct power of 1 ÷ √2.

The section that begins at the *pwr_ok* label checks the sign of SE+15. If it is negative, MR contains the correct value of √s. This value is multiplied by the square root approximation of the scaled input, which was stored in MY0. The product is shifted right six bits to put it in 8.8 *unsigned* format and returned. If SE+15 is positive, the reciprocal of the value in MR is calculated to yield the correct value of √s. The product of √s and the square root approximation in MY0 is calculated and added to H#2000 in MR0 to round the low order bits. This result is shifted right two bits to form an *unsigned* result in 8.8 format.

```
.MODULE Square_root;

{
     Square Root
          y = √x

     Calling Parameters
          MR1 = MSW of x (16.0 portion)
          MR0 = LSW of x (0.16 portion)
          M3 = 1
          L3 = 0

     Return Values
          SR1 = y in 8.8 UNSIGNED format

     Altered Registers
          AX0,AY0,AY1,AR,AF,MY0,MY1,MX0,MF,MR,SE,SR,I3

     Computation Time
          75 cycles (maximum)
}

.CONST   base=H#0D49, sqrt2=H#5A82;

.VAR/DM sqrt_coeff[5];

.INIT    sqrt_coeff : H#5D1D, H#A9ED, H#46D6, H#DDAA, H#072D;

.ENTRY   sqrt;
```

(listing continues on next page)

```
sqrt:   I3=^sqrt_coeff;                          {Pointer to coeff. buffer}
        SE=EXP MR1 (HI);                         {Check for redundant bits}
        SE=EXP MR0 (LO);
        AX0=SE, SR=NORM MR1 (HI);                {Remove redundant bits}
        SR=SR OR NORM MR0 (LO);
        MY0=SR1, AR=PASS SR1;
        IF EQ RTS;
        MR=0;
        MR1=base;                                {Load constant value}
        MF=AR*MY0 (RND), MX0=DM(I3,M3);          {MF = x²}
        MR=MR+MX0*MY0 (SS), MX0=DM(I3,M3);       {MR = base + C₁x}
        CNTR=4;
        DO approx UNTIL CE;
           MR=MR+MX0*MF (SS), MX0=DM(I3,M3);
approx:    MF=AR*MF (RND);
        AY0=15;
        MY0=MR1, AR=AX0+AY0;                     {SE + 15 = 0?}
        IF NE JUMP scale;                        {No, compute √s}
        SR=ASHIFT MR1 BY -6 (HI);
        RTS;
scale:  MR=0;
        MR1=sqrt2;                               {Load 1÷√2}
        MY1=MR1, AR=ABS AR;
        AY0=AR;
        AR=AY0-1;
        IF EQ JUMP pwr_ok;
        CNTR=AR;                                 {Compute (1÷√2)^(SE+15)}
        DO compute UNTIL CE;
compute:    MR=MR1*MY1 (RND);
pwr_ok: IF NEG JUMP frac;
        AY1=H#0080;                              {Load a 1 in 9.23 format}
        AY0=0;                                   {Compute reciprocal of MR}
        DIVS AY1, MR1;
        DIVQ MR1; DIVQ MR1; DIVQ MR1;
        DIVQ MR1; DIVQ MR1; DIVQ MR1;
        DIVQ MR1; DIVQ MR1; DIVQ MR1;
        DIVQ MR1; DIVQ MR1; DIVQ MR1;
        DIVQ MR1; DIVQ MR1; DIVQ MR1;
        MX0=AY0;
        MR=0;
        MR0=H#2000;
        MR=MR+MX0*MY0 (US);
        SR=ASHIFT MR1 BY 2 (HI);
        SR=SR OR LSHIFT MR0 BY 2 (LO);
        RTS;
frac:   MR=MR1*MY0 (RND);

        SR=ASHIFT MR1 BY -6 (HI);
        RTS;
.ENDMOD;
```

Listing 4.3 Square Root Approximation

4.5 LOGARITHM APPROXIMATION

The common logarithm (base ten) of any number *x* between one and two can be approximated using the following equation:

$$2\log_{10}(x) = 0.8678284(x-1) - 0.4255677(x-1)^2 + 0.2481384(x-1)^3 - 0.1155701(x-1)^4 + 0.0272522(x-1)^5$$

The natural logarithm (base *e*) of any number *x* between one and two can be approximated using the following equation:

$$\ln_e(x) = 0.9991150(x-1) - 0.4899597(x-1)^2 + 0.2856751(x-1)^3 - 0.1330566(x-1)^4 + 0.03137207(x-1)^5$$

To calculate the logarithm of a number greater than two or less than one using these formulas, you must scale the value to a number within the valid input range. The exponent detector of the shifter in the ADSP-2100 can be used to scale the input number. It determines the amount of left shifting needed to remove redundant sign bits of the input value, if any exist, and stores a number that represents the shift amount in the SE register. Because the format of the input number is 16.16, a left shift of 14 bits indicates that the input number is already between one and two; no scaling is needed. If the shift is more than 14 bits, the input must be scaled by a value greater than one. If the shift is fewer than 14 bits, the input must be scaled by a value less than one. The exponent will shift the number into the range $0.5 < x < 1$, so the number must be shifted to the left one bit more to place it in the range $1 < x < 2$.

The logarithm of the scaled input must be adjusted to produce the logarithm of the original unscaled input. The adjustment is determined as follows:

$$Y = \log(X)$$
$$Z = \log(sX)$$
$$\log(sX) = \log(s) + \log(X)$$
$$Y = Z - \log(s)$$

Therefore, the logarithm of the unscaled input is equal to the logarithm of the scaled input less the logarithm of the scaling factor *s*. Computation of log(*s*) is simplified by the fact that *s* is a power of two.

$$s = 2^{SE+14}$$
$$\log(s) = \log(2^{SE+14}) = (SE+14)\log(2)$$

Computation of the $\log_{10}$ and $\ln_e$ can be accomplished using a single routine initialized with one of two sets of coefficients (for either $\log_{10}$ or $\ln_e$). The routine shown in Listing 4.4 has two entry points: *ln*, to compute the natural log, and *log*, to compute the common log. The entry point sets I3 to the start of the appropriate coefficient buffer and MY1 to either ln(2) or $\log_{10}(2)$, as appropriate. This routine yields results accurate to within two LSBs.

At the *compute* label, the input number is adjusted to the range from 0.5 to 1 by the exponent detector. Because the negative of the value in SE is used by the NORM command, the value in SE is decremented by one, so that the number is normalized to the range from one to two in unsigned 1.15 format.

The value of SE+14 is stored in AR. After the input number is normalized, ln(2) or $\log_{10}(2)$ (stored in MY1) is multiplied by SE+14 in AR. The value in AR is in 16.0 format and the value in MY1 is in either 1.15 format (common log) or 2.14 format (natural log). The product is stored in MR and shifted left 16 bits to match the format of the terms to be accumulated in the MR register during the approximation.

In the natural log computation, ln(2) is in 2.14 format. When it is multiplied by SE+14, a 25.15 formatted number is produced. Shifting this value left 16 bits yields a number in MR in 9.31 format. Each term of the approximation is produced in 1.31 format and added to the MR value. The final result in MR is shifted left 12 bits (right four bits) to place it in 5.11 format in SR1.

```
.MODULE Logarithms;

{
   Logarithm Approximations
        y=log10(x)
        y=lne(x)

   Calling Parameters
        MR1 = Integer Portion of x in 16.0 twos complement
        MR0 = Fractional Portion of x in 0.16 unsigned
        M3 = 1
        L3 = 0

   Return Values
        SR1 = Y (4.12 format for log; 5.11 format for ln)
```

```
   Altered Registers
        AX0,AY0,AR,MY1,MX1,MF,MR,SE,SR,I3

   Computation Time
        33 cycles (maximum)
}

.CONST  log_2=H#2688,ln_2=H#2C5D;

.VAR/DM ln_coeffs[5];
.VAR/DM log_coeffs[5];
.INIT   ln_coeffs : H#7FE3, H#C149, H#2491, H#EEF0, H#0404;
.INIT   log_coeffs : H#6F15, H#C987, H#1FC3, H#F135, H#037D;

.ENTRY  log,ln;

ln:     MY1=ln_2;                                {Natural log start here}
        I3=^ln_coeffs;
        JUMP compute;
log:    I3=^log_coeffs;                          {Common log start here}
        MY1=log_2;
compute:SE=EXP MR1 (HI);                         {Check for redundant bits}
        SE=EXP MR0 (LO);
        AY0=SE;
        AR=AY0-1;
        AX0=14;
        SE=AR, AR=AX0+AY0;
        SR=NORM MR1 (HI);                        {Remove redundant bits}
        SR=SR OR NORM MR0 (LO);
        MR=AR*MY1 (SS);                          {(SE+14) x log(2)}
        MY1=MR1;                                 {Shift left 16 bits}
        MR1=MR0;
        MR2=MY1;
        MR0=0;
        AY0=H#8000;
        AR=SR1-AY0;
        MY1=AR;
        MF=AR*MY1 (RND), MX1=DM(I3,M3);          {MF = x^2}
        MR=MR+MX1*MY1 (SS), MX1=DM(I3,M3);       {MR = C_1x}
        CNTR=3;
        DO approx UNTIL CE;
            MR=MR+MX1*MF (SS);
approx:     MF=AR*MF (RND), MX1=DM(I3,M3);
        MR=MR+MX1*MF (RND);
        SR=ASHIFT MR2 BY 12 (HI);                {Shift to correct format}
        SR=SR OR LSHIFT MR1 BY 12 (LO);
        RTS;
.ENDMOD;
```

Listing 4.4 Logarithm Approximation

4.6 UNIFORM RANDOM NUMBER GENERATION

Although the generation of a random number is not, strictly speaking, a function, it is a useful operation for many applications. One such application is in high-speed modems, in which it can be used as a training signal for the adaptive equalizer (see Chapter 13). The means for generating random numbers on a digital computer, of course, is by the computation of a function that approximates the random number. Many of such functions have been proposed (Knuth, 1969). The implementation presented here is based on the linear congruence method, which uses the following equation.

$$x(n+1) = (ax(n) + c) \bmod m$$

The initial value of *x*, x(0), is called the seed value and is generally not important, because with a good choice of *a* and *c* all *m* values are generated before the output sequence repeats. The random number sequence produced by the above equation is thus uniform in the sense that the output is uniformly distributed between 0 and $m-1$. Of course, different seed values should be used at different times if different sequences are desired. By choosing the modulus $m=2^{32}$, we ensure a long sequence and have a convenient modulus for the ADSP-2100. The values of *a* and *c* that are used in the following program (a=1664525 and c=32767) were chosen according to the rules in Knuth, 1969.

Listing 4.5 (on the next page) shows the ADSP-2100 routine used to compute random numbers based on the linear congruence method. The first number produced by this routine is the initial seed value in SR1. Note that, although only the most significant 16 bits of the 32-bit *x* value are used as random numbers in this routine, any or all of the bits can be used. However, as stated in Knuth, 1969, when using a value of *m* equal to the word size of the machine, the least significant bits of x(n) are much less random than the most significant bits. Thus, one should always use the *b* most significant bits when only a *b*-bit random number is desired.

The routine requires 10N+4 cycles to execute, where N is the number of random numbers desired. For example, computing 2^{16} (65,536) random numbers using an 8 MHz ADSP-2100 takes 81.9 milliseconds. Computing all $m=2^{32}$ numbers in the sequence requires almost one and a half hours.

```
.MODULE urand_sub;

{
   Linear Congruence Uniform Random Number Generator

   Calling Parameters
        I0 -> Output buffer                       L0 = 0
        M0 = 1
        SR1 = MSW of seed value
        SR0 = LSW of seed value
        CNTR = desired number of random numbers

   Return Values
        Desired number of random numbers in output buffer
        SR1 = MSW of updated seed value
        SR0 = LSW of updated seed value

   Altered Registers
        MY0,MY1,MR,SI,SR

   Computation Time
        10 × N + 4 cycles
}

.ENTRY  urand;

urand:  MY1=25;                                 {Upper half of a}
        MY0=26125;                              {Lower half of a}
        DO randloop UNTIL CE;
            DM(I0,M0)=SR1, MR=SR0*MY1(UU);      {a(hi) × x(lo)}
            MR=MR+SR1*MY0(UU);                  {a(hi) × x(lo) + a(lo) × x(hi)}
            SI=MR1;
            MR1=MR0;
            MR2=SI;
            MR0=H#FFFE;                         {c=32767, left-shifted by 1}
            MR=MR+SR0*MY0(UU);                  {(above) + a(lo) × x(lo) + c}
            SR=ASHIFT MR2 BY 15 (HI);
            SR=SR OR LSHIFT MR1 BY -1 (HI);     {right-shift by 1}
randloop:   SR=SR OR LSHIFT MR0 BY -1 (LO);
        RTS;
.ENDMOD;
```

Listing 4.5 Random Number Generator

4.7 REFERENCES

Burrington, R.S. 1973. *Handbook of Mathematical Tables and Formulas*. Fifth Edition. New York: McGraw-Hill Book Company.

Knuth, D. E. 1969. *The Art of Computer Programming: Volume 2 / Seminumerical Algorithms*. Second Edition. Reading, MA: Addison-Wesley Publishing Company.

Digital Filters ■ 5

5.1 OVERVIEW

The digital computation of filter transfer functions has always been an important area of digital signal processing. Apart from the obvious advantages of virtually eliminating errors in the filter associated with voltage and temperature drift, component aging, and EMI-induced power supply noise, digital filters are capable of performance specifications that would, at best, be extremely difficult to achieve with an analog implementation. Digital filters are able to realize sharp cutoff characteristics, tight passband and stopband specifications, exactly linear phase responses, and even arbitrary magnitude responses.

Many of the routines in this chapter make use of circular buffers for storing data and coefficients. To implement circular addressing, the length register (Ln) that corresponds to the circular buffer pointer register (In) must be set to the buffer length. See the discussion in Chapter 2 of the *ADSP-2100 User's Manual* for more information.

The bibliography at the end of this manual provides several excellent sources of introductions to digital filter theory, design, and implementation.

5.2 FINITE IMPULSE RESPONSE (FIR) FILTERS

A finite impulse response (FIR) filter is a discrete linear time-invariant system whose output is based on the weighted summation of a finite number of past inputs. FIR filters, unlike infinite impulse response (IIR) filters, are nonrecursive and require no feedback loops in their computation. This property allows simple analysis and implementation on microprocessors such as the ADSP-2100. A graphic representation of an FIR filter is shown in Figure 5.1, on the next page.

5 Digital Filters

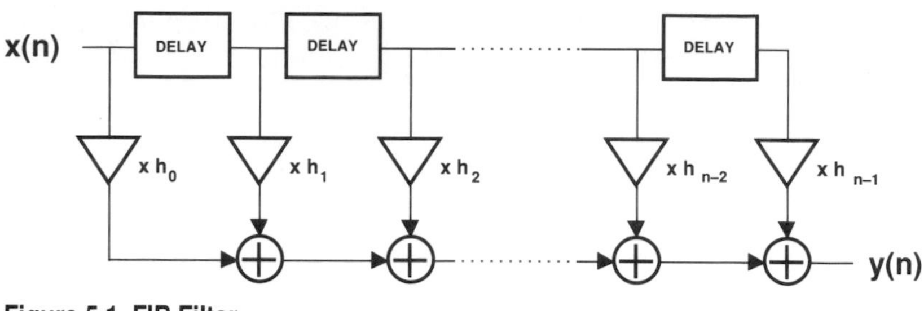

Figure 5.1 FIR Filter

5.2.1 Single-Precision FIR Transversal Filter

The realization of an FIR filter can take many forms, although the most useful in practice are generally the transversal and lattice structures. The FIR lattice filter is described later in this chapter. Another implementation of the transversal filter is given in Chapter 13. An FIR transversal filter structure can be obtained directly from the equation for discrete-time convolution.

$$y(n) = \sum_{k=0}^{N-1} h_k(n)\, x(n-k)$$

In this equation, x(n) and y(n) represent the input to and output from the filter at time *n*. The output y(n) is formed as a weighted linear combination of the current and past input values of x, x(n–k). The weights, $h_k(n)$, are the transversal filter coefficients at time *n*. (For a nonadaptive filter, the coefficients do not change with *n*. Adaptive filters are described later in this chapter.) In the equation, x(n–k) represents the past value of the input signal "contained" in the (k+1)th tap of the transversal filter. For example, x(n), the present value of the input signal, would correspond to the first tap, while x(n–42) would correspond to the forty-third filter tap.

The subroutine that realizes the sum-of-products operation used in computing the transversal filter is shown in Listing 5.1. The first instruction sets up the computation by clearing MR and loading MX0 and MY0 with the first data and coefficient values from data and program memory. The multiply/accumulate with dual data fetch in the *sop* loop is then executed N–1 times in N cycles to compute the sum of the first N–1 products. The final multiply/accumulate instruction is performed with the rounding mode enabled to round the result to the upper 24 bits of MR. MR1 is then conditionally saturated to its most positive or negative value based on the status of the overflow flag MV. In this manner, results are accumulated to the full 40-bit resolution of MR, with saturation of the output only if the final result overflowed beyond the least significant 32 bits of MR.

The limit on the number of filter taps attainable for a real-time implementation of the transversal filter subroutine is determined primarily by the processor cycle time, the sampling rate, and the number of other computations required. The transversal filter subroutine presented here requires a total of N+6 cycles for a filter of length N; at an 8 kHz sampling rate and an instruction cycle time of 125 nanoseconds, this permits a filter of 900 taps with 94 instruction cycles for other operations.

```
.MODULE fir_sub;

{
   FIR Transversal Filter Subroutine

   Calling Parameters
        I0 -> Oldest input data value in delay line
        L0 = Filter length (N)
        I4 -> Beginning of filter coefficient table
        L4 = Filter length (N)
        M1,M5 = 1
        CNTR = Filter length - 1 (N-1)

   Return Values
        MR1 = Sum of products (rounded and saturated)
        I0 -> Oldest input data value in delay line
        I4 -> Beginning of filter coefficient table

   Altered Registers
        MX0,MY0,MR

   Computation Time
        N - 1 + 5 + 2 cycles

   All coefficients and data values are assumed to be in 1.15 format.
}

.ENTRY  fir;

fir:    MR=0, MX0=DM(I0,M1), MY0=PM(I4,M5);
        DO sop UNTIL CE;
sop:       MR=MR+MX0*MY0(SS), MX0=DM(I0,M1), MY0=PM(I4,M5);
        MR=MR+MX0*MY0(RND);
        IF MV SAT MR;
        RTS;
.ENDMOD;
```

Listing 5.1 FIR Filter Single-Precision

5.2.2 Double-Precision FIR Transversal Filter

Many digital filters require a sum-of-products computation using operands that are greater than 16 bits in magnitude. The following subroutine implements a sum-of-products calculation using coefficients and data that are both represented in double precision. On the ADSP-2100, this is accomplished through the use of the mixed-mode multiply instructions, in much the same manner as described in Chapter 2.

The subroutine that realizes the sum-of-products operation used in computing the transversal filter is shown in Listing 5.2. First, the sum of the products of the low halves of the coefficients and the high halves of the data values is computed; this sum is accumulated with the sum of the products of the high halves of the coefficients and the low halves of the data values. This sum is then shifted right 16 bits and then accumulated with the sum of the products of the high halves of the coefficients and the high halves of the data values. A conditional saturation is then performed on the final 32-bit result before storage to data memory. Note that because the result is only the most significant 32 bits, the products of the low-order coefficients and the low-order data affect only the least significant bit of the result and are therefore not computed.

The above routine is easily extended to applications requiring other multiprecision formats or even those requiring mixed precision. For example, to use 32-bit coefficients and 16-bit data values, you would eliminate the *lhloop* loop and make corresponding changes in the data memory pointer values and the size of the circular buffer. Chapter 2 describes the basic techniques for performing multiprecision multiplications, which are directly applicable to multiprecision multiply/accumulate operations.

```
.MODULE dfir_sub;

{
   Double-Precision Transversal Filter Subroutine

   Calling Parameters
        I0 -> Oldest input data value in delay line
        L0 = 2 × Filter length (N)
        I4 -> 2nd location (LSW of 1st value) of filter coefficient table
        L4 = 2 × Filter length (N)
        M0,M4 = 1
        M1,M5 = 2
        M2,M6 = 3
        AX0 = Filter length - 2 (N-2)
        CNTR = Filter length - 2 (N-2)

   Return Values
        MR1,MR0 = sum of products
        (conditionally saturated to 32 bits)
        I0 -> Oldest input data value in delay line
        I4 -> 2nd location (LSW of 1st value) of filter coefficient table

   Altered Registers
        MX0,MY0,MR

   Computation Time
        3 × (N - 2) + 16 + 9

   All coefficients and data values are assumed to be in 1.15 format.
}

.ENTRY  dfir;

dfir:   MR=0, MX0=DM(I0,M1), MY0=PM(I4,M5);
        DO hlloop UNTIL CE;
hlloop:    MR=MR+MX0*MY0(SU), MX0=DM(I0,M1), MY0=PM(I4,M5);
        MR=MR+MX0*MY0(SU), MX0=DM(I0,M2), MY0=PM(I4,M4);
        MR=MR+MX0*MY0(SU), MX0=DM(I0,M1), MY0=PM(I4,M5);
        CNTR=AX0;
        DO lhloop UNTIL CE;
lhloop:    MR=MR+MX0*MY0(US), MX0=DM(I0,M1), MY0=PM(I4,M5);
        MR=MR+MX0*MY0(US), MX0=DM(I0,M0), MY0=PM(I4,M5);
        MR=MR+MX0*MY0(US), MX0=DM(I0,M1), MY0=PM(I4,M5);
        MR0=MR1;                          {downshift 16 places}
        MR1=MR2;
        CNTR=AX0;
        DO hhloop UNTIL CE;
hhloop:    MR=MR+MX0*MY0(SS), MX0=DM(I0,M1), MY0=PM(I4,M5);
        MR=MR+MX0*MY0(SS), MX0=DM(I0,M1), MY0=PM(I4,M6);
        MR=MR+MX0*MY0(SS);
        IF MV SAT MR;
        RTS;
.ENDMOD;
```

Listing 5.2 Double-Precision FIR Filter

5.2.3 Two-Dimensional FIR Filter

The two-dimensional FIR filter is used in a variety of applications, including smoothing and edge detection in image processing, in which the input is a matrix that represents a digitized image. The routine presented in this section is a two-dimensional version of the single-precision FIR filter presented earlier in this chapter. Instead of performing a sum-of-products operation on a one-dimensional input signal, it convolves a two-dimensional (QxR) coefficient matrix with a two-dimensional (SxT) input matrix using the equation:

$$G(x,y) = \sum_{i=0}^{Q-1} \sum_{j=0}^{R-1} [H(i,j)\ F(x-i,\ y-j)] \qquad \text{(Oppenheim, 1978)}$$

The two-dimensional FIR filter is computed by multiplying and accumulating a section of each row of the input matrix by each row of the coefficient matrix. The value of a point in the output matrix is equal to a sum-of-products operation of the input matrix with the coefficient matrix.

The routine, shown in Listing 5.3, assumes that the first (data memory) address of the input matrix is stored in I0, the first output matrix address in I1, and the first coefficient matrix address in I4. The length registers L0 and L1 should each be set to zero, and L4 should be set to the length of (total number of elements in) the QxR coefficent matrix. The number of rows of the output buffer (S–Q) should be stored in the CNTR, and the number of columns of the output buffer (T–R) should be stored in AX0. AX1 should store Q, the number rows in the coefficient matrix. AY0 should store R–2. The modify registers M0 and M4 should both be set to one. M1 should store T–R+1, M2 should be set to $-(Q \times T+1)$, and M3 should store R–1. All of these values must be initialized before the routine is called.

Convolution at the edges of the input matrix can yield meaningless results because the edge values do not have valid adjacent data. This situation can be remedied in several ways; one way is to set any value outside the input matrix to zero. Another way, used in this example, is to perform the convolution only if the coefficient matrix is completely enclosed by the input matrix. To use a coefficient matrix that is not enclosed by the input matrix, you must call the routine with CNTR set to S (the number of input matrix rows), AX0 set to T (the number input matrix columns), and M3 set to zero.

The routine begins by reading the first coefficient into MY0. The *in_row* loop is executed once for each row of the output matrix. In this loop, the CNTR register is loaded and the *in_col* loop is executed, generating a column of the output matrix on each pass. The *row_loop* loop executes the *col_loop* loop once for each column of the coefficient matrix.

The last two multiply/accumulate and data read instructions are removed from the *col_loop* loop in order to provide efficient pointer manipulation. The first multiply/accumulate operation outside the loop is performed in parallel with moving I0 to point to the first element of the next convolution row. The second multiply/accumulate operation is performed in parallel with reading in the values that will be used for the next sequence of execution of the *row_loop* loop. When all iterations of the *row_loop* loop have been executed, the value in MR1 is stored in the appropriate location of the output matrix. The last instruction in the *in_row* loop modifies I0 to point to the first element of the next row of input data.

The output matrix, stored by rows, is smaller than the input matrix if the input matrix fully encloses the coefficient matrix. In this case, the number of cycles the routine requires is:

$$(((R-2+4) \times Q+5) \times (T-R)+3) \times (S-Q)+3+4$$

If the coefficient matrix is not fully enclosed by the input matrix, the number is:

$$(((R-2+4) \times Q+5) \times T+3) \times S+3+4$$

```
.MODULE two_dimensional_FIR_filter;

{            Q-1 R-1
   G(x,y) =  Σ   Σ  [H(i,j) F(x-i, y-j)]
             i=0 j=0

   Calling Parameters
        I0 -> F, SxT Input Matrix stored by rows                   L0 = 0
        I1 -> G, (S-Q)x(T-R) Output Matrix stored by rows          L1 = 0
        I4 -> H, QxR Coefficient Matrix stored by rows             L4 = Q × R
        M0,M4 = 1
        M1 = T-R+1
        M2 = -(Q × T+1)                        AX1 = Q
        M3 = R-1                               AY0 = R-2
        CNTR = S-Q                             AX0 = T-R

   Return Values
        G(x,y) filled [Output Matrix]

   Altered Registers
        MX0,MY0,MR,I0,I1,I4,L4

   Computation Time
        (((R-2+4) × Q + 5) × (T-R) + 3) × (S-Q) + 3 + 4 cycles
}

.ENTRY  tdfir;

tdfir:  MY0=PM(I4,M4);                         {Get first coefficient}
        DO in_row UNTIL CE;                    {Loop through output rows}
           CNTR=AX0;
           DO in_col UNTIL CE;                 {Loop through output columns}
              CNTR=AX1;
              MR=0, MX0=DM(I0,M0);             {Clear MR, get input data}
              DO row_loop UNTIL CE;            {Loop through coefficient rows}
                 CNTR=AY0;
                 DO col_loop UNTIL CE;         {Loop through coefficient cols}
col_loop:           MR=MR+MX0*MY0 (SS), MX0=DM(I0,M0), MY0=PM(I4,M4);
                 MR=MR+MX0*MY0 (SS), MX0=DM(I0,M1), MY0=PM(I4,M4);
                 {Move pointer to next convolution window row}
row_loop:        MR=MR+MX0*MY0 (SS), MX0=DM(I0,M0), MY0=PM(I4,M4);
                 {Read values for next loop}
              DM(I1,M0)=MR1;                   {Save output points}
in_col:       MODIFY(I0,M2);                   {Get next conv. start same row}
in_row:    MODIFY(I0,M3);                      {Point to next input row}
        RTS;
.ENDMOD;
```

Listing 5.3 Two-Dimensional FIR Filter

5.3 INFINITE IMPULSE RESPONSE (IIR) FILTERS

Compared to the FIR filter, an IIR filter can often be much more efficient in terms of attaining certain performance characteristics with a given filter order. This is because the IIR filter incorporates feedback and is capable of realizing both poles and zeroes of a system transfer function, whereas the FIR filter is only capable of realizing the zeroes (although the FIR filter is still more desirable in many applications, because of features such as stability and the ability to realize exactly linear phase responses).

5.3.1 Direct Form IIR Filter

The IIR filter can realize both the poles and zeroes of a system because it has a rational transfer function, described by polynomials in *z* in both the numerator and the denominator:

$$H(z) = \frac{\sum_{k=0}^{M} b_k z^{-k}}{1 - \sum_{k=1}^{N} a_k z^{-k}}$$

The difference equation for such a system is described by the following:

$$y(n) = \sum_{k=0}^{M} b_k x(n-k) + \sum_{k=1}^{N} a_k y(n-k)$$

In most applications, the order of the two polynomials M and N are the same.

The roots of the denominator determine the pole locations of the filter, and the roots of the numerator determine the zero locations. There are, of course, several means of implementing the above transfer function with an IIR filter structure. The "direct form" structure presented in Listing 5.4 implements the difference equation above.

Note that there is a single delay line buffer for the recursive and non-recursive portions of the filter (Oppenheim and Schafer's Direct Form II). The sum-of-products of the *a* values and the delay line values are first computed, followed by the sum-of-products of the *b* values and the delay line values.

```
.MODULE diriir_sub;

{
   Direct Form II IIR Filter Subroutine

   Calling Parameters
        MR1 = Input sample (x[n])
        MR0 = 0
        I0 -> Delay line buffer current location (x[n-1])
        L0 = Filter length
        I5 -> Feedback coefficients (a[1], a[2], ... a[N])
        L5 = Filter length - 1
        I6 -> Feedforward coefficients (b[0], b[1], ... b[N])
        L6 = Filter length
        M0 = 0
        M1,M4 = 1
        CNTR = Filter length - 2
        AX0 = Filter length - 1

   Return Values
        MR1 = output sample (y[n])
        I0 -> delay line current location (x[n-1])
        I5 -> feedback coefficients
        I6 -> feedforward coefficients

   Altered Registers
        MX0,MY0,MR

   Computation Time
        (N - 2) + (N - 1)) + 10 + 4 cycles  (N = M = Filter order)

   All coefficients and data values are assumed to be in 1.15 format.
}

.ENTRY  diriir;

diriir: MX0=DM(I0,M1), MY0=PM(I5,M4);
        DO poleloop UNTIL CE;
poleloop:  MR=MR+MX0*MY0(SS), MX0=DM(I0,M1), MY0=PM(I5,M4);
        MR=MR+MX0*MY0(RND);
        CNTR=AX0;
        DM(I0,M0)=MR1;
        MR=0, MX0=DM(I0,M1), MY0=PM(I6,M4);
        DO zeroloop UNTIL CE;
zeroloop:  MR=MR+MX0*MY0(SS), MX0=DM(I0,M1), MY0=PM(I6,M4);
        MR=MR+MX0*MY0(RND);
        MODIFY (I0,M2);
        RTS;
.ENDMOD;
```

Listing 5.4 Direct Form IIR Filter

5.3.2 Cascaded Biquad IIR Filter

A second-order biquad IIR filter section is shown on Figure 5.2. Its transfer function in the z-domain is:

$$H(z) = Y(z)/X(z) = (B_0 + B_1z^{-1} + B_2z^{-2})/(1 + A_1z^{-1} + A_2z^{-2})$$

where A_1, A_2, B_0, B_1 and B_2 are coefficients that determine the desired impulse response of the system H(z). Furthermore, the corresponding difference equation for a biquad section is:

$$Y(n) = B_0X(n) + B_1X(n–1) + B_2X(n–2) – A_1Y(n–1) – A_2Y(n–2)$$

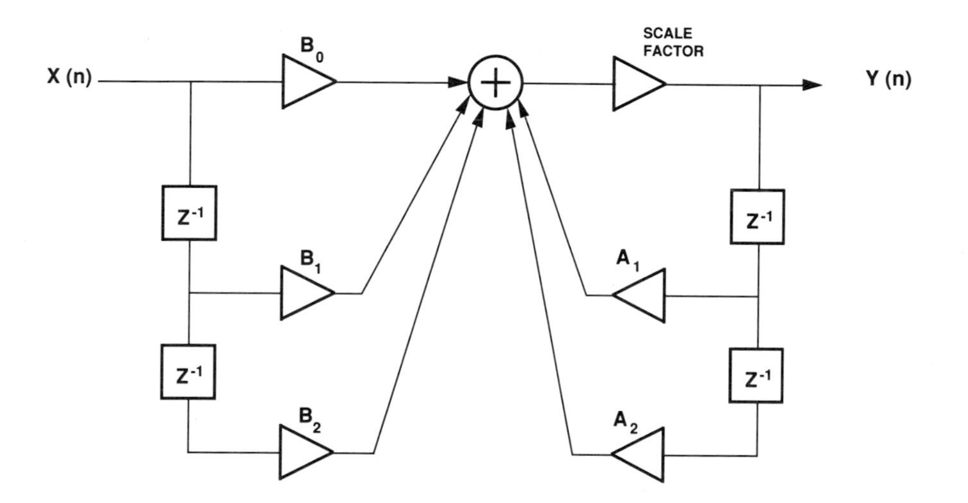

Figure 5.2 Second-order Biquad IIR Filter Section

Higher-order filters can be obtained by cascading several biquad sections with appropriate coefficients. Another way to design higher-order filters is to use only one complicated single section. This approach is called the direct form implementation. The biquad implementation executes slower but generates smaller numerical errors than the direct form implementation. The biquads can be scaled separately and then cascaded in order to minimize the coefficient quantization and the recursive accumulation errors. The coefficients and data in the direct form implementation must be scaled all at once, which gives rise to larger errors. Another disadvantage of the direct form implementation is that the poles of such single-stage high-order polynomials get increasingly sensitive to quantization errors. The second-order polynomial sections (i.e., biquads) are less sensitive to quantization effects.

An ADSP-2100 subroutine that implements a high-order filter is shown in Listing 5.5. The subroutine is arranged as a module and is labeled *biquad_sub*. There are a number of registers that need to be initialized in order to execute this subroutine. It may be sufficient to do this initialization only once (e.g., at powerup) if other executed algorithms do not need these registers. In most typical cases, however, some of these registers may need to be set every time the *biquad_sub* routine is called. It may sometimes be beneficial, from a modular software point of view, to always initialize all the setup registers as a part of this subroutine.

The *biquad_sub* routine takes its input from the SR1 register. This register must contain the 16-bit input X(n). X(n) is assumed to be already computed before this subroutine is called. The output of the filter is also made available in the SR1 register.

After the initial design of a high order filter, all coefficients must be scaled down in each biquad stage separately. This is necessary in order to conform to the 16-bit fixed-point fractional number format as well as to ensure that overflows will not occur in the final multiply-accumulate operations in each stage. The scaled-down coefficients are the ones that get stored in the processor's memory. The operations in each biquad are performed with scaled data and coefficients and are eventually scaled up before being output to the next one. The choice of a proper scaling factor depends greatly on the design objectives, and in some cases it may even be unnecessary. The filter coefficients are usually designed with a commercial software package in higher than 16-bit precision arithmetic. System performance deviates from ideal when such high precision coefficients are quantized to 16 bits and further scaled down. In systems that require stringent specifications, careful simulations of quantization and scaling effects must be performed.

During the initialization of the *biquad_sub* routine, the index register I0 points to the data memory buffer that contains the previous error inputs and the previous biquad section outputs. This buffer must be initialized to zero at powerup unless some nonzero initial condition is desired. The index register I1 points to another buffer in data memory that contains the individual scale factors for each biquad. The buffer length register L1 is set to zero if the filter has only one biquad section. In the case of multiple biquads, L1 is initialized with the number of biquad sections. The index register I4, on the other hand, points to the circular program memory buffer that contains the scaled biquad coefficients. These coefficients are stored in the order: B_2, B_1, B_0, A_2 and A_1 for each biquad. All of the individual biquad coefficient groups must be stored in the same order that the biquads are cascaded in, such as: B_2, B_1, B_0, A_2, A_1, B_2^*, B_1^*, B_0^*, A_2^*, A_1^*,B_2^{**}, etc. The buffer length register L4 must be set to the value of (5 x number of biquad sections). Finally, the loop counter register CNTR must be set to the number of biquad sections since the filter code will be executed as a loop.

The core of the *biquad_sub* routine starts its execution at the *biquad* label. The routine is organized in a looped fashion where the end of the loop is the instruction labeled *sections*. Each iteration of the loop executes the computations for one biquad. The number of loops to be executed is determined by the CNTR register contents. The SE register is loaded with the appropriate scaling factor for the particular biquad at the beginning of each loop iteration. After this operation, the coefficients and the data values are fetched from memory in the sequence that they have been stored. These numbers are multiplied and accumulated until all of the values for a particular biquad have been accessed. The result of the last multiply/accumulate is rounded to 16 bits and upshifted by the scaling value. At this point the *biquad* loop is executed again, or the filter computations are completed by doing the final update to the delay line. The delay lines for data values are always being updated within the *biquad* loop as well as outside of it.

The filter coefficients must be scaled appropriately so that no overflows occur after the upshifting operation between the biquads. If this is not ensured by design, it may be necessary to include some overflow checking between the biquads.

The execution time for an Nth order *biquad_sub* routine can be calculated as follows (assuming that the appropriate registers have been initialized and N is a power of 2):

ADSP-2101/2102 : (8 x N/2) + 4 processor cycles
ADSP-2100/2100A : (8 x N/2) + 4 + 5 processor cycles

It may take up to a maximum of 12 cycles to initialize the appropriate registers every time the filter is called, but typically this number will be lower.

```
.MODULE       biquad_sub;

{             Nth order cascaded biquad filter subroutine

              Calling Parameters:

                 SR1=input X(n)
                 I0 -> delay line buffer for X(n-2), X(n-1),
                     Y(n-2), Y(n-1)
                 L0 = 0
                 I1 -> scaling factors for each biquad section
                 L1 = 0  (in the case of a single biquad)
                 L1 = number of biquad sections
                      (for multiple biquads)
                 I4 -> scaled biquad coefficients
                 L4 = 5 x [number of biquads]
                 M0, M4 = 1
                 M1 = -3
                 M2 = 1 (in the case of multiple biquads)
                 M2 = 0 (in the case of a single biquad)
                 M3 = (1 - length of delay line buffer)

              Return Value:
                 SR1 = output sample Y(n)

              Altered Registers:
                 SE, MX0, MX1, MY0, MR, SR

              Computation Time (with N even):
                 ADSP-2101/2102: (8 x N/2) + 5 cycles
                 ADSP-2100/2100A: (8 x N/2) + 5 + 5 cycles

              All coefficients and data values are assumed to be in 1.15 format
}

.ENTRY        biquad;

biquad:       CNTR = number_of_biquads
              DO sections UNTIL CE;
                 SE=DM(I1,M2);
                 MX0=DM(I0,M0), MY0=PM(I4,M4);
                 MR=MX0*MY0(SS), MX1=DM(I0,M0), MY0=PM(I4,M4);
                 MR=MR+MX1*MY0(SS), MY0=PM(I4,M4);
                 MR=MR+SR1*MY0(SS), MX0=DM(I0,M0), MY0=PM(I4,M4);
                 MR=MR+MX0*MY0(SS), MX0=DM(I0,M1), MY0=PM(I4,M4);
                 DM(I0,M0)=MX1, MR=MR+MX0*MY0(RND);
sections:        DM(I0,M0)=SR1, SR=ASHIFT MR1 (HI);
              DM(I0,M0)=MX0;
              DM(I0,M3)=SR1;
              RTS;
.ENDMOD;
```

Listing 5.5 Cascaded Biquad IIR Filter

5.4 LATTICE FILTERS

The lattice filter is used often in the analysis and synthesis of speech, most commonly to simulate the vocal tract. Its physical analogue is a series of cylinders of different radii; each of the filter coefficients represents the amount of energy reflected at a boundary of two cylinders. The all-pole lattice filter is used in voice synthesis (see Chapter 10).

5.4.1 All-Zero Lattice Filter

The all-zero lattice filter (Bellanger, 1984) is the FIR representation of the lattice filter whose structure is shown in Figure 5.3.

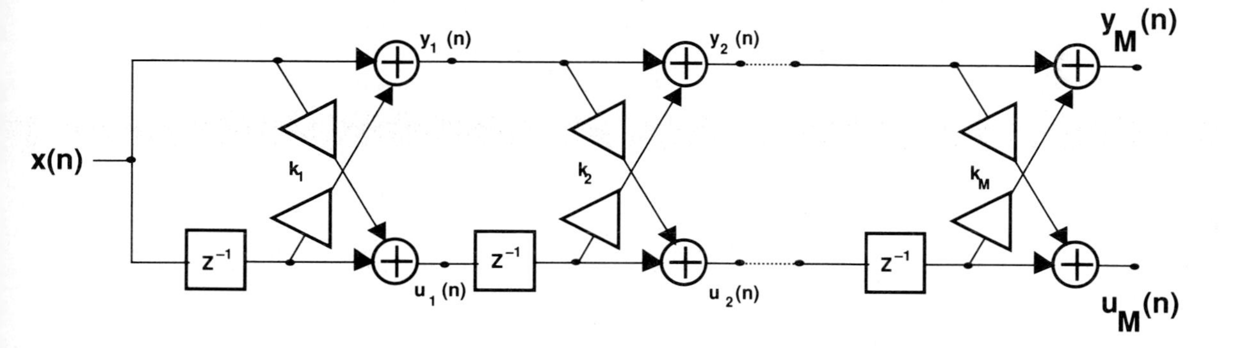

Figure 5.3 All-Zero Lattice Filter

Each stage of the filter has an input and output that are related by the equations

$$y_z(n) = y_{z-1}(n) + k_z u_{z-1}(n-1)$$

$$u_z(n) = k_z y_{z-1}(n) + u_{z-1}(n-1)$$

The initial values of $y_z(n)$ and $u_z(n)$ are both the value of the filter input, x(n).

$$y_0(n) = x(n)$$

$$u_0(n) = x(n)$$

For example

$y_1(n) = x(n) + k_1\, x(n-1)$

$u_1(n) = k_1\, x(n) + x(n-1)$

and

$y_2(n) = x(n) + k_1\,(1+k_2)\, x(n-1) + k_2\, x(n-2)$

$u_2(n) = k_2\, x(n) + k_1\,(1+k_2)\, x(n-1) + x(n-2)$

The filter output is the output of the last stage.

The ADSP-2100 implementation of the all-zero lattice filter is shown in Listing 5.6. Various registers must be preloaded before this routine is called. The index register I0 should contain the starting address of the input buffer, and I2 should hold the starting address of the output buffer. I3 should contain the starting address of the filter delay line, and I4 should contain the starting address of the coefficient buffer. The length registers L0 and L2 should be set to zero, but L3 and L4 should be set to the order of the filter (number of sections) to make the delay line and coefficient buffers circular. The modify register M3 should be set to one, and the SE register should contain the value needed to maintain a valid output data format. (For example, if two 4.12 numbers are multiplied, the product is a 7.23 number. To obtain a product in 9.21 format, the SE register must be set to –2.) MF, the multiplier feedback register, should contain the value one in the output format. Multiplication by MF is an alternative method of converting output to the correct format. The CNTR register should contain the number of locations in the output buffer.

The *out_loop* loop is executed once for each output data point. CNTR is loaded with the order of the filter, and the first input data point is loaded into MX0. The *latt_loop* loop performs the filtering operation on the input data point.

The first multiplication in the *latt_loop* loop formats the $y_{z-1}(n)$ value into the MR register and also reads in values for $u_{z-1}(n-1)$ and k_z. These values are then multiplied and accumulated to produce $y_z(n)$, at the same time the value $u_{z-1}(n)$ is stored in the delay line for the next pass. The value $y_z(n)$ is reformatted in the shifter for use by the multiplier in the next pass of the *latt_loop* loop.

Next, $u_{z-1}(n-1)$ is formatted into the multiplier to compute the value of $u_z(n)$. This value is then accumulated with the product of k_z and $y_z(n)$. Again, the shifter reformats the value before storage.

```
.MODULE all_zero_lattice_filter;

{
   All Zero Lattice

   Calling Parameters
        CNTR = Length of Output Buffer
        I0 -> Input Buffer                          L1 = 0
        I2 -> Output Buffer                         L2 = 0
        I3 -> Delay Line Buffer (circular)          L3 = Filter Order
        I4 -> Coefficient Buffer (circular)         L4 = Filter Order
        M0 = 1
        M2 = 0
        M3 = 1
        M4 = 1
        SE = Appropriate Scale Factor
        MF = Formatted 1

   Return Values
        Output Buffer Filled

   Altered Registers
        MX0,MX1,MY0,MF,MR,SR,I2,I3,I4

   Computation Time
        (8 × Filter Order + 4) × Output Buffer Length + 3 + 1 cycles
}

.ENTRY z_latt;

z_latt: SR1=0;                                     {Clear SR1 for first pass}
        DO out_loop UNTIL CE;                      {Loop output length}
           CNTR=L3;
           MX0=DM(I0,M0);                          {Get excitation signal}
           DO latt_loop UNTIL CE;                  {Loop through filter}
              MR=MX0*MF (SS), MX1=DM(I3,M2), MY0=PM(I4,M4);   {Get U,K}
              MR=MR+MX1*MY0 (SS), DM(I3,M3)=SR1;              {Compute Yz store U}
              SR=ASHIFT MR1 (HI);                  {Reformat Yz}
              SR=SR OR LSHIFT MR0 (LO);
              MR=MX1*MF (SS);                      {Format Uz-1}
              MX0=SR1, MR=MR+MX0*MY0 (SS);         {Compute Uz and Hold Yz}
              SR=ASHIFT MR1 (HI);                  {Reformat Uz}
latt_loop:    SR=SR OR LSHIFT MR0 (LO);
out_loop:  DM(I2,M0)=MX0;                          {Save output}
        RTS;
.ENDMOD;
```

Listing 5.6 All-Zero Lattice Filter

5.4.2 All-Pole Lattice Filter

The all-pole lattice filter, shown in Figure 5.4, relates the variables $x_z(n)$, $u_z(n)$, and y(n) by the following equations (Bellanger, 1984):

$$x_{z-1}(n) = x_z(n) - k_z u_{z-1}(n-1)$$

$$u_z(n) = k_z x_{z-1}(n) + u_{z-1}(n-1)$$

Therefore

$$y(n) = x_1(n) - k_1 y(n-1)$$

$$u_1(n) = k_1 y(n) + y(n-1)$$

and

$$x_1(n) = x_2(n) - k_2 u_1(n-1)$$

$$u_2(n) = k_2 x_1(n) + u_1(n-1)$$

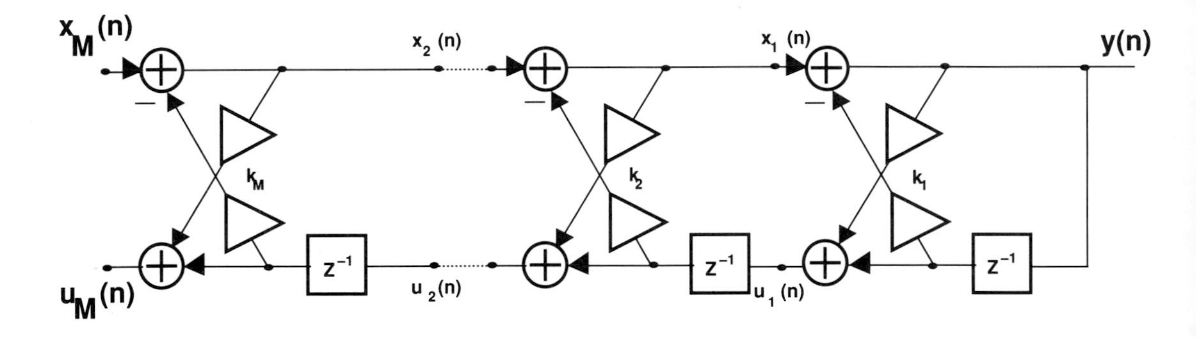

Figure 5.4 All-Pole Lattice Filter Structure

The all-pole lattice filter routine is shown in Listing 5.7. Various registers must be preloaded before the routine is called. The index register I0 should point to the start of the input buffer, I1 to the start of the coefficient buffer, I2 to the start of the output buffer, and I4 to the start of the filter delay line. The length registers L0 and L2 should both be set to zero, and L1 and L4 should be set to the filter order to make the coefficient and delay line buffers circular. The modify registers M0 and M4 should both be set to one; M1 and M5 should both be set to –1. M6 should be set to three and M7 to –2. The SE register, which controls data scaling, should be set to an appropriate value, and AX0 should be set to the order of the filter less one.

The routine loads the first input data value into MY0. The *outloop* loop is executed once for each output data value. The MR register is loaded with the scaled value of $x_z(n)$ at the same time the coefficient k_z and delay line value $u_{z-1}(n-1)$ are loaded. The next instruction computes the value $x_{z-1}(n)$ and also loads the next multiplier operands. The *dataloop* loop performs the remainder of the filtering operation on the data point.

In the *dataloop* loop, $x_{z-1}(n)$ is computed and then shifted to the proper format for the next multiplication. Then the value of $u_z(n)$ is computed and stored in the delay line. After the *dataloop* loop has been executed, the pointers to the delay line and coefficient buffer are moved to the tops of their buffers at the same time the output of the filter and the last delayed point $u_1(n)$ are saved.

```
.MODULE all_pole_lattice_filter;

{  All-Pole Lattice Filter Routine

   Calling Parameters
        CNTR = Length of Excitation Signal
        I0 -> Excitation Signal                  L0 = 0
        I1 -> Coefficient Buffer (circular)      L1 = Filter Order
        I2 -> Output Buffer                      L2 = 0
        I4 -> Delay Line Buffer (circular)       L4 = Filter Order
        AR = Formatted 1
        M0, M4 = 1                  M1,M5 = -1
        M6 = 3                      M7 = -2
        SE = Appropriate scale value
        AX0 = Filter Order - 1

   Return Values
        Output Buffer Filled

   Altered Registers
        MX0,MY0,MY1,MR,SR,I0,I1,I2,I4

   Computation Time
        (6 × (Filter Length-1) +8) × Output Buffer Length + 3 + 6 cycles
}
.ENTRY  p_latt;

p_latt: MY0=DM(I0,M0);                          {Get Input data}
        DO outloop UNTIL CE;                    {Loop through output}
           CNTR=AX0;
           MR=AR*MY0 (SS), MX0=DM(I1,M0), MY0=PM(I4,M4);          {Get U,K}
           MR=MR-MX0*MY0 (SS), MX0=DM(I1,M0), MY0=PM(I4,M4);      {MR=X10}
           DO dataloop UNTIL CE;                {Loop through filter}
              MR=MR-MX0*MY0 (SS);               {Compute Xz}
              SR=ASHIFT MR1 (HI);               {Reformat Xz}
              MY1=SR1, MR=AR*MY0 (SS);          {Format Uz+1}
              MR=MR+MX0*MY1 (SS), MX0=DM(I1,M0),MY0=PM(I4,M7);     {MR=Uz}
              SR=ASHIFT MR1 (HI);               {Reformat Uz}
dataloop:     PM(I4,M6)=SR1, MR=AR*MY1 (SS);    {Save Uz format Xz}
           MY0=PM(I4,M7), MX0=DM(I1,M1);        {Reset Pointers}
           MY0=DM(I0,M0), SR=ASHIFT MR1 (HI);   {Get new data point}
           DM(I2,M0)=MY1, SR=SR OR LSHIFT MR0 (LO);{Store output}
outloop:   PM(I4,M4)=SR1;                       {Save Y}
        RTS;
.ENDMOD;
```

Listing 5.7 All-Pole Lattice Filter

5.5 MULTIRATE FILTERS

Multirate filters are digital filters that change the sampling rate of a digitally-represented signal. These filters convert a set of input samples to another set of data that represents the same analog signal sampled at a different rate. A system incorporating multirate filters (a multirate system) can process data sampled at various rates.

Some examples of applications for multirate filters are:

- Sample-rate conversion between digital audio systems
- Narrow-band low-pass and band-pass filters
- Sub-band coding for speech processing in vocoders
- Transmultiplexers for TDM (time-division multiplexing) to FDM (frequency-division multiplexing) translation
- Quadrature modulation
- Digital reconstruction filters and anti-alias filters for digital audio, and
- Narrow-band spectra calculation for sonar and vibration analysis.

For additional information on these topics, see *References* at the end of this chapter.

The two types of multirate filtering processes are decimation and interpolation. Decimation reduces the sample rate of a signal. It eliminates redundant or unnecessary information and compacts the data, allowing more information to be stored, processed, or transmitted in the same amount of data. Interpolation increases the sample rate of a signal. Through calculations on existing data, interpolation fills in missing information between the samples of a signal. Decimation reduces a sample rate by an integer factor M, and interpolation increases a sample rate by an integer factor L. Non-integer rational (ratio of integers) changes in sample rate can be achieved by combining the interpolation and decimation processes.

The ADSP-2100 programs in this chapter demonstrate decimation and interpolation as well as efficient rational changes in sample rate. Cascaded stages of decimation and interpolation, which are required for large rate changes (large values of L and M) and are useful for implementing narrow-band low-pass and band-pass filters, are also demonstrated.

5.5.1 Decimation

Decimation is equivalent to sampling a discrete-time signal. Continuous-time (analog) signal sampling and discrete-time (digital) signal sampling are analogous.

5.5.1.1 Continuous-Time Sampling

Figure 5.5 shows the periodic sampling of a continuous-time signal, $x_c(t)$, where t is a continuous variable. To sample $x_c(t)$ at a uniform rate every T seconds, we modulate (multiply) $x_c(t)$ by an impulse train, s(t):

$$s(t) = \sum_{n=-\infty}^{+\infty} \partial(t-nT)$$

The resulting signal is a train of impulses, spaced at intervals of T, with amplitudes equal to samples of $x_c(t)$. This impulse train is converted to a set of values x(n), where n is a discrete-time variable and $x(n)= x_c(nT)$. Thus, $x_c(t)$ is quantized both in time and in amplitude to form digital values x(n). The modulation process is equivalent to a track-and-hold circuit, and the quantization process is equivalent to an analog-to-digital (A/D) converter.

Figure 5.6 shows a frequency-domain interpretation of the sampling process. $X_c(w)$ is the spectrum of the continuous-time signal $x_c(t)$. S(w), a train of impulses at intervals of the sampling frequency, F_s or 1/T, is the frequency transform of s(t). Because modulation in the time domain is equivalent to convolution in the frequency domain, the convolution of $X_c(w)$ and S(w) yields the spectrum of x(n). This spectrum is a sequence of periodic repetitions of $X_c(w)$, called *images* of $X_c(w)$, each centered at multiples of the sampling frequency, F_s.

The frequency that is one-half the sampling frequency ($F_s/2$) is called the Nyquist frequency. The analog signal $x_c(t)$ must be bandlimited before sampling to at most the Nyquist frequency. If $x_c(t)$ is not bandlimited, the images created by the sampling process overlap each other, mirroring the spectral energy around $nF_s/2$, and thus corrupting the signal representation. This phenomenon is called aliasing. The input $x_c(t)$ must pass through an analog anti-alias filter to eliminate any frequency component above the Nyquist frequency.

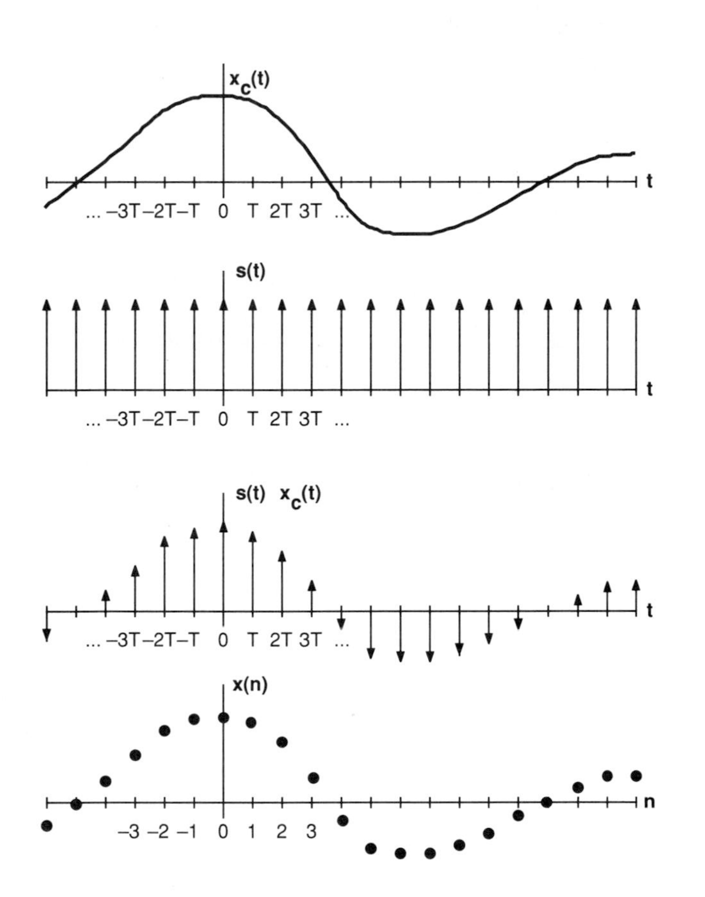

Figure 5.5 Sampling Continuous-Time Signal

5.5.1.2 Discrete-Time Sampling

Figure 5.7 shows the sampling of a discrete-time signal, x(n). The signal x(n) is multiplied by s(n), a train of impulses occurring at every integer multiple of M. The resulting signal consists of every Mth sample of x(n) with all other samples zeroed out. In this example, M is 4; the decimated version of x(n) is the result of discarding three out of every four samples. The original sample rate, F_s, is reduced by a factor of 4; $F_s'=F_s/4$.

5 Digital Filters

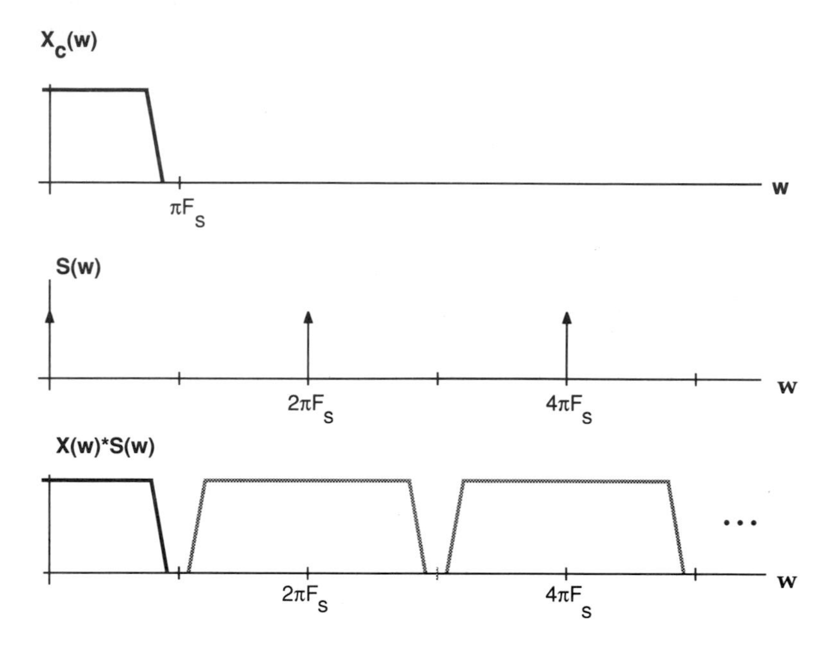

Figure 5.6 Spectrum of Continuous-Time Signal Sampling

Figure 5.8 shows the frequency-domain representation of sampling the discrete-time signal. The spectrum of the waveform to be sampled is X(w). The original analog signal was bandlimited to $w_s/2$ before being sampled, where w_s is the original sample rate. The shaded lines indicate the images of X(w) above $w_s/2$. Before decimation, X(w) must be bandlimited to one-half the *final* sample rate, to eliminate frequency components that could alias. H(w) is the transfer function of the low-pass filter required for a decimation factor (M) of four. Its cutoff frequency is $w_s/8$. This digital anti-alias filter performs the equivalent function as the analog anti-alias filter described in the previous section.

W(w) is a version of X(w) filtered by H(w), and W(w)*S(w) is the result of convolving W(w) with the sampling function S(w), the transform of s(n). The shaded lines in W(w)*S(w) represent the images of W(w) formed by this convolution. These images show the energy in the original signal that would alias if we decimated X(w) without bandlimiting it. X(w′) is the result of decimating W(w)*S(w) by four. Decimation effectively spreads out the energy of the original sequence and eliminates unwanted information located above w_s/M.

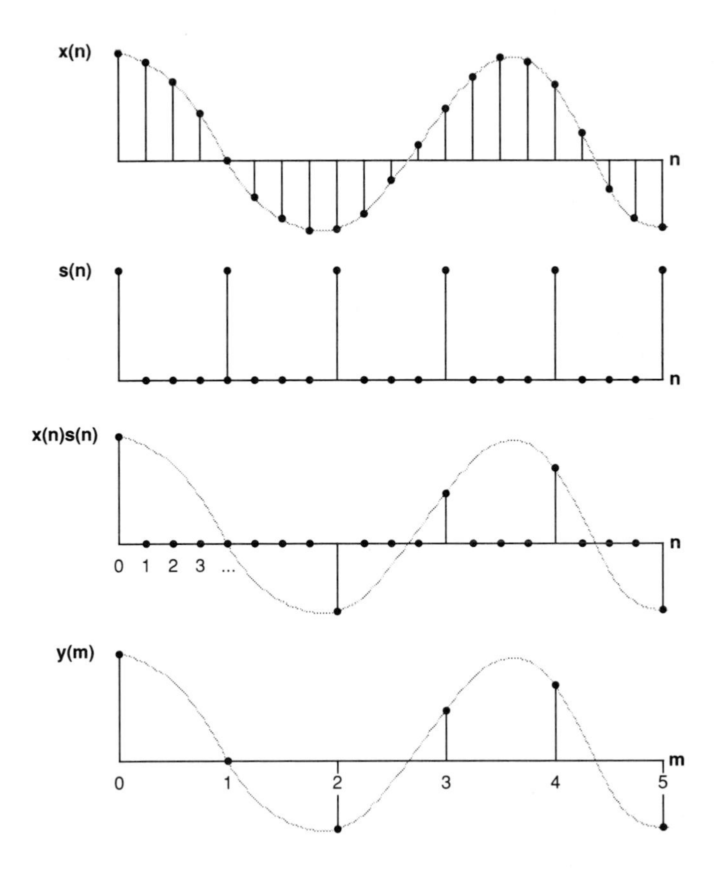

Figure 5.7 Sampling Discrete -Time Signal

The decimation and anti-alias functions are usually grouped together into one function called a decimation filter. Figure 5.9 shows a block diagram of a decimation filter. The input samples x(n) are put through the digital anti-alias filter, h(k). The box labeled with a down-arrow and M is the sample rate compressor, which discards M–1 out of every M samples. Compressing the filtered input w(n) results in y(m), which is equal to w(Mm).

Data acquisition systems such as the digital audio tape recorder can take advantage of decimation filters to avoid using expensive high-

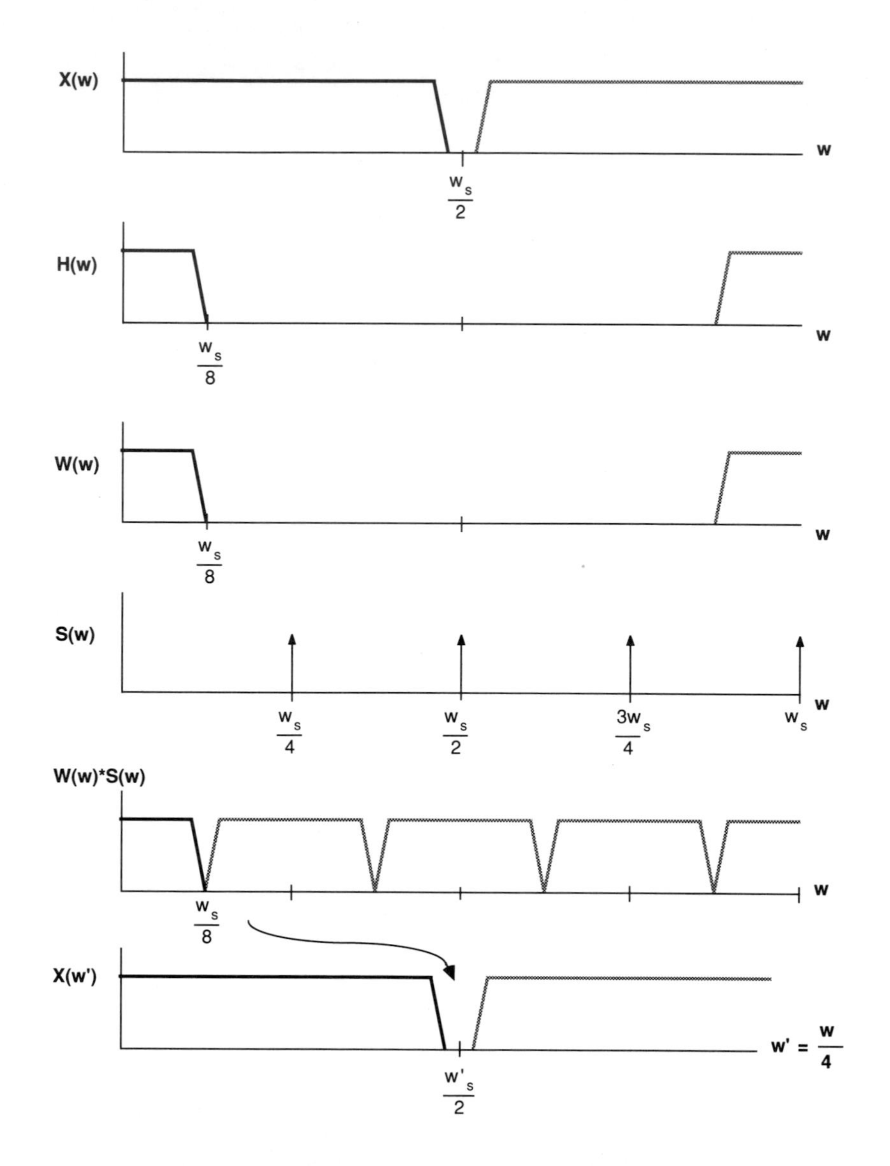

Figure 5.8 Spectrum of Discrete-Time Signal Sampling

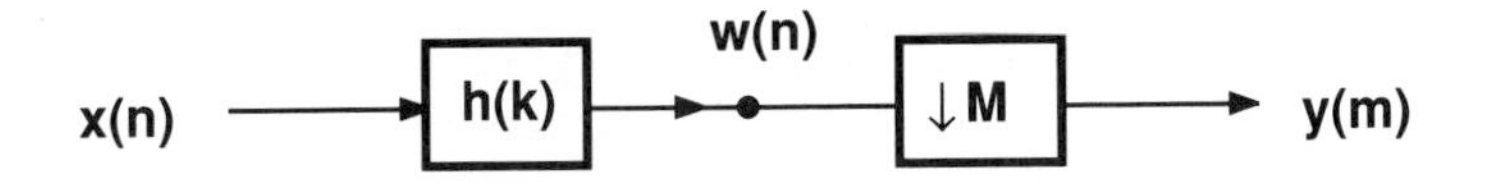

Figure 5.9 Decimation Filter Block Diagram

performance analog anti-aliasing filters. Such a system over-samples the input signal (usually by a factor of 2 to 8) and then decimates to the required sample rate. If the system over-samples by two, the transition band of the front end filter can be twice that required in a system without decimation, thus a relatively inexpensive analog filter can be used.

5.5.1.3 Decimation Filter Structure

The decimation algorithm can be implemented in an FIR (Finite Impulse Response) filter structure. The FIR filter has many advantages for multirate filtering including: linear phase, unconditional stability, simple structure, and easy coefficient design. Additionally, the FIR structure in multirate filters provides for an increase in computational efficiency over IIR structures. The major difference between the IIR and the FIR filter is that the IIR filter must calculate all outputs for all inputs. The FIR multirate filter calculates an output for every Mth input. For a more detailed description of the FIR and IIR filters, refer to Crochiere and Rabiner, 1983; see *References* at the end of this chapter.

The impulse response of the anti-imaging low-pass filter is h(n). A time-series equation filtering x(n) is the convolution

$$w(n) = \sum_{k=0}^{N-1} h(k)\, x(n-k)$$

where N is the number of coefficients in h(n). N is the order, or number of taps, in the filter. The application of this equation to implement the filter response $H(e^{jw})$ results in an FIR filter structure.

Figure 5.10a, on the next page, shows the signal flowgraph of an FIR decimation filter. The N most recent input samples are stored in a delay line; z–1 is a unit sample delay. N samples from the delay line are multiplied by N coefficients and the resulting products are summed to form a single output sample w(n). Then w(n) is down-sampled by M using the rate compressor.

(a)

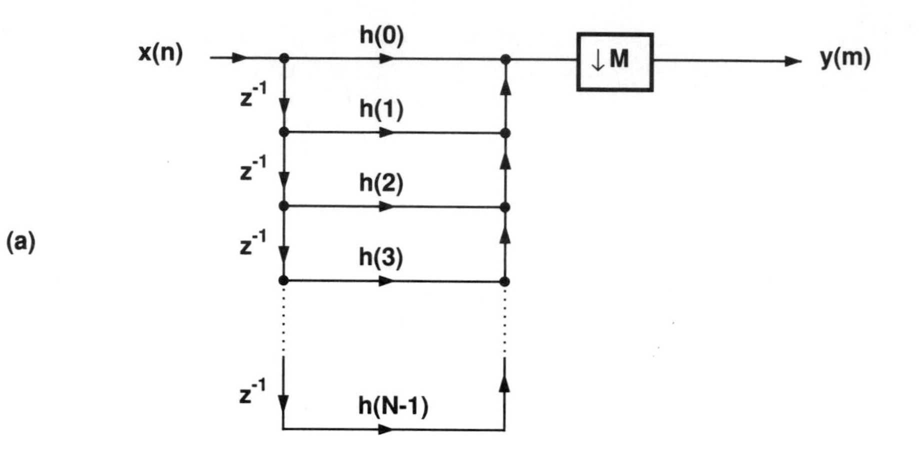

(b)

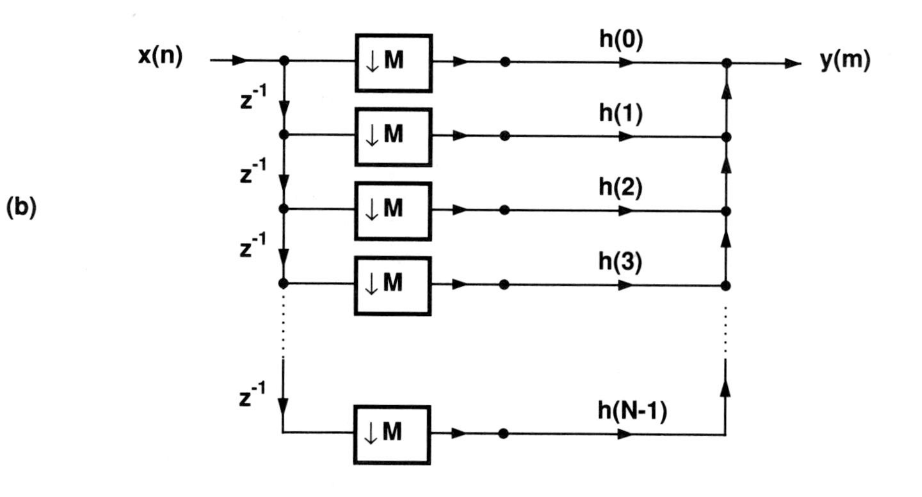

Figure 5.10 FIR Form Decimation Filter

It is not necessary to calculate the samples of w(n) that are discarded by the rate compressor. Accordingly, the rate compressor can be moved in front of the multiply/accumulate paths, as shown in Figure 5.10b. This change reduces the required computation by a factor of M. This filter structure can be implemented by updating the delay line with M inputs before each output sample is calculated.

Substitution of the relationship between w(n) and y(m) into the convolution results in the decimation filtering equation

$$y(m) = \sum_{k=0}^{N-1} h(k)\, x(Mm-k)$$

Some of the implementations shown in textbooks on digital filters take advantage of the symmetry in transposed forms of the FIR structure to reduce the number of multiplications required. However, such a reduction of multiplications results in an increased number of additions. In this application, because the ADSP-2100 is capable of both multiplying and accumulating in one cycle, trading off multiplication for addition is a useless technique.

5.5.1.4 ADSP-2100 Decimation Algorithm

Figure 5.11 shows a flow chart of the decimation filter algorithm used for the ADSP-2100 routine. The decimator calculates one output for every M inputs to the delay line.

External hardware causes an interrupt at the input sample rate F_s, which triggers the program to fetch an input data sample and store it in the *data* circular buffer. The index register that points into this data buffer is then incremented by one, so that the next consecutive input sample is written to the next address in the buffer. The counter is then decremented by one and compared to zero. If the counter is not yet zero, the algorithm waits for another input sample. If the counter has decremented to zero, the algorithm calculates an output sample, then resets the counter to M so that the next output will be calculated after the next M inputs.

The output is the sum of products of N data buffer samples in a circular buffer and N coefficients in another circular buffer. Note that M input samples are written into the data buffer before an output sample is calculated. Therefore, the resulting output sample rate is equal to the input rate divided by the decimation factor: $F_s'=F_s/M$.

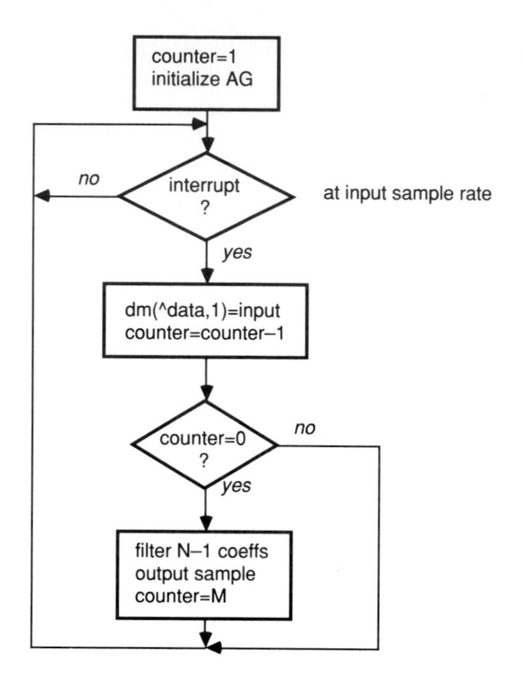

Figure 5.11 Decimator Flow Chart

For additional information on the use of the ADSP-2100's address generators for circular buffers, see the *ADSP-2100 User's Manual*, Chapter 2.

The ADSP-2100 program for the decimation filter is shown in Listing 5.8. Inputs to this filter routine come from the memory-mapped port *adc*, and outputs go to the memory-mapped port *dac*. The filter's I/O interfacing hardware is described in more detail later in this chapter.

```
{DECIMATE.dsp
Real time Direct Form FIR Filter, N taps, decimates by M for a
decrease of 1/M times the input sample rate.

   INPUT: adc
   OUTPUT: dac
}

.MODULE/RAM/ABS=0    decimate;
.CONST               N=300;
.CONST               M=4;                  {decimate by factor of M}
.VAR/PM/RAM/CIRC     coef[N];
.VAR/DM/RAM/CIRC     data[N];
.VAR/DM/RAM          counter;
.PORT                adc;
.PORT                dac;
.INIT                coef:<coef.dat>;

                RTI;                {interrupt 0}
                RTI;                {interrupt 1}
                RTI;                {interrupt 2}
                JUMP sample;        {interrupt 3= input sample rate}

initialize:     IMASK=b#0000;       {disable all interrupts}
                ICNTL=b#01111;      {edge sensitive interrupts}
                SI=M;               {set decimation counter to M}
                DM(counter)= SI;    {for first input data sample}
                I4=^coef;           {setup a circular buffer in PM}
                L4=%coef;
                M4=1;               {modifier for coef is 1}
                I0=^data;           {setup a circular buffer in DM}
                L0=%data;
                M0=1;               {modifier for data is 1}
                IMASK=b#1000;       {enable interrupt 3}
wait_interrupt: JUMP wait_interrupt;      {infinite wait loop}

{__________Decimator, code executed at the sample rate________________}
sample:         AY0=DM(adc);
                DM(I0,M0)=AY0;      {update delay line with newest}
                AY0=DM(counter);
                AR=AY0-1;           {decrement and update counter}
                DM(counter)=AR;
                IF NE RTI;          {test and return if not M times}
```

(listing continues on next page)

```
{_________code below executed at 1/M times the sample rate____________}
do_fir:          AR=M;                      {reset the counter to M}
                 DM(counter)=AR;
                 CNTR=N - 1;
                 MR=0, MX0=DM(I0,M0), MY0=PM(I4,M4);
                 DO taploop UNTIL CE;       {N-1 taps of FIR}
taploop:            MR=MR+MX0*MY0(SS), MX0=DM(I0,M0), MY0=PM(I4,M4);
                 MR=MR+MX0*MY0(RND);        {last tap with round}
                 IF MV SAT MR;              {saturate result if overflow}
                 DM(dac)=MR1;               {output data sample}
                 RTI;
.ENDMOD;
```

Listing 5.8 Decimation Filter

The routine uses two circular buffers, one for data samples and one for coefficients, that are each N locations long. The *coef* buffer is located in program memory and stores the filter coefficients. Each time an output is calculated, the decimator accesses all these coefficients in sequence, starting with the first location in *coef*. The I4 index register, which points to the coefficient buffer, is modified by one (from modify register M0) each time it is accessed. Therefore, I4 is always modified back to the beginning of the coefficient buffer after the calculation is complete.

The FIR filter equation starts the convolution with the most recent data sample and accesses the oldest data sample last. Delay lines implemented with circular buffers, however, access data in the opposite order. The oldest data sample is fetched first from the buffer and the newest data sample is fetched last. Therefore, to keep the data/coefficient pairs together, the coefficients must be stored in memory in reverse order.

The relationship between the address and the contents of the two circular buffers (after N inputs have occurred) is shown in the table below. The *data* buffer is located in data memory and contains the last N data samples input to the filter. Each pass of the filter accesses the locations of both buffers sequentially (the pointer is modified by one), but the first address accessed is not always the first location in the buffer, because the decimation filter inputs M samples into the delay line before starting each filter pass. For each pass, the first fetch from the data buffer is from an address M greater than for the previous pass. The data delay line moves forward M samples for every output calculated.

Data		*Coefficient*	
DM(0)	= x(n–(N–1)) oldest	PM(0)	= h(N–1)
DM(1)	= x(n–(N–2))	PM(1)	= h(N–2)
DM(2)	= x(n–(N–3))	PM(2)	= h(N–3)
	•		•
	•		•
	•		•
DM(N–3)	= x(n–2)	PM(N–3)	= h(2)
DM(N–2)	= x(n–1)	PM(N–2)	= h(1)
DM(N–1)	= x(n–0) newest	PM(N–1)	= h(0)

A variable in data memory is used to store the decimation counter. One of the processor's registers could have been used for this counter, but using a memory location allows for expansion to multiple stages of decimation (described in *Multistage Implementations*, later in this chapter).

The number of cycles required for the decimation filter routine is shown below. The ADSP-2100 takes one cycle to calculate each tap (multiply and accumulate), so only 18+N cycles are necessary to calculate one output sample of an N-tap decimator. The 18 cycles of overhead for each pass is just six cycles greater than the overhead of a non-multirate FIR filter.

Interrupt response	2 cycles
Fetch input	1 cycle
Write input to data buffer	1 cycle
Decrement and test counter	4 cycles
Reload counter with M	2 cycles
FIR filter pass	7 + N cycles
Return from interrupt	1 cycle
Maximum total	18 + N cycles/output

5.5.1.5 A More Efficient Decimator

The routine in Listing 5.8 requires that the 18+N cycles needed to calculate an output occur during the first of the M input sample intervals. No calculations are done in the remaining M–1 intervals. This limits the number of filter taps that can be calculated in real time to:

$$N = \frac{1}{F_s\, t_{CLK}} - 18$$

where t_{CLK} is the instruction cycle time of the processor.

An increase in this limit by a factor of M occurs if the program is modified so that the M data inputs overlap the filter calculations. This more efficient version of the program is shown in Listing 5.9.

In this example, a circular buffer *input_buf* stores the M input samples. The code for loading *input_buf* is placed in an interrupt routine to allow the input of data and the FIR filter calculations to occur simultaneously.

A loop waits until the input buffer is filled with M samples before the filter output is calculated. Instead of counting input samples, this program determines that M samples have been input when the input buffer's index register I0 is modified back to the buffer's starting address. This strategy saves a few cycles in the interrupt routine.

After M samples have been input, a second loop transfers the data from *input_buf* to the data buffer. An output sample is calculated. Then the program checks that at least one sample has been placed in *input_buf*. This check prevents a false output if the output calculation occurs in less than one sample interval. Then the program jumps back to wait until the next M samples have been input.

This more efficient decimation filter spreads the calculations over the output sample interval $1/F_s'$ instead of the input interval $1/F_s$. The number of taps that can be calculated in real time is:

$$N = \frac{M}{F_s\, t_{CLK}} - 20 - 2M - 6(M-1)$$

which is approximately M times greater than for the first routine.

```
{DEC_EFF.dsp
Real time Direct Form FIR Filter, N taps, decimates by M for a decrease of 1/M times
the input sample rate. This version uses an input buffer to allow the filter
computations to occur in parallel with inputs. This allows larger order filter for a
given input sample rate. To save time, an index register is used for the input buffer
as well as for a decimation counter.

   INPUT: adc
   OUTPUT: dac
}

.MODULE/RAM/ABS=0    eff_decimate;
.CONST               N=300;
.CONST               M=4;                    {decimate by factor of M}
.VAR/PM/RAM/CIRC     coef[N];
.VAR/DM/RAM/CIRC     data[N];
.VAR/DM/RAM/CIRC     input_buf[M];
.PORT                adc;
.PORT                dac;
.INIT                coef:<coef.dat>;

        RTI;                    {interrupt 0}
        RTI;                    {interrupt 1}
        RTI;                    {interrupt 2}
        JUMP sample;            {interrupt 3= input sample rate}

initialize:  IMASK=b#0000;            {disable all interrupts}
             ICNTL=b#01111;           {edge sensitive interrupts}
             I4=^coef;                {setup a circular buffer in PM}
             L4=%coef;
             M4=1;              {modifier for coef is 1}
             I0=^data;                {setup a circular buffer in DM}
             L0=%data;
             M0=1;              {modifier for data is 1}
             I1=^input_buf;           {setup a circular buffer in DM}
             L1=%input_buf;
             IMASK=b#1000;            {enable interrupt 3}

wait_M:      AX0=I1;                  {wait for M inputs}
             AY0=^input_buf;
             AR=AX0-AY0;        {test if pointer is at start}
             IF NE JUMP wait_M;
```

(listing continues on next page)

```
{____________code below executed at 1/M times the sample rate__________}
             CNTR=M;
             DO load_data UNTIL CE;
                  AR=DM(I1,M0);
load_data:        DM(I0,M0)=AR;

fir:         CNTR=N - 1;
             MR=0, MX0=DM(I0,M0), MY0=PM(I4,M4);
             DO taploop UNTIL CE;     {N-1 taps of FIR}
taploop:          MR=MR+MX0*MY0(SS), MX0=DM(I0,M0), MY0=PM(I4,M4);
             MR=MR+MX0*MY0(RND);     {last tap with round}
             IF MV SAT MR;           {saturate result if overflow}
             DM(dac)=MR1;            {output data sample}

wait_again:  AX0=I1;
             AY0=^input_buf;
             AR=AX0-AY0;        {test and wait if i1 still}
             IF EQ JUMP wait_again; {points to start of input_buf}
             JUMP wait_M;

{__________sample input, code executed at the sample rate____________}
sample:      ENA SEC_REG;            {so no registers will get lost}
             AY0=DM(adc);            {get input sample}
             DM(I1,M0)=AY0;          {load in M long buffer}
             RTI;
.ENDMOD;
```

Listing 5.9 Efficient Decimation Filter

5.5.2 Decimator Hardware Configuration

Both decimation filter programs assume an ADSP-2100 system with the I/O hardware configuration shown in Figure 5.12. The processor is interrupted by an interval timer at a frequency equal to the input sample rate F_s and responds by inputting a data value from the A/D converter. The track/hold (sampler) and the A/D converter (quantizer) are also clocked at this frequency. The D/A converter on the filter output is clocked at a rate of F_s/M, which is generated by dividing the interval timer frequency by M.

To keep the output signal jitter-free, it is important to derive the D/A converter's clock from the interval timer and not from the ADSP-2100. The sample period of the analog output should be disassociated from writes to the D/A converter. If an instruction-derived clock is used, any conditional

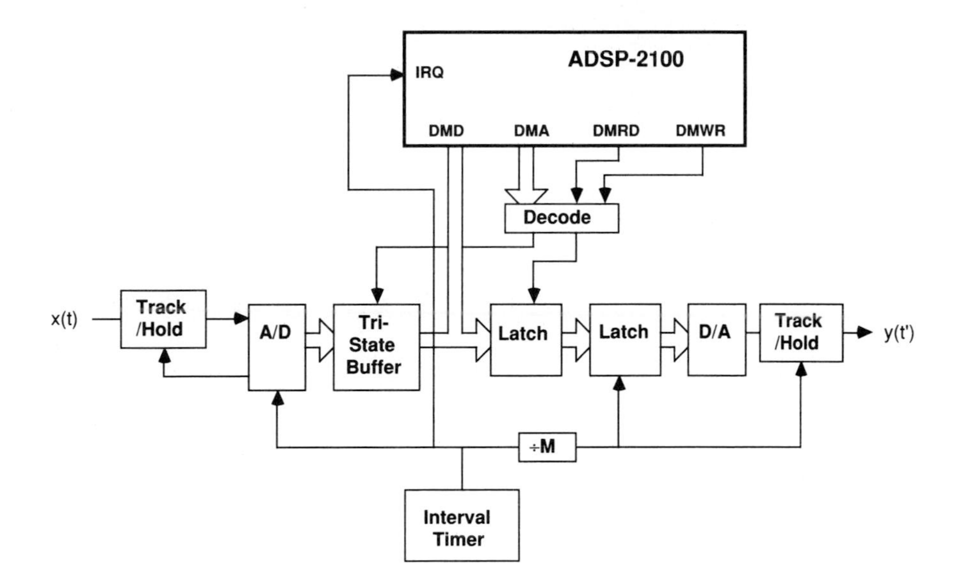

Figure 5.12 Decimator Hardware

instructions in the program could branch to different length program paths, causing the output samples to be spaced unequally in time. The D/A converter must be double-buffered to accommodate the interval-time-derived clock. The ADSP-2100 outputs data to one latch. Data from this latch is fed to a second latch that is controlled by an interval-timer-derived clock.

5.5.3 Interpolation

The process of recreating a continuous-time signal from its discrete-time representation is called reconstruction. Interpolation can be thought of as the reconstruction of a discrete-time signal from another discrete-time signal, just as decimation is equivalent to sampling the samples of a signal. Continuous-time (analog) signal reconstruction and discrete-time (digital) signal reconstruction are analogous.

Figure 5.13, on the following page, illustrates the reconstruction of a continuous-time signal from a discrete-time signal. The discrete-time signal x(n) is first made continuous by forming an impulse train with amplitudes at times nT equal to the corresponding samples of x(n). In a

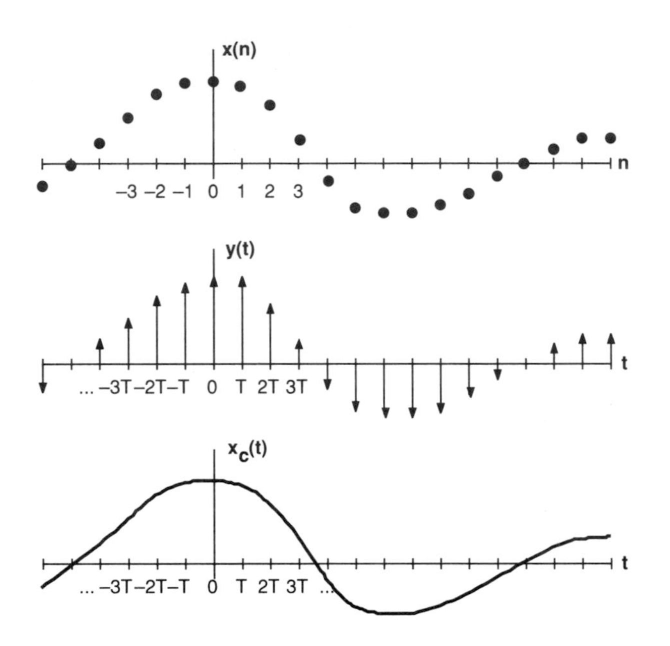

Figure 5.13 Reconstruction of a Continuous-Time Signal

real system, a D/A converter performs this operation. The result is a continuous signal y(t), which is smoothed by a low-pass anti-imaging filter (also called a reconstruction filter), to produce the reconstructed analog signal $x_c(t)$. The frequency domain representation of y(t) in Figure 5.14 shows that images of the original signal appear in the discrete- to continuous-time conversion. These images are eliminated by the anti-imaging filter, as shown in the spectrum of the resulting signal $X_c(w)$.

5.5.3.1 Reconstruction of a Discrete-Time Signal

Figure 5.15 shows the interpolation by a factor of L (4 in this example) of the discrete-time signal x(n). This signal is expanded by inserting L–1 zero-valued samples between its data samples. The resulting signal w(m) is low-pass filtered to produce y(m). The insertion of zeros and the smoothing filter fill in the data missing between the samples of x(n). Because one sample of x(n) corresponds to L samples of y(m), the sample rate is increased by a factor of L.

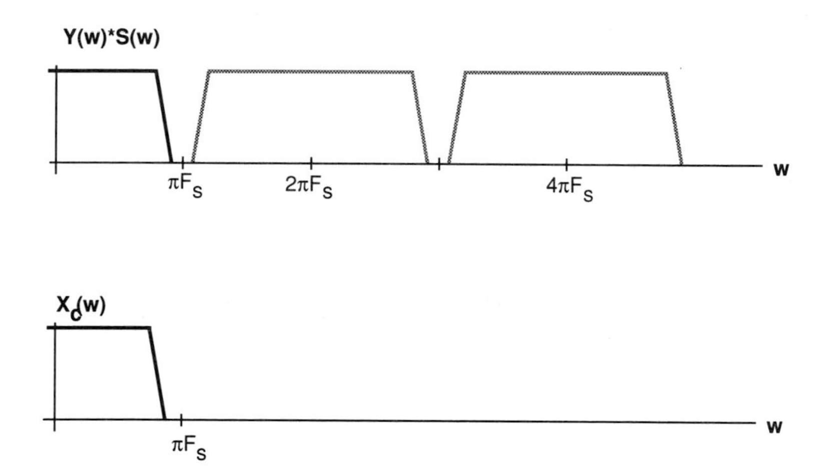

Figure 5.14 Spectrum of Continuous-Time Signal Reconstruction

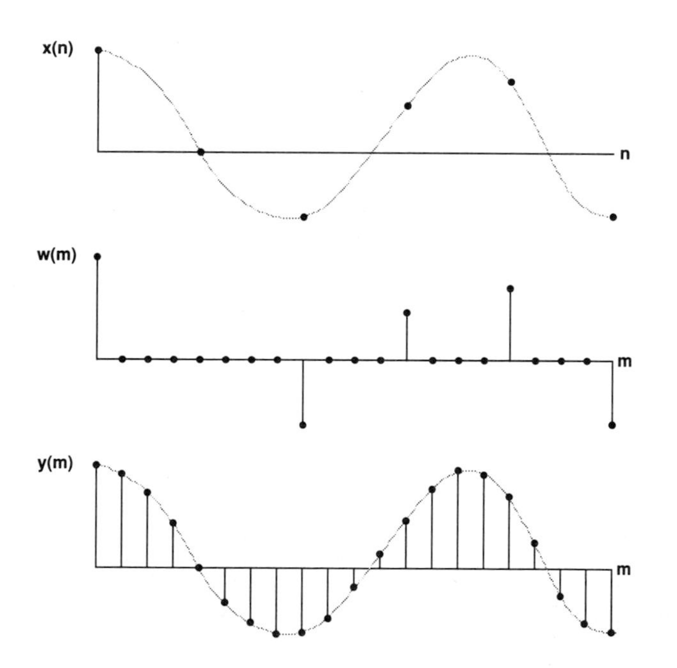

Figure 5.15 Interpolation of Discrete-Time Signal

Figure 5.16 shows the frequency interpretation of interpolation. F_s' is the input sample frequency and F_s'' is the output sample frequency. F_s'' is equal to F_s' multiplied by the interpolation factor L (3 in this example). H(w) is the response of the filter required to eliminate the images in W(w). The lower stopband frequency of H(w) must be less than $F_s''/2L$, which is the Nyquist frequency of the original signal. Thus filtering by H(w) accomplishes the function of a digital anti-imaging filter.

Digital audio systems such as compact disk and digital audio tape players frequently use interpolation (oversampling) techniques to avoid using costly high performance analog reconstruction filters. The anti-imaging function in the digital interpolator allows these systems to use inexpensive low-order analog filters on the D/A outputs.

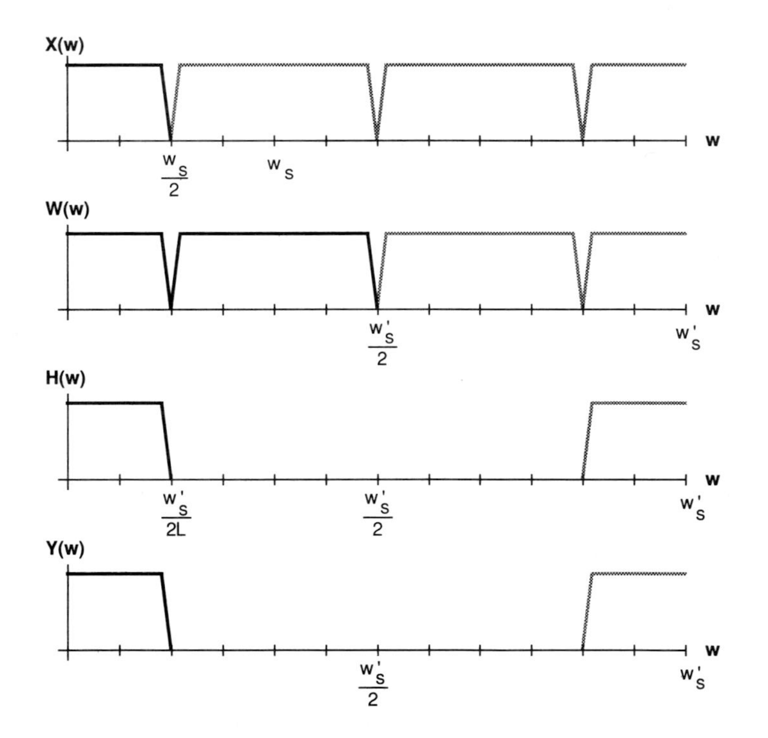

Figure 5.16 Spectrum of Interpolation

5.5.3.2 Interpolation Filter Structure

Figure 5.17a shows a block diagram of an interpolation filter. The two major differences from the decimation filter are that the interpolator uses a sample rate expander instead of the sample rate compressor and that the interpolator's low-pass filter is placed after the rate expander instead of before the rate compressor. The rate expander, which is the block labeled with an up-arrow and L, inserts L–1 zero-valued samples after each input sample. The resulting w(m) is low-pass filtered to produce y(m), a smoothed, anti-imaged version of w(m). The transfer function of the interpolator H(k) incorporates a gain of 1/L because the L 1 zeros inserted by the rate expander cause the energy of each input to be spread over L output samples.

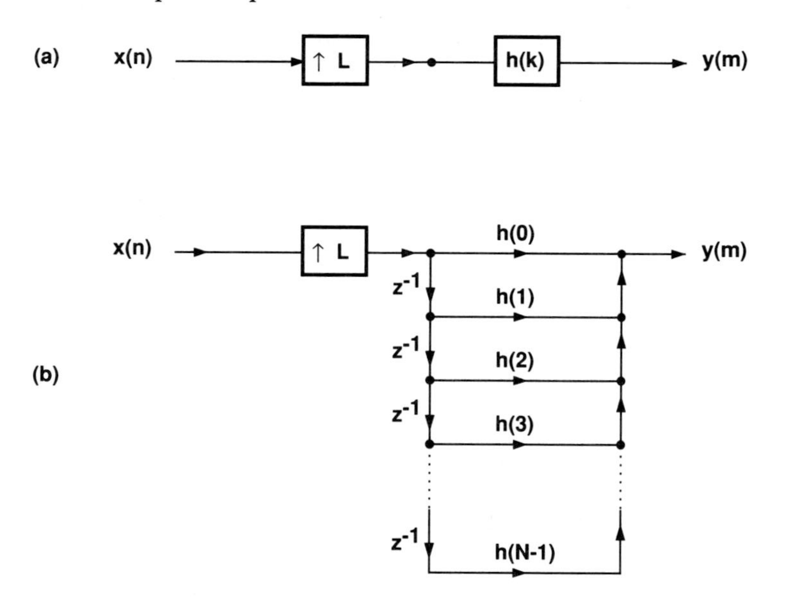

Figure 5.17 Interpolation Filter Block Diagram

The low-pass filter of the interpolator uses an FIR filter structure for the same reasons that an FIR filter is used in the decimator, notably computational efficiency. The convolution equation for this filter is

$$y(m) = \sum_{k=0}^{N-1} h(k)\, w(m-k)$$

N–1 is the number of filter coefficients (taps) in h(k), w(m–k) is the rate expanded version of the input x(n), and w(m–k) is related to x(n) by

$$w(m-k) = \begin{cases} x((m-k)/L)) & \text{for } m-k = 0, \pm L, \pm 2L, \ldots \\ 0 & \text{otherwise} \end{cases}$$

The signal flow graph that represents the interpolation filter is shown in Figure 5.17b, on previous page. A delay line of length N is loaded with an input sample followed by L–1 zeros, then the next input sample and L–1 zeros, and so on. The output is the sum of the N products of each sample from the delay line and its corresponding filter coefficient. The filter calculates an output for every sample, zero or data, loaded into the delay line.

An example of the interpolator operation is shown in the signal flowgraph in Figure 5.18. The contents of the delay line for three consecutive passes of the filter are highlighted. In this example, the interpolation factor L is 3. The delay line is N locations long, where N is the number of coefficients of the filter; N=9 in this example. There are N/L or 3 data samples in the delay line during each pass. The data samples x(1), x(2), and x(3) in the first pass are separated by L–1 or 2 zeros inserted by the rate expander. The zero-valued samples contribute (L–1)N/L or 6 zero-valued products to the output result. These (L–1)N/L multiplications are unnecessary and waste processor capacity and execution time.

A more efficient interpolation method is to access the coefficients and the data in a way that eliminates wasted calculations. This method is accomplished by removing the rate expander to eliminate the storage of the zero-valued samples and shortening the data delay line from N to N/L locations. In this implementation, the data delay line is updated only after L outputs are calculated. The same N/L (three) data samples are accessed for each set of L output calculations. Each output calculation

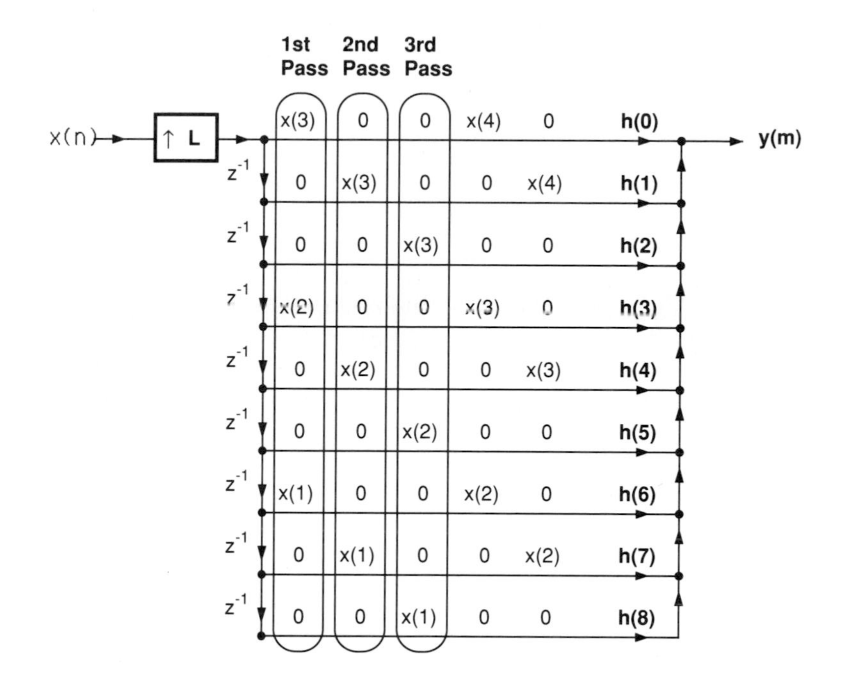

Figure 5.18 Example Interpolator Flowgraph

accesses every Lth (third) coefficient, skipping the coefficients that correspond to zero-valued data samples.

Crochiere and Rabiner (see *References* at the end of this chapter) refer to this efficient interpolation filtering method as *polyphase filtering*, because a different phase of the filter function h(k) (equivalent to a set of interleaved coefficients) is used to calculate each output sample.

5.5.3.3 ADSP-2100 Interpolation Algorithm

A circular buffer of length N/L located in data memory forms the data delay line. Although the convolution equation accesses the newest data sample first and the oldest data sample last, the ADSP-2100 fetches data samples from the circular buffer in the opposite order: oldest data first, newest data last. To keep the data/coefficient pairs together, the coefficients are stored in program memory in reverse order, e.g., h(N–1) in PM(0) and h(0) in PM(N–1).

Figure 5.19 shows a flow chart of the interpolation algorithm. The processor waits in a loop and is interrupted at the output sample rate (L times the input sample rate). In the interrupt routine, the coefficient address pointer is decremented by one location so that a new set of interleaved coefficients will be accessed in the next filter pass. A counter tracks the occurrence of every Lth output; on the Lth output, an input sample is taken and the coefficient address pointer is set forward L locations, back to the first set of interleaved coefficients. The output is then calculated with the coefficient address pointer incremented by L locations to fetch every Lth coefficient. One restriction in this algorithm is that the number of filter taps must be an integral multiple of the interpolation factor; N/L must be an integer.

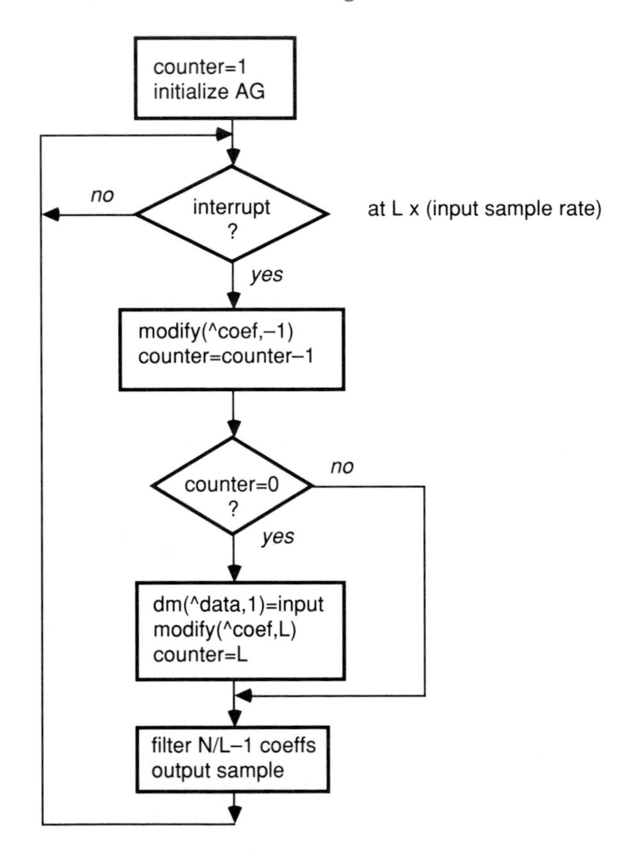

Figure 5.19 Interpolation Flow Chart

Listing 5.10 is an ADSP-2100 program that implements this interpolation algorithm. The ADSP-2100 is capable of calculating each filter pass in ((N/L)+17) processor instruction cycles. Each pass must be calculated within the period between output samples, equal to $1/F_s L$. Thus the maximum number of taps that can be calculated in real time is:

$$N = \frac{1}{F_s t_{CLK}} - 17L$$

where t_{CLK} is the processor cycle time and F_s is the input sampling rate.

```
{INTERPOLATE.dsp
Real time Direct Form FIR Filter, N taps, uses an efficient algorithm
to interpolate by L for an increase of L times the input sample rate. A
restriction on the number of taps is that N/L be an integer.

   INPUT: adc
   OUTPUT: dac
}

.MODULE/RAM/ABS=0 interpolate;
.CONST            N=300;
.CONST            L=4;                       {interpolate by factor of L}
.CONST            NoverL=75;
.VAR/PM/RAM/CIRC  coef[N];
.VAR/DM/RAM/CIRC  data[NoverL];
.VAR/DM/RAM       counter;
.PORT             adc;
.PORT             dac;
.INIT             coef:<coef.dat>;

        RTI;                           {interrupt 0}
        RTI;                           {interrupt 1}
        RTI;                           {interrupt 2}
        JUMP sample;                   {interrupt 3 at (L*input rate)}

initialize:     IMASK=b#0000;          {disable all interrupts}
                ICNTL=b#01111;         {edge sensitive interrupts}
                SI=1;                  {set interpolate counter to 1}
                DM(counter)=SI;        {for first data sample}
                I4=^coef;              {setup a circular buffer in PM}
                L4=%coef;
```

(listing continues on next page)

```
                M4=L;                {modifier for coef is L}
                M5=-1;               {modifier to shift coef back -1}
                I0=^data;            {setup a circular buffer in DM}
                L0=%data;
                M0=1;
                IMASK=b#1000;        {enable interrupt 3}
wait_interrupt: JUMP wait_interrupt;{infinite wait loop}

{_______________________Interpolate___________________________________}

sample:      MODIFY(I4,M5);          {shifts coef pointer back by -1}
             AY0=DM(counter);
             AR=AY0-1;               {decrement and update counter}
             DM(counter)=AR;
             IF NE JUMP do_fir;      {test and input if L times}

{________input data sample, code executed at the sample rate__________}

do_input:    AY0=DM(adc);            {input data sample}
             DM(I0,M0)=AY0;          {update delay line with newest}
             MODIFY(I4,M4);          {shifts coef pointer up by L}
             DM(counter)=M4;         {reset counter to L}

{_________filter pass, occurs at L times the input sample rate________}

do_fir:      CNTR=NoverL - 1;        {N/L-1 since round on last tap}
             MR=0, MX0=DM(I0,M0), MY0=PM(I4,M4);
             DO taploop UNTIL CE;    {N/L-1 taps of FIR}
taploop:         MR=MR+MX0*MY0(SS), MX0=DM(I0,M0), MY0=PM(I4,M4);
             MR=MR+MX0*MY0(RND);     {last tap with round}
             IF MV SAT MR;           {saturate result if overflowed}
             DM(dac)=MR1;            {output sample}
             RTI;
.ENDMOD;
```

Listing 5.10 Efficient Interpolation Filter

The interpolation filter has a gain of 1/L in the passband. One method to attain unity gain is to premultiply (offline) all the filter coefficients by L. This method requires the maximum coefficient amplitude to be less than 1/L, otherwise the multiplication overflows the 16-bit coefficient word length. If the maximum coefficient amplitude is not less than 1/L, then you must multiply each output result by 1/L instead. The code in Listing

5.11 performs the 16-by-32 bit multiplication needed for this gain correction. This code replaces the saturation instruction in the interpolation filter program in Listing 3.3. The MY1 register should be initialized to L at the start of the routine, and the last multiply/accumulate of the filter should be performed with (SS) format, not the rounding option. This code multiplies a filter output sample in 1.31 format by the gain L, in 16.0 format, and produces in a 1.15 format corrected output in the SR0 register.

```
MX1= MR1;
MR= MR0*MY1 (UU);
MR0= MR1;
MR1= MR2;
MR= MR+MX1*MY1 (SU);
SR= LSHIFT MR0 BY -1 (LO);
SR= SR OR ASHIFT MR1 BY -1 (HI);
```

Listing 5.11 Extended Precision Multiply

5.5.3.4 Interpolator Hardware Configuration

The I/O hardware required for the interpolation filter is the same as that for the decimation filter with the exception that the interval timer clocks the output D/A converter, and the input A/D converter is clocked by the interval counter signal divided by L. The interval timer interrupts the ADSP-2100 at the output sample rate. This configuration is shown in Figure 5.20, on the following page.

5.5.4 Rational Sample Rate Changes

The preceding sections describe processes for decreasing or increasing a signal's sample rate by an integer factor (M for decreasing, L for increasing). In real systems, the integer factor restriction is frequently unacceptable. For instance, if two sequences of different sample rates are analyzed in operations such as cross-correlation, cross-spectrum, and transfer function determination, the sequences must be converted to a common sample rate. However, the two sample rates are not generally related by an integer factor. Another instance in which the integer factor restriction is unacceptable is in transferring data between two storage media that use different sampling rates, such as from a compact disk at 44.1kHz to a digital audio tape at 48.0kHz. The compact disk data must have its sampling rate increased by a factor of 48/44.1.

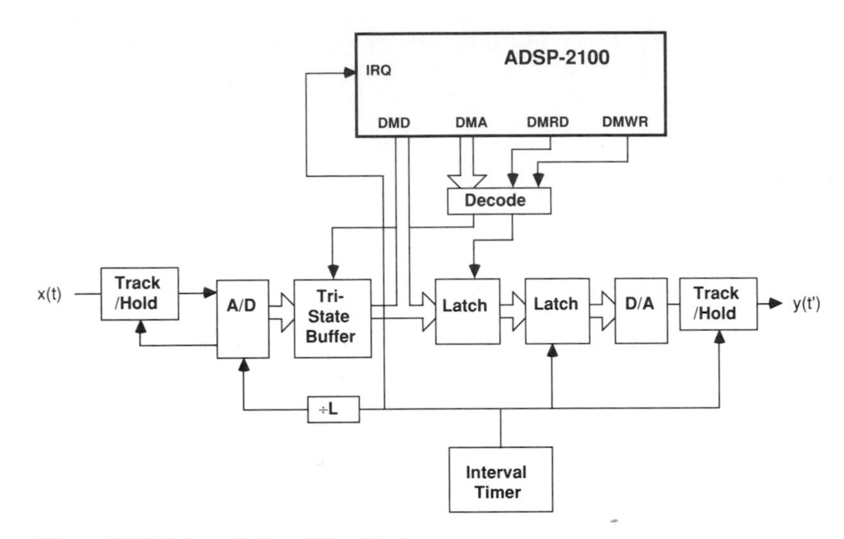

Figure 5.20 Interpolation Filter Hardware

5.5.4.1 L/M Change in Sample Rate

A noninteger rate change factor can be approximated by a rational number closest to the desired factor. This number should be the least common multiple of the two sample rates. A rational number is the ratio of two integers and can be expressed as the fraction L/M. For an L/M sample rate change, the signal is first interpolated to increase its sample rate by L. The resulting signal is decimated to decrease its sample rate by M. The overall change in the sample rate is L/M. Thus, a rational rate change can be accomplished by cascading interpolation and decimation filters. For example, to increase a signal's sample rate by a ratio of 48/44.1 requires interpolation by L=160 and decimation by M=147.

Figure 5.21a shows the cascading of the interpolation and decimation processes. The cascaded filters are the interpolation and decimation filters discussed in the two previous sections of this chapter. The rate expander and the low-pass filter h′(k) make up the interpolator, and the low-pass filter h″(k) and the rate compressor make up the decimator. The interpolator has the same restriction that N/L must be an integer. The input signal x(n) is interpolated to an intermediate sample rate that is a common multiple of the input and output sample rates. To maintain the maximum bandwidth in the intermediate signal x(k), the interpolation must precede the decimation; otherwise, some of the desired frequency content in the original signal would be filtered out by the decimator.

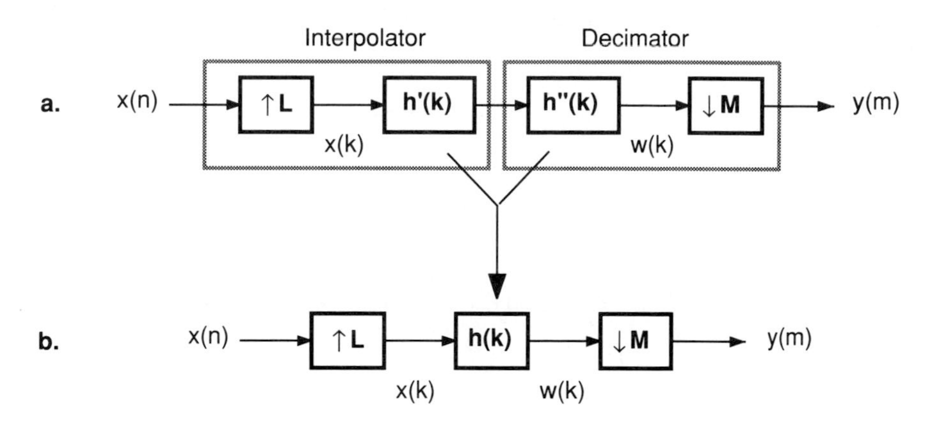

Figure 5.21 Combined Interpolation and Decimation

A significant portion of the computations can be eliminated because the two filters, h′(k) and h″(k), are redundant. Both filters have low-pass transfer functions, and thus the filter with the highest cutoff frequency is unnecessary. The interpolation and decimation functions can be combined using one low-pass filter, h(k), as shown in Figure 5.21b. This rate changer incorporates a gain of L to compensate for the 1/L gain of the interpolation filter, as described earlier in this chapter.

Figure 5.22, on the next page, shows the frequency representation for a sample rate increase of 3/2. The input sample frequency of 4kHz is first increased to 12kHz which is the least common multiple of 4 and 6kHz. This intermediate signal X(k) is then filtered to eliminate the images caused by the rate expansion and to prevent any aliasing that the rate compression could cause. The filtered signal W(k) is then rate compressed by a factor of 2 to result in the output Y(k) at a sample rate of 6kHz. Figure 5.23, also on the next page, shows a similar example that decreases the sample rate by a factor of 2/3.

Figure 5.24 shows the relationship between the sample periods used in the 3/2 and 2/3 rate changes. The intermediate sample period is one-Lth of the input period, $1/F_s$, and one-Mth of the output period.

5.5.4.2 Implementation of Rate Change Algorithm

The rational rate change algorithm applies the same calculation-saving techniques used in the decimation and interpolation filters. In the interpolation, the rate expander is incorporated into the filter so that all

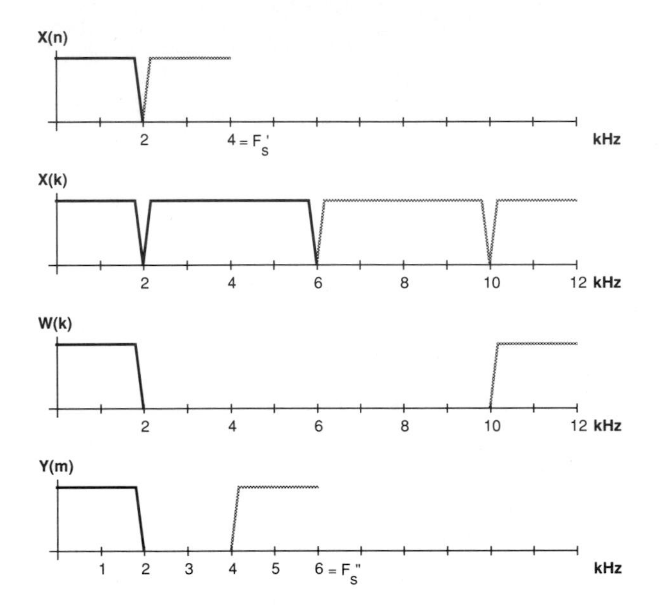

Figure 5.22 3/2 Sample Rate Change

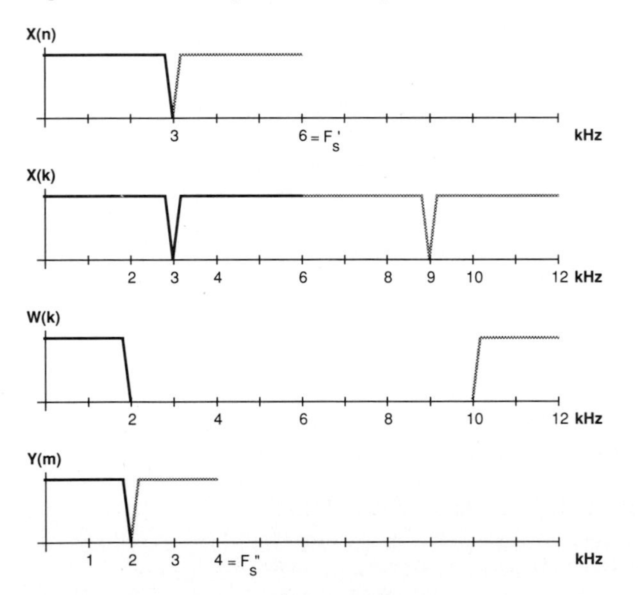

Figure 5.23 2/3 Sample Rate Change

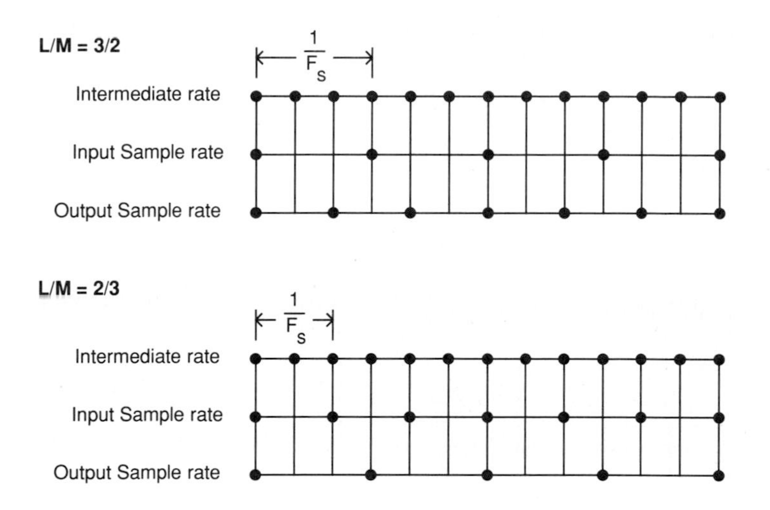

Figure 5.24 Intermediate, Input and Output Sample Rates

zero-valued multiplications are skipped. In the decimation, the rate compressor is incorporated into the filter so that discarded results are not calculated, and a data buffer stores input samples that arrive while the filter output is being calculated. Thus, an entire output period is allocated for calculating one output sample.

The flow charts in Figures 5.25 and 5.26 show two implementations of the rate change algorithm. The first one uses software counters to derive the input and output sample rates from the common sample rate. In this algorithm, the main routine is interrupted at the common sample rate. Depending on whether one or both of the counters has decremented to zero, the interrupt routine reads a new data sample into the input buffer and/or sets a flag that causes the main routine to calculate a new output sample.

For some applications, the integer factors M and L are so large that the overhead for dividing the common sample rate leaves too little time for the filter calculations. The second implementation of the rate change algorithm solves this problem by using two external hardware dividers, ÷L and ÷M, to generate the input and output sample rates from the common rate. The ÷L hardware generates an interrupt that causes the processor to input a data sample. The ÷M hardware generates an interrupt (with a lower priority) that starts the calculation of an output sample.

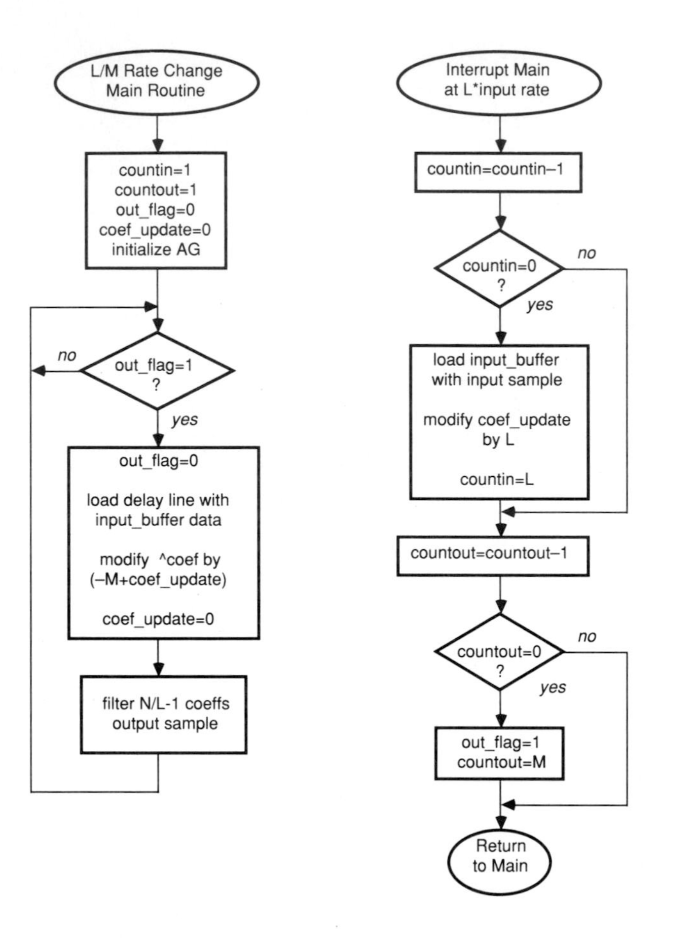

Figure 5.25 L/M Rate Change with Software Division

5.5.4.3 ADSP-2100 Rational Rate Change Program

Listings 5.12 and 5.13 contain the ADSP-2100 programs for the two implementations of the rational factor sample rate change discussed above. The programs are identical except that the first uses two counters to derive the input and output sample periods from IRQ1, whereas the second relies on external interrupts IRQ0 and IRQ1 to provide these periods. In the second program, the input routine has the higher priority interrupt so that inputs always precede outputs when both interrupts coincide.

To implement the calculation-saving techniques, the programs must update the coefficient pointer with two different modification values. First, the algorithm must update the coefficient pointer by L each time an input sample is read. This modification moves the coefficient pointer back to the first set of the polyphase coefficients. The variable *coef_update* tracks these updates. The second modification to the coefficient pointer is to set it back by one location for each interpolated output, even the outputs that not calculated because they are discarded in the decimator. The modification constant is –M because M–1 outputs are discarded by the rate compressor. The total value that updates the coefficient pointer is –M + *coef_update*.

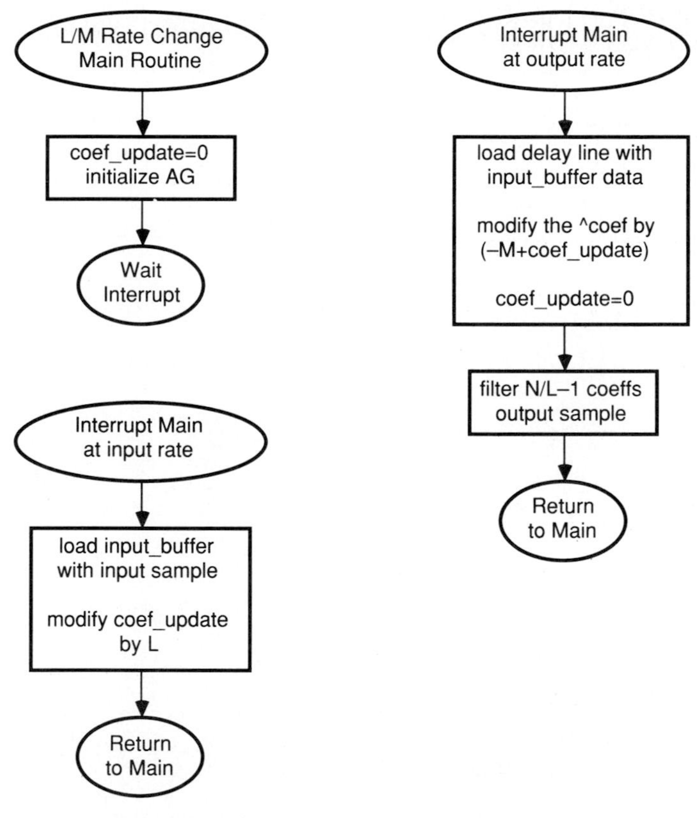

Figure 5.26 L/M Rate Change with Hardware Division

The rate change program in Listing 5.12 can calculate the following number of filter taps within one output period:

$$N = \frac{M}{F_s t_{CLK}} - 30L - 11ML - 6L\lceil M/L \rceil$$

where t_{CLK} is the ADSP-2100 instruction cycle time, and the notation $\lceil u \rceil$ means the smallest integer greater than or equal to u. The program in Listing 3.6 can execute

$$N = \frac{M}{F_s t_{CLK}} - 22L - 9L\lceil M/L \rceil$$

taps in one output period. These equations determine the upper limit on N, the order of the low-pass filter.

```
{RATIO_BUF.dsp
Real time Direct Form FIR Filter, N taps. Efficient algorithm
to interpolate by L and decimate by M for a L/M change in the input sample rate. Uses
an input buffer so that the filter computations to occur in parallel with inputting and
outputting data. This allows a larger number of filter taps for a given input sample
rate.

        INPUT: adc
        OUTPUT: dac
    }

    .MODULE/RAM/ABS=0 Ratio_eff;
    .CONST              N=300;                   {N taps, N coefficients}
    .CONST              L=3;                     {decimate by factor of L}
    .CONST              NoverL=100;              {NoverL must be an integer}
    .CONST              M=2;                     {decimate by factor of M}
    .CONST              intMoverL=2;             {smallest integer GE M/L}
    .VAR/PM/RAM/CIRC    coef[N];                 {coefficient circular buffer}
    .VAR/DM/RAM/CIRC    data[NoverL];            {data delay line}
    .VAR/DM/RAM         input_buf[intMoverL];    {input buffer is not circular}
    .VAR/DM/RAM         countin;
    .VAR/DM/RAM         countout;
    .VAR/DM/RAM         out_flag;                {true when time to calc. output}
    .PORT               adc;
    .PORT               dac;
    .INIT               coef:<coef.dat>;
```

```
                RTI;                 {interrupt 0}
                JUMP interrupt;      {interrupt 1= L * input rate}
                RTI;                 {interrupt 2}
                RTI;                 {interrupt 3}

initialize:     IMASK=b#0000;        {disable all interrupts}
                ICNTL=b#01111;       {edge sensitive interrupts}
                SI=0;                {variables initial conditions}
                DM(out_flag)=SI;     {true if time to calc. output}
                SI=1;
                DM(countin)=SI;      {input every L interrupts}
                DM(countout)=SI;     {output every M interrupts}
                I4=^coef;            {setup a circular buffer in PM}
                L4=%coef;
                M4=L;                {modifier for coef buffer}
                I5=0;                {i5 tracks coefficient updates}
                L5=0;
                M5=-M;               {coef modify done each output}
                I0=^data;            {setup delay line in DM}
                L0=%data;
                M0=1;                {modifier for data is 1}
                I1=^input_buf;       {setup input data buffer in DM}
                L1=0;                {input buffer is not circular}
                IMASK=b#0010;        {enable interrupt 3}

{___________wait for time to calculate an output sample______________}
wait_out:       AX0=DM(out_flag);
                AR=PASS AX0;         {test if true}
                IF EQ JUMP wait_out;

{____________code below occurs at the output sample rate_____________}
                DM(out_flag)=L5;     {reset the output flag to 0}
                AX0=I1;              {calculate ammount in in buffer}
                AY0=^input_buf;
                AR=AX0-AY0;
                IF EQ JUMP modify_coef;     {skip do loop if buffer empty}
                CNTR=AR;             {dump in buffer into delay line}
```

(listing continues on next page)

```
                I1=^input_buf;
                DO load_data UNTIL CE;
                   AR=DM(I1,M0);
load_data:         DM(I0,M0)=AR;
                I1=^input_buf;          {fix pointer to input_buf}

modify_coef:    MODIFY(I5,M5);          {modify coef update by -M}
                M6=I5;
                MODIFY(I4,M6);          {modify ^coef by coef update}
                I5=0;                   {reset coef update}

fir:            CNTR=NoverL-1;
                MR=0, MX0=DM(I0,M0), MY0=PM(I4,M4);
                DO taploop UNTIL CE;    {N/L-1 taps of FIR}
taploop:           MR=MR+MX0*MY0(SS), MX0=DM(I0,M0), MY0=PM(I4,M4);
                MR=MR+MX0*MY0(RND);     {last tap with round}
                IF MV SAT MR;           {saturate result if overflow}
                DM(dac)=MR1;            {output data sample}
                JUMP wait_out;

{__________interrupt, code executed at L times the input rate_________}
interrupt:      ENA SEC_REG;            {so no registers will get lost}
                AY0=DM(countin);        {test if time for input}
                AR=AY0-1;
                DM(countin)=AR;
                IF NE JUMP skipin;

input:          AY0=DM(adc);            {get input sample}
                DM(I1,M0)=AY0;          {load in M long buffer}
                MODIFY(I5,M4);          {modify the coef update by L}
                DM(countin)=M4;         {reset the input count to L}

skipin:         AY0=DM(countout);       {test if time for output}
                AR=AY0-1;
                DM(countout)=AR;
                IF NE RTI;

output:         DM(out_flag)=M0;        {set output flag to true 1}
                AR=M;                   {reset output counter to M}
                DM(countout)=AR;
                RTI;
.ENDMOD;
```

Listing 5.12 Rational Rate Change Program with Software Division

```
{RATIO_2INT.dsp
Real time Direct Form FIR Filter, N taps.  Efficient algorithm to interpolate by L and
decimate by M for a L/M change in the input sample rate. Uses an input buffer so that
the filter computations to occur in parallel with inputting and outputting data. This
allows a larger number of filter taps for a given input sample rate. This version uses
two interrupts and external divide by L and divide by M to eliminate excessive overhead
for large values of M and L.

   INPUT: adc
   OUTPUT: dac
}

.MODULE/RAM/ABS=0 Ratio_2int;
.CONST             N=300;                     {N taps, N coefficients}
.CONST             L=3;                       {decimate by factor of L}
.CONST             NoverL=100;                {NoverL must be an integer}
.CONST             M=2;                       {decimate by factor of M}
.CONST             intMoverL=2;               {smallest integer GE M/L}
.VAR/PM/RAM/CIRC coef[N];                     {coefficient circular buffer}
.VAR/DM/RAM/CIRC   data[NoverL];              {data delay line}
.VAR/DM/RAM        input_buf[intMoverL];      {input buffer is not circular}
.PORT              adc;
.PORT              dac;
.INIT              coef:<coef.dat>;

        JUMP output;                {interrupt 0= L*Fin/M}
        JUMP input;                 {interrupt 1= L*Fin/L}
        RTI;                        {interrupt 2}
        RTI;                        {interrupt 3}

initialize: IMASK=b#0000;           {disable all interrupts}
            ICNTL=b#11111;          {edge sens. nested interrupts}
            I4=^coef;               {setup a circular buffer in PM}
            L4=%coef;
            M4=L;                   {modifier for coef buffer}
            I5=0;                   {i5 tracks coefficient updates}
            L5=0;
            M5=-M;                  {coef modify done each output}
            I0=^data;               {setup delay line in DM}
            L0=%data;
            M0=1;                   {modifier for data is 1}
            I1=^input_buf;          {setup input data buffer in DM}
            L1=0;                   {input buffer is not circular}
            IMASK=b#0011;           {enable interrupt 3}
```

(llsting continues on next page)

```
{___________wait for time to output or input a sample________________}
wait_out:     JUMP wait_out;

{______interrupt code below executed at the output sample rate________}
output:       AX0=I1;                  {calculate ammount in in buffer}

              AY0=^input_buf;
              AR=AX0-AY0;
              IF EQ JUMP modify_coef;{skip do loop if buffer empty}
              CNTR=AR;                 {dump in buffer into delay line}

              I1=^input_buf;
              DO load_data UNTIL CE;
                 AR=DM(I1,M0);
load_data:       DM(I0,M0)=AR;
              I1=^input_buf;           {fix pointer to input_buf}
modify_coef: MODIFY(I5,M5);            {modify coef update by -M}
              M6=I5;
              MODIFY(I4,M6);           {modify ^coef by coef update}
              I5=0;                    {reset coef update}

fir:          CNTR=NoverL-1;
              MR=0, MX0=DM(I0,M0), MY0=PM(I4,M4);
              DO taploop UNTIL CE;     {N/L-1 taps of FIR}
taploop:         MR=MR+MX0*MY0(ss), MX0=DM(I0,M0), MY0=PM(I4,M4);
              MR=MR+MX0*MY0(RND);      {last tap with round}
              IF MV SAT MR;            {saturate result if overflow}
              DM(dac)=MR1;             {output data sample}
              RTI;

{_______interrupt code below executed at the input sample rate_________}

input:        ENA SEC_REG;             {context save}
              AY0=DM(adc);             {get input sample}
              DM(I1,M0)=AY0;           {load in M long buffer}
              MODIFY(I5,M4);           {modify the coef update by L}
              RTI;
.ENDMOD;
```

Listing 5.13 Rational Rate Change Program with Hardware Division

5.5.5 Multistage Implementations

The previous examples of sample rate conversion in this chapter use a single low-pass filter to prevent the aliasing or imaging associated with the rate compression and expansion. One method for further improving the efficiency of rate conversion is to divide this filter into two or more cascaded stages of decimation or interpolation. Each successive stage reduces the sample rate until the final sample rate is equal to twice the bandwidth of the desired data. The product of all the stages' rate change factors should equal the total desired rate change, M or L.

Crochiere and Rabiner (see *References* at the end of this chapter) show that the total number of computations in a multi-stage design can be made substantially less than that for a single-stage design because each stage has a wider transition band than that of the single-stage filter. The sample rate at which the first stage is calculated may be large, but because the transition band is wide, only a small number of filter taps (N) is required. In the last stage, the transition band may be small, but because the sample rate is small also, fewer taps and thus a reduced number of computations are needed. In addition to computational efficiency, multistage filters have the advantage of a lower round-off noise due to the reduced number of taps.

Figure 5.27, on the following page, shows the frequency spectra for an example decimation filter implemented in two stages. The bandwidth and transition band of the desired filter is shown in Figure 5.27a and the frequency response of the analog anti-alias filter required is shown in Figure 5.27b. The shaded lines indicate the frequencies that alias into the interstage transition bands. These frequencies are sufficiently attenuated so as not to disturb the final pass or transition bands. The frequency responses of the first and final stage filters are shown in Figure 5.27c and d. The example in Figure 5.28 is the same except that aliasing is allowed in the final transition band. This aliasing is tolerable, for instance, when the resulting signal is used for spectral analysis and the aliased band can be ignored. All the transition bands are wide and therefore the filter stages require fewer taps than a single-stage filter.

Crochiere and Rabiner (see *References* at the end of this chapter) contains some design curves that help to determine optimal rate change factors for the intermediate stages. A two- or three-stage design provides a substantial reduction in filter computations; further reductions in computations come from adding more stages. Because the filters are cascaded, each stage must have a passband ripple equal to final passband

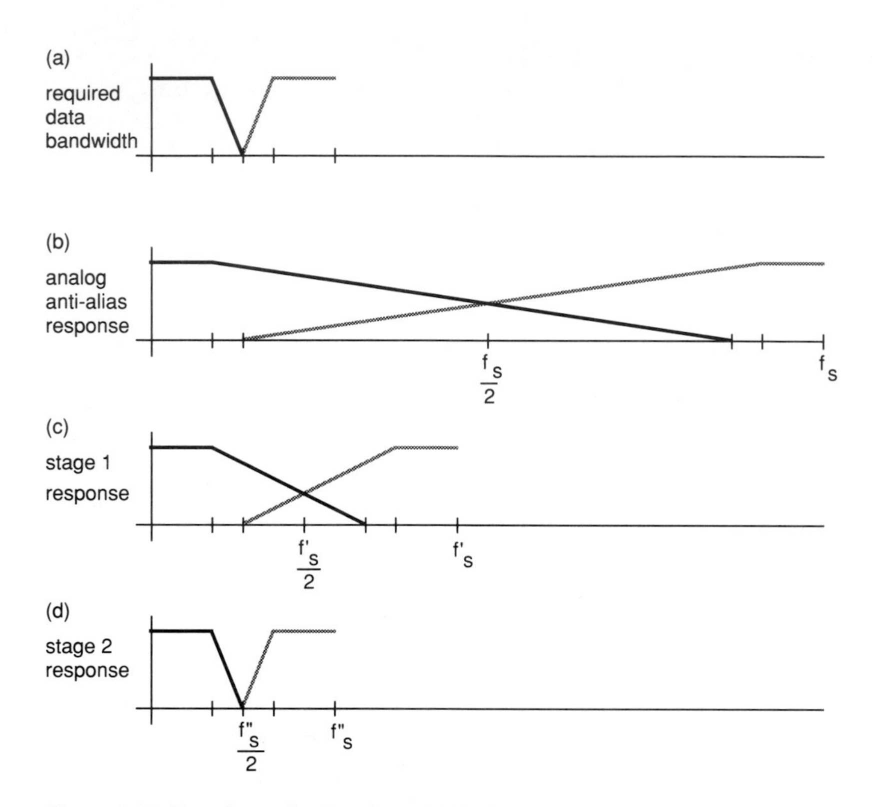

Figure 5.27 Two-Stage Decimation with No Transition Band Aliasing

ripple divided by the number of stages used. The stopband ripple does not have to be less than the final stopband ripple because each successive stage attenuates the stopband ripple of the previous stage.

Listing 5.14 is the ADSP-2100 program for a two-stage decimation filter. This two-stage decimation program is similar to the program for the buffered decimation listed in *Decimation*, earlier in this chapter. The main difference is that two *ping-ponged* buffers are used to store input samples. While one buffer is filled with input data, the other is dumped into the delay line of the first filter stage. The result of the first stage filter is written to the second stage's delay line. After M samples have been input and one final result has been calculated, the two input buffers swap functions, or *ping-pong*.

Two buffers are needed because only a portion (M–1 samples) of the input buffer can be dumped into the first stage delay line at once. The single-stage decimation algorithm used only one buffer because it could dump all input data into the buffer at once. The two ping-ponged buffers are implemented as one double-length circular buffer, 2M locations long, indexed by two pointers offset from each other by M. Because the pointers follow each other and wrap around the buffer, the two halves switch between reading and writing (ping-pong) after every M inputs.

Listing 5.15 is the ADSP-2100 program for a two-stage interpolation filter. This routine is essentially a cascade of the program for the single-stage interpolator. The required interrupt rate is the product of the interpolation factors of the individual stages, (L–1)(L–2). The program can be expanded easily if more than two stages of interpolation or decimation are required.

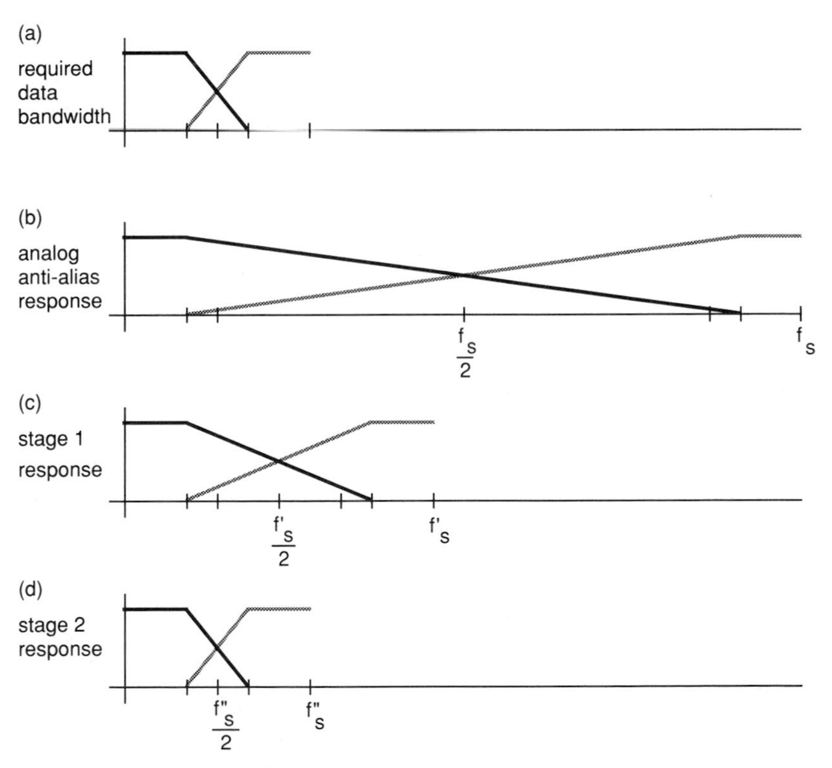

Figure 5.28 Two-Stage Decimation with Transition Band Aliasing

```
{DEC2STAGEBUF.dsp
Real time Direct Form Finite Impulse Filter, N1,N2 taps, uses two cascaded stages to
decimate by M1 and M2 for a decrease of 1/(M1*M2) times the input sample rate. Uses an
input buffer to increase efficiency.

   INPUT: adc
   OUTPUT: dac
}

.MODULE/RAM/ABS=0 dec2stagebuf;
.CONST            N1=32;
.CONST            N2=148;
.CONST            M_1=2;                {decimate by factor of M}
.CONST            M_2=2;                {decimate by factor of M}
.CONST            M1xM2=4;              {(M_1 * M_2)}
.CONST            M1xM2x2=8;            {(M_1 * M_2 * 2)}
.VAR/PM/RAM/CIRC  coef1[N1];
.VAR/PM/RAM/CIRC  coef2[N2];
.VAR/DM/RAM/CIRC  data1[N1];
.VAR/DM/RAM/CIRC  data2[N2];
.VAR/DM/RAM/CIRC  input_buf[m1xm2x2];
.VAR/DM/RAM       input_count;
.PORT             adc;
.PORT             dac;
.INIT             coef1:<coef1.dat>;
.INIT             coef2:<coef2.dat>;

        RTI;                        {interrupt 0}
        RTI;                        {interrupt 1}
        RTI;                        {interrupt 2}
        JUMP sample;                {interrupt 3 at input rate}

initialize: IMASK=b#0000;           {disable all interrupts}
            ICNTL=b#01111;          {edge sensitive interrupts}
            SI=M1xM2;
            DM(input_count)=SI;
            I4=^coef1;              {setup a circular buffer in PM}
            L4=%coef1;
            M4=1;                   {modifier for coef is 1}
            I5=^coef2;              {setup a circular buffer in PM}
            L5=%coef2;
```

```
           I0=^data1;                {setup a circular buffer in DM}
           L0=%data1;
           M0=1;                     {modifier for data is 1}
           I1=^data2;                {setup a circular buffer in DM}
           L1=%data2;
           I2=^input_buf;            {setup input buffer in DM}
           L2=%input_buf;
           I3=^input_buf;            {setup second in buffer in DM}
           L3=%input_buf;
           IMASK=b#1000;             {enable interrupt 3}

wait_full: AX0=DM(input_count);      {wait until input buffer is full}
           AR=PASS AX0;
           IF NE JUMP wait_full;

           AR=M1xM2;                 {reinitialize input counter}
           DM(input_count)=AR;
           CNTR=M_2;
           DO stage_1 UNTIL CE;
              CNTR=M_1;
              DO dump_buf UNTIL CE;
                 AR=DM(I3,M0);
dump_buf:        DM(I0,M0)=AR;
              CNTR=N1 - 1;
              MR=0, MX0=DM(I0,M0), MY0=PM(I4,M4);
              DO taploop1 UNTIL CE;{N-1 taps of FIR}
taploop1:        MR=MR+MX0*MY0(SS), MX0=DM(I0,M0), MY0=PM(I4,M4);
              MR=MR+MX0*MY0(RND);    {last tap with round}
              IF MV SAT MR;          {saturate result if overflow}
stage_1:      DM(I1,M0)=MR1;         {pass to next filter stage}

           CNTR=N2 - 1;
           MR=0, MX0=DM(I1,M0), MY0=PM(I5,M4);
           DO taploop2 UNTIL CE;     {N-1 taps of FIR}
taploop2:     MR=MR+MX0*MY0(SS), MX0=DM(I1,M0), MY0=PM(I5,M4);
           MR=MR+MX0*MY0(RND);       {last tap with round}
           IF MV SAT MR;             {saturate result if overflow}
           DM(dac)=MR1;              {output data sample}
           JUMP wait_full;
```

(listing continues on next page)

```
{_______________interrupt code executed at the sample rate_____________}
sample:      ENA SEC_REG;               {context save}
             AY0=DM(adc);               {input data sample}
             DM(I2,M0)=AY0;             {load input buffer}
             AY0=DM(input_count);
             AR=AY0-1;                  {decrement and update counter}
             DM(input_count)=AR;
             RTI;
.ENDMOD;
```

Listing 5.14 ADSP-2100 Program for Two-Stage Decimation

```
{INT2STAGE.dsp
Two stage cascaded real time Direct Form Finite Impulse Filter, N1, N2 taps, uses an
efficient algorithm to interpolate by L1*L2 for an increase of L1*L2 times the input
sample rate. A restriction on the number of taps is that N divided by L be an integer.

   INPUT: adc
   OUTPUT: dac
}

.MODULE/RAM/ABS=0 interpolate_2stage;
.CONST            N1=32;
.CONST            N2=148;
.CONST            L_1=2;                {stage one factor is L1}
.CONST            L_2=2;                {stage two factor is L2}
.CONST            N1overL1=16;
.CONST            N2overL2=74;
.VAR/PM/RAM/CIRC  coef1[N1];
.VAR/PM/RAM/CIRC  coef2[N2];
.VAR/DM/RAM/CIRC  data1[N1overL1];
.VAR/DM/RAM/CIRC  data2[N2overL2];
.VAR/DM/RAM       counter1;
.VAR/DM/RAM       counter2;
```

```
.PORT           adc;
.PORT           dac;
.INIT           coef1:<coef1.dat>;
.INIT           coef2:<coef2.dat>;

                RTI;                 {interrupt 0}
                RTI;                 {interrupt 1}
                RTI;                 {interrupt 1}
                JUMP sample;         {interrupt 3= L1*L2 output rate}

initialize:     IMASK=b#0000;        {disable all interrupts}
                ICNTL=b#01111;       {edge sensitive interrupts}
                SI=1;                {set interpolate counters to 1}
                DM(counter1)=SI;     {for first data sample}
                DM(counter2)=SI;
                I4=^coef1;           {setup a circular buffer in PM}
                L4=%coef1;
                M4=L_1;              {modifier for coef is L1}
                M6=-1;               {modifier to shift coef back -1}

                I5=^coef2;           {setup a circular buffer in PM}
                L5=%coef2;
                M5=L_2;              {modifier for coef is L2}
                I0=^data1;           {setup a circular buffer in DM}
                L0=%data1;
                M0=1;
                I1=^data2;           {setup a circular buffer in DM}
                L1=%data2;
                IMASK=b#1000;        {enable interrupt 3}
wait_interrupt: JUMP wait_interrupt;       {infinite wait loop}


{_________________________Interpolate__________________________________}
sample:     MODIFY(I5,M6);           {shifts coef pointer back by -1}

            AY0=DM(counter2);
            AR=AY0-1;                {decrement and update counter}
            DM(counter2)=AR;
            IF NE JUMP do_fir2;      {test, do stage 1 if L_2 times}
```

(listing continues on next page)

```
            MODIFY(I4,M6);            {shifts coef pointer back by -1}
            AY0=DM(counter1);
            AR=AY0-1;                 {decrement and update counter.}
            DM(counter1)=AR;
            IF NE JUMP do_fir1;       {test and input if L_1 times}

{_________input data sample, code occurs at the input sample rate____}
do_input:   AY0=DM(adc);              {input data sample}
            DM(I0,M0)=AY0;            {update delay line with newest}
            MODIFY(I4,M4);            {shifts coef1 pointer by L1}
            DM(counter1)=M4;          {reset counter1}

{_________filter pass, occurs at L1 times the input sample rate_______}
do_fir1:    CNTR=N1overL1 - 1;        {N1/L1-1 because round last tap}

            MR=0, MX0=DM(I0,M0), MY0=PM(I4,M4);
            DO taploop1 UNTIL CE;     {N1/L_1-1 taps of FIR}
taploop1:      MR=MR+MX0*MY0(SS), MX0=DM(I0,M0), MY0=PM(I4,M4);
            MR=MR+MX0*MY0(RND);       {last tap with round}
            IF MV SAT MR;             {saturate result if overflow}
            DM(I1,M0)=MR1;            {update delay line with newest}
            MODIFY(I5,M5);            {shifts coef2 pointer by L2}
            DM(counter2)=M5;          {reset counter2}

{_______filter pass, executed at (L1*L2) times the input sample rate__}
do_fir2:    CNTR=N2overL2 - 1;        {N2/L2-1 because round last tap}

            MR=0, MX0=DM(I1,M0), MY0=PM(I5,M5);
            DO taploop2 UNTIL CE;     {N2/L_2-1 taps of FIR}
taploop2:      MR=MR+MX0*MY0(SS), MX0=DM(I1,M0), MY0=PM(I5,M5);
            MR=MR+MX0*MY0(RND);       {last tap with round}
            IF MV SAT MR;             {saturate result if overflow}
            DM(dac)=MR1;              {output sample}
            RTI;
.ENDMOD;
```

Listing 5.15 ADSP-2100 Program for Two-Stage Interpolation

5.5.6 Narrow-Band Spectral Analysis

The computation of the spectrum of a signal is a fundamental DSP operation that has a wide range of applications. The spectrum can be calculated efficiently with the fast Fourier transform (FFT) algorithm. An N-point FFT results in N/2 bins of spectral information spanning zero to the Nyquist frequency. The frequency resolution of this spectrum is F_s/N Hz per bin, where F_s is the sample rate of the data. The number computations required is on the order of $N\log_2N$. Often, applications such as sonar, radar, and vibration analysis need to determine only a narrow band of the entire spectrum of a signal. The FFT would require calculating the entire spectrum of the signal and discarding the unwanted frequency components.

Multirate filtering techniques let you translate a frequency band to the baseband and reduce the sample rate to twice the width of the narrow band. An FFT performed on reduced-sample-rate data allows either greater resolution for about the same amount of computations or an equivalent resolution for a reduced amount of computations. Thus, the narrow band can be calculated more efficiently. In addition to computation savings, this frequency translation technique has the advantage of eliminating the problem of an increased noise floor caused by the bit growth of data in a large-N FFT.

One method of frequency translation takes advantage of the aliasing properties of the rate compressor. As discussed in *Decimation*, earlier in this chapter, rate compression (discrete-time sampling) in the frequency domain results in images, a sequence of periodic repetitions of the baseband signal. These images are spaced at harmonics of the sampling frequency.

The modulation and the sample rate reduction can be done simultaneously, as shown in Figure 5.29a, on the next page. The input signal is band-pass-filtered to eliminate all frequencies but the narrow band of interest (between w_1 and w_2). This signal is rate-compressed by a factor of M to get the decimated and frequency-translated output. An (N/M)-point FFT is performed on the resulting signal, producing the spectrum of the signal, which now contains only the narrow band.

Figure 5.29b shows the modulation of the band-pass signal. The modulating impulses are spaced at intervals of $2\pi/M$. Figure 5.29c shows that the narrow band is translated to baseband because it is forced to alias. The spectrum of the final signal y(m) is shown in Figure 5.29d. Zero corresponds to w_1, and π or the Nyquist frequency corresponds to w_2. If

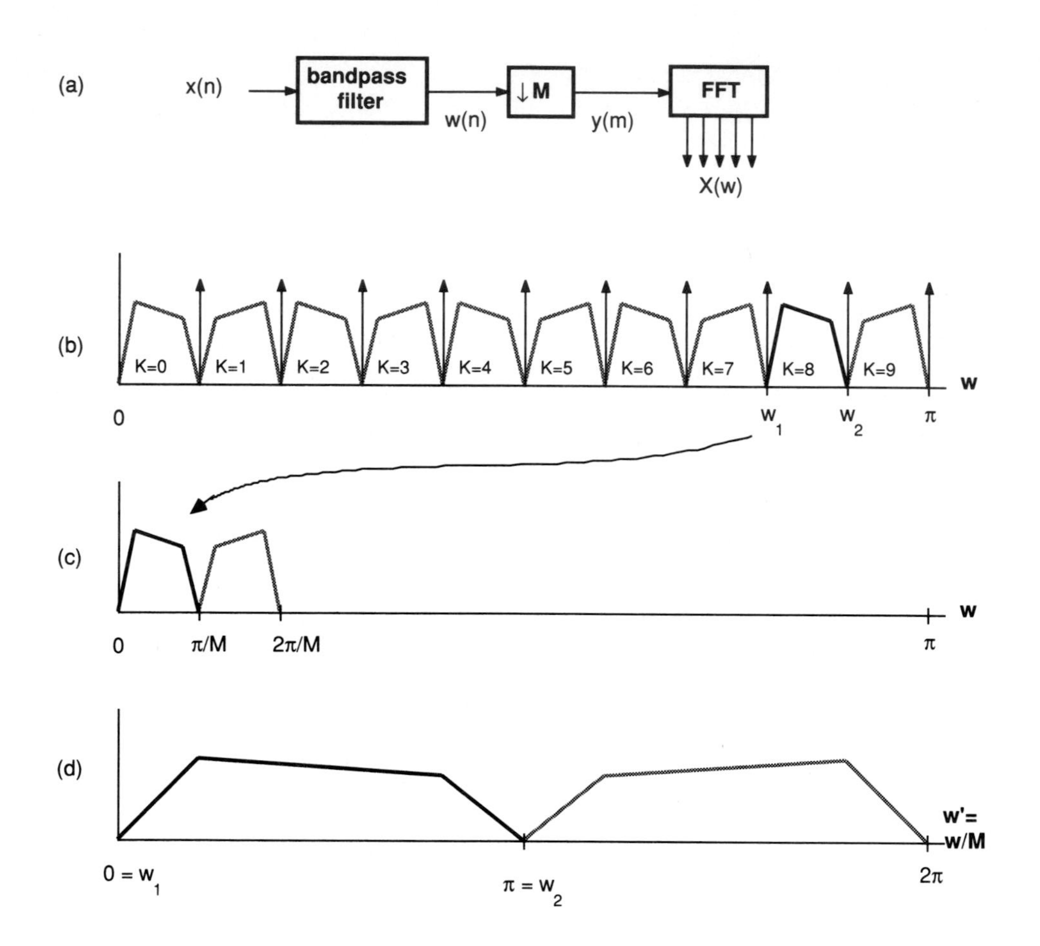

Figure 5.29 Integer Band Decimator for High Resolution Spectral Analysis

an odd integer band is chosen for the band-pass filter, e.g., K=9, then the translated signal is inverted in frequency. This situation can be corrected by multiplying each output sample y(m) by $(-1)^m$, i.e., inverting the sign of all odd samples.

The entire narrow band filtering process can be accomplished using the single- or multi-stage decimation program listed in this chapter. A listing of an ADSP-2100 FFT implementation can be found in Chapter 6.

5.6 ADAPTIVE FILTERS

The stochastic gradient (SG) adaptive filtering algorithm, developed in its present form by Widrow and Hoff (Widrow and Hoff, 1960), provides a powerful and computationally efficient means of realizing adaptive filters. It is used to accomplish a variety of applications, including

- echo cancellation in voice or data signals,
- channel equalization in data communication to minimize the effect of intersymbol interference, and,
- noise cancellation in speech, audio, and biomedical signal processing.

The SG algorithm is the most commonly used adaptation algorithm for transversal filter structures. Honig and Messerschmitt, 1984, provides an excellent and thorough treatment of the SG transversal filter as well as other algorithms and filter structures. Using the notation developed therein, the estimation error $e_c(T)$ between the two input signals y(T) and d(T) of a joint process estimator implemented with a transversal filter structure is given by the following equation:

$$e_c(T) = d(T) - \sum_{j=1}^{n} c_j(T)\, y(T - j + 1)$$

The estimation error of the joint process estimator is thus formed by the difference between the signal it is desired to estimate, d(T), and a weighted linear combination of the current and past input values y(T). The weights, $c_j(T)$, are the transversal filter coefficients at time T. The adaptation of the jth coefficient, $c_j(T)$, is performed according to the following equation:

$$c_j(T + 1) = c_j(T) + \beta e_c(T)\, y(T - j + 1)$$

In this equation, y(T–j+1) represents the past value of the input signal "contained" in the jth tap of the transversal filter. For example, y(T), the present value of the input signal, corresponds to the first tap and y(T–42) corresponds to the forty-third filter tap. The step size β controls the "gain" of the adaptation and allows a tradeoff between the convergence rate of the algorithm and the amount of random fluctuation of the coefficients after convergence.

5.6.1 Single-Precision Stochastic Gradient

The transversal filter subroutine that performs the sum-of-products operation to calculate $e_c(T)$, the estimation error, is given in *FIR Filters*, earlier in this chapter. The subroutine that updates the filter coefficients is shown in Listing 5.16. This subroutine is based on the assumption that all n data values used to calculate the coefficient are real.

The first instruction multiplies $e_c(T)$ (the estimation error at time T, stored in MX0) by β (the step size, stored in MY1) and loads the product into the MF register. In parallel with this multiplication, the data memory read which transfers y(T–n+1) (pointed to by I2) to MX0 is performed. The nth coefficient update value, $\beta e_c(T)$ y(T–n+1), is computed by the next instruction in parallel with the program memory read which transfers the nth coefficient (pointed to by I6) to the ALU input register AY0. The *adapt* loop is then executed n times in 2n+2 cycles to update all n coefficients. The first time through the loop, the nth coefficient is updated in parallel with a dual fetch of y(T–n+2) and the (n–1)th coefficient. The updated nth coefficient value is then written back to program memory in parallel with the computation of the (n–1)th coefficient update value, $e_c(T)$ y(T–n+2). This continues until all n coefficients have been updated and execution falls out of the loop. If desired, the saturation mode of the ALU may be enabled prior to entering the routine so as to automatically implement a saturation capability on the coefficient updates.

The maximum allowable filter order when using the stochastic gradient algorithm for an adaptive filtering application is determined primarily by the processor cycle time, the sampling rate, and the number of other computations required. The transversal filter subroutine requires a total of n+7 cycles for a filter of length n, while the gradient update subroutine requires 2n+9 cycles to update all n coefficients. At an 8-kHz sampling rate and an instruction cycle time of 125 nanoseconds, the ADSP-2100 can implement an SG transversal filter of approximately 300 taps. This implementation would also have 84 instruction cycles for miscellaneous operations.

```
.MODULE rsg_sub;

{
   Real SG Coefficient Adaptation Subroutine

   Calling Parameters
        MX0 = Error
        MY1 = Beta
        I2 -> Oldest input data value in delay line
        L2 = Filter length
        I6 -> Beginning of filter coefficient table
        L6 = Filter length
        M1,M5 = 1
        M6 = 2
        M3,M7 = -1
        CNTR = Filter length

   Return Values
        Coefficients updated
        I2 -> Oldest input data value in delay line
        I6 -> Beginning of filter coefficient table

   Altered Registers
        AY0,AR,MX0,MF,MR

   Computation Time
        (2 × Filter length) + 6 + 3 cycles

   All coefficients and data are assumed to be in 1.15 format.
}

.ENTRY  rsg;

rsg:    MF=MX0*MY1(RND), MX0=DM(I2,M1);         {MF=Error × Beta}
        MR=MX0*MF(RND), AY0=PM(I6,M5);
        DO adapt UNTIL CE;
           AR=MR1+AY0, MX0=DM(I2,M1), AY0=PM(I6,M7);
adapt:     PM(I6,M6)=AR, MR=MX0*MF(RND);
        MODIFY (I2,M3);                         {Point to oldest data}
        MODIFY (I6,M7);                         {Point to start of table}
        RTS;
.ENDMOD;
```

Listing 5.16 Single-Precision Stochastic Gradient

5.6.2 Double-Precision Stochastic Gradient

In some adaptive filtering applications, such as the local echo cancellation required in high-speed data transmission systems, the precision afforded by 16-bit filter coefficients is not adequate. In such applications it is desirable to perform the coefficient adaptation (and generally the filtering operation as well) using a higher-precision representation for the coefficient values. The subroutine in Listing 5.17 implements a stochastic gradient adaptation algorithm that is again based on the equations in the previous section but performs the coefficient adaptation in double precision. Data values, of course, are still maintained in single precision.

The 16-bit coefficients are stored in program memory LSW first, so that the LSWs of all coefficients are stored at even addresses, and the MSWs at odd addresses. The coefficients thus require a circular buffer length (specified by L6) that is twice the length of the filter. As in the single-precision SG program in the previous section, the first instruction is used to compute the product of $e_c(T)$, the estimation error, and β, the step size, in parallel with the data memory read that transfers the oldest input data value in the delay line, y(T–n+1), to MX0. Upon entering the *adaptd* loop, y(T–n+1) is multiplied by $\beta e_c(T)$ to yield the nth coefficient update value. This is performed in parallel with the fetch of the LSW of the nth coefficient (to AY0). The next instruction computes the sum of the update value LSW (in MR0) and the LSW of the nth coefficient, while performing a dual fetch of the MSW of the nth coefficient (again to AY0) and the next data value in the delay line (to MX0). The LSW of the updated nth coefficient is then written back to program memory in parallel with the update of the MSW of the nth coefficient, and the final instruction of the loop writes this updated MSW to program memory. The *adaptd* loop continues execution in this manner until all *n* double-precision coefficients have been updated. If you want saturation capability on the coefficient update, you must enable and disable the saturation mode of the ALU within the update loop. The updates of the LSWs of the coefficients should be performed with the saturation mode disabled; the update of the MSWs of the coefficients should be performed with the saturation mode enabled.

To determine whether an application can benefit from double-precision adaptation, you should evaluate the performance of the associated filtering routine using both single-precision and double-precision coefficients. In some instances, maintaining the coefficients in double precision while performing the filtering operation on only the MSWs of the coefficients may result in the desired amount of cancellation. This adaptation can be achieved using the single-precision transversal filter routine with an M5 value of two. If more cancellation is needed, the routine in Listing 5.17 must be used.

```
.MODULE drsg_sub;

{
   Double-Precision SG Coefficient Adaptation Subroutine

   Calling Parameters
        MX0 = Error
        MY1 = Beta
        I2 -> Oldest input data value in delay line
        L2 = Filter length
        I6 -> Beginning of filter coefficient table
        L6 = 2 × Filter length
        M1,M5 = 1
        M3,M7 = -1
        CNTR = Filter length

   Return Values
        Coefficients updated
        I2 -> Oldest input data value in delay line
        I6 -> Beginning of filter coefficient table

   Altered Registers
        AY0,AR,MX0,MF,MR

   Computation Time
        (4 × Filter length) + 5 + 4 cycles

   All coefficients are assumed to be in 1.31 format.
   All data are assumed to be in 1.15 format.
}

.ENTRY  drsg;

drsg:   MF=MX0*MY1(RND), MX0=DM(I2,M1);           {MF=Error × Beta}
        MR=MX0*MF(SS);
        DO adaptd UNTIL CE;
               MX0=DM(I2,M1), AY0=PM(I6,M5);
               AR=MR0+AY0, AY0=PM(I6,M7);
               PM(I6,M5)=AR, AR=MR1+AY0+C;
adaptd:        MR=MX0*MF(SS), PM(I6,M5)=AR;
        MODIFY (I2,M3);                           {Point to oldest data}
        RTS;
.ENDMOD;
```

Listing 5.17 Double-Precision Stochastic Gradient

5.7 REFERENCES

Bellanger, M. 1984. *Digital Processing of Signals: Theory and Practice*. New York: John Wiley and Sons.

Bloom, P.J. October 1985. *High Quality Digital Audio in the Entertainment Industry: An Overview of Achievements and Changes*. IEEE ASSP Magazine, Vol. 2, Num. 4, pp. 13-14.

Crochiere, Ronald E. and Lawrence R. Rabiner. 1983. *Multirate Digital Signal Processing*. Englewood Cliffs, N.J.: Prentice-Hall.

Hamming, R. W. 1977. *Digital Filters*. Englewood Cliffs, N.J.: Prentice-Hall, Inc.

Honig, M. and Messerschmitt, D. 1984. *Adaptive Filters: Structures, Algorithms, and Applications*. Boston: Kluwer Academic Publishers.

Jackson, L. B. 1986. *Digital Filters and Signal Processing*. Boston: Kluwer Academic Publishers.

Liu, Bede and Abraham Peled. 1976. *Theory Design and Implementation of Digital Signal Processing*, pp. 77-88. John Wiley & Sons.

Liu, Bede and Fred Mintzer. December 1978. *Calculation of Narrow Band Spectra by Direct Decimation*. IEEE Trans. Acoust. Speech Signal Process., Vol. ASSP-26, No. 6, pp. 529-534.

Oppenheim, A. V., and Schafer, R. W. 1975. *Digital Signal Processing*. Englewood Cliffs, N.J.: Prentice-Hall, Inc.

Oppenheim, A.V. ed. 1978. *Applications of Digital Signal Processing*. Englewood Cliffs, N.J.: Prentice-Hall, Inc.

Otnes, R.K and L.E Enochson. 1978. *Applied Time Series Analysis*, pp. 202-212. Wiley-Interscience.

Rabiner, L. R. and Gold, B. 1975. *Theory and Applications of Digital Signal Processing*. Englewood Cliffs, N.J.: Prentice-Hall, Inc.

Schafer, Ronald W. and Lawrence R. Rabiner. June 1973. *A Digital Signal Processing Approach to Interpolation*. Proc. IEEE, vol. 61, pp. 692-702.

Widrow, B., and Hoff, M., Jr. 1960. *Adaptive Switching Circuits*. IRE WESCON Convention Record, Pt. 4., pp. 96-104.

One-Dimensional FFTs ■ 6

6.1 OVERVIEW

In many applications, frequency analysis is necessary and desirable. Applications ranging from radar to spread-spectrum communications employ the Fourier transform for spectral analysis and frequency domain processing.

The discrete Fourier transform (DFT) is the discrete-time equivalent of the continuous-time Fourier transform. Whereas the continuous-time Fourier transform operates on a continuous time signal x(t), the DFT operates on samples of x(t). If X(f) is the continuous Fourier transform of x(t), then the DFT of x(t) (sampled) is a sequence of samples of X(f), equally spaced in frequency. Equation (1) computes the DFT (Oppenheim and Schafer, 1975). X(k) is the discrete or sampled Fourier transform, and x(n) is a sequence of samples of the input signal, x(t). The term W_N, defined as $e^{-j2\pi/N}$, corresponds to the term $e^{-j2\pi ft}$ used to compute the continuous Fourier transform. Various powers of W_N are used for each multiplication in the DFT calculation.

(1) $$X(k) = \sum_{n=0}^{N-1} x(n)\, W_N^{nk} \qquad k = 0 \text{ to } N-1$$

where $W_N = e^{-j2\pi/N}$

A complex summation of N complex multiplications is required for each of the N output samples. In all, N^2 complex multiplications and N^2 complex additions compute an N-point DFT. The time burden created by this large number of calculations limits the usefulness of the DFT in many applications. For this reason, tremendous effort was devoted to developing more efficient ways of computing the DFT. This effort produced the fast Fourier transform (FFT). The FFT uses mathematical shortcuts to reduce the number of calculations the DFT requires.

This chapter describes three variations of the FFT algorithm: the radix-2 decimation-in-time FFT, the radix-2 decimation-in-frequency FFT and the radix-4 decimation-in-frequency FFT. Optimization, to make the code run faster, and block floating-point scaling, to increase data precision, are also addressed. In addition, bit- and digit-reversal and windowing, operations related to the FFT, are described. An FFT in two dimensions is presented in the next chapter.

6.2 RADIX-2 FAST FOURIER TRANSFORMS

Suppose an N-point DFT is accomplished by performing two N/2-point DFTs and combining the outputs of the smaller DFTs to give the same output as the original DFT. The original DFT requires N^2 complex multiplications and N^2 complex additions. Each DFT of N/2 input samples requires $(N/2)^2 = N^2/4$ multiplications and additions, a total of $N^2/2$ calculations for the complete DFT. Dividing the DFT into two smaller DFTs reduces the number of computations by 50 percent. Each of these smaller DFTs can be divided in half, yielding four N/4-point DFTs. If we continue dividing the N-point DFT calculation into smaller DFTs until we have only two-point DFTs, the total number of complex multiplications and additions is reduced to $N\log_2 N$. For example, a 1024-point DFT requires over a million complex additions and multiplications. A 1024-point DFT divided down into two-point DFTs needs fewer than ten thousand complex additions and multiplications, a reduction of over 99 percent.

Dividing the DFT into smaller DFTs is the basis of the FFT. A radix-2 FFT divides the DFT into two smaller DFTs, each of which is divided into two smaller DFTs, and so on, resulting in a combination of two-point DFTs. In a similar fashion, a radix-4 FFT divides the DFT into four smaller DFTs, each of which is divided into four smaller DFTs, and so on. Two types of radix-2 FFTs are described in this section: the decimation-in-time FFT and the decimation-in-frequency FFT. The radix-4 decimation-in-frequency FFT is described in a later section.

6.2.1 Radix-2 Decimation-In-Time FFT Algorithm

The decimation-in-time (DIT) FFT divides the input (time) sequence into two groups, one of even samples and the other of odd samples. N/2-point DFTs are performed on these sub-sequences, and their outputs are combined to form the N-point DFT.

Decimation-in-time is illustrated by the following equations (Oppenheim and Schafer, 1975). First, x(n), the input sequence in equation (1), is divided into even and odd sub-sequences:

(2)
$$X(k) = \sum_{n=0}^{N/2-1} x(2n)\, W_N^{2nk} + \sum_{n=0}^{N/2-1} x(2n+1)\, W_N^{(2n+1)k}$$

$$= \sum_{n=0}^{N/2-1} x(2n)\, W_N^{2nk} + W_N^{k} \sum_{n=0}^{N/2-1} x(2n+1)\, W_N^{2nk}$$

for $k = 0$ to $N-1$

By the substitutions

$W_N^{2nk} = (e^{-j2\pi/N})^{2nk} = (e^{-j2\pi/(N/2)})^{nk} = W_{N/2}^{nk}$
$x_1(n) = x(2n)$
$x_2(n) = x(2n+1)$

this equation becomes

(3)
$$X(k) = \sum_{n=0}^{N/2-1} x_1(n)\, W_{N/2}^{nk} + W_N^{k} \sum_{n=0}^{N/2-1} x_2(n)\, W_{N/2}^{nk}$$

$$= Y(k) + W_N^{k} Z(k)$$

for $k = 0$ to $N-1$

Equation (3) is the sum of two N/2-point DFTs (Y(k) and Z(k)) performed on the sub-sequences of even and odd samples, respectively, of the input sequence, x(n). Multiples of W_N (called "twiddle factors") appear as coefficients in the FFT calculation. In equation (3), Z(k) is multiplied by the twiddle factor W_N^{k}.

Because $W_N^{k+N/2} = (e^{-j2\pi/N})^k \times (e^{-j2\pi/N})^{N/2} = -W_N^{k}$, equation (3) can also be expressed as two equations:

(4)
$$X(k) = Y(k) + W_N^{k} Z(k)$$

$$X(k+N/2) = Y(k) - W_N^{k} Z(k)$$

for $k = 0$ to $N/2-1$

Together these equations form an N-point FFT. Figure 6.1 illustrates this first decimation of the DFT.

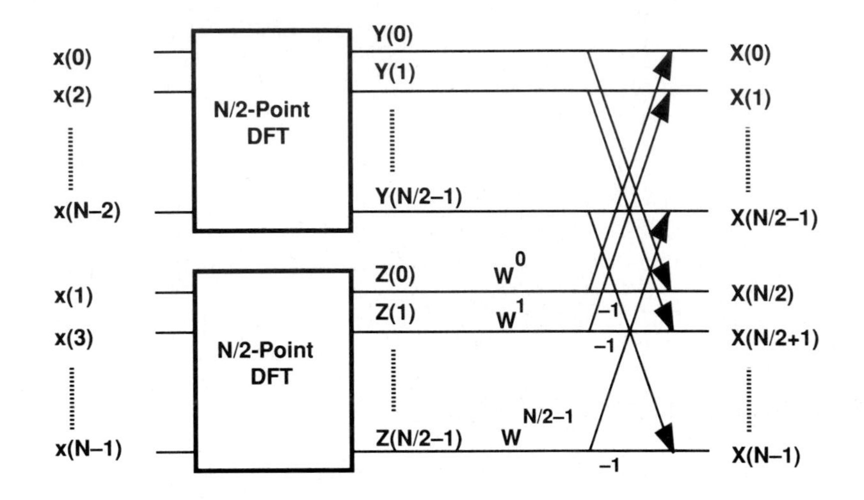

Figure 6.1 First Decimation of DIT FFT

The two N/2-point DFTs (Y(k) and Z(k)) can be divided to form four N/4-point DFTs, yielding equation pairs (5) and (6).

(5) $Y(k) = U(k) + W_N^{2k} V(k)$

$Y(k+N/4) = U(k) - W_N^{2k} V(k)$

for $k = 0$ to $N/4-1$

(6) $Z(k) = R(k) + W_N^{2k} S(k)$

$Z(k+N/4) = R(k) - W_N^{2k} S(k)$

for $k = 0$ to $N/4-1$

U(k) and V(k) are N/4-point DFTs whose input sequences are created by dividing $x_1(n)$ into even and odd sub-sequences. Similarly, R(k) and S(k) are N/4-point DFTs performed on the even and odd sub-sequences of $x_2(n)$. Each of these four equations can be divided to form two more. The final decimation occurs when each pair of equations together computes a two-point DFT (one point per equation). The pair of equations that make up the two-point DFT is called a radix-2 "butterfly." The butterfly is the core calculation of the FFT. The entire FFT is performed by combining butterflies in patterns determined by the FFT algorithm.

A complete eight-point DIT FFT is illustrated graphically in Figure 6.2. Each pair of arrows represents a butterfly. Notice that the entire FFT computation is made up of butterflies organized in different patterns, called groups and stages. The first stage consists of four groups of one butterfly each. The second stage has two groups of two butterflies, and the last has one group of four butterflies. Every stage contains N/2 (four) butterflies. Each butterfly has two input points, called the dual node and the primary node. The spacing between the nodes in the sequence is called the dual-node spacing. Associated with each butterfly is a twiddle factor whose exponent depends on the group and stage of the butterfly.

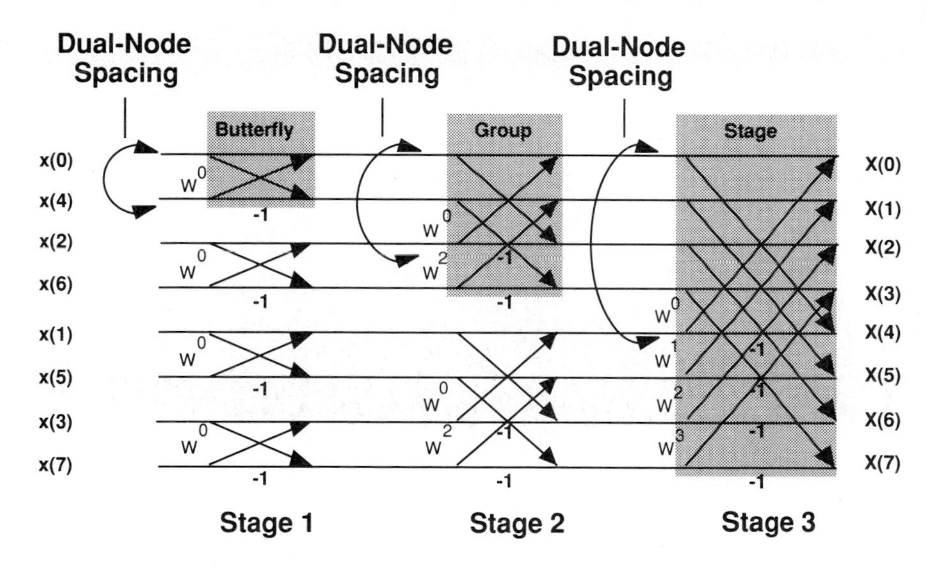

Figure 6.2 Eight-Point DIT FFT

Notice that whereas the output sequence is sequentially ordered, the input sequence is not. This is an effect of repeatedly dividing the input sequence into sub-sequences of even and odd samples. It is possible to perform an FFT using input and output sequences in other orders, but these approaches generally complicate addressing in the FFT program and can require a different butterfly. In this section, we have opted to scramble the input sequence of the DIT FFT because this approach uses twiddle factors in sequential order, produces the output sequence in sequential order, and requires a relatively simple butterfly. The scrambling of the inputs is achieved by a process called bit reversal, which is described later in this chapter.

The characteristics of an N-point DIT FFT with bit-reversed inputs are summarized below.

	Stage 1	*Stage 2*	*Stage 3*	*Stage Log_2N*
Number of Groups	N/2	N/4	N/8	1
Butterflies per Group	1	2	4	N/2
Dual-Node Spacing	1	2	4	N/2
Twiddle Factor Exponents	(N/2)k, k=0	(N/4)k, k=0, 1	(N/8)k, k=0, 1, 2, 3	k, k=0 to N/2–1

A generalized butterfly flow graph is shown in Figure 6.3. The variables *x* and *y* represent the real and imaginary parts, respectively, of a sample. The twiddle factor can be divided into real and imaginary parts because $W_N = e^{-j2\pi/N} = \cos(2\pi/N) - j\sin(2\pi/N)$. In the program presented later in this section, the twiddle factors are initialized in memory as cosine and –sine values (not +sine). For this reason, the twiddle factors are shown in Figure 6.3 as C + j(–S). C represents cosine and –S represents –sine.

The dual node (x_1+jy_1) is multiplied by the twiddle factor C+j(–S). The result of this multiplication is added to the primary node (x_0+jy_0) to produce $x_0'+jy_0'$ and subtracted from the primary node to produce $x_1'+jy_1'$. Equations (7) through (10) calculate the real and imaginary parts of the butterfly outputs.

(7) $$x_0' = x_0 + [(C)x_1 - (-S)y_1]$$

(8) $$y_0' = y_0 + [(C)y_1 + (-S)x_1]$$

(9) $x_1' = x_0 - [(C)x_1 - (-S)y_1]$

(10) $y_1' = y_0 - [(C)y_1 + (-S)x_1]$

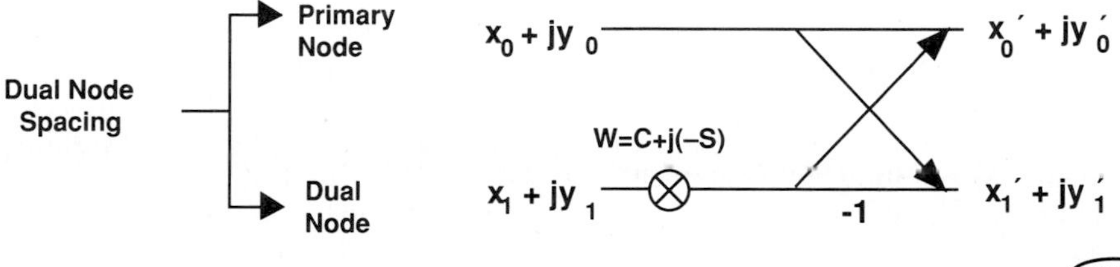

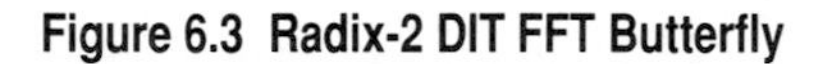

Figure 6.3 Radix-2 DIT FFT Butterfly

The butterfly produces two complex outputs that become butterfly inputs in the next stage of the FFT. Because each stage has the same number of butterflies (N/2), the number of butterfly inputs and outputs remains the same from one stage to the next. An "in-place" implementation writes each butterfly output over the corresponding butterfly input (x_0' overwrites x_0, etc.) for each butterfly in a stage. In an in-place implementation, the FFT results end up in the same memory range as the original inputs.

6.2.2 Radix-2 Decimation-In-Time FFT Program

The flow chart for the DIT FFT program is shown in Figure 6.4. The FFT program is divided into three subroutines. The first subroutine scrambles the input data. The next subroutine computes the FFT, and the third scales the output data.

Four modules are created. The main module declares and initializes data buffers and calls subroutines. The other three modules contain the FFT, bit reversal, and block floating-point scaling subroutines. The main module and FFT module are described in this section. The bit reversal and block floating-point scaling modules are described in later sections.

6.2.2.1 Main Module

The *dit_fft_main* module is shown in Listing 6.1. *N* is the number of points in the FFT (in this example, N=1024) and *N_div_2* is used for specifying the lengths of buffers. To change the number of points in the FFT, you

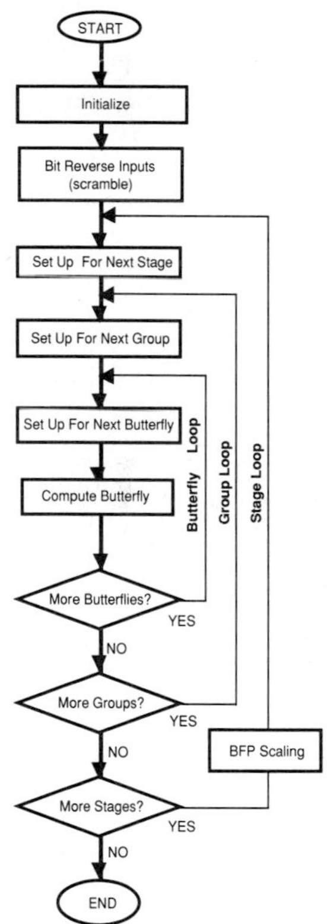

Figure 6.4 Radix-2 DIT FFT Flow Chart

change the value of these constants and the twiddle factors. The data buffers *twid_real* and *twid_imag* in program memory hold the twiddle factor cosine and sine values. The *inplacereal, inplaceimag, inputreal* and *inputimag* buffers in data memory store real and imaginary data values. Sequentially ordered input data is stored in *inputreal* and *inputimag*. This data is scrambled and written to *inplacereal* and *inplaceimag*, which are the data buffers used by the in-place FFT. A four-location buffer called *padding* is placed at the end of *inplaceimag* to allow data accesses to exceed the buffer length. If no *padding* was placed after *inplaceimag* and the program attempted to read undefined memory locations, the ADSP-2100 Simulator would signal an error. This buffer assists in debugging but is not necessary in a real system. Variables (one-location buffers) *groups, bflys_per_group, blk_exponent* and *node_space* are declared last.

The real part (cosine values) of the twiddle factors in the *twid_real.dat* file are placed in the buffer *twid_real*. Likewise, *twid_imag.dat* is placed in *twid_imag*. The variable *groups* is initialized to *N_div_2*, and *bflys_per_group* and *node_space* are initialized to one because there are N/2 groups of one butterfly in the first stage of the FFT. The *blk_exponent* is initialized to zero. This exponent value is updated when the output data is scaled.

Two subroutines are called. The first subroutine places the input sequence in bit-reversed order. The second performs the FFT and calls the block floating-point scaling routine.

6.2.2.2 DIT FFT Module

The FFT routine uses three nested loops. The inner loop computes butterflies, the middle loop controls the grouping of these butterflies, and the outer loop controls the FFT stage characteristics. These loops are described separately in the following sections. The complete routine is presented at the end.

Butterfly Loop

The butterfly calculation involves a complex multiplication, a complex addition, and a complex subtraction. These operations can potentially cause the butterfly data to grow by two bits from input to output. For example, if x_0 is H#07FF (five sign bits), x_0' could be H#100F (three sign bits). Because of this bit growth, precautions must be taken to ensure that 16-bit data never overflows.

One-Dimensional FFTs 6

```
.MODULE/ABS=4          dit_fft_main;
.CONST                 N=1024, N_div_2=512;      {Const. for 1024 points}
.VAR/PM/RAM/CIRC       twid_real [N_div_2];
.VAR/PM/RAM/CIRC       twid_imag [N_div_2];
.VAR/DM/RAM/ABS=0      inplacereal [N], inplaceimag [N];
.VAR/DM/RAM/ABS=H#1000 inputreal [N], inputimag [N], padding [4];
.VAR/DM/RAM            groups, bflys_per_group,
                       node_space, blk_exponent;

.INIT       twid_real: <twid_real.dat>;
.INIT       twid_imag: <twid_imag.dat>;
.INIT       inputreal: <inputreal.dat>;
.INIT       inputimag: <inputimag.dat>;
.INIT       inplaceimag: <inputimag.dat>;
.INIT       groups: N_div_2;
.INIT       bflys_per_group: 1;
.INIT       node_space: 1;
.INIT       blk_exponent: 0;
.INIT       padding: 0,0,0,0;                    {Zeros after inplaceimag}

.GLOBAL     inplacereal, inplaceimag;
.GLOBAL     inputreal, inputimag;
.GLOBAL     twid_real, twid_imag;
.GLOBAL     groups, bflys_per_group, node_space, blk_exponent;

.EXTERNAL   scramble, fft_strt;

            CALL scramble;                       {Subroutine calls}
            CALL fft_strt;
            TRAP;                                {Halt program}
.ENDMOD;
```

Listing 6.1 Main Module, Radix-2 DIT FFT

An example of bit growth and overflow is shown below.

Bit Growth:
Input to butterfly H#0F00 = 0000 1111 0000 0000
Output from butterfly H#1E00 = 0001 1110 0000 0000

Overflow:
Input to butterfly H#7000 = 0111 0000 0000 0000
Output from butterfly H#E000 = 1110 0000 0000 0000

In overflow, the positive number H#7000 is multiplied by a positive number, resulting in H#E000, which is too large to represent as a positive, signed 16-bit number. H#E000 is erroneously interpreted as a negative number.

To avoid errors caused by overflow, one of three methods of compensating for bit growth can be applied:

- Input data scaling
- Unconditional block floating-point scaling (output data)
- Conditional block floating-point scaling (output data)

Three different code segments for the butterfly calculation are presented in this section; each uses a different method of compensating for bit growth.

One way to ensure that overflow never occurs is to include enough extra sign bits, called guard bits, in the FFT input data to ensure that bit growth never results in overflow (Rabiner and Gold, 1975). Data can grow by a maximum factor of 2.4 from butterfly input to output (two bits of growth). However, a data value cannot grow by this maximum amount in two consecutive stages. The number of guard bits necessary to compensate for the maximum possible bit growth in an N-point FFT is $\log_2 N+1$. For example, each of the input samples of a 32-point FFT (requiring five stages) must contain six guard bits, so ten bits are available for data (one sign bit, nine magnitude bits). This method requires no data shifting and is therefore the fastest of the three methods discussed in this section. However, for large FFTs the resolution of the input data is greatly limited. For small, low-precision FFTs, this is the fastest and most efficient method.

The code segment for a butterfly with no shifting is shown in Listing 6.2. This section of code computes one butterfly equation while setting up values for the next butterfly. The butterfly outputs (x_0', y_0' x_1' and y_1') are written over the inputs to the butterfly (x_0, y_0, x_1 and y_1) in the boldface instructions. The input and output parameters of the butterfly loop are shown below.

Initial Conditions

MX0 = x_1
MY0 = C
MY1 = (–S)
I0 --> x_0
I1 --> x_1
I2 --> y_0
I3 --> y_1
I4 --> next C
I5 --> next (–S)
I6 --> y_1
CNTR = butterfly count
M0 = 0
M1 = 1
M4 = twiddle factor modify value
M5 = 1

Final Conditions

MX0 = next x_1
MY0 = next C
MY1 = next (–S)
I0 --> next x_0
I1 --> next x_1
I2 --> next y_0
I3 --> next y_1
I4 --> C after next C
I5 --> (–S) after next (–S)
I6 --> next y_1
CNTR = butterfly count –1

```
MR=MX0*MY0(SS),MX1=DM(I6,M5);        {MR=x1(C),MX1=y1}
MR=MR-MX1*MY1(RND),AY0=DM(I0,M0);    {MR=x1(C)-y1(-S),AY0=x0}
AR-MR1+AY0,AY1=DM(I2,M0);            {AR=x0+[x1(C)-y1(-S)],AY1=y0}
AR=AY0-MR1,DM(I0,M1)=AR;             {AR=x0-[x1(C)-y1(-S)],x0´=x0+[x1(C)-y1(-S)]}
MR=MX0*MY1(SS),DM(I1,M1)=AR;         {MR=x1(-S),x1´=x0-[x1(C)-y1(-S)]}
MR=MR+MX1*MY0(RND),MX0=DM(I1,M0),MY1=PM(I5,M4);
                                     {MR=x1(-S)+y1(C),MX0=next x1,MY1=next (-S)}
AR=AY1-MR1,MY0=PM(I4,M4);            {AR=y0-[x1(-S)+y1(C)],MY0=next C}
AR=MR1+AY1,DM(I3,M1)=AR;             {AR=y0+[x1(-S)+y1(C)],y1´=y0-[x1(-S)+y1(C)]}
DM(I2,M1)=AR;                        {y0´=y0+[x1(-S)+y1(C)]}
```

Listing 6.2 DIT FFT Butterfly, Input Data Scaled

Another way to compensate for bit growth is to scale the outputs down by a factor of two unconditionally after each stage. This approach is called unconditional block floating-point scaling. Initially, two guard bits are included in the input data to accommodate the maximum bit growth in the first stage. In each butterfly of a stage calculation, the data can grow into the guard bits. To prevent overflow in the next stage, the guard bits are replaced before the next stage is executed by shifting the entire block of data one bit to the right and updating the block exponent. This shift is necessary after every stage except the last, because no overflow can occur after the last stage.

The input data to an unconditional block floating-point FFT can have at most 14 bits (one sign bit and 13 magnitude bits). In the FFT calculation, the data loses a total of $(\log_2 N)-1$ bits because of shifting. Unconditional block floating-point scaling results in the same number of bits lost as in input data scaling. However, it produces more precise results because the FFT starts with more precise input data. The tradeoff is a slower FFT calculation because of the extra cycles needed to shift the output of each stage.

The code for the unconditional block floating-point butterfly is shown in Listing 6.3. Instructions that write butterfly results to memory are boldface. After the last stage of the FFT, no compensation for bit growth is needed, so a butterfly with no shifting can be used in the last stage.

Initial Conditions

SR0 = last y_0'
MX0 = x_1
MX1 = y_1
MY0 = C
MY1 = (–S)
I0 --> x_0
I1 --> x_1
I2 --> last y_0'
I3 --> y_1
I4 --> next C
I5 --> next (–S)
I6 --> next y_1
CNTR = butterfly count
M0 = 0
M1 = 1
M4 = twiddle factor modify value
M5 = 1
SE = –1

Final Conditions

SR0 = y_0'
MX0 = next x_1
MX1 = next y_1
MY0 = next C
MY1 = next (–S)
I0 --> next x_0
I1 --> next x_1
I2 --> y_0'
I3 --> next y_1
I4 --> C after next C
I5 --> (–S) after next (–S)
I6 --> y_1 after next y_1
CNTR = butterfly count –1

```
MR=MX0*MY0(SS),DM(I2,M1)=SR0;          {MR=x1(C),last y0=last y0´}
MR=MR-MX1*MY1(RND),AY0=DM(I0,M0); {MR=x1(C)-y1(-S),AY0=x0}
AR=MR1+AY0,AY1=DM(I2,M0);              {AR=x0+[x1(C)-y1(-S)],AY1=y0}
SR=ASHIFT AR(LO);                      {Shift result right 1 bit}
DM(I0,M1)=SR0,AR=AY0-MR1;              {x0´=x0-[x1(C)-y1(-S)],AR=x0-[x1(C)-y1(-S)]}
SR=ASHIFT AR(LO);                      {Shift result right 1 bit}
DM(I1,M1)=SR0,MR=MX0*MY1(SS);          {x1´=x0-[x1(C)-y1(-S)],MR=x1(-S)]}
MR=MR+MX1*MY0(RND),MX0=DM(I1,M0),MY1=PM(I5,M4);
                                       {MR=x1(-S)-y1(C),MX0=next x1,MY1=next(-S)}
AR=AY1-MR1,MY0=PM(I4,M4);              {AR=y0-[x1(-S)-y1(C)],MY0=next C}
SR=ASHIFT AR(LO),MX1=DM(I6,M5);        {Shift result right 1 bit,MX1=next y1}
DM(I3,M1)=SR0,AR=MR1+AY1;              {y1´=y0-[x1(-S)-y1(C),AR=y0+[x1(-S)-y1(C)]}
SR=ASHIFT AR(LO);                      {Shift result right 1 bit}
```

Listing 6.3 DIT FFT Butterfly, Unconditional Block Floating-Point Scaling

In conditional block floating-point scaling, data is shifted *only* if bit growth occurs. If one or more outputs grows, the entire block of data is shifted to the right and the block exponent is updated. For example, if the original block exponent is 0 and data is shifted three positions, the resulting block exponent is +3.

The code segment for the conditional block floating-point butterfly is shown in Listing 6.4. As in the other types of butterflies, one butterfly equation is calculated and its outputs (x_0', y_0', x_1' and y_1') are written over its inputs (x_0, y_0, x_1 and y_1) in the boldface instructions.

The conditional block floating-point butterfly checks each butterfly output for growth with the EXPADJ instruction. This instruction does no shifting; instead, it monitors the output data and updates the SB register if bit growth is detected. (See the *ADSP-2100 User's Manual* for a complete description of this instruction.) If shifting is necessary it is performed after the entire stage is complete (in the block floating-point scaling routine). The butterfly code computes one butterfly equation while setting up values for the next butterfly. The input and output parameters of the butterfly loop are as follows:

Initial Conditions

MX0 = x_1
MX1 = y_1
MY0 = C
MY1 = (–S)
I0 --> x_0
I1 -> x_1
I2 --> y_0
I3 --> y_1
I4 --> next C
I5 --> next (–S)
CNTR = butterfly count
M1 = 1
M4 = twiddle factor modify value
M0 = 0
SB = monitored block exponent for this stage

Final Conditions

MX0 = next x_1
MX1 = next y_1
MY0 = next C
MY1 = next (–S)
I0 --> next x_0
I1 --> next x_1
I2 --> next y_0
I3 --> next y_1
I4 --> C after next C
I5 --> (–S) after next (–S)
CNTR = butterfly count –1

```
MR=MX0*MY1(SS),AX0=DM(I0,M0);        {MR=x1(-S),AX0=x0}
MR=MR+MX1*MY0(RND),AX1=DM(I2,M0); {MR=[y1(C)+x1(-S)];AX1=y0}
AY1=MR1,MR=MX0*MY0(SS);              {AY1=[y1(C)+x1(-S)];MR=x1(C)}
MR=MR-MX1*MY1(RND);                  {MR=[x1(C)-y1(-S)]}
AY0=MR1,AR=AX1-AY1;                  {AY0=[x1(C)-y1(-S)],AR=y0-[y1(C)+x1(-S)]}
SB=EXPADJ AR,DM(I3,M1)=AR;           {check for bit growth,y1´=y0-[y1(C)+x1(-S)]}
AR=AX0-AY0,MX1=DM(I3,M0),MY1=PM(I5,M4);
                                     {AR=x0-[x1(C)-y1(-S)],MX1=next y1,MY1=next S}
SB=EXPADJ AR,DM(I1,M1)=AR;           {check for bit growth,x1´=x0-[x1(C)-y1(-S)]}
AR=AX0+AY0,MX0=DM(I1,M0),MY0=PM(I4,M4);
                                     {AR=x0+[x1(C)-y1(-S)],MX0=next x1,MY0=next C}
SB=EXPADJ AR,DM(I0,M1)=AR;           {check for bit growth,x0´=x0+[x1(C)-y1(-S)]}
AR=AX1+AY1;                          {AR=y0+[y1(C)+x1(-S)]}
SB=EXPADJ AR,DM(I2,M1)=AR;           {check for bit growth,y0´=y0+[y1(C)+x1(-S)]}
```

Listing 6.4 DIT FFT Butterfly, Conditional Block Floating-Point Scaling

Group Loop

The group loop controls the grouping of butterflies. It sets pointers to the input data and twiddle factors of the first butterfly in the group, initializes the butterfly counter and sets up the butterfly loop for each group.

The code segment for the group loop is shown in Listing 6.5. This code is designed for the conditional block floating-point butterfly and thus requires slight modification for use with the other types (input scaling, unconditional block floating-point) of butterflies. The first butterfly of every group in the first stage of the DIT FFT has a twiddle factor of W^0. Thus, I4 and I5 are initialized to point to the cosine and sine values of W^0 before the butterfly loop is entered. In the group loop, the butterfly counter is initialized and initial butterfly data is fetched. The butterfly loop is executed *bflys_per_group* times to compute all butterflies in the group. After the butterfly loop is complete, pointers I0, I1, I2 and I3 are updated with the MODIFY instruction to point to x_0, x_1, y_0 and y_1 of the first butterfly in the next group. The group loop is executed *groups* times.

The input and output parameters of the group loop are as follows:

Initial Conditions	*Final Conditions*
I0 --> x_0 of first butterfly in group	I0 --> x_0 of first butterfly in next group
I1 --> x_1 of first butterfly in group	I1 --> x_1 of first butterfly in next group
I2 --> y_0 of first butterfly in group	I2 --> y_0 of first butterfly in next group
I3 --> y_1 of first butterfly in group	I3 --> y_1 of first butterfly in next group
CNTR = group count	CNTR = group count –1
M2 = node_space	

```
            I4=^twid_real;
            I5=^twid_imag;                  {Initialize twiddle factor pointers}
            CNTR=DM(bflys_per_group);       {Initialize butterfly counter}
            MY0=PM(I4,M4),MX0=DM(I1,M0);    {MY0=C,MX0=x1}
            MY1=PM(I5,M4),MX1=DM(I3,M0);    {MY1=(-S),MX1=y1}
            DO bfly_loop UNTIL CE;

bfly_loop:      {Calculate All Butterflies in Group}

            MODIFY(I0,M2);                  {I0 -->first x0 in next group}
            MODIFY(I1,M2);                  {I1 -->first x1 in next group}
            MODIFY(I2,M2);                  {I2 -->first y0 in next group}
group_loop: MODIFY(I3,M2);                  {I3 -->first y1 in next group}
```

Listing 6.5 Radix-2 DIT FFT Group Loop

Stage Loop

The stage loop controls the grouping characteristics of the FFT. These include the number of groups in a stage, the number of butterflies in each group, and the node spacing. The stage loop also calls a subroutine which performs conditional block floating-point scaling on the outputs of a stage calculation. Note that if unconditional block floating-point scaling or input data scaling were used, this call would be omitted.

The stage loop code for a conditional block floating-point FFT is shown in Listing 6.6. The stage loop sets up the group loop by initializing I0, I1, I2 and I3 to point to x_0, x_1, y_0 and y_1, respectively, for the first butterfly in the first group. It also initializes the group loop counter and node space modifier so that pointers can be updated for new groups. The value of the twiddle factor exponent is increased by *groups* for each butterfly. M4, initialized to *groups*, is the modifier for the twiddle factor pointers.

The group loop calculates all groups in the stage. After the group loop is complete, a block floating-point subroutine is called to check the stage outputs for bit growth and scale the data if necessary. The stage characteristics are then updated for the next stage; *bflys_per_group* and *node_space* are doubled and *groups* is divided by two.

The input and output parameters for the stage loop are as follows. Note that all the parameters except the stage count are passed in memory.

Initial Conditions	*Final Conditions*
groups=# groups current stage	*groups*=# groups next stage
bflys_per_group=# butterflies/group	*bflys_per_group*=# butterflies/ group next stage
node_space=node spacing current stage	*node_space*=node spacing next stage
CNTR=stage count	CNTR=stage count –1
inplacereal=real stage input data	*inplacereal*=real stage output data
inplaceimag=imag. stage input data	*inplaceimag*=imag. stage output data

```
           I0=^inplacereal;          {I0 -->first x0 in first group of stage}
           I2=^inplaceimag;          {I2 -->first y0 in first group of stage}
           SB=-2                     {SB = -(number of guard bits)}
           SI=DM(groups);            {SI = groups}
           CNTR=SI;                  {Initialize group counter}
           M4=SI;                    {Initialize twiddle factor modifier}
           M2=DM(node_space);        {Initialize node spacing modifier}
           I1=I0;
           MODIFY(I1,M2);            {I1 -->first x1 of first group in stage}
           I3=I2;
           MODIFY(I3,M2);            {I3 -->first y1 of first group in stage}
           DO group_loop UNTIL CE;
group_loop:    {Compute All Groups in Stage}

           CALL bfp_adj;             {Adjust stage output for bit growth}
           SI=DM(bflys_per_group);
           SR=ASHIFT SI BY 1(LO);
           DM(node_space)=SR0;       {node_space=node_space × 2}
           DM(bflys_per_group)=SR0;  {bflys_per_group=bflys_per_group × 2}
           SI=DM(groups);
           SR=ASHIFT SI BY -1(LO);
           DM(groups)=SR0;           (groups = groups ÷ 2}
```

Listing 6.6 Radix-2 DIT FFT Stage Loop

DIT FFT Subroutine

The complete conditional block floating-point radix-2 DIT FFT routine is shown in Listing 6.7. The constants N and log_2N are the number of points and the number of stages in the FFT, respectively. To change the number of points in the FFT, you modify these constants. Notice that the length and modify registers (that retain the same values throughout the FFT calculation) and the stage counter are initialized before the stage loop is executed. Instructions that write butterfly results to memory are boldface.

```
.MODULE      fft;

{            Performs Radix-2 DIT FFT

             Calling Parameters
                inplacereal = Real input data in scrambled order
                inplaceimag = All zeroes (real input assumed)
                twid_real = Twiddle factor cosine values
                twid_imag = Twiddle factor sine values
                groups = N/2
                bflys_per_group = 1
                node_space = 1

             Return Values
                inplacereal = Real FFT results in sequential order
                inplaceimag = Imaginary FFT results in sequential order

             Altered Registers
                I0,I1,I2,I3,I4,I5,L0,L1,L2,L3,L4,L5
                M0,M1,M2,M3,M4,M5
                AX0,AX1,AY0,AY1,AR,AF
                MX0,MX1,MY0,MY1,MR,SB,SE,SR,SI

             Altered Memory
                inplacereal, inplaceimag, groups, node_space,
                bflys_per_group, blk_exponent
}
.CONST       log2N=10, N=1024;                  {Set constants for N-point FFT}
.EXTERNAL    twid_real, twid_imag;
.EXTERNAL    inplacereal, inplaceimag;
.EXTERNAL    groups, bflys_per_group, node_space;
.EXTERNAL    bfp_adj;
.ENTRY       fft_strt;

fft_strt:    CNTR=log2N;                        {Initialize stage counter}
             M0=0;
             M1=1;
             L1=0;
             L2=0;
             L3=0;
             L4=%twid_real;
             L5=%twid_imag;
             DO stage_loop UNTIL CE;            {Compute all stages in FFT}
                I0=^inplacereal;                {I0 -->x0 in 1st grp of stage}
                I2=^inplaceimag;                {I2 -->y0 in 1st grp of stage}
                SB=-2                           {SB to detect data > 14 bits}
                SI=DM(groups);
                CNTR=SI;                        {CNTR = group counter}
                M4=SI;                          {M4=twiddle factor modifier}
                M2=DM(node_space);              {M2=node space modifier}
                I1=I0;
```

```
            MODIFY(I1,M2);                    {I1 -->x1 of 1st grp in stage}
            I3=I2;
            MODIFY(I3,M2);                    {I3 -->y1 of 1st grp in stage}
            DO group_loop UNTIL CE;
               I4=^twid_real;                 {I4 --> C of W⁰}
               I5=^twid_imag;                 {I5 --> (-S) of W⁰}
               CNTR=DM(bflys_per_group);      {CNTR = butterfly counter}
               MY0=PM(I4,M4),MX0=DM(I1,M0);      {MY0=C,MX0=x1 }
               MY1=PM(I5,M4),MX1=DM(I3,M0);      {MY1=-S,MX1=y1}
               DO bfly_loop UNTIL CE;
                  MR=MX0*MY1(SS),AX0=DM(I0,M0); {MR=x1(-S),AX0=x0}
                  MR=MR+MX1*MY0(RND),AX1=DM(I2,M0);
                                                {MR=(y1(C)+x1(-S)),AX1=y0}
                  AY1=MR1,MR=MX0*MY0(SS);       {AY1=y1(C)+x1(-S),MR=x1(C)}
                  MR=MR-MX1*MY1(RND);           {MR=x1(C)-y1(-S)}
                  AY0=MR1,AR=AX1-AY1;           {AY0=x1(C)-y1(-S),}
                                                {AR=y0-[y1(C)+x1(-S)]}
                  SB=EXPADJ AR,DM(I3,M1)=AR;    {Check for bit growth,}
                                                {y1´=y0-[y1(C)+x1(-S)]}
                  AR=AX0-AY0,MX1=DM(I3,M0),MY1=PM(I5,M4);
                                                {AR=x0-[x1(C)-y1(-S)],}
                                                {MX1=next y1,MY1=next (-S)}
                  SB=EXPADJ AR,DM(I1,M1)=AR;    {Check for bit growth,}
                                                {x1´=x0-[x1(C)-y1(-S)]}
                  AR=AX0+AY0,MX0=DM(I1,M0),MY0=PM(I4,M4);
                                                {AR=x0+[x1(C)-y1(-S)],}
                                                {MX0=next x1,MY0=next C}
                  SB=EXPADJ AR,DM(I0,M1)=AR;    {Check for bit growth,}
                                                {x0´=x0+[x1(C)-y1(-S)]}
                  AR=AX1+AY1;                   {AR=y0+[y1(C)+x1(-S)]}
bfly_loop:        SB=EXPADJ AR,DM(I2,M1)=AR;    {Check for bit growth,}
                                                {y0´=y0+[y1(C)+x1(-S)]}
               MODIFY(I0,M2);                 {I0 -->1st x0 in next group}
               MODIFY(I1,M2);                 {I1 -->1st x1 in next group}
               MODIFY(I2,M2);                 {I2 -->1st y0 in next group}
group_loop:    MODIFY(I3,M2);                 {I3 -->1st y1 in next group}
            CALL bfp_adj;                     {Compensate for bit growth}
            SI=DM(bflys_per_group);
            SR=ASHIFT SI BY 1(LO);
            DM(node_space)=SR0;               {node_space=node_space × 2}
            DM(bflys_per_group)=SR0;          {bflys_per_group= }
                                                 {bflys_per_group × 2}
            SI=DM(groups);
            SR=ASHIFT SI BY -1(LO);
stage_loop: DM(groups)=SR0;                   {groups=groups ÷ 2}
         RTS;
.ENDMOD;
```

Listing 6.7 Radix-2 DIT FFT Routine, Conditional Block Floating-Point

6.2.3 Radix-2 Decimation-In-Frequency FFT Algorithm

In the DIT FFT, each decimation consists of two steps. First, a DFT equation is expressed as the sum of two DFTs, one of even samples and one of odd samples. This equation is then divided into two equations, one that computes the first half of the output (frequency) samples and one that computes the second half. In the decimation-in-frequency (DIF) FFT, a DFT equation is expressed as the sum of two calculations, one on the first half of the samples and one on the second half of the samples. This equation is then expressed as two equations, one that computes even output samples and one that computes odd output samples. Decimation in time refers to grouping the input sequence into even and odd samples, whereas decimation in frequency refers to grouping the output (frequency) sequence into even and odd samples. Decimation-in-frequency can thus be visualized as repeatedly dividing the output sequence into even and odd samples in the same way that decimation in time divides down the input sequence (Oppenheim, 1975). The following equations illustrate radix-2 decimation in frequency.

The DIF FFT divides an N-point DFT into two summations, shown in (11).

$$(11) \qquad X(k) = \sum_{n=0}^{N-1} x(n)\, W_N^{nk}$$

$$= \sum_{n=0}^{N/2-1} x(n)\, W_N^{nk} + \sum_{n=N/2}^{N-1} x(n) W_N^{nk}$$

$$= \sum_{n=0}^{N/2-1} x(n)\, W_N^{nk} + \sum_{n=0}^{N/2-1} x(n+N/2) W_N^{(n+N/2)k}$$

Because $W_N^{(n+N/2)k} = W_N^{nk} \times W_N^{(N/2)k}$ and $W_N^{(N/2)k} = (-1)^k$, equation (11) can also be expressed as

$$(12) \qquad X(k) = \sum_{n=0}^{N/2-1} x(n)\, W_N^{nk} + (-1)^k \sum_{n=0}^{N/2-1} x(n+N/2)\, W_N^{nk}$$

$$= \sum_{n=0}^{N/2-1} [x(n) + (-1)^k x(n+N/2)] W_N^{nk}$$

for k = 0 to N–1

The decimation of the output (frequency) sequence is accomplished by dividing X(k) into two equations, one that computes even output samples and one that computes odd output samples. For even values of X(k), k=2r.

$$(13) \qquad X(2r) = \sum_{n=0}^{N/2-1} [x(n) + (-1)^{2r} x(n+N/2)] W_N^{2nr}$$

$$= \sum_{n=0}^{N/2-1} [x(n) + x(n+N/2)] W_{N/2}^{nr}$$

r = 0 to N/2–1

For odd values of X(k), k=2r+1.

$$(14) \qquad X(2r+1) = \sum_{n=0}^{N/2-1} [x(n) + (-1)^{2r+1} x(n+N/2)] W_N^{(2r+1)n}$$

$$= \sum_{n=0}^{N/2-1} [[x(n) - x(n+N/2)] W_N^{n}] W_{N/2}^{nr}$$

r = 0 to N/2–1

Note that X(2r) and X(2r+1) are the results of N/2-point DFTs performed on the sum and difference of the first and second halves of the input sequence. In equation (14), the difference of the two halves of the input sequence is multiplied by a twiddle factor, W_N^n. Figure 6.5, on the next page, illustrates the first decimation of the DIF FFT, which eliminates half ($N^2/2$) of the DFT calculations.

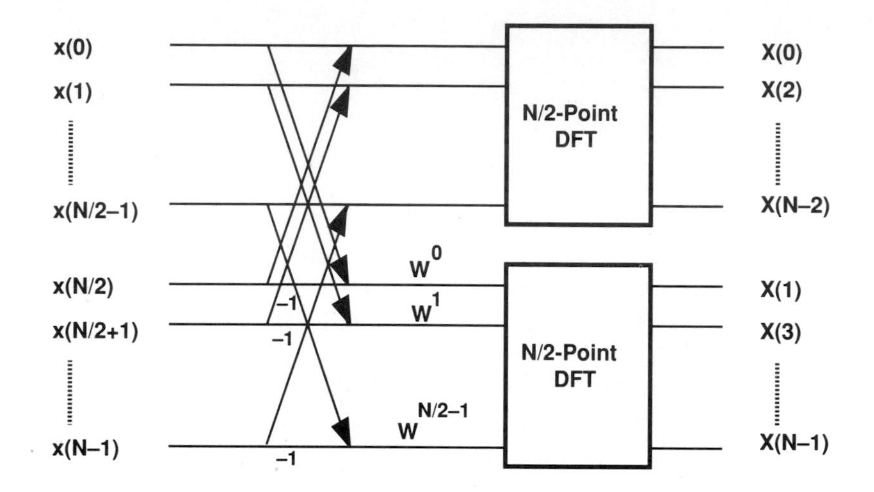

Figure 6.5 First Decimation of DIF FFT

Each of the two N/2-point DFTs (X(2r) and X(2r+1)) are divided into two N/4-point DFTs in the same way as the N-point DFT is divided into two N/2-point DFTs. By the substitutions

$$X_1(r) = X(2r) \qquad r = 0 \text{ to } N/2-1$$

$$x_1(n) = x(n) + x(n+N/2) \qquad n = 0 \text{ to } N/2-1$$

the sequence of even samples in equation (13) becomes

$$(15) \qquad X_1(r) = \sum_{n=0}^{N/2-1} x_1(n)\, W_{N/2}^{rk}$$

This N/2-point sequence has the same form as the original N-point sequence in equation (11) and can be divided in half in the same manner to yield

$$(16) \qquad X_1(r) = \sum_{n=0}^{N/4-1} [\, x_1(n) + (-1)^r x_1(n+N/4)]\, W_{N/2}^{nr}$$

For even output samples, let r=2s.

$$X_1(2s) = \sum_{n=0}^{N/4-1} [x_1(n) + x_1(n+N/4)]\, W_{N/4}^{sn} \quad (17)$$

For odd output samples, let r=2s+1.

$$X_1(2s+1) = \sum_{n=0}^{N/4-1} [(x_1(n) - x_1(n+N/4))\, W_N^{2n}]\, W_{N/4}^{sn} \quad (18)$$

X(2r+1) is also divided into two equations, one that computes even output samples and one that computes odd output samples, in the same way that X(2r) is divided into $X_1(2s)$ and $X_1(2s+1)$. Thus we have four N/4-point sequences.

If we make the substitutions

$$X_2(s) = X_1(2s)$$

$$x_2(n) = x_1(n) + x_1(n+N/4)$$

equation (17) becomes

$$X_2(s) = \sum_{n=0}^{N/4-1} x_2(n)\, W_{N/4}^{sn} \quad (19)$$

The four N/4-point sequences that result from the decimation of X(2r) and X(2r+1) are divided to form eight N/8-point sequences in the third decimation. This process is repeated until the division of a sequence results in a pair of equations that together compute a two-point DFT. In this pair, the summation variable n (see equations 17 and 18) is equal to zero only, so no summation is performed. The two-point DFT computed by this pair of equations is the core calculation (butterfly) for the radix-2 DIF FFT.

Figure 6.6, on the next page, shows the complete decimation for an eight-point DIF FFT. Notice that the inputs of the DIF FFT are in sequential order and the outputs are in scrambled order. The DIF FFT can also be performed with inputs in bit-reversed order, resulting in outputs in sequential order. In this case, however, the twiddle factors must be in bit-reversed order. In this chapter, both the DIT FFT and the DIF FFT are presented with twiddle factors in sequential order to simplify programming.

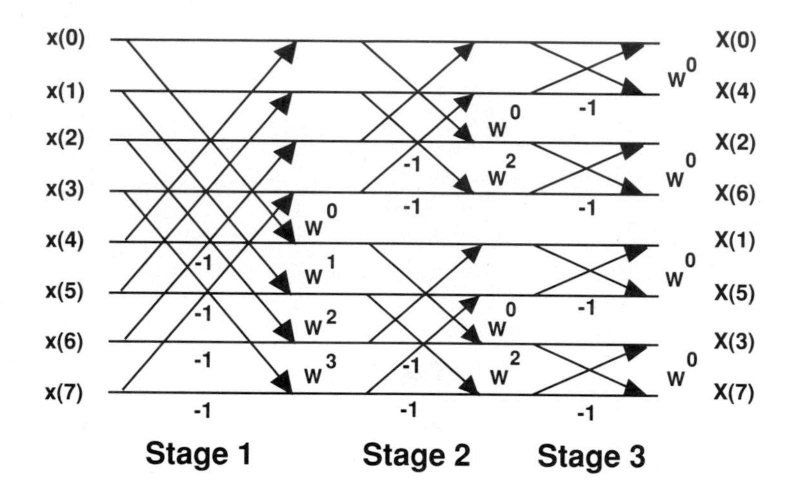

Figure 6.6 Eight-Point DIF FFT

As in the DIT FFT, the DIF FFT butterflies are organized into groups and stages. In the eight-point FFT, the first stage has one group of four (N/2) butterflies. The second stage has two groups of two (N/4) butterflies, and the last has four groups of one butterfly. In general, an N-point DIF FFT has the characteristics summarized below.

	Stage 1	*Stage 2*	*Stage 3*	*Stage Log_2N*
Number of Groups	1	2	4	N/2
Butterflies per Group	N/2	N/4	N/8	1
Dual-Node Spacing	N/2	N/4	N/8	1
Twiddle Factor Exponents	n, n=0 to N/2–1	2n, n=0 to N/4–1	4n, n=0 to N/8–1	(N/2)n, n=0

The DIF FFT butterfly is similar to that of the DIT FFT except that the twiddle factor multiplication occurs after rather than before the primary-node and dual-node subtraction. The DIF butterfly is illustrated graphically in Figure 6.7. The variables x and y represent the real and imaginary parts, respectively, of a sample. The twiddle factor can be divided into real and imaginary parts because $W_N = e^{-j2\pi/N} = \cos(2\pi/N) - j\sin(2\pi/N)$. In the program presented later in this section, the twiddle factors are initialized in memory as cosine and –sine values (not +sine). For this reason, the twiddle factors are shown in Figure 6.7 as C + j(–S). C represents cosine and –S represents –sine.

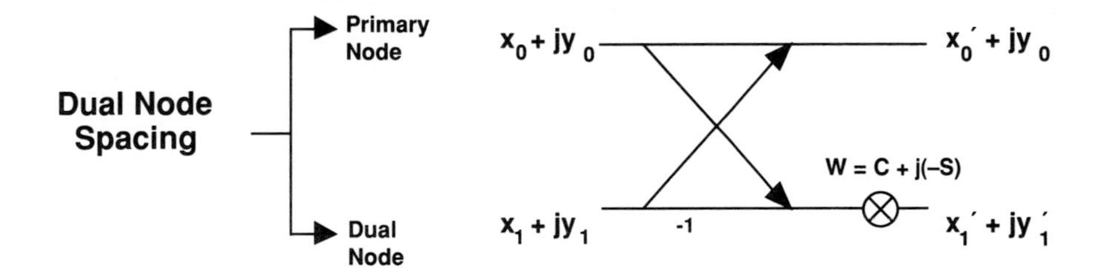

Figure 6.7 Radix-2 DIF FFT Butterfly

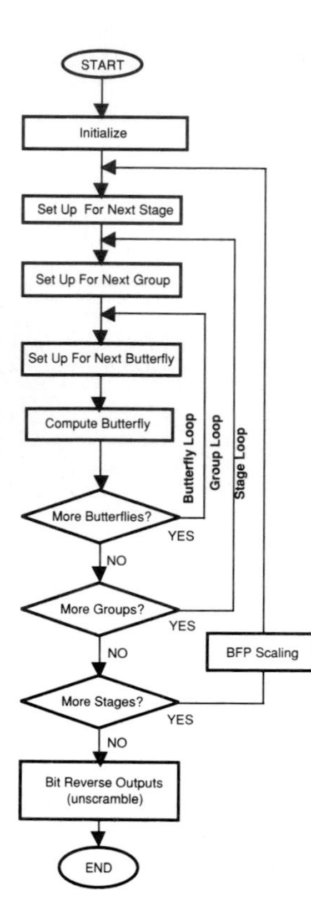

Figure 6.8 Radix-2 DIF FFT Flow Chart

Equations (20) through (23) describe the DIF FFT butterfly outputs.

(20) $x_0' = x_0 + x_1$

(21) $y_0' = y_0 + y_1$

(22) $x_1' = C(x_0 - x_1) - (-S)(y_0 - y_1)$

(23) $y_1' = (-S)(x_0 - x_1) + C(y_0 - y_1)$

As in the DIT FFT, the butterfly is performed in-place; that is, the results of each butterfly are written over the corresponding inputs. For example, x_0' is written over x_0.

6.2.4 Radix-2 Decimation-In-Frequency FFT Program

The DIF flow chart is shown in Figure 6.8. Like the DIT FFT, the DIF FFT uses three subroutines. The first subroutine computes the FFT. The second subroutine performs conditional block floating-point scaling at the end of each stage (except the last). The third subroutine bit-reverses the locations of the FFT output data to "unscramble" the data. The DIF FFT subroutine is described in this section.The block floating-point and bit reversal routines are described in later sections.

6.2.4.1 Main Module

The module *dif_fft_main* is shown in Listing 6.8. The FFT calculation is performed in one buffer (*inplacedata*). In this program, the real and imaginary input data are interleaved in the buffer. The length of *inplacedata* is thus twice the number of points in the FFT and is specified by the constant *N_x_2* (N_x_2 = 2048 for a 1024-point FFT). Unlike the DIT FFT, the DIF FFT is performed on sequentially ordered input data and produces data in bit-reversed order; therefore, no additional buffers for scrambling the input data are needed.

When the output data is unscrambled, it is separated into real and imaginary values and placed in two buffers (*real_results, imaginary_results*). Twiddle-factor buffers are defined and initialized as in the DIT FFT.

The DIF FFT uses the variables *groups*, *bflys_per_group* and *blk_exponent*. Because the first stage of the DIF FFT contains one group of N/2 butterflies, *groups* is initialized to one and *bflys_per_group* is initialized to *N_div_2*. The node spacing (*node_space*) is N instead of N/2 because the real and imaginary input data are interleaved.

Two subroutines are called. The first performs the DIF FFT and calls the block floating-point scaling routine. The second bit-reverses the FFT outputs to unscramble them.

```
.MODULE/ABS=4             dif_fft_main;

.CONST                    N=1024,N_x_2=2048;        {Const. for 1024 points}
.CONST                    N_div_2=512,log2N=10;
.VAR/DM/RAM               inplacedata[N_x_2];
.VAR/DM/RAM               real_results[N];
.VAR/DM/RAM               imaginary_results[N];
.VAR/PM/ROM/CIRC          twid_imag[N_div_2];
.VAR/PM/ROM/CIRC          twid_real[N_div_2];

.VAR/DM/RAM               groups,node_space,bflys_per_group,blk_exponent;

.INIT                     inplacedata: <inplacedata.dat>;
.INIT                     twid_imag: <twid_imag.dat>;
.INIT                     twid_real: <twid_real.dat>;
.INIT                     groups: 1;
.INIT                     node_space: N;
.INIT                     bflys_per_group: N_div_2;
.INIT                     blk_exponent: 0;

.GLOBAL                   inplacedata, real_results, imaginary_results;
.GLOBAL                   twid_real, twid_imag;
.GLOBAL                   groups, bflys_per_group, node_space, blk_exponent;

.EXTERNAL                 unscramble, fft_start;

                          CALL fft_start;
                          CALL unscramble;
                          TRAP;
.ENDMOD;
```

Listing 6.8 Main Module, Radix-2 DIF FFT

6.2.4.2 DIF FFT Module

The conditional block floating-point DIF FFT program is described in this section. The butterfly loop is described first, then the group and stage loops. The complete FFT program is presented at the end of this section.

Butterfly Loop

The code segment for the DIF butterfly (with conditional block floating-point scaling) is shown in Listing 6.9, on the next page. The primary-node outputs x_0' and y_0' are calculated first and written over x_0 and y_0. Complex subtraction for the dual-node calculation is then performed, followed by the twiddle factor multiplication. The outputs x_1' and y_1' are written over x_1 and y_1. Instructions that write butterfly results to memory are boldface. Each butterfly output is checked for bit growth using the EXPADJ instruction. This loop is repeated *bflys_per_group* times.

The input and output parameters for the butterfly loop are as follows:

Initial Conditions	*Final Conditions*
AX0 = x_0	AX0 = next x_0
AY0 = x_1	AY0 = next x_1
AY1 = y_1	AY1 = next y_1
I0 --> y_0	I0 --> next y_0
I1 --> next x_1	I1 --> x_1 after next
I2 --> x_1	I2 --> next x_1
I4 --> C	CNTR = butterfly count –1
I5 --> (-S)	
M0 = –1	
M1 = 1	
M5 = twiddle factor modifier	
CNTR = butterfly count	

```
AR=AX0+AY0,AX1=DM(I0,M0),MY0=PM(I4,M5);  {AR=x0+x1,AX1=y0,MY0=C,I0 -->x0}
SB=EXPADJ AR;                            {Check for bit growth}
DM(I0,M1)=AR,AR=AX1+AY1;                 {x0´=x0+x1,AR=y0+y1,I0 -->y0}
SB=EXPADJ AR;                            {Check for bit growth}
DM(I0,M1)=AR,AR=AX0-AY0;                 {y0´=y0+y1,AR=x0-x1,I0 -->next x0}
MX0=AR,AR=AX1-AY1;                       {MX0=x0-x1,AR=y0-y1}
MR=MX0*MY0(SS),AX0=DM(I0,M1),MY1=PM(I5,M5);
                                         {MR=(x0-x1)C,AX0=next x0,MY1=(-S)}
MR=MR-AR*MY1(RND),AY0=DM(I1,M1);         {MR=(x0-x1)C-(y0-y1)(-S),AY0=next x1}
SB=EXPADJ MR1;                           {Check for bit growth}
DM(I2,M1)=MR1,MR=AR*MY0 (SS);            {x1´=(x0-x1)C-(y0-y1)(-S),MR=(y0-y1)C}
MR=MR+MX0*MY1(RND),AY1=DM(I1,M1);        {MR=(y0-y1)C+(x0-x1)(-S),AY1=next y1}
DM(I2,M1)=MR1,SB=EXPADJ MR1;             {y1´=(y0-y1)C+(x0-x1)(-S),check}
                                         {for bit growth}
```

Listing 6.9 Radix-2 DIF FFT Butterfly, Conditional Block Floating-Point

Group Loop

The group loop code is shown in Listing 6.10. The group loop sets up the butterfly loop by fetching initial data and initializing the butterfly loop counter. When all the butterflies in a group have been calculated, data pointers are updated to point to the inputs for the first butterfly of the next group. This loop is repeated until all groups in a stage are complete.

The input and output parameters of the group loop are as follows:

Initial Conditions	*Final Conditions*
I0 --> x_0 of first butterfly in group	I0 --> x_0 of first butterfly in next group
I1 --> x_1 of first butterfly in group	I1 --> x_1 of first butterfly in next group
I2 --> x_1 of first butterfly in group	I2 --> x_1 of first butterfly in next group
CNTR = group count	CNTR = group count –1
M1 = 1	
M2 = *node_space*	
M3 = *node_space*–2	

```
            CNTR=DM(bflys_per_group); {Initialize butterfly counter}
            AX0=DM(I0,M1);            {AX0=x0}
            AY0=DM(I1,M1);            {AY0=x1}
            AY1=DM(I1,M1);            {AY1=y1}
            DO bfly_loop UNTIL CE;
bfly_loop:      {Calculate All Butterflies in Group}

            MODIFY(I2,M2);            {I2 ->x1 of 1st butterfly in next group}
            MODIFY(I1,M3);            {I1 ->x1 of 1st butterfly in next group}
            MODIFY(I0,M3);
group_loop: MODIFY(I0,M1);            {I0 ->x0 of 1st butterfly in next group}
```

Listing 6.10 Radix-2 DIF FFT Group Loop

Stage Loop

The stage loop code is shown in Listing 6.11, on the next page. This code segment sets up and computes all groups in a stage and controls stage characteristics, such as the number of groups in a stage. Pointers I0 and I1 are set to point to x_0 and x_1 of the first butterfly in the first group of the stage. Pointer I2 also points to x_1 and is used to write the dual-node butterfly results to data memory. M3 is set to *node_space*–2 and is used to modify pointers for the next group. The group counter is initialized to *groups*, the number of groups in the stage. The twiddle factor modifier stored in M5 is also *groups*. This value is the exponent increment value for the twiddle factors of consecutive butterflies in a group.

SB is set to –2 to detect any bit growth into the guard bits of any butterfly output. When all the groups in a stage are computed, the *bfp_adjustment* routine is called to check for bit growth and adjust the output data if necessary. Then parameters for the next stage are updated; *groups* is doubled and *node_space* and *bflys_per_group* are divided in half. The stage loop is repeated log_2N times.

The input and output parameters of the stage loop are summarized below. Note that the parameters are passed in memory locations.

Initial Conditions	*Final Conditions*
groups=# groups in stage	*groups*=# groups in next stage
bflys_per_group=# butterflies/group	*bflys_per_group*=#butterflies/ group (next stage)
node_space=node spacing current stage	*node_space*=node spacing next stage
inplacedata=stage input data	*inplacedata*=stage output data

```
        I0=^inplacedata;            {I0 ->x0 in 1st butterfly of stage}
        I1=^inplacedata;
        AY0=DM(node_space);
        M2=AY0;                     {M2=dual node spacing}
        MODIFY(I1,M2);              {I1 ->x1 in 1st butterfly of stage}
        I2=I1;
        AX0=2;
        AR=AY0-AX0;
        M3=AR;                      {M3=node_space-2}
        CNTR=DM(groups);            {Initialize group counter}
        SB=-2;                      {Set minimum allowable sign bits to two}
        M5=DM(groups);              {M5=twiddle factor modifier}
        DO group_loop UNTIL CE;

group_loop:     {Calculate All Groups in Stage}

        CALL bfp_adjustment;        {Adjust block data for bit growth}
        SI=DM(groups);
        SR=LSHIFT SI BY 1 (LO);
        DM(groups)=SR0;             {groups=groups × 2}
        SI=DM(node_space);
        SR=LSHIFT SI BY -1 (LO);
        DM(node_space)=SR0;         {node_space=node_space ÷ 2}
        SR=LSHIFT SR0 BY -1(LO);
        DM(bflys_per_group)=SR0;    {bflys_per_group=bflys_per_group ÷ 2}
```

Listing 6.11 Radix-2 DIF FFT Stage Loop

DIF FFT Subroutine

The complete block floating-point DIF FFT subroutine is shown in Listing 6.12. Initializations of index, modifier and length registers that retain the same values throughout the FFT calculation are performed before the stage loop is entered. Instructions that write butterfly results to memory are boldface.

```
.MODULE       dif_fft;

.CONST        N=1024, N_div_2=512, log2N=10;

.EXTERNAL     inplacedata, twid_real, twid_imag;
.EXTERNAL     groups,bflys_per_group,node_space;
.EXTERNAL     bfp_adjust;

.ENTRY        fft_start;

fft_start:  I4=^twid_real;          {I4 -> C OF W0}
            L4=N_div_2;
            I5=^twid_imag;          {I5 -> (-S) OF W0}
            L5=N_div_2;
            M0=-1;
            M1=1;
            CNTR=log2N;             {Initialize stage counter}
            L0=0;
            L1=0;
            L2=0;
            DO stage_loop UNTIL CE;
               I0=^inplacedata;          {I0 -> x0}
               I1=^inplacedata;
               AY0=DM(node_space);
               M2=AY0;
               MODIFY(I1,M2);            {I1 -> x1}
               I2=I1;
               AX0=2;
               AR=AY0-AX0;
               M3=AR;                    {M3=node_space-2}
               CNTR=DM(groups);          {Initialize group counter}
               SB=-2;
               M5=CNTR;                  {Init. twiddle factor modifier}
               DO group_loop UNTIL CE;
                  CNTR=DM(bflys_per_group);      {Init. butterfly counter}
                  AX0=DM(I0,M1);                 {AX0=x0}
                  AY0=DM(I1,M1);                 {AY0=x1}
                  AY1=DM(I1,M1);                 {AY1=y1}
                  DO bfly_loop UNTIL CE;
                     AR=AX0+AY0, AX1=DM(I0,M0), MY0=PM(I4,M5);
                                                 {AR=x0+x1,AX1=y0,MY0=C}
                     SB=EXPADJ AR;               {Check for bit growth}
```

(lisitng continues on next page)

```
                    DM(I0,M1)=AR,AR=AX1+AY1;     {x0´=x0+x1,AR=y0+y1}
                    SB=EXPADJ AR;                {Check for bit growth}
                    DM(I0,M1)=AR,AR=AX0-AY0;     {y0´=y0+y1,AR=x0-x1}
                    MX0=AR, AR=AX1-AY1;          {MX0=x0-x1,AR=y0-y1}
                    MR=MX0*MY0 (SS), AX0=DM(I0,M1), MY1=PM(I5,M5);
                                       {MR=(x0-x1)C,AX0=next x0,MY1=(-S)}
                    MR=MR-AR*MY1 (RND), AY0=DM(I1,M1);
                                       {MR=(x0-x1)C-(y0-y1)(-S),AY0=next x1}
                    SB=EXPADJ MR1;               {Check for bit growth}
                    DM(I2,M1)=MR1, MR=AR*MY0 (SS);
                                  {x1´=(x0-x1)C-(y0-y1)(-S),MR=(y0-y1)C}
                    MR=MR+MX0*MY1 (RND), AY1=DM(I1,M1);
                                  {MR=(y0-y1)C+(x0-x1)(-S),AY1=next y1}
bfly_loop:          DM(I2,M1)=MR1,SB=EXPADJ MR1;
                                  {y1´=(y0-y1)C+(x0-x1)(-S),check bit growth}
                 MODIFY(I2,M2);   {I2->x1 of first butterfly in next group}
                 MODIFY(I1,M3);   {I1->x1 of first butterfly in next group}
                 MODIFY(I0,M3);
group_loop:      MODIFY(I0,M1);   {I0->x0 of first butterfly in next group}
              CALL bfp_adjust;    {Adjust block data for bit growth}
              SI=DM(groups);
              SR=LSHIFT SI BY 1 (LO);
              DM(groups)=SR0;              {groups=groups × 2}
              SI=DM(node_space);
              SR=LSHIFT SI BY -1 (LO);
              DM(node_space)=SR0;          {node_space=node_space ÷ 2}
              SR=LSHIFT SR0 BY -1 (LO);
stage_loop:   DM(bflys_per_group)=SR0;
                                  {bflys_per_group=bflys_per_group ÷ 2}
          RTS;
.ENDMOD;
```

Listing 6.12 Radix-2 DIF FFT Routine, Conditional Block Floating-Point

6.2.5 Bit Reversal

Bit reversal is an addressing technique used in FFT calculations to obtain results in sequential order. Because the FFT repeatedly subdivides data sequences, the data and/or twiddle factors may be scrambled (in bit-reversed order). All radix-2 FFTs can be calculated with either the input sequence or the output sequence scrambled. The twiddle factors may also need to be scrambled, depending on the order of the input and output sequences. In this chapter, however, input and output sequences are set up so that twiddle factors are never scrambled. This simplifies the FFT explanation as well as the program.

As described earlier, the input sequence to the radix-2 DIT FFT is scrambled before the FFT is performed. Similarly, the output sequence of the radix-2 DIF FFT is unscrambled after the FFT is performed. Scrambling and unscrambling are both accomplished through bit reversal.

An example of bit reversal is shown below. Bit reversal operates on the binary number that represents the position of a sample within an array of samples. Each position is shown in decimal and binary. For example, the position of x(4) in sequential order is four (binary 100). Note that this position does not necessarily correspond to the location of the sample in memory. The bit-reversed position is the transpose of the bits of the binary number about its center; the transpose of the binary number 100 is 001. In this example, three bits are needed to represent eight positions, so bits zero and two are interchanged. Four bits are needed to represent 16 positions, so in a 16-point sequence, bits zero and three and bits one and two would be interchanged. A 1024-point sequence requires the reversal of ten bits.

Sample, Sequential Order	*Sequential Location*		*Bit-Reversed Location*		*Sample, Bit-Reversed Order*
	decimal	binary	decimal	binary	
x(0)	0	000	0	000	x(0)
x(1)	1	001	4	100	x(4)
x(2)	2	010	2	010	x(2)
x(3)	3	011	6	110	x(6)
x(4)	4	100	1	001	x(1)
x(5)	5	101	5	101	x(5)
x(6)	6	110	3	011	x(3)
x(7)	7	111	7	111	x(7)

When the samples in sequential order are scrambled, x(0) remains in location zero, x(1) moves to location four, x(2) remains in location two, x(3) moves to location six, etc. Conversely, if samples are already scrambled, bit reversal unscrambles them.

The ADSP-2100 has a bit-reverse capability built into its data address generator #1 (DAG1). When a mode bit is enabled (through software), the 14-bit address generated by DAG1 is automatically bit-reversed for any data memory read or write. The two address generators of the ADSP-2100 greatly simplify bit reversal. One address generator can be used to read sequentially ordered data, and the other can be used to write the same data to its bit-reversed location. Because the address generators are independent, intermediate enabling and disabling of the bit-reverse mode is not needed.

The base (starting) address of an array being accessed with bit-reversed addressing must be an integer multiple of the length (N) of the transform (i.e., base address=0, N, 2N, ...).

In many cases, fewer than 14 bits must be reversed (for example, an eight-point FFT needing only three bits reversed). Reversal of fewer than 14 bits is accomplished by adding the correct modify value to the address pointer after each memory access. The following example demonstrates bit reversal of ten bits using I0 to store the address to be reversed and M0 to store the modify value.

First, we determine the first bit-reversed address. This address is the first 14-bit address with the ten least significant bits reversed. For the DIT FFT subroutine, the first address in the *inplacereal* buffer is H#0000. If we reverse the ten least significant bits of H#0000, we still have H#0000. Thus, we want to output H#0000 as the first bit-reversed address. To do so, I0 must be initialized to the number that, when bit-reversed by the ADSP-2100 (all 14 bits), is H#0000. In this case, that number is also H#0000.

The second bit-reversed address must be H#0200 (H#0001 with ten least significant bits reversed). We must modify I0 to the value that, when bit-reversed (all 14 bits) is H#0200. This value is H#0010. Since I0 contains H#0000, we must add H#0010 to it. Thus, H#0010 is loaded into M0. After the first data memory read or write, which outputs H#0000, M0 is added to the (non-bit-reversed) address in I0 so that I0 contains H#0010. On the second data memory read or write, I0 is bit-reversed (14 bits) and the resulting address is H#0200, the correct second bit-reversed address.

In general, the modify value is determined by raising two to the difference between 14 and the number of bits to be reversed. In this ten-bit example, the value is $2^{(14-10)}$ = H#0010. Adding this value to I0 after each memory access and reversing all 14 bits on the next memory access yields the correct bit-reversed addresses for ten bits. The first four bit-reversed addresses are shown below.

Sequence	*I0, Non-Bit-Reversed*		*I0, Bit-Reversed*	
	H#	B#	H#	B#
0	0000	00 0000 0000 0000	0000	00 00**00 0000 0000**
1	0010	00 0000 0001 0000	0200	00 00**10 0000 0000**
2	0020	00 0000 0010 0000	0100	00 00**01 0000 0000**
3	0030	00 0000 0011 0000	0300	00 00**11 0000 0000**

Only the ten least significant bits (boldface) are bit-reversed. Each time a data memory write is performed, I0 is modified by M0. Note that the modified I0 value is not bit-reversed. Bit reversal only occurs when a data memory read or write is executed.

Two bit reversal routines are shown in Listings 6.13 and 6.14 (*scramble* and *unscramble*, respectively). The *scramble* routine places the inputs to the DIT FFT in bit-reversed order. The *unscramble* routine places the output data of the DIF FFT in sequential order. Both modules begin by initializing two constants. The first constant (*N*) is the number of input points in the FFT. The second constant (*mod_value*) is the modify value for the pointer which outputs the bit-reversed addresses. Pointers to the data buffers are initialized, and the bit-reverser is enabled for DAG1. In bit-reverse mode, any addresses output from registers I0, I1, I2, or I3 will be bit-reversed. I0 is used in *scramble*, and I1 is used in *unscramble*.

The *scramble* routine assumes real input data. In this case, the imaginary data is all zeros and can be initialized directly into the *inplaceimag* buffer. The *brev* loop consists of two instructions. First, the sequentially ordered data is read from the *input_real* buffer using I4 (from DAG2). Then, the same data is written to the bit-reversed location in the *inplacereal* buffer using I0 (from DAG1). After all the real input data has been placed in bit-reversed order in the *inplacereal* buffer, the bit-reverser is disabled for the rest of the FFT calculation.

The *unscramble* routine uses two loops: one to unscramble the real FFT output data, the other to unscramble the imaginary output data. I4 points to the first of the scrambled real data values in the *inplacedata* buffer. I4 is modified by two (in M4) after each read. Because the real and imaginary data in *inplacedata* are interleaved, this ensures that only real data is read for the first loop. I1 contains the (bit-reversed) address of the first location in the *real_results* buffer (for unscrambled real data). The appropriate modify value (stored in M1) is added to I1 upon each data memory write. Before entering the second loop, I4 is then updated to point to the first imaginary data in *inplacedata* and I1 is set to the first address (bit-reversed) of the *imag_results* buffer (for sequentially ordered imaginary data).

```
.MODULE      dit_scramble;

{            Calling Parameters
                Sequentially ordered input data in inputreal
                The base (starting) address of an array being accessed with
                bit-reversed addressing must be an integer multiple of the
                length (N) of the transform (i.e., base address=0,N,2N,…).

             Return Values
                Scrambled input data in inplacereal

             Altered Registers
                I0,I4,M0,M4,AY1

             Altered Memory
                inplacereal
}

.CONST       N=1024,mod_value=H#0010;       {Initialize constants}
.EXTERNAL    inputreal, inplacereal;
.CONST       N=1024,mod_value=H#0010;       {Initialize constants}
.ENTRY       scramble;

scramble:    I4=^inputreal;                 {I4-->sequentially ordered data}
             I0=^inplacereal;               {I0-->scrambled data}
             M4=1;
             M0=mod_value;                  {M0=modifier for reversing N bits}
             L4=0;
             L0=0;
             CNTR = N;
             ENA BIT_REV;                   {Enable bit-reversed outputs on DAG1}
             DO brev UNTIL CE;
                AY1=DM(I4,M4);              {Read sequentially ordered data}
brev:           DM(I0,M0)=AY1;              {Write data in bit-reversed location}
             DIS BIT_REV;                   {Disable bit-reverse}
             RTS;                           {Return to calling program}
.ENDMOD;
```

Listing 6.13 Bit-Reverse (Scramble) Routine

```
.MODULE         dif_unscramble;

{               Calling Parameters
                   Real and imaginary scrambled output data in inplacedata
                   Output data is stored with bit-reversed addressing, starting
                   at address 0.

                Return Values
                   Sequentially ordered real output data in real_results
                   Sequentially ordered imag. output data in imaginary_results

                Altered Registers
                   I1,I4,M1,M4,L1,AY1,CNTR

                Altered Memory
                   real_results, imaginary_results
}

.CONST          N=1024,mod_value=H#0010;             {Initialize constants}

.EXTERNAL       inplacedata;

.ENTRY          unscramble;          {Declare entry point into module}

unscramble:     I4=^inplacedata;     {I4-->real part of 1st data point}
                M4=2;                {Modify by 2 to fetch only real data}
                L1=0;
                L4=0;
                I1=H#4;              {I1=1st real output addr, bit-reversed}
                M1=mod_value;        {Modifier for 10-bit reversal}
                CNTR=N;              {N=number of real data points}
                ENA BIT_REV;         {Enable bit-reverse}
                DO bit_rev_real UNTIL CE;
                   AY1=DM(I4,M4);    {Read real data}
bit_rev_real:      DM(I1,M1)=AY1;    {Place in sequential order}
                I4=^inplacedata+1;   {I4-->imag. part of 1st data point}
                I1=H#C;              {I1=1st imag. output addr, bit-reversed}
                CNTR=N;              {N=number of imaginary data points}
                DO bit_rev_imag UNTIL CE;
                   AY1=DM(I4,M4);    {Read imag. data}
bit_rev_imag:      DM(I1,M1)=AY1;    {Place in sequential order}
                DIS BIT_REV;         {Disable bit-reverse}
                RTS;
.ENDMOD;
```

Listing 6.14 Bit-Reverse (Unscramble) Routine

6.3 BLOCK FLOATING-POINT SCALING

Block floating-point scaling is used to maximize the dynamic range of a fixed-point data field. The block floating-point system is a hybrid between fixed-point and floating-point systems. Instead of each data word having its own exponent, block floating-point format assumes the same exponent for an entire block of data.

The initial input data contains enough guard bits to ensure that no overflow occurs in the first FFT stage. During each stage of the FFT calculation, bit growth can occur. This bit growth can result in magnitude bits replacing guard bits. Because the stage output data is used as input data for the next stage, these guard bits must be replaced; otherwise, an output of the next stage might overflow. In a conditional block floating-point FFT, bit growth is monitored in each stage calculation. When the stage is complete, the output data of the entire stage is shifted to replace any lost guard bits.

Because a radix-2 butterfly calculation has the potential for two bits of growth, SB (the block floating-point exponent register) is initialized to –2. This sets up the ADSP-2100 block floating-point compare logic to detect any data with more than 13 bits of magnitude (or fewer than three sign bits). After each butterfly calculation, the EXPADJ instruction determines if bit growth occurred by checking the number of guard bits. For example, the value 1111 0000 0000 0000 has an exponent of –3. The value 0111 1111 1111 1111 has an exponent of zero (no guard bits). If a butterfly result has an exponent larger than the value in SB, bit growth into the guard bits has occurred, and SB is assigned the larger exponent (if it has not already been changed by bit growth in a previous butterfly of the same stage). Therefore, at the end of each stage, SB contains the exponent of the largest butterfly result(s). If no bit growth occurred, SB is not changed. An example of how the EXPADJ instruction affects SB is shown below (assume SB = –2 initially).

Butterfly Output Data	*Value of SB After EXPADJ Performed on Data*
1111 0000 0000 0000	SB= –2
1110 0000 0000 0000	SB= –2
1100 0000 0000 0000	SB= –1
1110 0000 0000 0000	SB= –1
1000 0000 0000 0000	SB= 0

The *dit_radix-2_bfp_adjust* routine is shown in Listing 6.15, on the next page. This routine performs block floating-point scaling on the outputs of each stage except the last of the DIT FFT. It can be modified for the DIF FFT subroutine by replacing *inplacereal* references with *inplacedata.*

Because guard bits only need to be replaced to ensure that an output of the next stage does not overflow, the subroutine first checks to see if the block of data is the output of the last stage. If it is, no shifting is needed and the subroutine returns. If the data block is not the output of the last stage, shifting is necessary only if SB is not –2 (indicating that bit growth into guard bits occurred). If SB is –2, no bit growth occurred, so the subroutine returns.

If bit growth occurred, shifting is needed. The subroutine determines the amount to shift from the value of SB. The data can grow by either one or two bits for each stage; therefore, if bit growth occurred, SB must be either –1 or zero. If SB is –1, the data block is shifted right one bit. If SB is not –1, it must be zero. In this case, the data block is shifted right two bits. When shifting is complete, the block exponent is updated by the shifted amount (one or two).

In this routine, shifting to the right is performed through multiplication rather than shift instructions. Multiplication by an appropriate power of two gives a shifted result. For example, to shift a number two bits to the right, the number is multiplied by H#0200. In multiplication, the product can be rounded to preserve LSB information, whereas in shifting, this information is merely lost. Multiplication thus minimizes noise.

```
.MODULE      dit_radix_2_bfp_adjust;

{            Calling Parameters
                Radix-2 DIT FFT stage results in inplacereal and inplaceimag

             Return Parameters
                inplacereal and inplaceimag adjusted for bit growth

             Altered Registers
                I0,I1,AX0,AY0,AR,MX0,MY0,MR,CNTR

             Altered Memory
                inplacereal, inplaceimag, blk_exponent
}

.CONST       Ntimes2 = 2048;
.EXTERNAL    inplacereal, blk_exponent;    {Begin declaration section}

.ENTRY       bfp_adj;

bfp_adj:     AY0=CNTR;                     {Check for last stage}
             AR=AY0-1
             IF EQ RTS;                    {If last stage, return}
             AY0=-2;
             AX0=SB;
             AR=AX0-AY0;                   {Check for SB=-2}
             IF EQ RTS;                    {IF SB=-2, no bit growth, return}
             I0=^inplacereal;              {I0=read pointer}
             I1=^inplacereal;              {I1=write pointer}
             AY0=-1;
             MY0=H#4000;                   {Set MY0 to shift 1 bit right}
             AR=AX0-AY0,MX0=DM(I0,M1);     {Check if SB=-1; Get first sample}
             IF EQ JUMP strt_shift;        {If SB=-1, shift block data 1 bit}
             AX0=-2;                       {Set AX0 for block exponent update}
             MY0=H#2000;                   {Set MY0 to shift 2 bits right}
strt_shift:  CNTR=Ntimes2 - 1;             {initialize loop counter}
             DO shift_loop UNTIL CE;           {Shift block of data}
                MR=MX0*MY0(RND),MX0=DM(I0,M1); {MR=shifted data,MX0=next value}
shift_loop:     DM(I1,M1)=MR1;                 {Unshifted data=shifted data}
             MR=MX0*MY0(RND);              {Shift last data word}
             AY0=DM(blk_exponent);         {Update block exponent and}
             DM(I1,M1)=MR1,AR=AY0-AX0;     {store last shifted sample}
             DM(blk_exponent)=AR;
             RTS;
.ENDMOD;
```

Listing 6.15 Radix-2 Block Floating-Point Scaling Routine

6.4 OPTIMIZED RADIX-2 DIT FFT

Because the FFT is often just the first step in the processing of a signal, the execution speed is important. The faster the FFT executes, the more time the processor can devote to the remainder of the signal processing task. Improving the execution speed of the FFT in a given algorithm may allow a faster sampling rate for the system, a higher-order filter or a more detailed algorithm.

The radix-2 DIT FFT implemented in a looped form—the *stage* loop, the *group* loop and the *butterfly* loop—is the most compact form of the FFT, in terms of program memory storage requirements. The performance of a looped program which accommodates all stages of the FFT can be improved by mathematical optimization. The fully looped program does not exploit the unique mathematical characteristics of the first and last stage of the FFT. By breaking the first and last FFT stage out of the nested loop structure, a few tricks can be employed to improve the execution speed. The tradeoff for increased speed is a modest increase in program memory storage requirements.

This section focuses on the optimizing of the radix-2 DIT FFT. Similar modifications can be applied to the radix-2 DIF algorithm.

6.4.1 First Stage Modifications

The DIT FFT butterfly equations are as follows.

$$(7) \quad x'_0 = x_0 + [\mathbf{C}x_1 - (-\mathbf{S})y_1]$$

$$(8) \quad y'_0 = y_0 + [\mathbf{C}y_1 + (-\mathbf{S})x_1]$$

$$(9) \quad x'_1 = x_0 - [\mathbf{C}x_1 - (-\mathbf{S})y_1]$$

$$(10) \quad y'_1 = y_0 - [\mathbf{C}y_1 + (-\mathbf{S})x_1]$$

In the first stage, there are N/2 groups, each containing a single butterfly. Each butterfly uses a twiddle factor W^0, where

$$W^0 = e^{j0} = \cos(0) + j\sin(0) = 1 + j0$$

All of the multiplications in the first stage are by a value of either 0 or 1 and therefore can be removed. The first-stage butterflies do not need multiplications. The butterfly equations reduce to the following.

$$x'_0 = x_0 + x_1$$

$$y'_0 = y_0 + y_1$$

$$x'_1 = x_0 - x_1$$

$$y'_1 = y_0 - y_1$$

Because there is only one butterfly per group in the first stage, the butterfly loop (which would execute only once per group) and the group loop can be combined. The combination of the group and butterfly loops is shown in Listing 6.16. The elimination of the multiplications and the combination of the group and butterfly loops saves eleven clock cycles per group. In a 1024-point FFT, for example, the first stage contains 512 butterfly loops. These simple modifications save 5632 (512x11) clock cycles.

```
            DO group_lp UNTIL CE;
               AR=AX0+AY0, AX1=DM(I2,M0);    {AR=X1+Y1 , AX1=Y0}
               DM(I0,M2)=AR, AR=AX0-AY0;     {store X0', AR=X0-X1}
               DM(I1,M2)=AR, AR=AX1+AY1;     {store X1', AR=Y0+Y1}
               DM(I2,M2)=AR, AR=AX1-AY1;     {store Y0', AR=Y0-Y1}
               DM(I3,M2)=AR;                 {store Y1'}
               AX0=DM(I0,M0);                {AX0 = next X0}
               AY0=DM(I1,M0);                {AY0 = next X1}
group_lp:      AY1=DM(I3,M0);                {AY1 = next Y1}
```

Listing 6.16 First Stage Butterfly

Similar modifications can be performed on the block floating-point DIT FFT algorithm. First-stage butterfly multiplications can be reduced to additions, and the group and butterfly loops can be combined. Listing 6.17 contains a block floating-point first-stage butterfly.

```
          DO group_lp UNTIL CE;
             AR=AX0+AY0, AX1=DM(I2,M0);
             SB=EXPADJ AR, DM(I0,M2)=AR;
             AR=AX0-AY0;
             SB=EXPADJ AR;
             DM(I1,M2)=AR, AR=AX1+AY1;
             SB=EXPADJ AR, DM(I2,M2)=AR;
             AR=AX1-AY1, AX0=DM(I0,M0);
             SB=EXPADJ AR, DM(I3,M2)=AR;
             AY0=DM(I1,M0);
group_lp:    AY1=DM(I3,M0);
          CALL bfp_adj;
```

Listing 6.17 Block Floating-Point First Stage Butterfly

6.4.2 Last Stage Modifications

The last stage of the DIT FFT can also be modified to increase performance. This stage consists of a single group of N/2 butterflies. Because there is only one group in this stage, the group loop is not needed. The calculations in previous stages of the number of butterflies per group and the number of groups can also be removed. Furthermore, in the last stage, no setup for the next stage is needed. The code for the last stage with these modifications is shown in Listing 6.18.

```
I0=^inplacereal;               {I0 -> x0}
I1=^inplacereal+nover2;        {I1 -> x1}
I2=^inplaceimag;               {I2 -> y0}
I3=^inplaceimag+nover2;        {I3 -> y1}
CNTR=nover2;                   {# of butterflies}
M2=DM(node_space);             {node space modifier}
M4=1;
I6=I3;
I4=^twid_real;      {real twiddle pointer}
I5=^twid_imag;      {imag. twiddle pointer}
MY0=PM(I4,M4),MX0=DM(I1,M0);
MY1=PM(I5,M4);
DO bfly_lp UNTIL CE;
   MR=MX0*MY0(SS),MX1=DM(I6,M5);
   MR=MR-MX1*MY1(RND),AY0=DM(I0,M0);
   AR=MR1+AY0,AY1=DM(I2,M0);
   DM(I0,M1)=AR,AR=AY0-MR1;
```

(listing continues on next page)

```
                MR=MX0*MY1(SS),DM(I1,M1)=AR;
                MR=MR+MX1*MY0(RND),MX0=DM(I1,M0),MY1=PM(I5,M4);
                AR=AY1-MR1,MY0=PM(I4,M4);
                DM(I3,M1)=AR,AR=MR1+AY1;
    bfly_lp:    DM(I2,M1)=AR;
```

Listing 6.18 Last Stage DIT FFT

6.4.3 Optimized Radix-2 DIT FFT Program Listings

This section contains listings for two optimized FFT routines, one with no scaling (input data scaled) and one that performs block floating-point scaling. Listing 6.19 contains the main module, which performs initialization operations and places the input data in bit-reversed (scrambled) order. Listing 6.20 contains the *fft* module, which performs the FFT with input data scaling. Listing 6.21 contains a second *fft* module that performs the FFT with block floating-point scaling. Either FFT routine can be called from the main module.

The *scramble* and *bfp_adj* routines are the same as for the unoptimized radix-2 DIT FFT program.

```
.MODULE/ABS=4            dit_fft_main;
.CONST                   N=1024, N_div_2=512; {For 1024 points}
.VAR/PM/RAM/CIRC         twid_real [N_div_2];
.VAR/PM/RAM/CIRC         twid_imag [N_div_2], padding [4];
.VAR/DM/RAM/ABS=0        inplacereal [N], inplaceimag [N];
.VAR/DM/RAM/ABS=H#1000   inputreal [N], inputimag [N];
.VAR/DM/RAM              groups, bflys_per_group,
                         node_space, blk_exponent;

.INIT   twid_real: <twid_real.dat>;
.INIT   twid_imag: <twid_imag.dat>;
.INIT   inputreal: <inputreal.dat>;
.INIT   inputimag: <inputimag.dat>;
.INIT   inplaceimag: <inputimag.dat>;
.INIT   groups: N_div_2;
.INIT   bflys_per_group: 2;
.INIT   node_space: 2;
.INIT   blk_exponent: 0;
.INIT   padding: 0,0,0,0;            {Zeros after inplaceimag}

.GLOBAL twid_real, twid_imag;
.GLOBAL inplacereal, inplaceimag;
.GLOBAL inputreal, inputimag;
.GLOBAL groups, bflys_per_group, node_space, blk_exponent;
```

```
.EXTERNAL       scramble, fft_strt;

                CALL scramble;          {subroutine calls}
                CALL fft_strt;
                TRAP;                   {halt program}
.ENDMOD;
```

Listing 6.19 Main Module for Optimized Radix-2 DIT FFT

```
{1024-point DIT radix-2 FFT}
{No scaling: input data must be scaled}

.MODULE/ROM    fft;

.CONST         log2N=10, N=1024, nover2 =512, nover4 =256;

.EXTERNAL      twid_real,twid_imag,inplacereal,inplaceimag;
.EXTERNAL      inputreal,inputimag;
.EXTERNAL      groups,bflys_per_group,node_space;
.ENTRY         fft_strt;

fft_strt:      CNTR=log2N-2;           {CNTR=# stages in fft-2}
               M0=0;                   {in place modifier}
               M1=1;                   {advance 1 modifier}
               M3=-1;
               M5=1;
               L1=0;
               L2=0;
               L3=0;
               L4=%twid_real;          {length register for}
               L5=%twid_imag;          {twiddle factor tables}
               L6=0;

{First stage of 1024-point FFT}

               I0=^inplacereal;        {I0 -> x0}
               I1=^inplacereal+1;      {I1 -> x1}
               I2=^inplaceimag;        {I2 -> y0}
               I3=^inplaceimag+1;      {I3 -> y1}
               M2=2;                   {modifier for 1st stage}
                                       {node space}
               CNTR=nover2;            {number of groups}
```

(listing continues on next page)

```
                AX0=DM(I0,M0);          {AX0=x0}
                AY0=DM(I1,M0);          {AY0=x1}
                AY1=DM(I3,M0);          {AY1=y1}

                DO group_lp UNTIL CE;
                   AR=AX0+AY0, AX1=DM(I2,M0); {AR=x1+y1 , AX1=y0}
                   DM(I0,M2)=AR, AR=AX0-AY0;  {store x0', AR=x0-x1}
                   DM(I1,M2)=AR, AR=AX1+AY1;  {store x1', AR=y0+y1}
                   DM(I2,M2)=AR, AR=AX1-AY1;  {store y0', AR=y0-y1}
                   DM(I3,M2)=AR;              {store y1'}
                   AX0=DM(I0,M0);             {AX0=next x0}
                   AY0=DM(I1,M0);             {AY0=next x1}
group_lp:          AY1=DM(I3,M0);             {AY1=next y1}

{—— END STAGE 1 ——}

{—— STAGES 2 THRU 9 ——}

                DO stage_loop UNTIL CE;       {begin looping}
                                              {stage 2 thru stage 9}
                   I0=^inplacereal;           {I0 -> x0}
                   I2=^inplaceimag;           {I2 -> y0}
                   SI=DM(groups);             {SI=# groups in}
                                              {previous stage}
                   SR=ASHIFT SI BY -1(LO);    {SR0=groups/2}
                   CNTR=SR0;                  {CNTR=# groups in}
                                              {current stage}
                   DM(groups)=SR0;            {update "groups"}
                   M4=SR0;                    {Initialize twid}
                                              {pointer modifier}
                   M2=DM(node_space);         {M2=dual node spacing}
                   I1=I0;                     {points to 1st x_real}
                   MODIFY(I1,M2);             {points to 1st y_real}
                   I3=I2;                     {points to 1st x_imag}
                   MODIFY(I3,M2);             {points to 1st y_imag}
                   I6=I3;                     {I6->y1}

                   DO group_loop UNTIL CE;
                      I4=^twid_real;          {I4->real twid factor}
                      I5=^twid_imag;          {I5->imag twid factor}
                      CNTR=DM(bflys_per_group);    {# bflies/group}
                      MY0=PM(I4,M4),MX0=DM(I1,M0); {MY0=C, MX0=x1}
                      MY1=PM(I5,M4);               {MY1=(-S)}
```

```
                     DO bfly_loop UNTIL CE;
                        MR=MX0*MY0(SS),MX1=DM(I6,M5);
                                  {MR=x(C), MX1=y1}
                        MR=MR-MX1*MY1(SS),AY0=DM(I0,M0);
                                  {MR=x1(C)-y1(-S),AY0=x0}
                        AR=MR1+AY0,AY1=DM(I2,M0);
                                  {AR=x0+[x1(C)-y1(-S)],AY1=y0}
                        DM(I0,M1)=AR,AR=AY0-MR1;
                                  {store x0', calc. x1'}
                        MR=MX0*MY1(SS),DM(I1,M1)=AR;
                                  {MR=x1(-S), store x1'}
                        MR=MR+MX1*MY0(RND),MX0=DM(I1,M0),
                               MY1=PM(I5,M4);
                               {MR=x1(-S)+y1(C),MX0=x1, MY1=(-S)}
                        AR=AY1-MR1,MY0=PM(I4,M4);
                                             {calc y1', MY0=(C)}
                        DM(I3,M1)=AR,AR=MR1+AY1;
                                             {store y1',calc y0'}
bfly_loop:              DM(I2,M1)=AR;        {store y0'}


                     MODIFY(I0,M2);       {I0 -> x0 in next group}
                     MODIFY(I1,M2);       {I1 -> x1 in next group}
                     MODIFY(I2,M2);       {I2 -> y0 in next group}
group_loop:          MODIFY(I3,M2);       {I3 -> y1 in next group}

                  SI=DM(bflys_per_group);
                  SR=ASHIFT SI BY 1(LO); {SR=bflys_per_group * 2}
                  DM(node_space)=SR0;
stage_loop:       DM(bflys_per_group)=SR0; {update bflys_per_group}

{— LAST STAGE ———}

                I0=^inplacereal;             {I0 -> x0}
                I1=^inplacereal+nover2;      {I1 -> x1}
                I2=^inplaceimag;             {I2 -> y0}
                I3=^inplaceimag+nover2;      {I3 -> y1}
                CNTR=nover2;                 {# of butterflies}
                M2=DM(node_space);           {node space modifier}
                M4=1;
                I6=I3;
                I4=^twid_real;               {real twiddle pointer}
                I5=^twid_imag;               {imag. twiddle pointer}
```

(listing continues on next page)

```
{butterfly loop as above}
             MY0=PM(I4,M4),MX0=DM(I1,M0);
             MY1=PM(I5,M4);
             DO bfly_lp UNTIL CE;
                MR=MX0*MY0(SS),MX1=DM(I6,M5);
                MR=MR-MX1*MY1(RND),AY0=DM(I0,M0);
                AR=MR1+AY0,AY1=DM(I2,M0);
                DM(I0,M1)=AR,AR=AY0-MR1;
                MR=MX0*MY1(SS),DM(I1,M1)=AR;
                MR=MR+MX1*MY0(RND),MX0=DM(I1,M0),MY1=PM(I5,M4);
                AR=AY1-MR1,MY0=PM(I4,M4);
                DM(I3,M1)=AR,AR=MR1+AY1;
bfly_lp:        DM(I2,M1)=AR;

             RTS;                        {return to main}

.ENDMOD;
```

Listing 6.20 Optimized Radix-2 DIT FFT with Input Scaling

```
{1024 point DIT radix 2 FFT}
{Block Floating Point Scaling}

.MODULE       fft;

{     Calling Parameters
             inplacereal=real input data in scrambled order
             inplaceimag=all zeroes (real input assumed)
             twid_real=twiddle factor cosine values
             twid_imag=twiddle factor sine values
             groups=N/2
             bflys_per_group=1
             node_space=1

      Return Values
             inplacereal=real FFT results, sequential order
             inplaceimag=imag. FFT results, sequential order

      Altered Registers
             I0,I1,I2,I3,I4,I5,L0,L1,L2,L3,L4,L5
             M0,M1,M2,M3,M4,M5
             AX0,AX1,AY0,AY1,AR,AF
             MX0,MX1,MY0,MY1,MR,SB,SE,SR,SI
```

```
      Altered Memory
            inplacereal, inplaceimag, groups, node_space,
            bflys_per_group, blk_exponent
}

.CONST        log2N=10, N=1024, nover2=512, nover4=256;

.EXTERNAL     twid_real, twid_imag;
.EXTERNAL     inplacereal, inplaceimag;
.EXTERNAL     groups, bflys_per_group, node_space;
.EXTERNAL     bfp_adj;
.ENTRY        fft_strt;

fft_strt:     CNTR=log2N - 2;     {Initialize stage counter}
              M0=0;
              M1=1;
              L1=0;
              L2=0;
              L3=0;
              L4=%twid_real;
              L5=%twid_imag;
              L6=0;
              SB=-2;

{——— STAGE 1 ———}

              I0=^inplacereal;
              I1=^inplacereal + 1;
              I2=^inplaceimag;
              I3=^inplaceimag + 1;
              M2=2;

              CNTR=nover2;
              AX0=DM(I0,M0);
              AY0=DM(I1,M0);
              AY1=DM(I3,M0);

              DO group_lp UNTIL CE;
                 AR=AX0+AY0, AX1=DM(I2,M0);
                 SB=EXPADJ AR, DM(I0,M2)=AR;
                 AR=AX0-AY0;
                 SB=EXPADJ AR;
                 DM(I1,M2)=AR, AR=AX1+AY1;
```

(listing continues on next page)

```
                    SB=EXPADJ AR, DM(I2,M2)=AR;
                    AR=AX1-AY1, AX0=DM(I0,M0);
                    SB=EXPADJ AR, DM(I3,M2)=AR;
                    AY0=DM(I1,M0);
group_lp:           AY1=DM(I3,M0);

                CALL bfp_adj;

{————————————————————————————————————————}

                DO stage_loop UNTIL CE;{Compute all stages in FFT}
                    I0=^inplacereal; {I0 ->x0 in 1st grp of stage}
                    I2=^inplaceimag; {I2 ->y0 in 1st grp of stage}
                    SI=DM(groups);
                    SR=ASHIFT SI BY -1(LO); {groups / 2}
                    DM(groups)=SR0;         {groups=groups / 2}
                    CNTR=SR0;          {CNTR=group counter}
                    M4=SR0;            {M4=twiddle factor modifier}
                    M2=DM(node_space);  {M2=node space modifier}
                    I1=I0;
                    MODIFY(I1,M2);     {I1 ->y0 of 1st grp in stage}
                    I3=I2;
                    MODIFY(I3,M2);     {I3 ->y1 of 1st grp in stage}

                    DO group_loop UNTIL CE;
                        I4=^twid_real;     {I4 -> C of W0}
                        I5=^twid_imag;     {I5 -> (-S) of W0}
                        CNTR=DM(bflys_per_group);
                                           {CNTR=butterfly counter}
                        MY0=PM(I4,M4),MX0=DM(I1,M0); {MY0=C,MX0=x1 }
                        MY1=PM(I5,M4),MX1=DM(I3,M0); {MY1=-S,MX1=y1}
                        DO bfly_loop UNTIL CE;
                            MR=MX0*MY1(SS),AX0=DM(I0,M0);
                                           {MR=x1(-S),AX0=x0}
                            MR=MR+MX1*MY0(RND),AX1=DM(I2,M0);
                                           {MR=(y1(C)+x1(-S)),AX1=y0}
                            AY1=MR1,MR=MX0*MY0(SS);
                                       {AY1=y1(C)+x1(-S),MR=x1(C)}
                            MR=MR-MX1*MY1(RND);  {MR=x1(C)-y1(-S)}
                            AY0=MR1,AR=AX1-AY1;
                              {AY0=x1(C)-y1(-S),AR=y0-[y1(C)+x1(-S)]}
                            SB=EXPADJ AR,DM(I3,M1)=AR;
                            {Check for bit growth, y1=y0-[y1(C)+x1(-S)]}
```

```
                   AR=AX0-AY0,MX1=DM(I3,M0),MY1=PM(I5,M4);
              {AR=x0-[x1(C)-y1(-S)], MX1=next y1,MY1=next (-S)}
                   SB=EXPADJ AR,DM(I1,M1)=AR;
                   {Check for bit growth, x1=x0-[x1(C)-y1(-S)]}
                   AR=AX0+AY0,MX0=DM(I1,M0),MY0=PM(I4,M4);
                {AR=x0+[x1(C)-y1(-S)], MX0=next x1,MY0=next C}
                   SB=EXPADJ AR,DM(I0,M1)=AR;
                   {Check for bit growth, x0=x0+[x1(C)-y1(-S)]}
                   AR=AX1+AY1;   {AR=y0+[y1(C)+x1(-S)]}
bfly_loop:         SB=EXPADJ AR,DM(I2,M1)=AR;
                   {Check for bit growth, y0=y0+[y1(C)+x1(-S)]}
                MODIFY(I0,M2);   {I0 ->1st x0 in next group}
                MODIFY(I1,M2);   {I1 ->1st x1 in next group}
                MODIFY(I2,M2);   {I2 ->1st y0 in next group}
group_loop:     MODIFY(I3,M2);   {I3 ->1st y1 in next group}

           CALL bfp_adj;        {Compensate for bit growth}
           SI=DM(bflys_per_group);
           SR=ASHIFT SI BY 1(LO);
           DM(node_space)=SR0; {node_space=node_space / 2}
stage_loop:    DM(bflys_per_group)=SR0;  {bflys_per_group= }
                                         {bflys_per_group / 2}

{—— LAST STAGE ———}

         I0=^inplacereal;
         I1=^inplacereal+nover2;
         I2=^inplaceimag;
         I3=^inplaceimag+nover2;

         CNTR=nover2;
         M2=DM(node_space);
         M4=1;
         I4=^twid_real;
         I5=^twid_imag;

         MY0=PM(I4,M4),MX0=DM(I1,M0);       {MY0=C,MX0=x1}
         MY1=PM(I5,M4),MX1=DM(I3,M0);       {MY1=-S,MX1=y1}
         DO bfly_lp UNTIL CE;
            MR=MX0*MY1(SS),AX0=DM(I0,M0);{MR=x1(-S),AX0=x0}
            MR=MR+MX1*MY0(RND),AX1=DM(I2,M0);
                                 {MR=(y1(C)+x1(-S)),AX1=y0}
            AY1=MR1,MR=MX0*MY0(SS);
```

(listing continues on next page)

```
                                      {AY1=y1(C)+x1(-S),MR=x1(C)}
                MR=MR-MX1*MY1(RND); {MR=x1(C)-y1(-S)}
                AY0=MR1,AR=AX1-AY1;
                       {AY0=x1(C)-y1(-S), AR=y0-[y1(C)+x1(-S)]}
                SB=EXPADJ AR,DM(I3,M1)=AR;
                    {Check for bit growth, y1=y0-[y1(C)+x1(-S)]}
                AR=AX0-AY0,MX1=DM(I3,M0),MY1=PM(I5,M4);
                 {AR=x0-[x1(C)-y1(-S)], MX1=next y1,MY1=next (-S)}
                SB=EXPADJ AR,DM(I1,M1)=AR;
                    {Check for bit growth, x1=x0-[x1(C)-y1(-S)]}
                AR=AX0+AY0,MX0=DM(I1,M0),MY0=PM(I4,M4);
                    {AR=x0+[x1(C)-y1(-S)], MX0=next x1,MY0=next C}
                SB=EXPADJ AR,DM(I0,M1)=AR;
                    {Check for bit growth, x0=x0+[x1(C)-y1(-S)]}
                AR=AX1+AY1;  {AR=y0+[y1(C)+x1(-S)]}
bfly_lp:        SB=EXPADJ AR,DM(I2,M1)=AR;
                    {Check for bit growth}

            CALL bfp_adj;

            RTS;
.ENDMOD;
```

Listing 6.21 Optimized Radix-2 Block Floating-Point DIT FFT

6.5 RADIX-4 FAST FOURIER TRANSFORMS

Whereas a radix-2 FFT divides an N-point sequence successively in half until only two-point DFTs remain, a radix-4 FFT divides an N-point sequence successively in quarters until only four-point DFTs remain. An N-point sequence is divided into four N/4-point sequences; each N/4-point sequence is broken into four N/16-point sequences, and so on, until only four-point DFTs are left. The four-point DFT is the core calculation (butterfly) of the radix-4 FFT, just as the two-point DFT is the butterfly for a radix-2 FFT.

A radix-4 FFT essentially combines two stages of a radix-2 FFT into one, so that half as many stages are required. The radix-4 butterfly is consequently larger and more complicated than a radix-2 butterfly; however, fewer butterflies are needed. Specifically, N/4 butterflies are used in each of $(\log_2 N)/2$ stages, which is one quarter the number of butterflies in a radix-2 FFT. Although addressing of data and twiddle factors is more complex, a radix-4 FFT requires fewer calculations than a radix-2 FFT. The addressing capability of the ADSP-2100 can accommodate the added complexity, and so the it can compute a radix-4 FFT significantly faster than a radix-2 FFT. Like the radix-2 FFT, the radix-4 FFT requires data scrambling and/or unscrambling. However, radix-4 FFT sequences are scrambled and unscrambled through digit reversal, rather than bit reversal as in the radix-2 FFT. Digit reversal is described later in this section.

6.5.1 Radix-4 Decimation-In-Frequency FFT Algorithm

The radix-4 FFT divides an N-point DFT into four N/4-point DFTs, then into 16 N/16-point DFTs, and so on. In the radix-2 DIF FFT, the DFT equation is expressed as the sum of two calculations, one on the first half and one on the second half of the input sequence. Then the equation is divided to form two equations, one that computes even samples and the other that computes odd samples. Similarly, the radix-4 DIF FFT expresses the DFT equation as four summations, then divides it into four equations, each of which computes every fourth output sample. The following equations illustrate radix-4 decimation in frequency.

$$(24) \qquad X(k) = \sum_{n=0}^{N-1} x(n)\, W_N^{nk}$$

$$= \sum_{n=0}^{N/4-1} x(n)\, W_N^{nk} + \sum_{n=N/4}^{N/2-1} x(n)\, W_N^{nk} + \sum_{n=N/2}^{3N/4-1} x(n)\, W_N^{nk} + \sum_{n=3N/4}^{N-1} x(n)\, W_N^{nk}$$

$$= \sum_{n=0}^{N/4-1} x(n)\, W_N^{nk} + \sum_{n=0}^{N/4-1} x(n+N/4)\, W_N^{(n+N/4)k}$$

$$+ \sum_{n=0}^{N/4-1} x(n+N/2)\, W_N^{(n+N/2)k} + \sum_{n=0}^{N/4-1} x(n+3N/4) W_N^{(n+3N/4)k}$$

$$= \sum_{n=0}^{N/4-1} [\, x(n) + W_N^{k(N/4)}\, x(n+N/4) + W_N^{k(N/2)}\, x(n+N/2) + W_N^{k3N/4}\, x(n+3N/4)]\, W_N^{nk}$$

The three twiddle factor coefficients can be expressed as follows:

(25) $$W_N^{k(N/4)} = (e^{-j2\pi/N})^{k(N/4)} = (e^{-j\pi/2})^k = (\cos(\pi/2) - j\sin(\pi/2))^k = (-j)^k$$

(26) $$W_N^{k(N/2)} = (e^{-j2\pi/N})^{k(N/2)} = (e^{-j\pi})^k = (\cos(\pi) - j\sin(\pi))^k = (-1)^k$$

(27) $$W_N^{k3N/4} = (e^{-j2\pi/N})^{k3N/4} = (e^{-j3\pi/2})^k = (\cos(3/2\pi) - j\sin(3\pi/2))^k = j^k$$

Equation (23) can thus be expressed as

(28) $$X(k) = \sum_{n=0}^{N/4-1} [\, x(n) + (-j)^k x(n+N/4) + (-1)^k x(n+N/2) + (j)^k x(n+3N/4)\,]\, W_N^{nk}$$

Four sub-sequences of the output (frequency) sequence are created by setting k=4r, k=4r+1, k=4r+2 and k=4r+3:

(29) $$X(4r) = \sum_{n=0}^{N/4-1} [\, (\, x(n) + x(n+N/4) + x(n+N/2) + x(n+3N/4)\,)\, W_N^{0}\,]\, W_{N/4}^{nr}$$

$$(30)\ X(4r+1) = \sum_{n=0}^{N/4-1} [(x(n) - jx(n+N/4) - x(n+N/2) + jx(n+3N/4))W_N^{n}]\, W_{N/4}^{nr}$$

$$(31)\ X(4r+2) = \sum_{n=0}^{N/4-1} [(x(n) - x(n+N/4) + x(n+N/2) - x(n+3N/4))W_N^{2n}]\, W_{N/4}^{nr}$$

$$(32)\ X(4r+3) = \sum_{n=0}^{N/4-1} [(x(n) + jx(n+N/4) - x(n+N/2) - jx(n+3N/4))W_N^{3n}]\, W_{N/4}^{nr}$$

for r = 0 to N/4–1

X(4r), X(4r+1), X(4r+2), and X(4r+3) are N/4-point DFTs. Each of their N/4 points is a sum of four input samples (x(n), x(n+N/4), x(n+N/2) and x(n+3N/4)), each multiplied by either +1, –1, j, or –j. The sum is multiplied by a twiddle factor (W_N^{0}, W_N^{n}, W_N^{2n}, or W_N^{3n}).

These four N/4-point DFTs together make up an N-point DFT. Each of these N/4-point DFTs is divided into four N/16-point DFTs. Each N/16 DFT is further divided into four N/64-point DFTs, and so on, until the final decimation produces four-point DFTs (groups of four one-point DFT equations). The four one-point DFT equations make up the butterfly calculation of the radix-4 FFT. A radix-4 butterfly is shown graphically in Figure 6.9.

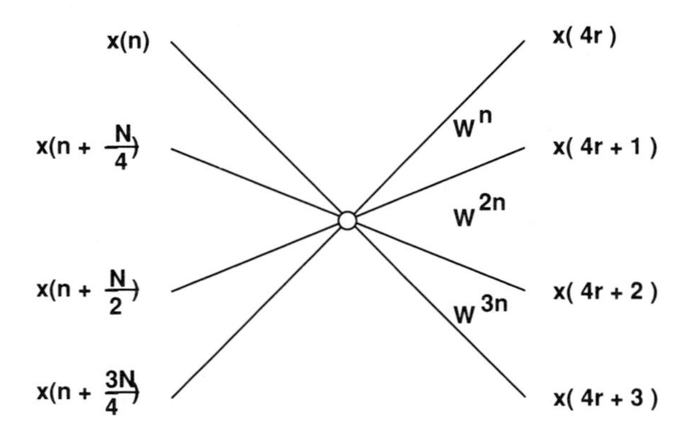

Figure 6.9 Radix-4 DIF FFT Butterfly

The output of each leg represents one of the four equations which are combined to make a four-point DFT. These four equations correspond to equations (29) through (32), for one point rather than N/4 points.

Each sample in the butterfly is complex. A butterfly flow graph with complex inputs and outputs is shown in Figure 6.10. The real part of each point is represented by *x*, and *y* represents the imaginary part. The twiddle factor can be divided into real and imaginary parts because $W_N = e^{-j2\pi/N} = \cos(2\pi/N) - j\sin(2\pi/N)$. In the program presented later in this section, the twiddle factors are initialized in memory as cosine and –sine values (not +sine). For this reason, the twiddle factors are shown in Figure 6.10 as C + j(–S). C represents cosine and –S represents –sine.

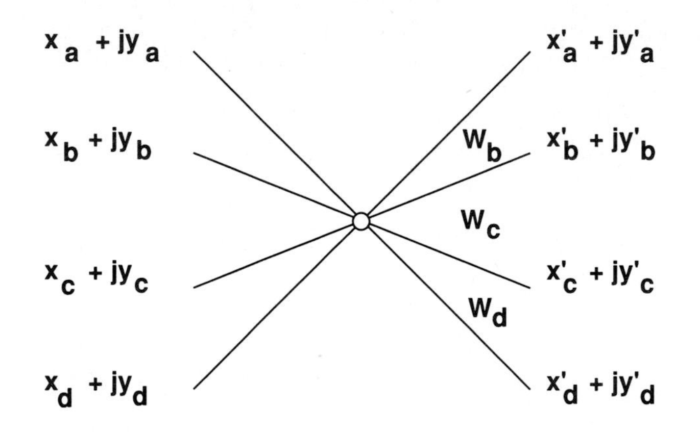

Figure 6.10 Radix-4 DIF FFT Butterfly, Complex Data

The real and imaginary output values for the radix-4 butterfly are given by equations (33) through (40).

(33) $x_a' = x_a + x_b + x_c + x_d$

(34) $y_a' = y_a + y_b + y_c + y_d$

(35) $x_b' = (x_a + y_b - x_c - y_d)C_b - (y_a - x_b - y_c + x_d)(-S_b)$

(36) $y_b' = (y_a - x_b - y_c + x_d)C_b + (x_a + y_b - x_c - y_d)(-S_b)$

(37) $x_c' = (x_a - x_b + x_c - x_d)C_c - (y_a - y_b + y_c - y_d)(-S_c)$

(38) $y_c' = (y_a - y_b + y_c - y_d)C_c + (x_a - x_b + x_c - x_d)(-S_c)$

(39) $x_d' = (x_a - y_b - x_c + y_d)C_d - (y_a + x_b - y_c - x_d)(-S_d)$

(40) $y_d' = (y_a + x_b - y_c - x_d)C_d + (x_a - y_b - x_c + y_d)(-S_d)$

A complete 64-point radix-4 FFT is shown in Figure 6.11, on the next page. As in the radix-2 FFT, butterflies are organized into groups and stages. The first stage has one group of 16 (N/4) butterflies, the next stage has four groups of four (N/16) butterflies, and the last stage has 16 groups of one butterfly. Notice that the twiddle factor values depend on the group and stage that are being performed. The table below summarizes the characteristics of an N-point radix-4 FFT.

Stage		1	2	3	$(\log_2 N)/2$
Butterfly Groups		1	4	16	N/4
Butterflies per Group		N/4	N/16	N/64	1
Dual-Node Spacing		N/4	N/16	N/64	1
Twiddle	*leg1*	0	0	0	0
Factor	*leg2*	n	4n	16n	(N/4)n
Exponents	*leg3*	2n	8n	32n	(N/2)n
	leg4	3n	12n	48n	(3N/4)n
		n=0 to N/4–1	n=0 to N/16–1	n=0 to N/32–1	n=0

A 64-point radix-4 FFT has half as many stages (three instead of six) and half as many butterflies in each stage (16 instead of 32) as a 64-point radix-2 FFT.

6 One-Dimensional FFTs

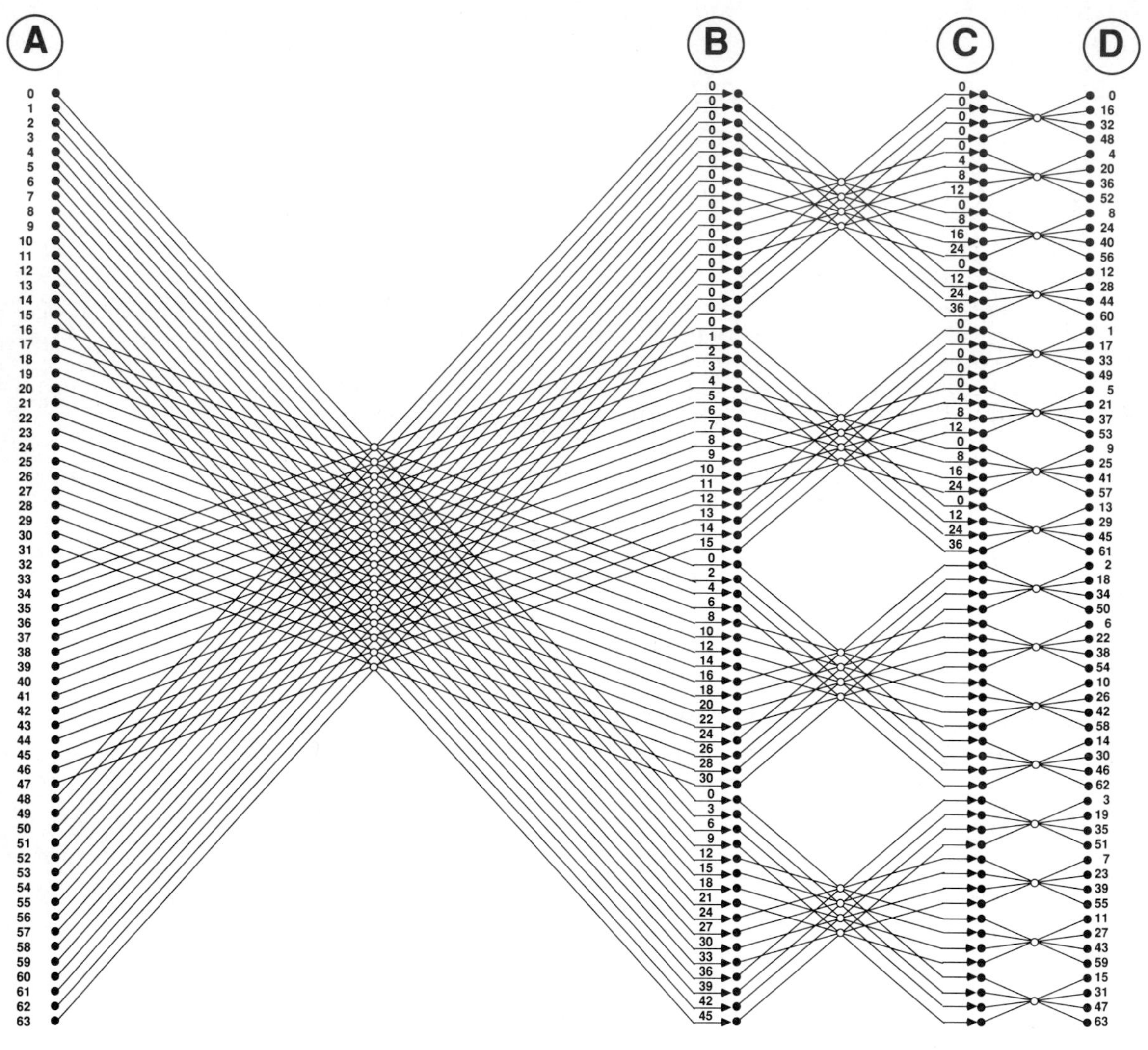

Figure 6.11 Sixty-Four-Point Radix-4 DIF FFT

Column a) indicates input sample; 44=x(44).
Column b) indicates twiddle factor exponent, stage one; 5=W_N^5.
Column c) indicates twiddle factor exponent, stage two.
Column d) indicates output sample; 51=X(51).

6.5.2 Radix-4 Decimation-In-Frequency FFT Program

A flow chart for the radix-4 DIF FFT program is shown in Figure 6.12. The program flow is identical to that of the radix-2 DIF FFT except that the outputs are unscrambled by digit reversal instead of bit reversal.

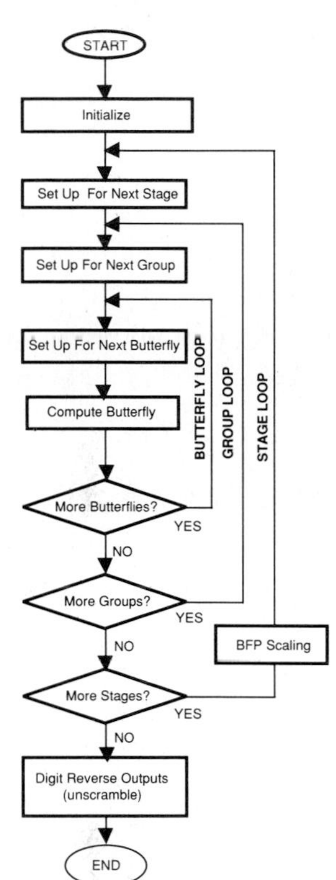

Figure 6.12 Radix-4 DIF FFT Flow Chart

The radix-4 DIF FFT routine uses three subroutines; the first computes the FFT, the second performs block floating-point scaling, and the third unscrambles the FFT results. The main routine (*rad4_main*) declares and initializes buffers and variables stored in external memory. It also calls the FFT and digit reversal subroutines. Three other modules contain the FFT, block floating-point scaling and digit reversal subroutines. The *rad4_main* and *rad4_fft* modules are described in this section. The block floating-point scaling and digit reversal routines are described later.

6.5.2.1 Main Module

The *rad4_main* module is shown in Listing 6.22. Constants *N*, *N_x_2*, *N_div_4*, and *N_div_2* are used throughout this module to specify buffer lengths as well as initial values for some variables. The in-place FFT calculation is performed in the *inplacedata* buffer. A small buffer called *padding* is placed at the end of the *inplacedata* buffer to allow memory accesses to exceed the buffer. The extra memory locations are necessary in a simulation because the ADSP-2100 Simulator does not allow undefined memory locations to be operated on; however, *padding* is not necessary in a real system.

The *input_data* buffer retains the initial FFT input data that is lost during the FFT calculation. This buffer allows you to look at the original input data after executing the program. However, *input_data* is also not needed in a real system.

The *digit_rev* subroutine unscrambles the FFT outputs and writes them in sequential order into *results*. The variables *groups*, *bflys_per_group*, *node_space*, and *blk_exponent* are declared to store stage characteristics and the block floating-point exponent, as in the radix-2 FFT routine.

Buffers *inplacedata*, *twids*, and *input_data* are initialized with data stored in external files. For example, *twids* is initialized with the external file *twids.dat*, which contains the twiddle factor values. Immediate zeros are placed in *padding*.

The variable *groups* is initialized to one and *bflys_per_group* to *N_div_4* because there is one group in the first stage of the FFT and N/4 butterflies in this first group. Node spacing for the radix-4 FFT in the first stage is N/4. However, because the *inplacedata* buffer is organized with real and imaginary data interleaved, the node spacing is doubled to N/2. Thus, the variable *node_space* is initialized to *N_div_2*.

The *rad4_fft* subroutine computes the FFT, and the *digit_rev* routine unscrambles the output using digit reversal. The TRAP instruction halts the ADSP-2100 when the FFT is complete.

6.5.2.2 DIF FFT Module

The conditional block floating-point radix-4 DIF FFT subroutine presented in this section consists of three nested loops. To simplify the explanation of this subroutine, each loop is described separately, starting with the innermost loop (the butterfly loop) and followed by the group loop and the stage loop. The entire subroutine is listed at the end of this section.

Butterfly Loop

The radix-4 butterfly equations (33-40) are repeated below.

$$(33)\quad x_a' = x_a + x_b + x_c + x_d$$

$$(34)\quad y_a' = y_a + y_b + y_c + y_d$$

$$(35)\quad x_b' = (x_a + y_b - x_c - y_d)C_b - (y_a - x_b - y_c + x_d)(-S_b)$$

$$(36)\quad y_b' = (y_a - x_b - y_c + x_d)C_b + (x_a + y_b - x_c - y_d)(-S_b)$$

$$(37)\quad x_c' = (x_a - x_b + x_c - x_d)C_c - (y_a - y_b + y_c - y_d)(-S_c)$$

$$(38)\quad y_c' = (y_a - y_b + y_c - y_d)C_c + (x_a - x_b + x_c - x_d)(-S_c)$$

$$(39)\quad x_d' = (x_a - y_b - x_c + y_d)C_d - (y_a + x_b - y_c - x_d)(-S_d)$$

$$(40)\quad y_d' = (y_a + x_b - y_c - x_d)C_d + (x_a - y_b - x_c + y_d)(-S_d)$$

The code segment to calculate these equations is shown in Listing 6.23. This code segment computes one radix-4 butterfly. The outputs (x_a', y_a', x_b', y_b', etc.) are written over the inputs (x_a, y_a, x_b, y_b, etc.) in the highlighted instructions. Each of the eight butterfly results is monitored for bit growth using the EXPADJ instruction and written to data memory in the same multifunction instruction. This code segment also sets up pointers and fetches the initial data for the next butterfly. The butterfly calculation is described in detail in the comments, and the instructions

```
.MODULE/ABS=4           rad4_main;

.CONST                  N=1024,N_x_2=2048,     {Define constants for N-point FFT}
                        N_div_4=256,N_div_2=512;
.VAR/DM/RAM/ABS=0       inplacedata[N_x_2], padding[4];
                                               {Pad end of inplacedata so memory}
.VAR/DM/RAM             twids[N_x_2];          {accesses can exceed end of buffer}
.VAR/DM/RAM             outputdata[N_x_2];
.VAR/DM/RAM             input_data[N_x_2];
.VAR/DM/RAM             groups,bflys_per_group,
                        node_space,blk_exponent;

.INIT                   inplacedata: <inplacedata.dat>;
.INIT                   input_data: <inplacedata.dat>;
.INIT                   twids: <twids.dat>;
.INIT                   groups: 1;
.INIT                   bflys_per_group: N_div_4;
.INIT                   node_space: N_div_2;
.INIT                   blk_exponent: 0;
.INIT                   padding: 0,0,0,0;

.GLOBAL                 inplacedata,twids, outputdata;
.GLOBAL                 groups,bflys_per_group,node_space,blk_exponent;

.EXTERNAL               rad4_fft,digit_rev;

                        CALL rad4_fft;
                        CALL digit_rev;
                        TRAP;                  {Stop program execution}
.ENDMOD;
```

Listing 6.22 Main Module, Radix-4 DIF FFT

that check for bit growth and write the butterfly results to data memory are boldface.

The input and output parameters of this code segment are shown below.

Initial Conditions	*Final Conditions*
I0 --> x_a	I0 --> next x_a
I1 --> x_b	I1 --> next x_b
I2 --> y_c	I2 --> next y_c
I3 --> x_d	I3 --> next x_d
I4 --> C_b	I4 --> next C_b
I5 --> S_c	I5 --> next S_c
I6 --> C_d	I6 --> next C_d
M0 = 0	AX0 = next x_a
M1 = 1	AY0 = next x_c
M3 = –1	MY0 = next C_c
CNTR = butterfly counter	CNTR = butterfly counter – 1
M4 = 1	
M5 = *groups* x 2 – 1	
M6 = *groups* x 4 – 1	
M7 = *groups* x 6 – 1	
AX0 = x_a	
AY0 = x_c	
MY0 = C_c	

```
AF=AX0+AY0,AX1=DM(I1,M1);           {AF=xa+xc; AX1=xb; I1 --> yb}
AR=AF-AX1,AY1=DM(I3,M1);            {AR=xa+xc-xb; AY1=xd; I3 --> yd}
AR=AR-AY1,SR1=DM(I1,M3);            {AR=xa-xb+xc-xd; SR1=yb; I1 --> xb}
MR=AR*MY0(SS),SR0=DM(I3,M3);        {MR=(xa-xb+xc-xd)Cc; SR0=yd; I3 --> xd}
MX0=AR,AR=AX1+AF;                   {AR=xa+xb+xc; MX0=(xa-xb+xc-xd)Cc}
AR=AR+AY1;                          {AR=xa+xb+xc+xd}
SB=EXPADJ AR,DM(I0,M1)=AR;          {xa´=xa+xb+xc+xd; I0 --> ya}
AF=AX0-AY0,AX0=DM(I0,M0);           {AF=xa-xc; AX0=ya; I0 --> ya}
AR=SR1+AF,AY0=SR0;                  {AR=xa+yb-xc; AY0=yd}
AF=AF-SR1;                          {AF=xa-yb-xc}
AR=AR-AY0,AY0=DM(I2,M3);            {AR=xa+yb-xc-yd; AY0=yc; I2 --> xc}
MX1=AR,AR=SR0+AF;                   {AR=xa-yb-xc+yd; MX1=xa+yb-xc-yd}
AF=AX0+AY0,DM(I3,M1)=AR;            {AR=ya+yc; location of xd=xa-yb-xc+yd}
                                    {I3 --> yd}
AY0=DM(I3,M3),AR=SR1+AF;            {AR=ya+yb+yc; AY0=yd; I3 --> xd}
AR=AR+AY0,MY1=DM(I5,M6);            {AR=ya+yb+yc+yd; MY1=(-Sc); I5 --> next Cc}
SB=EXPADJ AR,DM(I0,M1)=AR;          {ya´=ya+yb-yc+yd; I0 --> next xa}
AF=AF-SR1;                          {AF=ya-yb+yc}
AR=AF-SR0;                          {AR=ya-yb+yc-yd}
MR=MR-AR*MY1(SS);                   {MR=(xa-xb+xc-xd)Cc - (ya-yb+yc-yd)(-Sc)}
SB=EXPADJ MR1,DM(I2,M1)=MR1;        {xc´=(xa-xb+xc-xd)Cc - (ya-yb+yc-yd)(-Sc)}
                                    {I2 --> yc}
MR=AR*MY0(SS);                      {MR=(ya-yb+yc-yd)Cc}
MR=MR+MX0*MY1(SS),AY0=DM(I2,M0);    {MR=(ya-yb+yc-yd)Cc + (xa-xb+xc-xd)(-Sc)}
                                    {AY0=yc; I2 --> yc}
SB=EXPADJ MR1,DM(I2,M1)=MR1;        {yc´=(ya-yb+yc-yd)Cc + (xa-xb+xc-xd)(-Sc)}
                                    {I2 --> next xc}
AF=AX0-AY0,MY1=DM(I4,M4);           {AF=ya-yc; MY1=Cb; I4 -->(-Sb)}
AR=AF-AX1,AX0=DM(I0,M0);            {AR=ya-xb-yc; AX0=ya; I1 --> ya}
AR=AR+AY1,AY0=DM(I2,M1);            {AR=ya-xb-yc+xd; AY0=yc; I2 --> next xc}
MR=MX1*MY1(SS),MY0=DM(I4,M5);       {MR=(xa+yb-xc-yd)Cb; MY0=Sb; I4 --> next Cb}
MR=MR-AR*MY0(SS);                   {MR=(xa+yb-xc-yd)Cb - (ya-xb-yc+xd)(-Sb)}
SB=EXPADJ MR1,DM(I1,M1)=MR1;        {xb´=(xa+yb-xc-yd)Cb - (ya-xb-yc+xd)(-Sb)}
                                    {I1 --> yb}
MR=AR*MY1(SS);                      {MR=(ya-xb-yc+xd)Cb}
MR=MR+MX1*MY0(SS),MX1=DM(I3,M0);    {MR=(ya-xb-yc+xd)Cb + (xa+yb-xc-yd)(-Sb)}
                                    {MX1=xa-yb-xc+yd; I3 --> xd}
SB=EXPADJ MR1,DM(I1,M1)=MR1;        {yb´=(ya-xb-yc+xd)Cb + (xa+yb-xc-yd)(-Sb)}
                                    {I1 --> next xb}
AR=AX1+AF,MY0=DM(I6,M4);            {AR=ya+xb-yc; MY0=Cd; I6 -->-Sd}
AR=AR-AY1,MY1=DM(I6,M7);            {AR=ya+xb-yc-xd; MY1=-Sd; I6 -->Cd}
MR=MX1*MY0(SS);                     {MR=(xa-yb-xc+yd)Cd}
MR=MR-AR*MY1(SS);                   {MR=(xa-yb-xc+yd)Cd - (ya+xb-yc-xd)(-Sd)}
SB=EXPADJ MR1,DM(I3,M1)=MR1;        {xd´=(xa-yb-xc+yd)Cd - (ya+xb-yc-xd)(-Sd)}
                                    {I3 --> yd}
MR=AR*MY0(SS),MY0=DM(I5,M4);        {MR=(ya+xb-yc-xd)Cd; MY0=next Cc}
                                    {I5 --> next (-Sc)}
MR=MR+MX1*MY1(SS);                  {MR=(ya+xb-yc-xd)Cd + (xa-yb-xc+yd)(-Sd)}
SB=EXPADJ MR1,DM(I3,M1)=MR1;        {yd´=(ya+xb-yc-xd)Cd +(xa-yb-xc+yd)(-Sd)}
                                    {I3 --> next xd}
```

Listing 6.23 Radix-4 DIF FFT Butterfly, Conditional Block Floating-Point Scaling

Group Loop

The group loop is shown in Listing 6.24. This code segment sets up and computes one group of butterflies. Because each leg of the first butterfly in all groups in the FFT has the twiddle factor W^0, twiddle-factor pointers are initialized to point to the real part of W^0. Next, the butterfly loop is set up by initializing the butterfly loop counter and fetching initial data values (x_a, y_c and C_c). Notice that these are the initial conditions for the butterfly loop.

After all the butterflies in the group are calculated, pointers used in the butterfly are updated to point to x_a, x_b, x_c, and x_d for the first butterfly in the next group. For example, I0 points to the first x_a in the next group, I1 to the first x_b, etc. The group loop is executed *groups* times (the number of groups in a stage).

The input and output parameters of this code segment are as follows:

Initial Conditions	*Final Conditions*
I0 --> x_a	I0 --> first x_a of next group
I1 --> x_b	I1 --> first x_b of next group
I2 --> x_c	I2 --> first x_c of next group
I3 --> x_d	I3 --> first x_d of next group
M0 = 0	I4 --> invalid location for twiddle factor
M1 = 1	I5 --> invalid location for twiddle factor
M2 = 3 x node_space	I6 --> invalid location for twiddle factor
M3 = –1	CNTR = group count – 1
M4 = 1	
CNTR = group count	

```
           I4=^twids;                  {I4 --> Cb}
           I5=I4;                      {I5 --> Cc}
           I6=I5;                      {I6 --> Cd}
           CNTR=DM(bflys_per_group);   {Initialize butterfly counter}
           AX0=DM(I0,M0);              {AX0=xa; I0 --> xa}
           AY0=DM(I2,M1);              {AY0=xc; I2 --> yc}
           MY0=DM(I5,M4);              {MY0=Cc; I5 --> Sc}
           DO bfly_loop UNTIL CE;

bfly_loop: {Calculate All Butterflies}

           MODIFY(I0,M2);              {I0 --> first xa of next group}
           MODIFY(I1,M2);              {I1 --> first xb of next group}
           MODIFY(I2,M3);
           MODIFY(I2,M2);              {I2 --> first xc of next group}
           MODIFY(I3,M2);              {I3 --> first xd of next group}
```

Listing 6.24 Radix-4 DIF FFT Group Loop

Stage Loop

The stage characteristics of the FFT are controlled by the stage loop. For example, the stage loop controls the number of groups and the number of butterflies in each group. The stage loop code segment is shown in Listing 6.25. This code sets up and calculates all groups of butterflies in a stage and updates parameters for next stage.

The radix-4 butterfly data can potentially grow three bits from butterfly input to output (the worst case growth factor is 5.6). Therefore, each input value to the FFT contains three guard bits to prevent overflow. SB is initialized to –3, so any bit growth into the guard bits can be monitored. If bit growth occurs, it is compensated for in the block floating-point subroutine that is called after each stage is computed.

The variable *groups* is loaded into SI and used to calculate various stage parameters. These include *groups*x2–1, the leg b twiddle factor modifier, *groups*x4–1, the leg c twiddle factor modifier, and *groups*x6–1, the leg d modifier. Pointers are set to x_a, x_b, x_c, and x_d, the inputs to the first butterfly in the stage. The group loop counter is initialized and M2, which is used to update butterfly data pointers at the start of a new group, is set to three times the node spacing.

In the group loop, all groups in the stage are computed. After the groups are computed, the subroutine *bfp_adjust* is called to perform block floating-point scaling by checking for bit growth in the stage output data and adjusting all of the data in the block accordingly.

After the output data is scaled, parameters are adjusted for the next stage; *groups* is updated to *groups*x4, *node_space* to *node_space*/4, and *bflys_per_group* to *bflys_per_group*/4. The stage loop is repeated ($\log_2$N)/2 times (the number of stages in the FFT).

The input and output parameters for this code segment are as follows:

Initial Conditions	*Final Conditions*
groups = # groups/stage	*groups* =*groups* x 4
node_space = node spacing for stage	*node_space* = *node_space* /4
bflys_per_group = # butterflies/group	*bflys_per_group* =*bflys_per_group* /4
inplacedata=stage input data	*inplacedata*=stage output data
CNTR = stage count	CNTR = stage count – 1
	SB = –(number of guard bits remaining in data word(s) with largest magnitude)
	SI = # groups/stage
	I0 ->invalid location for data sample
	I1 ->invalid location for data sample
	I2 ->invalid location for data sample
	I3 ->invalid location for data sample
	M2 = *node_space* x 3
	M5 = *groups* x 2 – 1
	M6 = *groups* x 4 – 1
	M7 = *groups* x 6 – 1

```
          SB=-3;                        {SB detects growth into 3 guard bits}
          SI=DM(groups);                {SI=groups}
          SR=ASHIFT SI BY 1(HI);        {SR1=groups × 2}
          AY1=SR1;                      {AY1=groups × 2}
          AR=AY1-1;                     {AR=groups × 2 - 1}
          M5=AR;                        {M5=groups × 2 - 1}
          SR=ASHIFT SR1 BY 1(HI);       {SR1=groups × 4}
          AY1=SR1;                      {AY1=groups × 4}
          AR=AY1-1;                     {AR=groups × 4 - 1}
          M6=AR;                        {M6=groups × 4 - 1}
          AY0=SI;                       {AY0=groups}
          AR=AR+AY0;                    {AR=groups × 5 - 1}
          AR=AR+AY0;                    {AR=groups × 6 - 1}
          M7=AR;                        {M7=groups × 6 - 1}
          M2=DM(node_space);            {M2=node_space}
          I0=^inplacedata;              {I0 --> xa}
          I1=I0;
          MODIFY(I1,M2);                {I1 --> xb}
          I2=I1;
          MODIFY(I2,M2);                {I2 --> xc}
          I3=I2;
          MODIFY(I3,M2);                {I3 --> xd}
          CNTR=SI;                      {Initialize group counter}
          AY0=DM(node_space);
          M2=I3;                        {M2=node_space × 3}
          DO group_loop UNTIL CE;

group_loop: {Calculate All Groups in a Stage}

          CALL bfp_adjust;              {Check for bit growth}
          SI=DM(groups);                {SI=groups}
          SR=ASHIFT SI BY 2(HI);        {SR1=groups ×4 }
          DM(groups)=SR1;               {group count, next stage}
          SI=DM(bflys_per_group);       {SI=bflys_per_group}
          SR=ASHIFT SI BY -1(HI);       {SR1=bflys_per_group ÷ 2}
          DM(node_space)=SR1;           {node spacing, next stage}
          SR=ASHIFT SI BY -1(HI);       {SR1=node_space ÷ 2}
          DM(bflys_per_group)=SR1;      {butterfly count, next stage}
```

Listing 6.25 Radix-4 DIF FFT Stage Loop

Radix-4 DIF FFT Subroutine

The butterfly, group, and stage loop code segments are combined into the entire radix-4 DIF FFT subroutine, which is shown in Listing 6.26. Note that length and modify registers that retain the same value throughout the routine are initialized outside the stage loop. The stage loop counter is initialized to the number of stages in an N-point FFT ($log_2N\_div\_2$). Instructions that write butterfly results to memory are boldface.

```
.MODULE      radix_4_dif_fft;        {Declare and name module}

.CONST       log2N_div_2=5;          {Initial stage count}

.ENTRY       rad4_fft;

.EXTERNAL    groups,node_space,bflys_per_group;
.EXTERNAL    inplacedata,twids,bfp_adjust;

rad4_fft:    CNTR=log2N_div_2;       {Initialize stage counter}
             M0=0;                   {Set constant modifiers, length registers}
             M1=1;
             M3=-1;
             M4=1;
             L0=0;
             L1=0;
             L2=0;
             L3=0;
             L4=0;
             L5=0;
             L6=0;
             L7=0;
             DO stage_loop UNTIL CE;            {Compute all stages}
                SB=-4;                          {Detects bit growth into 4 MSBs}
                SI=DM(groups);                  {SI=groups}
                SR=ASHIFT SI BY 1(HI);          {SR1=groups × 2}
                AY1=SR1;                        {AY1=groups × 2}
                AR=AY1-1;                       {AR=groups × 2 - 1}
                M5=AR;                          {M5=groups × 2 - 1}
                SR=ASHIFT SR1 BY 1(HI);         {SR1=groups × 4}
                AY1=SR1;                        {AY1=groups × 4}
                AR=AY1-1;                       {AR=groups × 4 - 1}
                M6=AR;                          {M6=groups × 4 - 1}
                AY0=SI;                         {AY0=groups}
                AR=AR+AY0;                      {AR=groups × 5 - 1}
                AR=AR+AY0;                      {AR=groups × 6 - 1}
                M7=AR;                          {M7=groups × 6 - 1}
                M2=DM(node_space);              {M2=node_space}
                I0=^inplacedata;                {I0 -->xa}
                I1=I0;
                MODIFY(I1,M2);                  {I1 -->xb}
                I2=I1;
                MODIFY(I2,M2);                  {I2 -->xc}
                I3=I2;
                MODIFY(I3,M2);                  {I3 -->xd}
                CNTR=SI;                        {Initialize group counter}
                AY0=DM(node_space);
                M2=I3;                          {M2=node_space × 3}
                DO group_loop UNTIL CE;         {Compute all groups in stage}
```

```
                    I4=^twids;                   {I4 -->Cb}
                    I5=I4;                       {I5 -->Cc}
                    I6=I5;                       {I6 -->Cd}
                    CNTR=DM(bflys_per_group);    {Initialize butterfly counter}
                    AX0=DM(I0,M0);               {AX0=xa, I0 -->xa}
                    AY0=DM(I2,M1);               {AY0=xc, I2 -->yc}
                    MY0=DM(I5,M4);               {MY0=Cc, I5 -->(-Sc)}
                    DO bfly_loop UNTIL CE;       {Compute all butterflies in grp}
                       AF=AX0+AY0,AX1=DM(I1,M1);
                       AR=AF-AX1,AY1=DM(I3,M1);
                       AR=AR-AY1,SR1=DM(I1,M3);
                       MR=AR*MY0(SS),SR0=DM(I3,M3);
                       MX0=AR,AR=AX1+AF;
                       AR=AR+AY1;
                       SB=EXPADJ AR,DM(I0,M1)=AR;         {xa´=xa+xb+xc+xd}
                       AF=AX0+AY0,AX0=DM(I0,M0);
                       AR=SR1+AF,AY0=SR0;
                       AF=AF-SR1;
                       AR=AR-AY0,AY0=DM(I2,M3);
                       MX1=AR,AR=SR0+AF;
                       AF=AX0+AY0,DM(I3,M1)=AR;
                       AY0=DM(I3,M3),AR=SR1+AF;
                       AR=AR+AY0,MY1=DM(I5,M6);
                       SB=EXPADJ AR,DM(I0,M1)=AR;         {ya´=ya+yb+yc+yd}
                       AF=AF-SR1;
                       AR=AF-SR0;
                       MR=MR-AR*MY1(SS);
                       SB=EXPADJ MR1,DM(I2,M1)=MR1;       {xc´=(xa-xb+xc-xd)Cc}
                       MR=AR*MY0(SS);                     {-(ya-yb+yc-yd)(-Sc)}
                       MR=MR+MX0*MY1(SS),AY0=DM(I2,M0);
                       SB=EXPADJ MR1,DM(I2,M1)=MR1;       {yc´=(ya-yb+yc-yd)Cc}
                       AF=AX0-AY0,MY1=DM(I4,M4);          {+ (xa-xb+xc-xd)(-Sc)}
                       AR=AF-AX1,AX0=DM(I0,M0);
                       AR=AR+AY1,AY0=DM(I2,M1);
                       MR=MX1*MY1(SS),MY0=DM(I4,M5);
                       MR=MR-AR*MY0(SS);
                       SB=EXPADJ MR1,DM(I1,M1)=MR1;       {xb´=(xa+yb-xc-yd)Cb}
                       MR=AR*MY1(SS);                     {-(ya-xb-yc+yd)(-Sb)}
                       MR=MR+MX1*MY0(SS),MX1=DM(I3,M0);
                       SB=EXPADJ MR1,DM(I1,M1)=MR1;       {yb´=(ya-xb-yc+xd)Cb}
                       AR=AX1+AF,MY0=DM(I6,M4);           {+ (xa+yb-xc-yd)(-Sb)}
                       AR=AR-AY1,MY1=DM(I6,M7);
                       MR=MX1*MY0(SS);
                       MR=MR-AR*MY1(SS);
                       SB=EXPADJ MR1,DM(I3,M1)=MR1;       {xd´=(xa-yb-xc+yd)Cd}
                       MR=AR*MY0(SS),MY0=DM(I5,M4);       {- (ya+xb-yc-xd)(-Sd)}
                       MR=MR+MX1*MY1(SS);
bfly_loop:             SB=EXPADJ MR1,DM(I3,M1)=MR1;       {yd´= (ya+xb-yc-xd)Cd}
                                                          {+ (xa-yb-xc+yd)(-Sd)}
```

(listing continues on next page)

```
                 DO bfly_loop UNTIL CE;             {Compute all butterflies in grp}
                    AF=AX0+AY0,AX1=DM(I1,M1);
                    AR=AF-AX1,AY1=DM(I3,M1);
                    AR=AR-AY1,SR1=DM(I1,M3);
                    MR=AR*MY0(SS),SR0=DM(I3,M3);
                    MX0=AR,AR=AX1+AF;
                    AR=AR+AY1;
                    SB=EXPADJ AR,DM(I0,M1)=AR;      {xa´=xa+xb+xc+xd}
                    AF=AX0+AY0,AX0=DM(I0,M0);
                    AR=SR1+AF,AY0=SR0;
                    AF=AF-SR1;
                    AR=AR-AY0,AY0=DM(I2,M3);
                    MX1=AR,AR=SR0+AF;
                    AF=AX0+AY0,DM(I3,M1)=AR;
                    AY0=DM(I3,M3),AR=SR1+AF;
                    AR=AR+AY0,MY1=DM(I5,M6);
                    SB=EXPADJ AR,DM(I0,M1)=AR;      {ya´=ya+yb+yc+yd}
                    AF=AF-SR1;
                    AR=AF-SR0;
                    MR=MR-AR*MY1(SS);
                    SB=EXPADJ MR1,DM(I2,M1)=MR1;    {xc´=(xa-xb+xc-xd)Cc}
                    MR=AR*MY0(SS);                  {-(ya-yb+yc-yd)(-Sc)}
                    MR=MR+MX0*MY1(SS),AY0=DM(I2,M0);
                    SB=EXPADJ MR1,DM(I2,M1)=MR1;    {yc´=(ya-yb+yc-yd)Cc}
                    AF=AX0-AY0,MY1=DM(I4,M4);       {+ (xa-xb+xc-xd)(-Sc)}
                    AR=AF-AX1,AX0=DM(I0,M0);
                    AR=AR+AY1,AY0=DM(I2,M1);
                    MR=MX1*MY1(SS),MY0=DM(I4,M5);
                    MR=MR-AR*MY0(SS);
                    SB=EXPADJ MR1,DM(I1,M1)=MR1;    {xb´=(xa+yb-xc-yd)Cb}
                    MR=AR*MY1(SS);                  {-(ya-xb-yc+yd)(-Sb)}
                    MR=MR+MX1*MY0(SS),MX1=DM(I3,M0);
                    SB=EXPADJ MR1,DM(I1,M1)=MR1;    {yb´=(ya-xb-yc+xd)Cb}
                    AR=AX1+AF,MY0=DM(I6,M4);        {+ (xa+yb-xc-yd)(-Sb)}
                    AR=AR-AY1,MY1=DM(I6,M7);
                    MR=MX1*MY0(SS);
                    MR=MR-AR*MY1(SS);
                    SB=EXPADJ MR1,DM(I3,M1)=MR1;    {xd´=(xa-yb-xc+yd)Cd}
                    MR=AR*MY0(SS),MY0=DM(I5,M4);    {- (ya+xb-yc-xd)(-Sd)}
                    MR=MR+MX1*MY1(SS);
bfly_loop:          SB=EXPADJ MR1,DM(I3,M1)=MR1;    {yd´= (ya+xb-yc-xd)Cd}
                                                    {+ (xa-yb-xc+yd)(-Sd)}
                 MODIFY(I0,M2);            {I0 -->1st xa of next group}
                 MODIFY(I1,M2);            {I1 -->1st xb of next group}
                 MODIFY(I2,M3);
                 MODIFY(I2,M2);            {I2 -->1st xc of next group}
group_loop:   MODIFY(I3,M2);               {I3 -->1st xd of next group}
              CALL bfp_adjust;             {Check for bit growth}
              SI=DM(groups);               {SI=groups}
              SR=ASHIFT SI BY 2(HI);       {SR1=groups × 4}
```

```
                DM(groups)=SR1;                  {Group count, next stage}
                SI=DM(bflys_per_group);          {SI=bflys_per_group}
                SR=ASHIFT SI BY -1(HI);          {SR1=bflys_per_group ÷ 2}
                DM(node_space)=SR1;              {Node spacing, next stage}
                SR=ASHIFT SI BY -1(HI);          {SR1=node_space ÷ 2}
stage_loop:     DM(bflys_per_group)=SR1;         {Butterfly count, next stage}
            RTS;
.ENDMOD;
```

Listing 6.26 Radix-4 DIF FFT Routine, Conditional Block Floating-Point Scaling

A routine similar to the *dit_radix-2_bfp_adjust* routine is used to monitor bit growth in the radix-4 FFT. Because a radix-4 butterfly can cause data to grow by three bits from input to output, the radix-2 block floating-point routine is modified to adjust for three bits instead of two. The *dif_radix-4_bfp_adjust* routine is shown in Listing 6.27. This routine performs block floating-point adjustment on the radix-4 DIF FFT stage output.

The *dif_radix-4_bfp_adjust* routine checks for growth of three bits as well as for zero, one and two bits. This routine shifts data (by one, two or three bits to the right) using the shifter. As described above, shifting right by multiplication allows rounding of the shifted bit(s). However, multiplication is not always possible. This routine illustrates the use of the shifter.

```
.MODULE     dif_radix_4_bfp_adjust;

{           Calling Parameters
                Radix-4 DIF FFT stage results in inplacedata

            Return Values
                inplacedata adjusted for bit growth

            Altered Registers
                I0,I1,AX0,AY0,AR,SE,SI,SR

            Altered Memory
                inplacedata, blk_exponent
}

.CONST      N_x_2=2048;

.EXTERNAL   inplacedata, blk_exponent;

.ENTRY      bfp_adjust;
```

(listing continues on next page)

```
bfp_adjust: AY0=CNTR;
            AR=AY0-1;
            IF EQ RTS;                 {If last stage, return}
            AY0=-3;
            AX0=SB;
            AR=AX0-AY0;
            IF EQ RTS;                 {If SB=-3, no bit growth, return}
            AY0=-2;
            SE=-1;
            I0=^inplacedata;           {I0=read pointer}
            I1=^inplacedata;           {I1=write pointer}
            AR=AX0-AY0,SI=DM(I0,M1);   {Check SB, get 1st sample}
            IF EQ JUMP strt_shift;     {If SB=-2, shift block right 1 bit}
            AY0=-1;
            SE=-2;
            AR=AX0-AY0;
            IF EQ JUMP strt_shift;     {If SB=-1, shift block right 2 bits}
            SE=-3;                     {Otherwise, SB=0, shift right 3 bits}
strt_shift: CNTR=N_x_2-1;
            AY0=SE;
            DO shift_loop UNTIL CE;
               SR=ASHIFT SI(LO),SI=DM(I0,M1);  {SR=shifted data, SI=next data}
shift_loop:    DM(I1,M1)=SR0;                  {Unshifted data=shifted data}
            SR=ASHIFT SI(LO);          {Shift last data word}
            AX0=DM(blk_exponent);      {Update block exponent and}
            DM(I1,M1)=SR0,AR=AX0-AY0;  {store last shifted sample}
            DM(blk_exponent)=AR;
            RTS;
.ENDMOD;
```

Listing 6.27 Radix-4 Block Floating-Point Scaling Routine

6.5.3 Digit Reversal

Whereas bit reversal reverses the order of bits in binary (base 2) numbers, digit reversal reverses the order of digits in quarternary (base 4) numbers. Every two bits in the binary number system correspond to one digit in the quarternary number system. (For example, binary 1110 = quarternary 32.) The quarternary system is illustrated below for decimal numbers 0 through 15.

Decimal	*Binary*	*Quarternary*
0	0000	00
1	0001	01
2	0010	02
3	0011	03
4	0100	10
5	0101	11
6	0110	12
7	0111	13
8	1000	20
9	1001	21
10	1010	22
11	1011	23
12	1100	30
13	1101	31
14	1110	32
15	1111	33

The radix-4 DIF FFT successively divides a sequence into four subsequences, resulting in an output sequence in digit-reversed order. A digit-reversed sequence is unscrambled by digit-reversing the data positions. For example, position 12 in quarternary (six in decimal) becomes position 21 in quarternary (nine in decimal) after digit reversal. Therefore, data in position six is moved to position nine when the digit-reversed sequence is unscrambled. The digit-reversed positions for a 16-point sequence (samples X(0) through X(15)) are shown on the next page.

Sample, Sequential Order	*Sequential Location* decimal	quarternary	*Digit-Reversed Location* decimal	quarternary	*Sample, Digit-Reversed Order*
X(0)	0	00	0	00	X(0)
X(1)	1	01	4	10	X(4)
X(2)	2	02	8	20	X(8)
X(3)	3	03	12	30	X(12)
X(4)	4	10	1	01	X(1)
X(5)	5	11	5	11	X(5)
X(6)	6	12	9	21	X(9)
X(7)	7	13	13	31	X(13)
X(8)	8	20	2	02	X(2)
X(9)	9	21	6	12	X(6)
X(10)	10	22	10	22	X(10)
X(11)	11	23	14	32	X(14)
X(12)	12	30	3	03	X(3)
X(13)	13	31	7	13	X(7)
X(14)	14	32	11	23	X(11)
X(15)	15	33	15	33	X(15)

In an N-point radix-4 FFT, only the number of digits needed to represent N locations are reversed. Two digits are needed for a 16-point FFT, three digits for a 64-point FFT, and five digits for a 1024-point FFT.

The digit reversal subroutine that unscrambles the output sequence for the radix-4 DIF FFT is described later in the next section. This routine works with the optimized radix-4 FFT. A similar routine can be used for the unoptimized program.

6.6 OPTIMIZED RADIX-4 DIF FFT

This section explores changes to the radix-4 FFT program to increase its execution speed. Specifically, changes in the first and last stages, data structures and program flow are discussed.

6.6.1 First Stage Modifications

In the first stage, there are N/4 butterflies and only one group and therefore the group loop is not needed. The butterfly loop and the group loop can be combined into one. Because each loop requires several setup instructions to initialize the counter and other registers, combining the two loops enhances performance. The code for these combined loops is shown in Listing 6.28.

```
{———————— Stage 1 ————————————}


stage1:  I0=^inplace;                {in ->Xa,Xc}
         I1=^inplace+Nov2;           {in+N/2 ->Xb,Xd}
         I2=^inplace+1;              {in+1 ->Ya,Yc}
         I3=^inplace+Nov2+1;         {in+N/2+1 ->Yb,Yd}
         I5=^cos_table;
         I6=^cos_table;
         I7=^cos_table;

         M0=N;                       {N,    skip forward to dual node}
         M1=-N;                      {-N,   skip back to primary node}
         M2=-N+2;                    {-N+2, skip to next butterfly}
         M5=Nov4+1;          {N/4 + groups/stage*1, Cb Sb offset}
         M6=Nov4+2;          {N/4 + groups/stage*2, Cc Sc offset}
         M7=Nov4+3;          {N/4 + groups/stage*3, Cd Sc offset}
         AX0=DM(I0,M0);              {get first Xa}
         AY0=DM(I0,M1);              {get first Xc}
         AR=AX0-AY0, AX1=DM(I2,M0); {Xa-Xc,get first Ya}
         SR=LSHIFT AR(LO), AY1=DM(I2,M1); {SR1=Xa-Xc,get first Yc}
         CNTR=Nov4;                  {Bfly/group, stage one}
         DO stg1bfy UNTIL CE;

         <butterfly code here>

stg1bfy:

{———————— end Stage 1 ————————————}
```

Listing 6.28 First Stage with Combined Butterfly and Group Loops

6.6.2 Last Stage Modifications

In the last stage, all the twiddle factors are 1, therefore the multiplications can be removed from the butterfly. The butterfly equations reduce to the following:

$$x_a' = x_a + x_b + x_c + x_d$$

$$y_a' = y_a + y_b + y_c + y_d$$

$$x_b' = x_a - x_c + y_b - y_d$$

$$y_b' = y_a - y_c - x_b + x_d$$

$$x_c' = x_a - x_b + x_c - x_d$$

$$y_c' = y_a - y_b + y_c - y_d$$

$$x_d' = x_a - x_c - y_b + y_d$$

$$y_d' = y_a - y_c + x_b - x_d$$

These reduced equations can be computed using a simplified butterfly algorithm. This speed improvement entails a separate butterfly module for the last stage.

The general set of butterfly equations (including twiddle factor multiplications) can be implemented in 30 cycles using the code in Listing 6.29. On the other hand, the simplified equations of the last stage can be computed in 20 cycles using the code in Listing 6.30. A modest increase in program memory requirements is traded for a significant increase in the performance of this core loop.

```
DO stg1bfy UNTIL CE;
   AR=AX0+AY0, AX0=DM(I1,M0);          {AR=xa+xc, AX0=xb}
   MR0=AR, AR=AX1+AY1;                 {MR0=xa+xc, AR=ya+yc}
   MR1=AR, AR=AX1-AY1;                 {MR1=ya+yc, AR=ya-yc}
   SR=SR OR LSHIFT AR (HI), AY0=DM(I1,M1);
                                       {SR1=ya-yc, AY0=xd}
   AF=AX0+AY0, AX1=DM(I3,M0);          {AF=xb+xd, AX1=yb}
   AR=MR0+AF, AY1=DM(I3,M1);           {AR=xa+xb+xc+xd, AY1=yd}
   DM(I0,M0)=AR, AR=MR0-AF;
                     {output x'a=(xa+xb+xc+xd), AR=xa+xc-xb-xd}
   AF=AX1+AY1, MX0=AR;                 {AF=yb+yd, MX0=xa+xc-xb-xd}
   AR=MR1+AF, MY0=DM(I6,M4);           {AR=ya+yc+yb+yd, MY0=(Cc)}
```

```
               DM(I2,M0)=AR, AR=MR1-AF;          {output y'a, AR=ya+yc-yb-yd}
               MR=MX0*MY0(SS), MY1=DM(I6,M6);
                                           {MR=(xa+xc-xb-xd)(Cc), MY1=(Sc)}
               MR=MR+AR*MY1(RND);          {MR=(xa-xb+xc-xd)(Cc)+(ya-yb+yc-yd)(Sc)}
               DM(I0,M2)=MR1, MR=AR*MY0(SS);
                          {output x'c=xa-xb+xc-xd)(Cc)+(ya-yb+yc-yd)(Sc)}
                                                 {MR=(ya+yc-yb-yd)(Cc)}
               MR=MR-MX0*MY1(RND), MY0=DM(I5,M4);
                       {MR=(ya+yc-yb-yd)(Cc)-(xa+xc-xb-xd)(Sc), MY0=(Cb)}
               DM(I2,M2)=MR1, AR=AX0-AY0;
                       {output y'c=(ya+yc-yb-yd)(Cc)-(xa+xc-xb-xd)(Sc)}
                                                 {AR=xb-xd}
               AY0=AR, AF=AX1-AY1;               {AY0=xb-xd, AF=yb-yd}
               AR=SR0-AF, MY1=DM(I5,M5);         {AR=xa-xc-(yb-yd), MY1=(Sb)}
               MX0=AR, AR=SR0+AF;                {MX0=xa-xc-yb+yd, AR=xa-xc+yb-yd}
               SR0=AR, AR=SR1+AY0;         {SR0=xa-xc+yb-yd, AR=ya-yc+xb-xd}
               MX1=AR, AR=SR1-AY0;         {MX1=ya-yc+xb-xd, AR=ya-yc-(xb-xd)}
               MR=SR0*MY0(SS), AX0=DM(I0,M0);
                          {MR=(xa-xc+yb-yd)(Cb), AX0=xa of next bfly}
               MR=MR+AR*MY1(RND), AY0=DM(I0,M1);
                          {MR=(xa-xc+yb-yd)(Cb)+(ya-yc-xb+xd)(Sb)}
                                                 {AY0=xc of next bfly}
               DM(I1,M0)=MR1, MR=AR*MY0(SS);
                       {output x'b=(xa-xc+yb-yd)(Cb)+(ya-yc-xb+xd)(Sb)}
                                                 {MR=ya-yc-xb+xd)(Cb)}
               MR=MR-SR0*MY1(RND), MY0=DM(I7,M4);
                          {MR=(ya-yc-xb+xd)(Cb)-(xa-xc+yb-yd)(Sb)}
                                                 {MY0=(Cd)}
               DM(I3,M0)=MR1, MR=MX0*MY0(SS);
                       {output y'b=(ya-yc-xb+xd)(Cb)-(xa-xc+yb-yd)(Sb)}
                                                 {MR=(xa-yb-xc+yd)(Cd)}
               MY1=DM(I7,M7), AR=AX0-AY0;        {MY1=(Sd), AR=xa-xc}
               MR=MR+MX1*MY1(RND), AX1=DM(I2,M0);
                          {MR=(xa-yb-xc+yd)(Cd)+(ya+xb-yc-xd)(Sd)}
                                                 {AX1=ya of next bfly}
               DM(I1,M2)=MR1, MR=MX1*MY0(SS);
                       {output x'd=(xa-yb-xc+yd)(Cd)+(ya+xb-yc-xd)(Sd)}
                                                 {MR=(ya+yb-yc-yd)(Cd)}
               MR=MR-MX0*MY1(RND), AY1=DM(I2,M1);
                               {MR=(ya+yb-yc-yd)(Cd)-(xa-xc-yb+yd)Sd}
                                                 {yc of next bfly}
stg1bfy:       DM(I3,M2)=MR1, SR=LSHIFT AR(LO);
                          {output y'd=(ya+xb-yc-xd)Cd-(xa-xc-yb+yd)Sd}
                                                 {SR0=ya-yc of next bfly}
```

Listing 6.29 Unmodified Butterfly

```
            DO laststgbfy UNTIL CE;
               AR=AX0-AY0, AX1=DM(I2,M0);        {AR=xa-xc, AX1=ya}
               SR=LSHIFT AR(LO), AY1=DM(I2,M1); {SR0=xa-xc, AY1=yc}
               AR=AX0+AY0, AX0=DM(I1,M0);        {AR=xa+xc, AX0=xb}
               MR0=AR, AR=AX1+AY1;               {MR0=xa+xc, AR=ya+yc}
               MR1=AR, AR=AX1-AY1;               {MR1=ya+yc, AR=ya-yc}
               SR=SR OR LSHIFT AR (HI), AY0=DM(I1,M1); {SR1=ya-yc, AY0 xd}
               AF=AX0+AY0, AX1=DM(I3,M0);        {AF=xb+xd, AX1=yb}
               AR=MR0+AF, AY1=DM(I3,M1);         {AR=xa+xc+xb+xd, AY1=yd}
               DM(I0,M0)=AR, AR=MR0-AF;
                       {output x'a=xa+xc+xb+xd, AR=xa+xc-(xb+xd)}
               DM(I0,M0)=AR, AF=AX1+AY1; {output x'c=xa+xc-(xb+xd), AF=yb+yd}
               AR=MR1+AF;                        {AR=ya+yb+yc+yd}
               DM(I2,M0)=AR, AR=MR1-AF;
                       {output y'a=ya+yc+yb+yd, AR=ya+yc-(yb+yd)}
               DM(I2,M0)=AR, AR=AX0-AY0; {output y'c=ya+yc-(yb+yd), {AR=xb-xd}
               AX0=DM(I0,M0);                    {AX0=xa of next group}
               AF=AX1-AY1, AY1=AR;               {AF=yb-yd, AY1=xb-xd}
               AR=SR0+AF, AY0=DM(I0,M1);
                          {AR=xa-xc+yb-yd, AY0=xc of next group}
               DM(I1,M0)=AR, AR=SR0-AF;
                          {output x'b=xa-xc+yb-yd, AR=xa-xc-(yb-yd)}
               DM(I1,M0)=AR, AR=SR1-AY1;
                       {output x'd=xa-xc-(yb-yd), AR=ya-yc+(xb-xd)}
               DM(I3,M0)=AR, AR=SR1+AY1;
                       {output y'b=ya-yc+(xb-xd), AR=ya-yc-(xb-xd)}
laststgbfy:    DM(I3,M0)=AR;                     {output y'd=ya-yc-(xb-xd)}
```

Listing 6.30 Simplified Last Stage Butterfly

In addition to having its own butterfly algorithm, the last stage has only one butterfly in each of its N/4 groups, so the group and butterfly loops can be combined into one, reducing the loop setup overhead.

6.6.3 Program Structure Modifications

While the nested loop structure is efficient in terms of program storage requirements, it does not take advantage of the unique properties of the first and last stages as outlined above. In implementing the modifications to these stages it becomes convenient to restructure the algorithm. The outermost loop (stage loop) can be removed and the stage setup instructions can be done sequentially, in a nonrecursive manner. Each stage can have its own setup instructions followed by a call to a subroutine that executes the remaining two nested loops (group and butterfly loops). The program with the modified structure is shown in Listing 6.31.

```
{————————— Stage 1 —————————}

stage1:     I0=^inplace;                         {in ->Xa,Xc}
            I1=^inplace+Nov2;                    {in+N/2 ->Xb,Xd}
            I2=^inplace+1;                       {in+1 ->Ya,Yc}
            I3=^inplace+Nov2+1;                  {in+N/2+1 ->Yb,Yd}
            I5=^cos_table;
            I6=^cos_table;
            I7=^cos_table;

            M0=N;                         {N,    skip forward to dual node}
            M1=-N;                        {-N,   skip back to primary node}
            M2=-N+2;                      {-N+2, skip to next butterfly}
            M5=Nov4+1;                    {N/4+groups/stage*1, Cb Sb offset}
            M6=Nov4+2;                    {N/4+groups/stage*2, Cc Sc offset}
            M7=Nov4+3;                    {N/4+groups/stage*3, Cd Sc offset}
            AX0=DM(I0,M0);                {get first Xa}
            AY0=DM(I0,M1);                {get first Xc}
            AR=AX0-AY0, AX1=DM(I2,M0);    {Xa-Xc,get first Ya}
            SR=LSHIFT AR(LO), AY1=DM(I2,M1);    {SR1=Xa-Xc,get first Yc}
            CNTR=Nov4;                    {Bfly/group, stage one}
            DO stg1bfy UNTIL CE;

                < butterfly code here >

stg1bfy:

{————————— Stage 2 —————————}

stage2:     I0=^inplace;         {in -> Xa,Xc}
            I1=^inplace+128;     {in+N/8 -> Xb,Xd}
            I2=^inplace+1;       {in+1 -> Ya,Yc}
            I3=^inplace+129;     {in+N/8+1 -> Yb,Yd}
            M0=Nov4;             {N/4, skip forward to dual node}
            M1=-Nov4;            {-N/4, skip back to primary node}
            M2=-Nov4+2;          {-N/4+2, skip to next butterfly}
            M3=384;              {N*3/8,  skip to next group}
            M5=Nov4+4;           {N/4+groups/stage*1, Cb Sb offset}
            M6=Nov4+8;           {N/4+groups/stage*2, Cc Sc offset}
            M7=Nov4+12;          {N/4+groups/stage*3, Cd Sd offset}
            SI=64;               {Bfly/group, save counter for inner loop}
            DM(bfy_count)=SI;
            CNTR=4;              {groups/stage}
            CALL mid_stg;        {do stage 2}
```

(listing continues on next page)

```
{————————— Stage 3 —————————}

stage3:     I0=^inplace;            {in -> Xa,Xc}
            I1=^inplace+32;         {in+N/32 -> Xb,Xd}
            I2=^inplace+1;          {in+1 -> Ya,Yc}
            I3=^inplace+33;         {in+N/32+1 -> Yb,Yd}
            M0=64;                  {N/16, skip forward to dual node}
            M1=-64;                 {-N/16, skip back to primary node}
            M2=-62;                 {-N/16+2, skip to next butterfly}
            M3=96;                  {N*3/32, skip to next group}
            M5=Nov4+16;             {N/4+groups/stage*1, Cb Sb offset}
            M6=Nov4+32;             {N/4+groups/stage*2, Cc Sc offset}
            M7=Nov4+48;             {N/4+groups/stage*3, Cd Sd offset}
            SI=16;                  {Bfly/group, save counter for inner loop}
            DM(bfy_count)=SI;
            CNTR=16;                {groups/stage}
            CALL mid_stg;           {do stage 3}

{————————— Stage 4 —————————}

stage4:     I0=^inplace;            {in -> Xa,Xc}
            I1=^inplace+8;          {in+N/128 -> Xb,Xd}
            I2=^inplace+1;          {in+1 -> Ya,Yc}
            I3=^inplace+9;          {in+N/128+1 -> Yb,Yd}
            M0=16;                  {N/64, skip forward to dual node}
            M1=-16;                 {-N/64, skip back to primary node}
            M2=-14;                 {-N/64+2, skip to next butterfly}
            M3=24;                  {N*3/128, skip to next group}
            M5=Nov4+64;             {N/4 +groups/stage*1, Cb Sb offset}
            M6=Nov4+128;            {N/4 +groups/stage*2, Cc Sc offset}
            M7=Nov4+192;            {N/4 +groups/stage*3, Cd Sd offset}
            SI=4;                   {Bfly/group, save counter inner loop}
            DM(bfy_count)=SI;
            CNTR=64;                {groups/stage}
            CALL mid_stg;           {do stage 4}


{———— Last Stage (No Multiplies) ————}


laststage:  I0=^inplace;            {in ->Xa,Xc}
            I1=^inplace+2;          {in+N/512 ->Xb,Xd}
            I2=^inplace+1;          {in+1 ->Ya,Yc}
            I3=^inplace+3;          {in+N/512+1 ->Yb,Yd}
            M0=4;                   {N/256, skip forward to dual node}
            M1=-4;                  {-N/256, skip back to primary node}
```

```
        AX0=DM(I0,M0);          {first Xa}
        AY0=DM(I0,M1);          {first Xc}
        CNTR=Nov4;              {groups/stage}
        DO laststgbfy UNTIL CE;

            < last stage butterfly code here >

laststgbfy:
```

Listing 6.31 Modified Program Structure

Program size is increased in exchange for clarity and flexibility. The nonrecursive structure allows future modifications to be incorporated more readily than does the 3-nested-loop structure, because the changes can be applied to a particular stage without affecting the other stages.

6.6.4 Data Structure Modifications

6.6.4.1 Cosine Table

In the unoptimized FFT, real and imaginary parts of the twiddle factors are stored separately in two array structures, each of length N. In the optimized FFT, a single array of size N is used to store the real values (cosine values). Using an addressing modify value of –N/4 (–π/2 in terms of angular displacement), the imaginary values (sine values) can be derived from the cosine table. This is based on the trigonometric identity

$$\sin(x) = \cos(x-\pi/2)$$

This modification results in a 50% improvement in program data memory requirements.

For example, $W_{1024}^{256} = \cos[256] - j\sin[256]$
$= \cos[256] - j\cos[0]$

where cos[x] is defined as $\cos(2\pi x/N)$. The structure of the modified table is shown in Figure 6.13, on the next page.

6.6.4.2 In-Place Array

The computations of the FFT are performed in place, that is, the output data occupy the same memory locations as the input data (the *inplace* array). This approach simplifies indexing and reduces the amount of memory required. Input samples are stored in the 2N array *inplace;* the

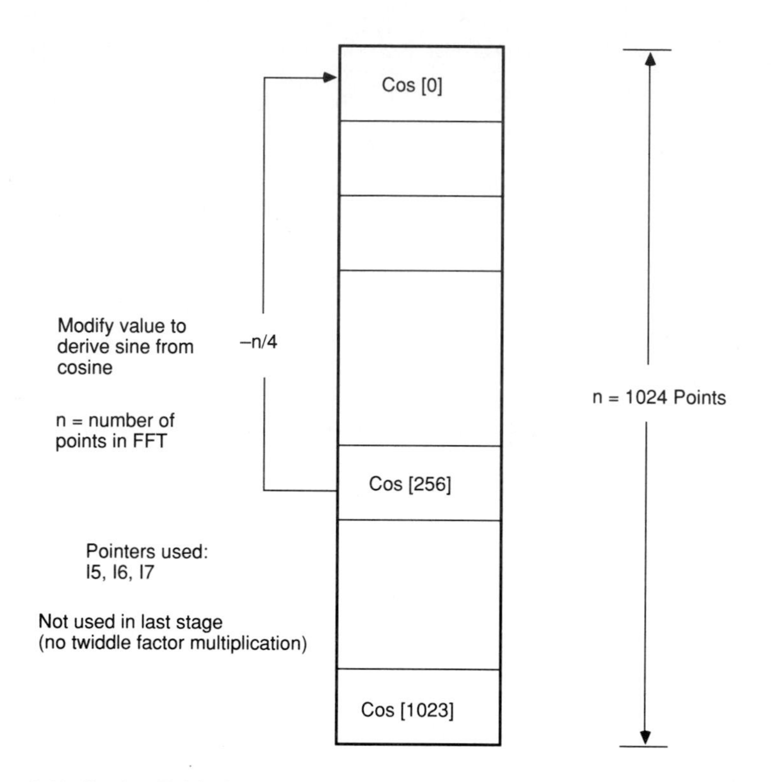

Figure 6.13 Cosine Table for 1024-Point FFT

real and imaginary values are interleaved, as shown in Figure 6.14. The program uses pointers I0, I1, I2 and I3 to read and write this array, except in the modified last stage, where I4, I5, I6 and I7 are used also (all pointers are used in last stage).

6.6.5 Digit-Reversing

In an in-place DIF FFT computation, the output is not in sequential order. Repeatedly subdividing the input data sequences produces a scrambled (nonsequentially-ordered) output. For a radix-4 FFT, the output is in digit-reversed order.

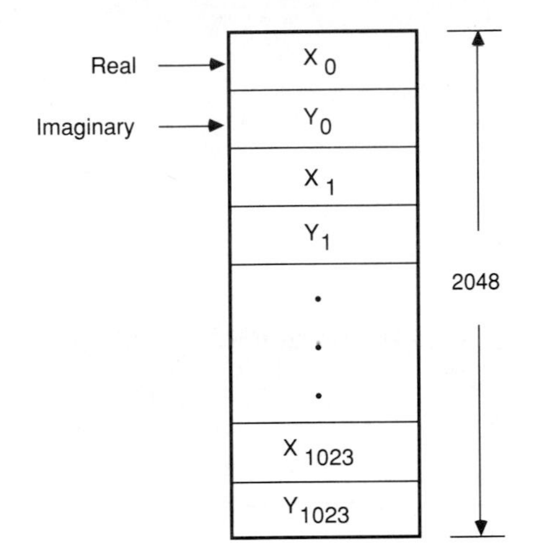

Figure 6.14 Interleaved Structure

Two methods to unscramble the digit-reversed output into sequential order are outlined below:

- A separate routine unscrambles the FFT output. This routine reads the digit-reversed output and writes it back in sequential order. It has the disadvantage of adding to the total execution time of the program.
- In a radix-4 algorithm, the butterfly can be modified to to produce a bit-reversed FFT output instead of a digit-reversed output. Combining this with the built-in bit-reverse mode of the ADSP-2100 processor family, the FFT routine can produce sequentially-ordered (unscrambled) output. This method does not affect program size or execution time.

6.6.5.1 Unscrambling Routine

The unscrambling routine in Listing 6.32 uses the bit-reverse mode of the ADSP-2100 with shift operations to digit-reverse an input array of size N. The routine uses a 7-instruction core loop and unscrambles a complex pair in seven cycles. This post-unscrambling approach adds 19% to 34% more computation time to the FFT, depending on the size of the input sequence.

```
.MODULE/BOOT=0  drev_1k;

{This routine unscrambles radix-4 dif fft results out of place}
{A 7-instruction core loop unscrambles a complex pair in 7 cycles}
{An additional 20 cycles for setup are required; total processing}
{time for 1024 points is 7188 cycles or 0.57504ms @ 80ns/cycle}

.CONST          N=1024;
.CONST          Nx2=2048;
.CONST          log2N=10;
.CONST          chkrbd_e=H#2AAA;     {even bit mask}
.CONST          chkrbd_o=H#1555;     {odd bit mask}
.CONST          base_adr=H#0004;     {results base, H#0800=2K bit rev}
.CONST          evenshift=14 - log2N - 2;
.CONST          oddshift=14 - log2N;
.VAR/ABS=2048   drev_out[Nx2];       {output at 2K}
.GLOBAL         drev_out;
.EXTERNAL       inplace;
.ENTRY          drev;

drev:   ENA BIT_REV;
        M1=H#2000;          {modify by one in b_rev mode}
        M3=base_adr;        {bit-rev base adr of output buffer}
        I0=base_adr;        {initialized, for first write is wrapped}
        L0=0;
        I4=^inplace;        {I4 points to scrambled fft results}
        M4=1;
        L4=0;
        AX0=chkrbd_e;       {used to isolate even index bits}
        AX1=chkrbd_o;       {used to isolate odd index bits}
        AY0=-1;             {initialize index for wrapped code}
        SE=evenshift;       {core loop even shift is not immediate}
        AF=PASS AY0;        {store index count in AF}
        CNTR=N;             {process N complex pairs}
        DO digit UNTIL CE;
           AF=AF+1,DM(I0,M1)=MX0;  {inc index, store b_rev real val}
           AR=AX0 AND AF,DM(I0,M1)=MX1;
                       {isolate even bits, store b_rev imag}
           SR=LSHIFT AR(HI),MX0=DM(I4,M4);
                       {shift for b_rev index, get next real}
           AR=AX1 AND AF,MX1=DM(I4,M4);
                       {isolate odd bits,get imag val}
           SR=SR OR LSHIFT AR BY oddshift (HI);   {OR for b_rev index}
           I0=SR1;            {store b_rev index}
digit:     MODIFY(I0,M3);     {compute b_rev adr}
```

```
        DM(I0,M1)=MX0;          {store last b_rev real val}
        DM(I0,M1)=MX1;          {store last b_rev imag val}
        DIS BIT_REV;
        RTS;
.ENDMOD;
```

Listing 6.32 Digit-Reverse (Unscrambling) Routine

6.6.5.2 Modified Butterfly

In a radix-4 in-place algorithm, interchanging the middle two branches of every butterfly computation results in a bit-reversed output (and not a digit-reversed output). Subsequently, the ENA BIT_REV instruction (enable bit-reverse mode) of the ADSP-2100 can be invoked to bit-reverse the output sequence as it is being written out. Since the output is already in bit-reversed order, invoking the bit-reverse mode in the last stage actually puts the output in sequential order. Interchanging the middle two branches of the butterfly is depicted in Figure 6.15.

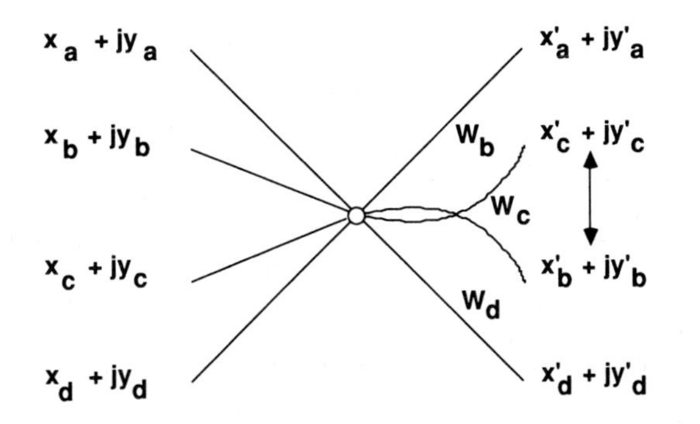

Figure 6.15 Butterfly with Interchange

In implementing this change, computations are done in the same order as before. However, in writing the results to memory we swap the middle two branches:

x_b' to position x_c', x_c' to position x_b'
y_b' to position y_c', y_c' to position y_b'

Listing 6.33 shows the butterfly code with the two branches interchanged.

```
{—————————— Stage 1 ——————————}

stage1: I0=^inplace;                          {in ->Xa,Xc}
        I1=^inplace+Nov2;                     {in+N/2 ->Xb,Xd}
        I2=^inplace+1;                        {in+1 ->Ya,Yc}
        I3=^inplace+Nov2+1;                   {in+N/2+1 ->Yb,Yd}
        I5=^cos_table;
        I6=^cos_table;
        I7=^cos_table;
        M0=N;                                 {N, skip forward to dual node}
        M1=-N;                          {-N, skip back to primary node}
        M2=-N+2;                        {-N+2, skip to next butterfly}
        M5=Nov4+1;                      {N/4 + groups/stage*1, Cb Sb offset}

        M3=-2;              {Because we have modified the middle branches}
                            {of bfly, pointers for I0 have to be treated a}
                            {little more carefully. This requires a more}
                            {complex pointer manipulation, using M3.}

        M6=Nov4+2;                      {N/4 + groups/stage*2, Cc Sc offset,
        M7=Nov4+3;                      {N/4 + groups/stage*3, Cd Sc offset}
        AX0=DM(I0,M0);                        {get first Xa}
        AY0=DM(I0,M1);                        {get first Xc}
        AR=AX0-AY0, AX1=DM(I2,M0);            {Xa-Xc,get first Ya}
        SR=LSHIFT AR(LO), AY1=DM(I2,M1);      {SR1=Xa-Xc,get first Yc}
        CNTR=Nov4;                            {Bfly/group, stage one}

{Middle 2 branches of butterfly are reversed.}
{This alteration, done in every stage, results in bit-reversed}
{outputs instead of digit-reversed outputs.}

        DO stg1bfy UNTIL CE;
           AR=AX0+AY0, AX0=DM(I1,M0);         {AR=xa+xc, AX0=xb}
           MR0=AR, AR=AX1+AY1;                {MR0=xa+xc, AR=ya+yc}
           MR1=AR, AR=AX1-AY1;                {MR1=ya+yc, AR=ya-yc}
           SR=SR OR LSHIFT AR (HI), AY0=DM(I1,M1); {SR1=ya-yc, AY0=xd}
           AF=AX0+AY0, AX1=DM(I3,M0);         {AF=xb+xd, AX1=yb}
           AR=MR0+AF, AY1=DM(I3,M1);          {AR=xa+xb+xc+xd, AY1=yd}
           DM(I0,M0)=AR, AR=MR0-AF;
                      {output x'a=(xa+xb+xc+xd), AR=xa+xc-xb-xd}
           AF=AX1+AY1, MX0=AR;                {AF=yb+yd, MX0=xa+xc-xb-xd}
           AR=MR1+AF, MY0=DM(I6,M4);          {AR=ya+yc+yb+yd, MY0=(Cc)}
           DM(I2,M0)=AR, AR=MR1-AF;           {output y'a, AR=ya+yc-yb-yd}
```

```
                MR=MX0*MY0(SS), MY1=DM(I6,M6); {MR=(xa+xc-xb-xd)(Cc), MY1=(Sc)}
                MR=MR+AR*MY1(RND),SI=DM(I0,M2);
                                   {MR=(xa-xb+xc-xd)(Cc)+(ya-yb+yc-yd)(Sc)}
                          {SI here is a dummy to perform a modify(I0,M2)}
                DM(I1,M0)=MR1, MR=AR*MY0(SS);     {output x'c to position x'b}
                                                  {MR=(ya+yc-yb-yd)(Cc)}
                MR=MR-MX0*MY1(RND), MY0=DM(I5,M4);
                                   {MR=(ya+yc-yb-yd)(Cc)-(xa+xc-xb-xd)(Sc)}
                                                  {MY0=(Cb)}
                DM(I3,M0)=MR1, AR=AX0-AY0;
                                   {output y'c=to position y'b, AR=xb-xd}
                AY0=AR, AF=AX1-AY1;               {AY0=xb-xd, AF=yb-yd}
                AR=SR0-AF, MY1=DM(I5,M5);         {AR=xa-xc-(yb-yd), MY1=(Sb)}
                MX0=AR, AR=SR0+AF;          {MX0=xa-xc-yb+yd, AR=xa-xc+yb-yd}
                SR0=AR, AR=SR1+AY0;         {SR0=xa-xc+yb-yd, AR=ya-yc+xb-xd}
                MX1=AR, AR=SR1-AY0;         {MX1=ya-yc+xb-xd, AR=ya-yc-(xb-xd)}
                MR=SR0*MY0(SS),AX0=DM(I0,M0);
                          {MR=(xa-xc+yb-yd)(Cb), AX0=xa of next bfly}
                MR=MR+AR*MY1(RND),AY0=DM(I0,M3);
                          {MR=(xa-xc+yb-yd)(Cb)+(ya-yc-xb+xd)(Sb)}
                                                  {AY0=xc of next bfly}
                DM(I0,M2)=MR1, MR=AR*MY0(SS);
                     {output x'b=to position x'c, MR=ya-yc-xb+xd)(Cb)}
                MR=MR-SR0*MY1(RND), MY0=DM(I7,M4);
                     {MR=(ya-yc-xb+xd)(Cb)-(xa-xc+yb-yd)(Sb), MY0=(Cd)}
                DM(I2,M2)=MR1, MR=MX0*MY0(SS);
                     {output y'b to position y'c, MR=(xa-yb-xc+yd)(Cd)}
                MY1=DM(I7,M7), AR=AX0-AY0;        {MY1=(Sd), AR=xa-xc}
                MR=MR+MX1*MY1(RND), AX1=DM(I2,M0);
                          {MR=(xa-yb-xc+yd)(Cd)+(ya+xb-yc-xd)(Sd)}
                                                  {AX1=ya of next bfly}
                DM(I1,M2)=MR1, MR=MX1*MY0(SS);
                          {output x'd=(xa-yb-xc+yd)(Cd)+(ya+xb-yc-xd)(Sd)}
                                                  {MR=(ya+yb-yc-yd)(Cd)}
                MR=MR-MX0*MY1(RND), AY1=DM(I2,M1);
                       {MR=(ya+yb-yc-yd)(Cd)-(xa-xc-yb+yd)Sd, yc of next bfly}
stglbfy:        DM(I3,M2)=MR1, SR=LSHIFT AR(LO);
                          {output y'd=(ya+xb-yc-xd)Cd-(xa-xc-yb+yd)Sd}
                                                  {SR0=ya-yc of next bfly}
```

Listing 6.33 Butterfly with Middle Two Branches Interchanged

Because index registers are modified after every memory access, it is important to make sure that interchanging these middle branches does not affect the remainder of the butterfly. This requires special attention to make sure that memory fetches directly following the interchanged pair are not affected.

In this new addressing sequence we define an additional modify value, M3, and use it to mask out the effect of this interchange. Figure 6.16 illustrates how, for example, index register I0 is modified through one iteration of the code in Listing 6.33. The arrow numbers correspond to the order in which I0 is sequenced.

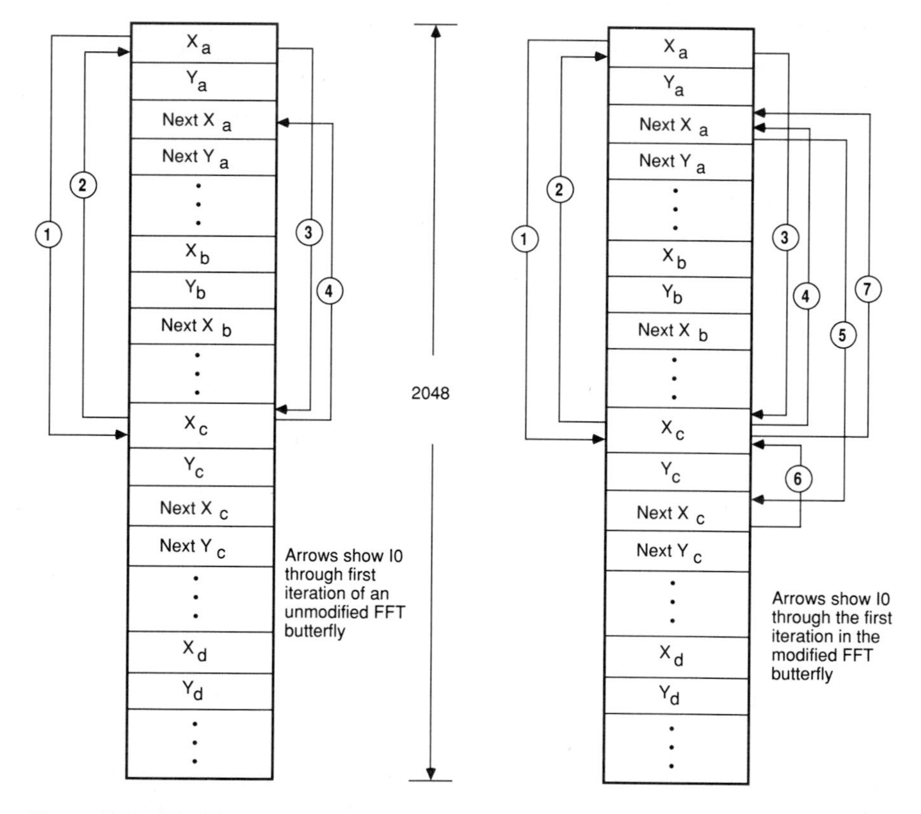

Figure 6.16 Modified Pointer Sequence

Since the algorithm calls two separate butterfly subroutines—one for the last stage (no multiplications) and one for the remainder of the stages—the modifications described here apply to both subroutines.

As a final step in the process, the output is sorted into sequential order using the bit-reverse mode of the ADSP-2100. This bit-reverse operation is concurrent with the execution of the last stage. The ENA BIT_REV instruction is executed at the start of the last stage. In this mode, the addresses of all memory accesses using I0, I1, I2 or I3 are bit-reversed and then placed on the address bus. Since this is done in the same clock cycle as the memory access itself, it generates no overhead.

The bit-reverse circuitry in the ADSP-2100 address generator reverses all 14 bits of the address value. Output sequences that are smaller than 2^{13} points require only $\log_2 2N$ (where N is the number of points in the FFT) bits to be reversed (2N because real and imaginary data are interleaved). For example, a 2K block of memory at H#0000 loaded with 1024 complex points, interleaved real and imaginary, in sequential order, requires reversing bits 10 through 1 of the 14-bit address value. Bit 0 is not reversed because data points are interleaved, so only every other address needs bit-reversing.

b13 b12 b11 **b10 b9 b8 b7 b6 b5 b4 b3 b2 b1** b0

becomes

b13 b12 b11 **b1 b2 b3 b4 b5 b6 b7 b8 b9 b10** b0

Bit-reversing fewer than 14 bits can be done on the ADSP-2100 by initializing I registers and an M register to appropriate values. The second column of Figure 6.17 (on the next page) shows the addresses that result from reversing bits 10 through 1 of the sequentially ordered addresses in the first column. The 14-bit addresses in the third column are the values (before bit-reverse) that the ADSP-2100 must generate in order to output the bit-reversed sequence in the second column. The addresses in the third column are obtained by initializing I0, I2, I1 and I3 to H#0000, H#2000, H#0008 and H#2008, respectively, and setting the M0 register (used to modify each of the I registers) to H#0010.

Sequentially ordered		*Bits 1 to 10 reversed*		*ADSP-2100 address (before bit-reverse)*	
data	address	data	address		
x0	00 0000 0000 0000	x0	00 0000 0000 0000	00 0000 0000 0000 = H#0000	I0
y0	00 0000 0000 0001	y0	00 0000 0000 0001	10 0000 0000 0000 = H#2000	I2
x1	00 0000 0000 0010	x512	00 0100 0000 0000	00 0000 0000 1000 = H#0008	I1
y1	00 0000 0000 0011	y512	00 0100 0000 0001	10 0000 0000 1000 = H#2008	I3
x2	00 0000 0000 0100	x256	00 0010 0000 0000	00 0000 0001 0000 = H#0010	I0+M0
y2	00 0000 0000 0101	y256	00 0010 0000 0001	10 0000 0001 0000 = H#2010	I2+M0
x3	00 0000 0000 0110	x768	00 0110 0000 0000	00 0000 0001 1000 = H#0018	I1+M0
y3	00 0000 0000 0111	y768	00 0110 0000 0001	10 0000 0001 1000 = H#2018	I3+M0
•		•		•	
•		•		•	
•		•		•	

Figure 6.17 Reversing Bits 10 Through 1

6.6.6 Variations

6.6.6.1 Inverse FFT

The inverse relationship for obtaining a sequence from its DFT is called the inverse DFT (IDFT). The transformation is described by the equation:

$$x(n) = 1/N \sum_{k=0}^{N-1} X(k)\, W_N^{-nk}$$

Although the FFT algorithms described in the chapter were presented in the context of computing the DFT efficiently, they may also be used in computing the IDFT.

The only difference between the two transformations is the normalization factor 1/N and the phase sign of the twiddle factor W_N. Consequently, an FFT algorithm for computing the DFT may be converted into an IFFT algorithm for computing the IDFT by using a reversed (upside down) twiddle factor table and by dividing the output of the algorithm by N.

6.6.6.2 Sizing the Program
Thus far, the discussion has been directed at the 1024-point FFT. It is possible with some changes to use the same code to transform 4s samples, where s is the number of stages required to do that. The modifications to resize the FFT program are as follows:

- Create input data files of new size.
- Generate twiddle factor table of new size.
- Modify symbolic constants in all modules used,
- Add/delete stages as required to satisfy the relation log4N=number of stages. Because the first and last stages are optimized for speed, additions or deletions are done to the center of the program.

In the program, all the stages with the exception of the optimized first and last have the same structure (see Listing 6.31). First, several instructions initialize the various registers, pointers and counters. Next, a subroutine that executes the group and butterfly loops is called and all the computations for that stage are carried out.

In reducing the size of the FFT, an entire stage block is deleted and the initial values of the remaining stages are recomputed using the new size. (The specific initial values for each stage as a function of size are documented in Listing 6.31.)

If the new size of the FFT is more than 1024 points, additional stages are required. This means another stage is written with setup instructions and a subroutine call identical to the other stages. This new stage is inserted in the middle of the FFT, and the initialization values for all the stages of the new program are recomputed following the documentation in Listing 6.31.

6.6.7 Programs and File Description
This section presents the files and the programs used in the optimized FFT example of this chapter. The flowchart below illustrates how the various files and modules are interrelated (see Figure 6.18 on the next page). Data files are shown in ovals and operations in rectangles.

6.6.7.1 Twiddle Factors
TWIDDLES.C is a C program that generates the data file for the cosine table (twiddle factor table). This hexadecimal data file (COS1024.DAT) is fully normalized in 1.15 format. The data on this file is loaded into memory through the .INIT directive during the linking process.

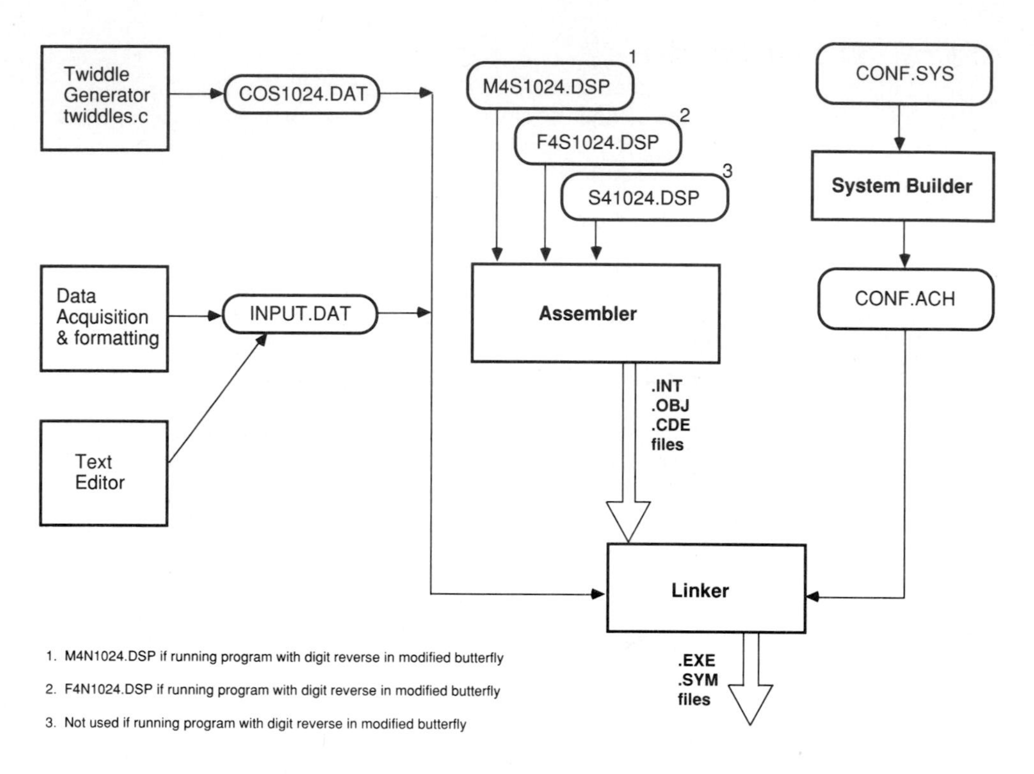

Figure 6.18 Program Generation Flowchart

6.6.7.2 Input Data

INPUT.DAT is a formatted (1.15 format, 11 guard bits) hexadecimal data file that contains the 1024 complex input samples (interleaved real and imaginary). The file is loaded into the *inplace* array through the .INIT directive during the linking process. In a real-time operation, the input samples would be loaded into RAM directly via a data acquisition board.

6.6.7.3 FFT Routines

The files F4S1024.DSP and F4N1024.DSP each contain a module fft that performs the FFT computations and forms the heart of the program. The module in F4S1024.DSP is called by the main routine in M4S1024.DSP and produces scrambled results which are passed to the *drev* routine in S41024.DSP to be unscrambled. The module in F4N1024.DSP is called by the main routine in M4N1024.DSP and produces normal (sequentially ordered) output at addresses H#0000-07FF.

6.6.7.4 FFT Program with Unscrambling Routine

Listing 6.34 shows the main module for the optimized radix-4 FFT, written for the ADSP-2101. This module is stored in the file M4S1024.DSP, which is the calling shell for the *fft* routine (in file F4S1024.DSP) and the *drev* unscrambling routine (in file S41024.DSP). The restart value for the PC is 0 and therefore the code starts immediately. The first three instructions

```
SI=0;
DM(H#3FFF)=SI;
DM(H#3FFE)=SI;
```

reset the system control register and the memory control register to allow zero wait state memory access (defaults to seven wait states).

```
.MODULE/RAM/BOOT=0/ABS=0            main;

{
Calling shell for the 1024-point radix-4 FFT with the unscrambling routine.

   fft  = FFT routine (in F4S1024.DSP)
   drev = unscrambling routine (in S41024.DSP)
}

.CONST      N=1024;
.CONST      Nx2=2048;
.VAR/DM     inplace[Nx2];      {inplace array contains original input}
                               {and also holds intermediate results}
.INIT       inplace:<input.dat>;      {load inplace array with data}
.GLOBAL     inplace;
.EXTERNAL   fft,drev;                 {2 routines used to perform FFT}



            SI=0;                     {These 3 lines reset the system}
            DM(H#3FFF)=SI;            {control register and the data}
            DM(H#3FFE)=SI;            {memory control register to allow}
                                      {zero state memory access}
            CALL fft;
            CALL drev;

trapper:    JUMP trapper;

.ENDMOD;
```

Listing 6.34 Main Module (ADSP-2101) for Radix-4 DIF FFT with Unscrambling Routine

In implementing the FFT on the ADSP-2100, the main module is modified in the following way. The ADSP-2100 has a PC restart value of 4 and an interrupt vector table in the 4-word address space 0000-0003. Therefore, the sequence of instructions below is required at the start of the module. The four RTI instructions ensure a NOP in case of a false interrupt signal.

```
RTI;    In the ADSP-2100 the first 4 PM locations
RTI;    are reserved for the interrupt vector
RTI;    table.
RTI;
```
< code starts here for ADSP-2100 >

The ADSP-2100 has no programmable wait states, so the first three instructions in Listing 6.34 should be removed. In addition, since the ADSP-2100 has no on-chip memory, the boot page select directive / BOOT=0 is not valid and should be removed.

6.6.7.5 FFT Program with Built-In Digit Reversal

Listing 6.35 contains the main routine for the optimized radix-4 FFT with built-in digit-reversal. Listing 6.36 contains the fft routine called by the main routine. Note that the unscrambling routine is not required here.

```
.MODULE/RAM/BOOT=0/ABS=0   main;

{Calling shell for 1024-point radix-4 FFT with built-in digit-reverse.

   fft = FFT routine (in F4N1024.DSP)}

.CONST              N=1024;            {number of samples in FFT}
.CONST              Nx2=2048;
.VAR/DM             inplace[Nx2]; {inplace array contains original input}
                                {and also holds intermediate results}
.VAR/DM/ABS=0       output[Nx2]; {output array holds results in order}
.INIT               inplace:<input.dat>;{load inplace array with data}
.GLOBAL             inplace,output;
.EXTERNAL           fft;               {FFT routine}

                    SI=0;              {these 3 lines reset the system}
                    DM(H#3FFF)=SI;     {control register and the data}
                    DM(H#3FFE)=SI;     {memory control register to allow}
                                       {zero wait state memory access}
                    CALL fft;
trapper:            JUMP trapper;

.ENDMOD;
```
Listing 6.35 Main Module for Radix-4 DIF FFT with Built-In Digit-Reversal

```
.MODULE/BOOT=0    fft_sub;

{
Optimized complex 1024-point radix-4 DIF FFT

This routine uses a modified radix-4 algorithm to unscramble results as
they are computed. The results are thus in sequential order.

Complex data is stored as x0 (real, imag), y0 (real, imag), x1 (real,
imag), y1 (real, imag),...

Butterfly terms
   xa = 1st real input leg          x'a = 1st real output leg
   xb = 2nd real input leg          x'b = 2nd real output leg
   xc = 3rd real input leg          x'c = 3rd real output leg
   xd = 4th real input leg          x'd = 4th real output leg
   ya = 1st imag input leg          y'a = 1st imag output leg
   yb = 2nd imag input leg          y'b = 2nd imag output leg
   yc = 3rd imag input leg          y'c = 3rd imag output leg
   yd = 4th imag input leg          y'd = 4th imag output leg

Pointers
   I0 -> xa,xc
   I1 -> xb,xd
   I2 -> ya,yc
   I3 -> yb,yd
   w0 (= Ca = Sa = 0)
   I5 -> w1 (1st Cb, - pi/4 for Sb)
   I6 -> w2 (2nd Cc, - pi/4 for Sc)
   I7 -> w3 (3rd Cd, - pi/4 for Sd)

Input
   inplace[2*N] normal order, interleaved real, imag.

Output
   inplace[2*N] normal order, interleaved real, imag.

Computation Time (cycles)
   setup   = 9
   stage 1 = 7700 = 20+256(30)
   stage 2 = 7758 = 18+4(15+64(30))
   stage 3 = 7938 = 18+16(15+16(30))
   stage 4 = 8658 = 18+64(15+4(30))
   stage 5 = 5140 = 20+256(20)

   Total 37203 cycles * 80ns/cycle = 2.97624ms}
```

(listing continues on next page)

```
.CONST          N = 1024;
.CONST          NX2 = 2048;
.CONST          Nov2 = 512;
.CONST          Nov4 = 256;
.CONST          N3ov8 = 384;
.VAR/DM/CIRC    cos_table[1024];

.VAR/DM         m3_space;   {memory space used to store M3 values}
                            {when M3 is loaded with its alternate value}

.VAR/DM         bfy_count;
.INIT           cos_table:<cos1024.dat>;   {N cosine values}
.EXTERNAL       inplace;

.ENTRY          fft;

fft:            M4=-Nov4;       {-N/4 = -90 degrees for sine}
                L0=0;
                L1=0;
                L2=0;
                L3=0;
                L5=%cos_table;
                L6=%cos_table;
                L7=%cos_table;
                SE=0;

{——————————— Stage 1 ———————————}

stage1:     I0=^inplace;            {in ->Xa,Xc}
            I1=^inplace+Nov2;       {in+N/2 ->Xb,Xd}
            I2=^inplace+1;          {in+1 ->Ya,Yc}
            I3=^inplace+Nov2+1;     {in+N/2+1 ->Yb,Yd}
            I5=^cos_table;
            I6=^cos_table;
            I7=^cos_table;
            M0=N;                   {N, skip forward to dual node}
            M1=-N;                  {-N,   skip back to primary node}
            M2=-N+2;                {-N+2, skip to next butterfly}

            M3=-2;      {Because we have modified the middle branches}
                        {of bfly, pointers for I0 require more}
                        {complex manipulation, using M3}

            M5=Nov4+1;              {N/4 + groups/stage*1, Cb Sb offset}
            M6=Nov4+2;              {N/4 + groups/stage*2, Cc Sc offset}
            M7=Nov4+3;              {N/4 + groups/stage*3, Cd Sc offset}
```

```
          AX0=DM(I0,M0);           {get first Xa}
          AY0=DM(I0,M1);           {get first Xc}
          AR=AX0-AY0, AX1=DM(I2,M0);     {Xa-Xc,get first Ya}
          SR=LSHIFT AR(LO), AY1=DM(I2,M1); {SR1=Xa-Xc,get first Yc}
          CNTR=Nov4;               {Bfly/group, stage one}

{Middle 2 branches of butterfly are reversed.}
{This alteration, dc .e in every stage, results in bit-reversed}
{outputs instead of digit-reversed outputs.}


          DO stg1bfy UNTIL CE;
             AR=AX0+AY0, AX0=DM(I1,M0); {AR=xa+xc, AX0=xb}
             MR0=AR, AR=AX1+AY1;        {MR0=xa+xc, AR=ya+yc}
             MR1=AR, AR=AX1-AY1;        {MR1=ya+yc, AR=ya-yc}
             SR=SR OR LSHIFT AR (HI), AY0=DM(I1,M1);
                                        {SR1=ya-yc, AY0=xd}
             AF=AX0+AY0, AX1=DM(I3,M0); {AF=xb+xd, AX1=yb}
             AR=MR0+AF, AY1=DM(I3,M1);  {AR=xa+xb+xc+xd, AY1=yd}
             DM(I0,M0)=AR, AR=MR0-AF;
                         {output x'a=(xa+xb+xc+xd), AR=xa+xc-xb-xd}
             AF=AX1+AY1, MX0=AR;        {AF=yb+yd, MX0=xa+xc-xb-xd}
             AR=MR1+AF, MY0=DM(I6,M4);  {AR=ya+yc+yb+yd, MY0=(Cc)}
             DM(I2,M0)=AR, AR=MR1-AF;   {output y'a, AR=ya+yc-yb-yd}
             MR=MX0*MY0(SS), MY1=DM(I6,M6);
                              {MR=(xa+xc-xb-xd)(Cc), MY1=(Sc)}
             MR=MR+AR*MY1(RND),SI=DM(I0,M2);
                         {MR=(xa-xb+xc-xd)(Cc)+(ya-yb+yc-yd)(Sc)}
                    {SI here is a dummy to perform a modify(I0,M2)}
             DM(I1,M0)=MR1, MR=AR*MY0(SS);
                                        {output x'c to position x'b}
                                        {MR=(ya+yc-yb-yd)(Cc)}
             MR=MR-MX0*MY1(RND), MY0=DM(I5,M4);
                   {MR=(ya+yc-yb-yd)(Cc)-(xa+xc-xb-xd)(Sc), MY0=(Cb)}
             DM(I3,M0)=MR1, AR=AX0-AY0;
                         {output y'c=to position y'b, AR=xb-xd}
             AY0=AR, AF=AX1-AY1;        {AY0=xb-xd, AF=yb-yd}
             AR=SR0-AF, MY1=DM(I5,M5);  {AR=xa-xc-(yb-yd), MY1=(Sb)}
             MX0=AR, AR=SR0+AF; {MX0=xa-xc-yb+yd, AR=xa-xc+yb-yd}
             SR0=AR, AR=SR1+AY0; {SR0=xa-xc+yb-yd, AR=ya-yc+xb-xd}
             MX1=AR, AR=SR1-AY0; {MX1=ya-yc+xb-xd, AR=ya-yc-(xb-xd)}
             MR=SR0*MY0(SS),AX0=DM(I0,M0);
                    {MR=(xa-xc+yb-yd)(Cb), AX0=xa of next bfly}
             MR=MR+AR*MY1(RND),AY0=DM(I0,M3);
                         {MR=(xa-xc+yb-yd)(Cb)+ (ya-yc-xb+xd)(Sb)}
                                        {AY0=xc of next bfly}
```

(listing continues on next page)

```
            DM(I0,M2)=MR1, MR=AR*MY0(SS);
                   {output x'b to position x'c, MR=ya-yc-xb+xd)(Cb)}
            MR=MR-SR0*MY1(RND), MY0=DM(I7,M4);
                   {MR=(ya-yc-xb+xd)(Cb)-(xa-xc+yb-yd)(Sb), MY0=(Cd)}
            DM(I2,M2)=MR1, MR=MX0*MY0(SS);
                   {output y'b to position y'c, MR=(xa-yb-xc+yd)(Cd)}
            MY1=DM(I7,M7), AR=AX0-AY0; {MY1=(Sd), AR=xa-xc}
            MR=MR+MX1*MY1(RND), AX1=DM(I2,M0);
                   {MR=(xa-yb-xc+yd)(Cd)+(ya+xb-yc-xd)(Sd)}
                                         {AX1=ya of next bfly}
            DM(I1,M2)=MR1, MR=MX1*MY0(SS);
                   {output x'd=(xa-yb-xc+yd)(Cd)+(ya+xb-yc-xd)(Sd)}
                                         {MR=(ya+yb-yc-yd)(Cd)}
            MR=MR-MX0*MY1(RND), AY1=DM(I2,M1);
                              {MR=(ya+yb-yc-yd)(Cd)-(xa-xc-yb+yd)Sd}
                                         {AY1=yc of next bfly}
stg1bfy:    DM(I3,M2)=MR1, SR=LSHIFT AR(LO);
                       {output y'd=(ya+xb-yc-xd)Cd-(xa-xc-yb+yd)Sd}
                                         {SR0=ya-yc of next bfly}

{——————— Stage 2 ———————}

stage2:     I0=^inplace;             {in -> Xa,Xc}
            I1=^inplace+128;         {in+N/8 -> Xb,Xd}
            I2=^inplace+1;           {in+1 -> Ya,Yc}
            I3=^inplace+129;         {in+N/8+1 -> Yb,Yd}
            M0=Nov4;                 {N/4, skip forward to dual node}
            M1=-Nov4;                {-N/4, skip back to primary node}
            M2=-Nov4+2;              {-N/4+2, skip to next butterfly}

            M3=384;                  {N*3/8, skip to next group}
            DM(m3_space)=M3;         {m3_space is temporary storage}
                                     {space needed because M3 is used}
                                     {in 2 contexts and will alternate}
                                     {in value}

            M5=Nov4+4;               {N/4+groups/stage*1, Cb Sb offset}
            M6=Nov4+8;               {N/4+groups/stage*2, Cc Sc offset}
            M7=Nov4+12;              {N/4+groups/stage*3, Cd Sd offset}
            SI=64;               {Bfy/group, save counter for inner loop}
            DM(bfy_count)=SI;  {SI is used as a temporary storage dummy}

            CNTR=4;                  {groups/stage}
            CALL mid_stg;            {do stage 2}
```

```
{——————————— Stage 3 ———————————}

stage3:     I0=^inplace;            {in -> Xa,Xc}
            I1=^inplace+32;         {in+N/32 -> Xb,Xd}
            I2=^inplace+1;          {in+1 -> Ya,Yc}
            I3=^inplace+33;         {in+N/32+1 -> Yb,Yd}
            M0=64;                  {N/16, skip forward to dual node}
            M1=-64;                 {-N/16, skip back to primary node}
            M2=-62;                 {-N/16+2, skip to next butterfly}
            M3=96;                  {N*3/32, skip to next group}

            DM(m3_space)=M3;        {M3_space is temporary storage}
                                    {space needed because M3 is used}
                                    {in 2 contexts and will alternate}
                                    {in value}

            M5=Nov4+16;             {N/4+groups/stage*1, Cb Sb offset}
            M6=Nov4+32;             {N/4+groups/stage*2, Cc Sc offset}
            M7=Nov4+48;             {N/4+groups/stage*3, Cd Sd offset}
            SI=16;             {Bfly/group, save counter for inner loop}
            DM(bfy_count)=SI;       {SI is a dummy temporary register}

            CNTR=16;                {groups/stage}
            CALL mid_stg;           {do stage 3}

{——————————— Stage 4 ———————————}

stage4:     I0=^inplace;            {in -> Xa,Xc}
            I1=^inplace+8;          {in+N/128 -> Xb,Xd}
            I2=^inplace+1;          {in+1 -> Ya,Yc}
            I3=^inplace+9;          {in+N/128+1 -> Yb,Yd}
            M0=16;                  {N/64, skip forward to dual node}
            M1=-16;                 {-N/64, skip back to primary node}
            M2=-14;                 {-N/64+2, skip to next butterfly}

            M3=24;                  {N*3/128, skip to next group}
            DM(m3_space)=M3;        {M3_space is temporary storage}
                                    {space needed because M3 is used}
                                    {in 2 contexts and will alternate}
                                    {in value}

            M5=Nov4+64;             {N/4+groups/stage*1, Cb Sb offset}
            M6=Nov4+128;            {N/4+groups/stage*2, Cc Sc offset}
            M7=Nov4+192;            {N/4+groups/stage*3, Cd Sd offset}
            SI=4;                 {Bfly/group, save counter inner loop}
            DM(bfy_count)=SI;       {SI is a dummy used for storage}
```

(listing continues on next page)

```
          CNTR=64;                {groups/stage}
          CALL mid_stg;           {do stage 4}

{———————— Last Stage, No Multiplies ————————}

laststage: I4=^inplace;           {in ->Xa,Xc}
          I5=^inplace+2;          {in+N/512 ->Xb,Xd}
          I6=^inplace+1;          {in+1 ->Ya,Yc}
          I7=^inplace+3;          {in+N/512+1 ->Yb,Yd}
          M4=4;                   {N/256, skip forward to dual node}

          M0=H#0010;      {This modify value is used to perform bit-}
                          {reverse as the final results are written}
                          {out. The derivation of this value}
                          {is explained in the text.}

          I0=H#0000;          {These base address values are derived}
          I2=H#2000;          {for output at address 0000}
          I1=H#0008;
          I3=H#2008;

          L4=0;               {This last stage has no twiddle factor}
          L5=0;               {multiplication}
          L6=0;               {Because the output addresses are bit-}
          L7=0;           {reversed, the I's M's & L's are reassigned}
                              {and reinitialized}

          AX0=DM(I4,M4);          {first Xa}
          AY0=DM(I4,M4);          {first Xc}
          CNTR=Nov4;              {groups/stage}
          ENA BIT_REV;            {all data accesses using I0..I3 are}
                                  {bit-reversed}


{Middle 2 branches of butterfly are reversed.}
{This alteration, done in every stage, results in bit-reversed}
{outputs instead of digit-reversed outputs.}

          DO laststgbfy UNTIL CE;
             AR=AX0-AY0, AX1=DM(I6,M4); {AR=xa-xc, AX1=ya}
             SR=LSHIFT AR(LO), AY1=DM(I6,M4); {SR0=xa-xc, AY1=yc}
             AR=AX0+AY0, AX0=DM(I5,M4); {AR=xa+xc, AX0=xb}
             MR0=AR, AR=AX1+AY1;        {MR0=xa+xc, AR=ya+yc}
             MR1=AR, AR=AX1-AY1;        {MR1=ya+yc, AR=ya-yc}
             SR=SR OR LSHIFT AR (HI), AY0=DM(I5,M4);
                                    {SR1=ya-yc, AY0 xd}
```

```
                AF=AX0+AY0, AX1=DM(I7,M4); {AF=xb+xd, AX1=yb}
                AR=MR0+AF, AY1=DM(I7,M4);  {AR=xa+xc+xb+xd, AY1=yd}
                DM(I0,M0)=AR, AR=MR0-AF;
                          {output x'a=xa+xc+xb+xd, AR=xa+xc-(xb+xd)}
                DM(I1,M0)=AR, AF=AX1+AY1;
                          {output x'c to position x'b, AF=yb+yd}
                AR=MR1+AF;                 {AR=ya+yb+yc+yd}
                DM(I2,M0)=AR, AR=MR1-AF;
                          {output y'a=ya+yc+yb+yd, AR=ya+yc-(yb+yd)}
                DM(I3,M0)=AR, AR=AX0-AY0;
                          {output y'c to position y'b, AR=xb-xd}
                AX0=DM(I4,M4);             {AX0=xa of next group}
                AF=AX1-AY1, AY1=AR;        {AF=yb-yd, AY1=xb-xd}
                AR=SR0+AF, AY0=DM(I4,M4);  {AR=xa-xc+yb-yd}
                                           {AY0=xc of next group}
                DM(I0,M0)=AR, AR=SR0-AF;
                      {output x'b to position x'c, AR=xa-xc-(yb-yd)}
                DM(I1,M0)=AR, AR=SR1-AY1;
                      {output x'd=xa-xc-(yb-yd), AR=ya-yc+(xb-xd)}
                DM(I2,M0)=AR, AR=SR1+AY1;
                      {output y'b to position y'c, AR=ya-yc-(xb-xd)}
laststgbfy: DM(I3,M0)=AR;              {output y'd=ya-yc-(xb-xd)}

            DIS BIT_REV;               {shut-off bit reverse mode}

            RTS;                       {end and exit from FFT subroutine}

{———————— Subroutine for middle stages ————————}


mid_stg:    DO midgrp UNTIL CE;
                I5=^cos_table;
                I6=^cos_table;
                I7=^cos_table;
                AX0=DM(I0,M0);             {get first Xa}
                AY0=DM(I0,M1);             {get first Xc}
                AR=AX0-AY0, AX1=DM(I2,M0); {Xa-Xc,get first Ya}
                SR=LSHIFT AR (LO), AY1=DM(I2,M1);{SR1=Xa-Xc,get first Yc}
                CNTR=DM(bfy_count);        {butterflies/group}

                M3=-2;                     {M3 is loaded with the value}
                                       {required for pointer manipulation}

{Middle 2 branches of butterfly are reversed.}
{This alteration, done in every stage, results in bit-reversed}
{outputs instead of digit-reversed outputs.}
```

(listing continues on next page)

```
DO midbfy UNTIL CE;
   AR=AX0+AY0, AX0=DM(I1,M0); {AR=xa+xc, AX0=xb}
MR0=AR, AR=AX1+AY1;         {MR0=xa+xc, AR=ya+yc}
MR1=AR, AR=AX1-AY1;         {MR1=ya+yc, AR=ya-yc}
SR=SR OR LSHIFT AR (HI), AY0=DM(I1,M1);
                            {SR1=ya-yc, AY0=xd}
AF=AX0+AY0, AX1=DM(I3,M0); {AF=xb+xd, AX1=yb}
AR=MR0+AF, AY1=DM(I3,M1);  {AR=xa+xb+xc+xd, AY1=yd}
DM(I0,M0)=AR, AR=MR0-AF;
            {output x'a=(xa+xb+xc+xd), AR=xa+xc-xb-xd}
AF=AX1+AY1, MX0=AR;         {AF=yb+yd, MX0=xa+xc-xb-xd}
AR=MR1+AF, MY0=DM(I6,M4);  {AR=ya+yc+yb+yd, MY0=(Cc)}
DM(I2,M0)=AR, AR=MR1-AF;    {output y'a, AR=ya+yc-yb-yd}
MR=MX0*MY0(SS), MY1=DM(I6,M6);
                      {MR=(xa+xc-xb-xd)(Cc), MY1=(Sc)}
MR=MR+AR*MY1(RND), SI=DM(I0,M2);
               {MR=(xa-xb+xc-xd)(Cc)+(ya-yb+yc-yd)(Sc)}
               {SI is a dummy to cause a modify(I0,M2)}
DM(I1,M0)=MR1, MR=AR*MY0(SS);
      {output x'c to position x'b, MR=(ya+yc-yb-yd)(Cc)}
MR=MR-MX0*MY1(RND), MY0=DM(I5,M4);
         {MR=(ya+yc-yb-yd)(Cc)-(xa+xc-xb-xd)(Sc)}
                            {MY0=(Cb)}
DM(I3,M0)=MR1, AR=AX0-AY0;
               {output y'c to position y'b, AR=xb-xd}
AY0=AR, AF=AX1-AY1;         {AY0=xb-xd, AF=yb-yd}
AR=SR0-AF, MY1=DM(I5,M5);  {AR=xa-xc-(yb-yd), MY1=(Sb)}
MX0=AR, AR=SR0+AF;    {MX0=xa-xc-yb+yd, AR=xa-xc+yb-yd}
SR0=AR, AR=SR1+AY0;   {SR0=xa-xc+yb-yd, AR=ya-yc+xb-xd}
MX1=AR, AR=SR1-AY0;   {MX1=ya-yc+xb-xd, AR=ya-yc-(xb-xd)}
MR=SR0*MY0(SS), AX0=DM(I0,M0);
            {MR=(xa-xc+yb-yd)(Cb), AX0=xa of next bfly}
MR=MR+AR*MY1(RND), AY0=DM(I0,M3);
            {MR=(xa-xc+yb-yd)(Cb)+(ya-yc-xb+xd)(Sb)}
                            {AY0=xc of next bfly}
DM(I0,M2)=MR1, MR=AR*MY0(SS);
                            {output x'b to position x'c}
                            {MR=ya-yc-xb+xd)(Cb)}
MR=MR-SR0*MY1(RND), MY0=DM(I7,M4);
                {MR=(ya-yc-xb+xd)(Cb)-(xa-xc+yb-yd)(Sb)}
                            {MY0=(Cd)}
DM(I2,M2)=MR1, MR=MX0*MY0(SS);
                            {output y'b to position y'c}
                            {MR=(xa-yb-xc+yd)(Cd)}
MY1=DM(I7,M7), AR=AX0-AY0; {MY1=(Sd), AR=xa-xc}
MR=MR+MX1*MY1(RND), AX1=DM(I2,M0);
               {MR=(xa-yb-xc+yd)(Cd)+(ya+xb-yc-xd)(Sd)}
                            {AX1=ya of next bfly}
```

```
                DM(I1,M2)=MR1, MR=MX1*MY0(SS);
                          {output x'd=(xa-yb-xc+yd)(Cd)+(ya+xb-yc-xd)(Sd)}
                                            {MR=(ya+yb-yc-yd)(Cd)}
                MR=MR-MX0*MY1(RND), AY1=DM(I2,M1);
                                 {MR=(ya+yb-yc-yd)(Cd)-(xa-xc-yb+yd)Sd}
                                            {yc of next bfly}
midbfy:         DM(I3,M2)=MR1, SR=LSHIFT AR(LO);
                            {output y'd=(ya+xb-yc-xd)Cd-(xa-xc-yb+yd)Sd}
                                            {SR0=ya-yc of next bfly}

             M3=DM(m3_space);      {modifier M3 is loaded with skip to}
                                   {next group_count and is used in the}
                                   {next four instructions}

             MODIFY (I0,M3);
             MODIFY (I1,M3);                {point to next group}
             MODIFY (I2,M3);                {of butterflies}
midgrp:      MODIFY (I3,M3);

             RTS;                  {return to middle stage calling code}

.ENDMOD;
```

Listing 6.36 Radix-4 DIF FFT Module with Built-In Digit-Reversal

6.7 LEAKAGE

The input to an FFT is not an infinite-time signal as in a continuous Fourier transform. Instead, the input is a section (a truncated version) of a signal. This truncated signal can be thought of as an infinite signal multiplied by a rectangular function. For a DFT, the product of the signal and the rectangular function is sampled (multiplied by a series of impulses). Because multiplication in the time domain corresponds to convolution in the frequency domain, the effect of truncating a signal is seen in the FFT results (Brigham, 1974). Figure 6.19 illustrates the effect truncation and sampling have on the Fourier transform.

Figure 6.19a shows z(t), a continuous cosine wave with a period of T_0. Its Fourier transform, Z(f) is two impulses, at $1/T_0$ and $-1/T_0$. Figure 6.19c shows the product of z(t) and u(t), a sampling function. The sampled signal z(t) x u(t) is truncated by multiplication with w(t), a rectangular function. Figure 6.19e shows the resulting signal, y(t), and its Fourier transform, Y(f) (the convolution of Z(f), U(f), and W(f)).

The DFT interprets its input as one complete cycle of a periodic signal. To create a periodic signal from the N samples (y(t)), we convolve y(t) with v(t), a series of impulses at intervals of T_0. T_0 is the length of the rectangular function as well as exactly one period of the input signal z(t). Notice that V(f), the Fourier transform of v(t), is a series of impulses located at multiples of $1/T_0$. Because the zero values of the side lobes of Y(f) are also located at multiples of $1/T_0$, multiplying Y(f) by V(f) in Figure 6.19g produces the same transform as in Figure 6.19c (the transform of the non-truncated signal).

If the length of the rectangular function (w(t)) is not equal to one period or a multiple of periods of the input signal, leakage effects appear in the DFT output. Figure 6.20 illustrates these effects. Notice that in this case T_1, the width of the rectangular function w(t), is not equal to T_0, the period of z(t). Because of the convolution of impulses at locations $-1/T_0$ and $1/T_0$ with W(f), which has zero values at multiples of $1/T_1$, Y(f) has zero values at frequencies other than multiples of $1/T_1$ or $1/T_0$. Convolution in the time domain of y(t) and v(t) in Figure 6.20g corresponds to multiplication of Y(f) and V(f) in the frequency domain. Because the samples in V(f) spaced at $1/T_1$ do not correspond to zero values in Y(f), noise (or leakage) in the DFT output is produced.

Another way to think of leakage is to conceptualize the DFT output as a series of bins at specific frequencies. If f_s is the sampling frequency and N

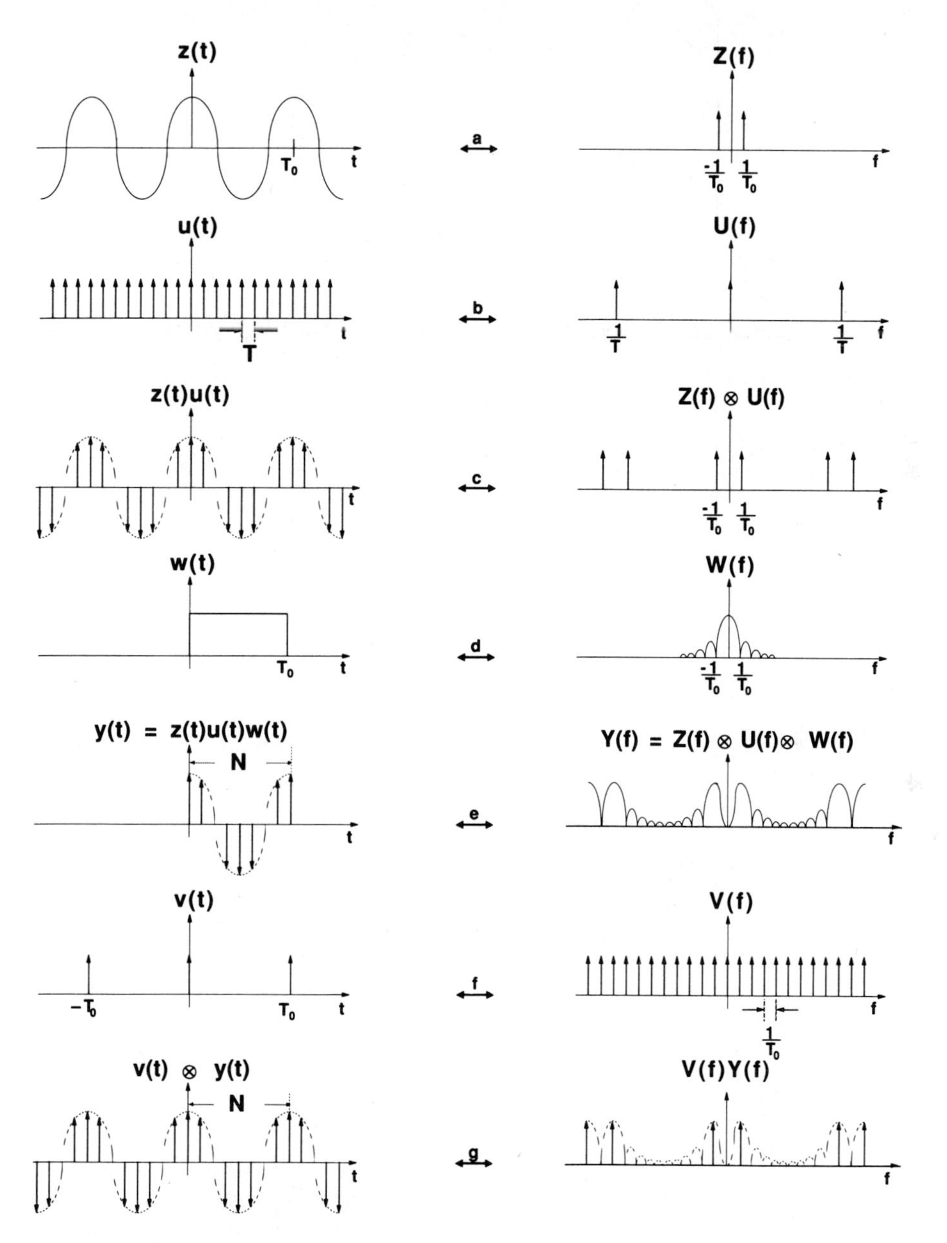

Figure 6.19 Development of DFT, Window = Input Period

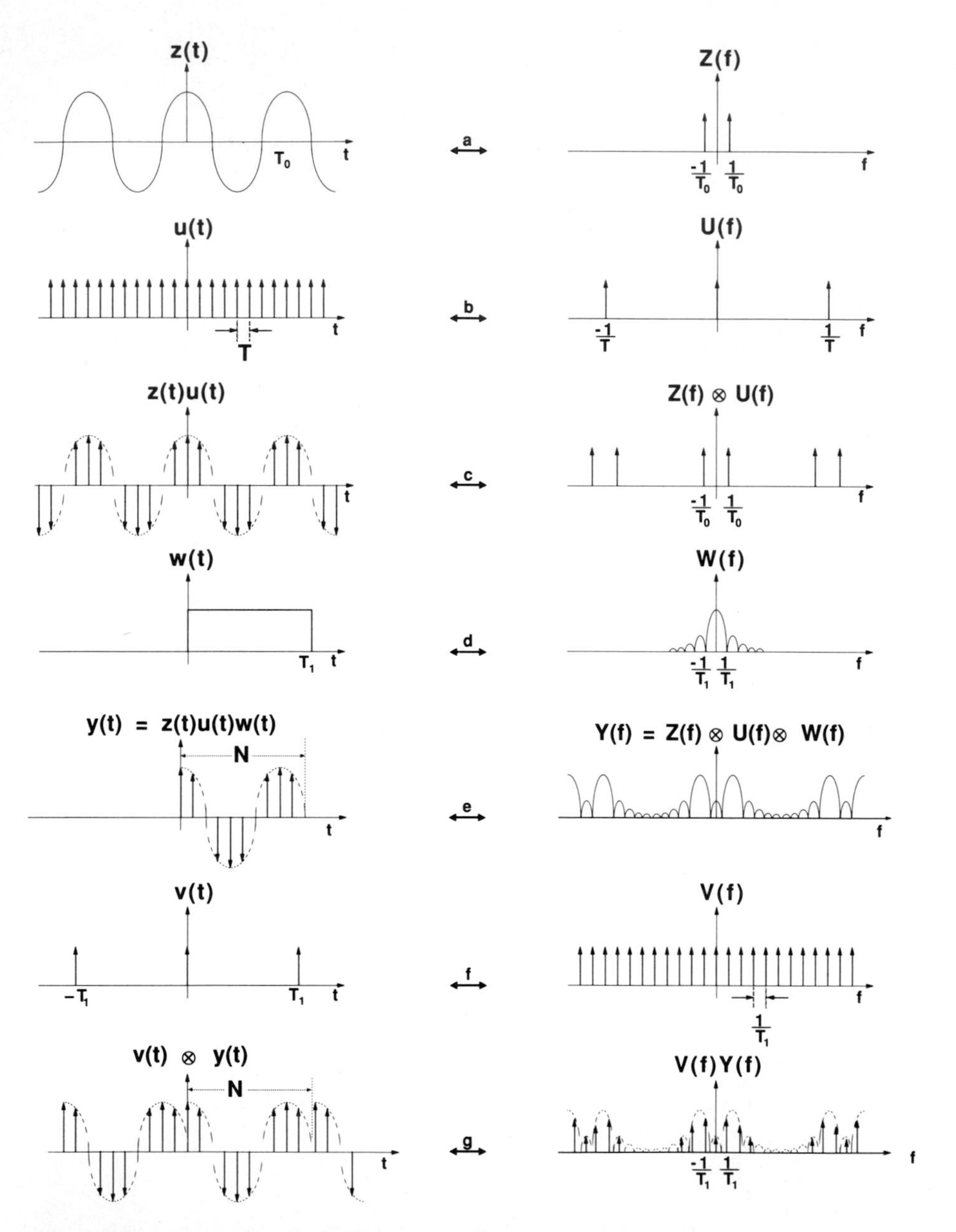

Figure 6.20 Development of DFT, Window ≠ Input Period

is the number of samples, the bins range from 0 Hz to $f_s/2$ Hz and are equally spaced in frequency f_s/N Hz apart. If the N input samples contain one period or a multiple of periods of the input signal, each of the frequency components falls into a frequency bin. If the N input samples do not contain one period or a multiple of periods, at least one of the frequency components falls between bins. The energy of this frequency component is distributed to the surrounding bins, producing a spectrum similar to Figure 6.20g (Brigham, 1974).

In a real system, it is difficult to capture exactly one period or a multiple of periods of a signal. In most cases, leakage in the FFT output will result. One method of reducing this leakage is called windowing.

Although windowing has many applications, we use it here to reduce leakage. Truncation of a signal is a form of windowing in which the window function is rectangular. Leakage caused by truncation can be reduced by selecting a non-rectangular window function with specific characteristics.

The window function is selected for two characteristics (Brigham, 1974). First, to reduce the effect of side lobe multiplication, the side lobes in the Fourier transform of the window function should be significantly smaller than those of the rectangular window function's Fourier transform. Second, the main lobe of the window function's Fourier transform should be sufficiently narrow so that important signal information is not lost. Two examples of window functions that exhibit these characteristics are the Hanning and the Hamming windows.

Hanning: $w(n) = 1/2\,[1 - \cos(2\pi n/(N-1))] \qquad 0 \le n \le N-1$

Hamming: $w(n) = 0.54 - 0.46\cos(2\pi n/(N-1)) \qquad 0 \le n \le N-1$

Noise reduction is accomplished by dividing the selected window function into N equally spaced samples (called window coefficients) and multiplying each FFT input sample by the corresponding coefficient. The module *window*, shown in Listing 6.37, performs this calculation. This module works with both the radix-2 and radix-4 DIF subroutines. It assumes that the *inplacedata* buffer contains sequentially ordered data organized with real and imaginary values interleaved. It is written for a 1024-point FFT, but the window size can be changed by changing the constant *N_x_2*.

The *window_coeffs* buffer is initialized with the window coefficients. This buffer is organized with real and imaginary values interleaved. The buffer is initialized with an external file *window_coeffs.dat* that contains the precalculated window coefficients.

Pointers are set to point to the *inplacedata* and *window_coeffs* buffers. The initial data fetch of a window coefficient and a sample is done before the loop. Inside the loop, a coefficient and sample are multiplied at the same time as the next coefficient and sample are read. After each multiplication, the product is written over the original FFT input sample. The loop is repeated for N samples (real and imaginary parts).

```
.MODULE         windowing;

{               Calling Parameters
                   FFT input data in the inplacedata buffer

                Return Values
                   Windowed FFT input data in the inplacedata buffer

                Altered Registers
                   I0,I1,I4,M0,M4,MX0,MY0,MR
}

.CONST          N_x_2=2048;

.VAR/PM         window_coeffs[N_x_2];

.INIT           window_coeffs:<window_coefs.dat>;

.ENTRY          window;


window:         I0=^inplacedata;          {I0 --> 1st sample in FFT input data}
                I1=I0;
                I4=^window_coeffs;        {I4 --> 1st window coefficient}
                M0=1;
                M4=1;
                CNTR=N_x_2-1;
                MX0=DM(I0,M0),MY0=PM(I4,M4);    {Read 1st sample and coefficient}
                DO window_loop UNTIL CE;        {Window N_x_2-1 samples}
                   MR=MX0*MY0(RND),MX0=DM(I0,M0),MY0=PM(I4,M4);
window_loop:       DM(I1,M0)=MR1;
                MR=MX0*MY0(RND);          {Multiply last sample and coefficient}
                DM(I1,M0)=MR1;            {Last sample updated}
                RTS;
.ENDMOD;
```

Listing 6.37 Windowing Routine

6.8 BENCHMARKS

Benchmarks for the optimized radix-2 DIT FFT and radix-4 DIF FFT routines are given in this section.

Straight-line code occupies much more memory than looped code, is hard to understand and is tedious to debug. All programs used to generate the benchmarks presented in this section are relatively short and uncomplicated looped programs.

The FFT benchmarks in Table 6.1 are worst-case. For example, the 1024-point radix-2 FFT benchmark assumes that the data grows by two bits in every stage.

Routine	*Number of Points*	*Number of Cycles*	*Execution Time* (12.5MHz ADSP-2100A)
Radix-2 DIT Input scaling	1024	52911	4.23 ms
Radix-2 DIT Conditional BFP*	1024	113482	9.08 ms
Radix-4 DIF	64	1381	0.11 ms
Input scaling	256	7372	0.59 ms
	1024	37021	2.96 ms
Radix-4 DIF	64	1405	0.11 ms
Input scaling	256	7423	0.59 ms
Built-in digit-reverse	1024	37203	2.98 ms

* BFP = Block Floating-Point Scaling

Table 6.1 Benchmarks for FFT Routines

Table 6.2 lists the benchmarks for the other routines presented in this chapter (bit reversal, digit reversal, and windowing).

Routine	*Number of Points*	*Number of Cycles*	*Time* (12MHz ADSP-2100A) *(μs)*
Bit-Reverse (scramble, real data)	64	138	11.04
	128	266	21.28
	256	522	41.76
	1024	2058	164.64
Bit-Reverse (unscramble, complex data)	64	270	21.60
	128	526	42.08
	256	1038	83.04
	1024	4110	328.80
Digit-Reverse (unscramble, complex data)	64	430	34.40
	256	1812	144.96
	1024	7188	575.04
Window (complex data)	64	267	21.36
	128	523	41.84
	256	1035	82.80
	1024	4107	328.56

Table 6.2 Benchmarks for Other Routines

6.9 REFERENCES

Brigham, E.O. 1974. *The Fast Fourier Transform*. Englewood Cliffs, N.J.: Prentice-Hall, Inc.

Burrus, C.S. and T.W. Parks. 1985. *DFT and Convolution Algorithms*. New York, NY: John Wiley and Sons.

Dudgeon, D. and Mersereau, R. 1984. *Multidimensional Digital Signal Processing*. Englewood Cliffs, N.J.: Prentice-Hall Inc.

Gonzalez, R. and Wintz, P. 1977. *Digital Image Processing*. Reading, MA: Addison-Wesley Publishing Company.

Hakimmashhadi, H. 1988. "Discrete Fourier Transform and FFT." *Signal Processing Handbook*. New York, NY: Marcel Dekker, Inc.

Haykin, S. 1983. *Communication Systems*. New York: John Wiley and Sons.

Oppenheim, A. V., and Schafer, R. W. 1975. *Digital Signal Processing*. Englewood Cliffs, N.J.: Prentice-Hall, Inc.

Proakis, J. G. and D. G. Manolakis. 1988. *Introduction to Digital Signal Processing*. New York, NY: Macmillan Publishing Company.

Rabiner, L. R. and Gold, B. 1975. *Theory and Applications of Digital Signal Processing*. Englewood Cliffs, N.J.: Prentice-Hall, Inc.

Rabiner, Lawrence R. and Gold, Bernard. 1975. *Theory and Applications of Digital Signal Processing*. Englewood Cliffs, N.J.: Prentice-Hall, Inc.

Two-Dimensional FFTs ■ 7

7.1 TWO-DIMENSIONAL FFTS

The two-dimensional discrete Fourier transform (2D DFT) is the discrete-time equivalent of the two-dimensional continuous-time Fourier transform. Operating on $x(n_1,n_2)$, a sampled version of a continuous-time two-dimensional signal, the 2D DFT is represented by the following equation:

$$X(k_1, k_2) = \sum_{n_1=0}^{N_1-1} \sum_{n_2=0}^{N_2-1} x(n_1, n_2)\, W_N^{\,n_1k_1}\, W_N^{\,n_2k_2}$$

where $W_N = e^{-j2\pi/N}$ and the signal is defined for the points $0 \le n_1, n_2 < N_1, N_2$.

Direct calculation of the 2D DFT is simple, but requires a very large number of complex multiplications. Assuming all of the exponential terms are precalculated and stored in a table, the total number of complex multiplications needed to evaluate the 2D DFT is $N_1^2N_2^2$. The number of complex additions required is also $N_1^2N_2^2$. Direct evaluation of the 2D DFT for a square image, 128 pixels by 128 pixels, requires over 268 million complex multiplications.

Two techniques can be employed to reduce the operation count of the two-dimensional transform. First, the row-column decomposition method (Dudgeon and Mersereau, 1984) partitions the two-dimensional DFT into many one-dimensional DFTs. Row-column decomposition reduces the number of complex multiplications from $N_1^2N_2^2$ (direct evaluation) to $N_1N_2\,(N_1+N_2)$ (row-column decomposition with direct DFT evaluation). For the 128-by-128-pixel example, the number of complex multiplications is reduced from 268 million to less than 4.2 million, a factor of 63.

The second technique to reduce the operation count of the two-dimensional transform is the fast Fourier transform (FFT). The FFT is a shortcut evaluation of the DFT. The FFT is used to evaluate the one-dimensional DFTs produced by the row-column decomposition.

7 Two-Dimensional FFTs

7.1.1 Row-Column Decomposition

Row-column decomposition is straightforward. The 2D DFT double summation

$$X(k_1, k_2) = \sum_{n_1=0}^{N_1-1} \sum_{n_2=0}^{N_2-1} x(n_1, n_2)\, W_{N_1}^{n_1k_1}\, W_{N_2}^{n_2k_2}$$

can be rewritten as

$$X(k_1, k_2) = \sum_{n_1=0}^{N_1-1} \left[\sum_{n_2=0}^{N_2-1} x(n_1, n_2)\, W_{N_2}^{n_2k_2} \right] W_{N_1}^{n_1k_1}$$

in which all occurrences of n_2 are grouped within the square brackets. The result is that the quantity within the brackets is a one-dimensional DFT. The equation can then be separated into the following two equations:

$$A(n_1, k_2) = \sum_{n_2=0}^{N_2-1} x(n_1, n_2)\, W_{N_2}^{n_2k_2}$$

$$X(k_1, k_2) = \sum_{n_1=0}^{N_1-1} A(n_1, k_2)\, W_{N_1}^{n_1k_1}$$

In the equations above, the function $A(n_1, k_2)$ has rows (n_1) and columns (k_2). The columns are one-dimensional DFTs of the corresponding columns of the original signal, $x(n_1, n_2)$. The second equation evaluates row DFTs. The DFT of each row is evaluated for the intermediate function $A(n_1, k_2)$. Therefore, in order to evaluate the DFT of a two-dimensional signal, $x(n_1, n_2)$, you first evaluate the DFT for each row. The results of these 1D DFTs are stored in an intermediate array, $A(n_1, k_2)$. Then 1D DFTs are evaluated for each column of the intermediate array. The result is the 2D DFT of the original signal. For an image of 128 by 128 pixels, 256 1D DFTs need to be evaluated.

When the FFT algorithm is employed to evaluate the 1D DFTs from the row-column decomposition, there are further significant computational savings. The number of complex multiplications is reduced from

$N_1N_2(N_1+N_2)$ for direct evaluation of each DFT to $N_1N_2(\log_2(N_1N_2/2))$ for FFT evaluation using the row-column decomposition method. For the 128-by-128 pixel example mentioned above, the number of complex multiplications necessary is less than 115 thousand. Therefore, using row-column decomposition and FFTs, there is a reduction by a factor of 2300 in the number of complex multiplications for a 128-by-128-point image.

The one-dimensional radix-2, decimation-in-frequency FFT algorithm from Chapter 6 is adapted for two dimensions in this chapter. Row-column decomposition facilitates the transition to two dimensions. To evaluate the FFT of a N-by-N-point image, you must simply evaluate 2N one-dimensional FFTs.

7.1.2 Radix-2 FFT

The FFT algorithm divides an input sequence into smaller sequences, evaluates the DFTs of these smaller sequences, and combines the outputs of the small DFTs to give the DFT of the original input sequence. The radix-2 FFT divides an N-point input sequence in half, into two N/2-point sequences. This division provides computational savings. Further computational savings are realized by dividing each N/2-point sequence into two N/4-point sequences. The strategy of a radix-2 FFT is to divide the input sequence successively by two until the resulting sequences contain only two points. For example, an 8-point DFT is reduced to four 2-point DFTs. The number of multiplications needed to evaluate a DFT is proportional to the square of the number of input points. Dividing the number of input points in half and evaluating two DFTs reduces the multiplication count from N^2 to $2(N/2)^2$. Each of these N/2-point sequences are again divided in half, and the operation count is reduced in the same fashion. With each decimation, the multiplication count is reduced by a factor of two. This scheme dramatically reduces the number of multiplications necessary to evaluate the DFT of the original sequence.

There are two basic varieties of radix-2 FFTs: decimation-in-time (DIT) and decimation-in-frequency (DIF). The DIT FFT divides the input sequence into even samples and odd samples. Each half-sequence is, in turn, divided into even and odd samples. Division into even and odd samples is facilitated by bit-reversing the addresses of the input sequence. The radix-2 DIT FFT takes a bit-reversed input sequence and outputs samples in normal order.

The DIF radix-2 FFT also successively divides the input sequence in half; however, the two N/2-point sequences consist of the first N/2 samples and the last N/2 samples. One sequence contains the points 0 through

N/2–1, and the other sequence contains the points N/2 through N–1. The inputs to the DIF FFT are addressed in normal order, and the results are output in bit-reversed order, which is the converse of the operation of the DIT FFT.

The row-column decomposition described in this chapter evaluates the 2D DFT of an N-by-N-point signal using 2N one-dimensional FFTs. The DIF radix-2 FFT is used to evaluate the rows and columns of an N-by-N-point input signal.

7.1.3 ADSP-2100 Implementation

This implementation of the 2D FFT uses nine software routines. The main routine declares and initializes data buffers and calls subroutines that perform the 2D FFT. Two subroutines perform initialization tasks. Row and column 1D FFTs are performed with block floating-point DIF FFTs. There are four subroutines that adjust the FFT output for bit growth. Two of these routines operate on row FFT output and two operate on column FFT output. There are also two bit reversal subroutines, one each for row and column FFTs.

Input data for this 2D FFT must be in 16 bit twos-complement fractional format. This input format is often called 1.15. Two guard bits is also necessary to prevent overflow in the first stage of the FFT. The implementation presented in this chapter is a block floating-point implementation of the 2D FFT. This implementation is computationally less intensive than full floating-point and provides more dynamic range than a fixed-point implementation. The ADSP-2100 family of DSP processors posess the necessary hardware to efficiently perform these block floating-point operations. There are four block floating-point adjust routines within the 2D FFT program. Two of these operate on row FFT data and two operate on column FFT data.

7.1.3.1 Main Module

Declaring variables, initializing data buffers and program variables, and calling subroutines are the major functions of the main module. The two-dimensional FFT is performed in place (using the same data buffers for inputs and outputs). The data buffers *realdata* and *imaginarydata* contain the real and imaginary parts of the input signal at the start of the program. As the program executes, all partial results are written to these data buffers, and at program completion, the buffers *realdata* and *imaginarydata* contain the resulting frequency samples. The in-place buffers are shown in Figure 7.1.

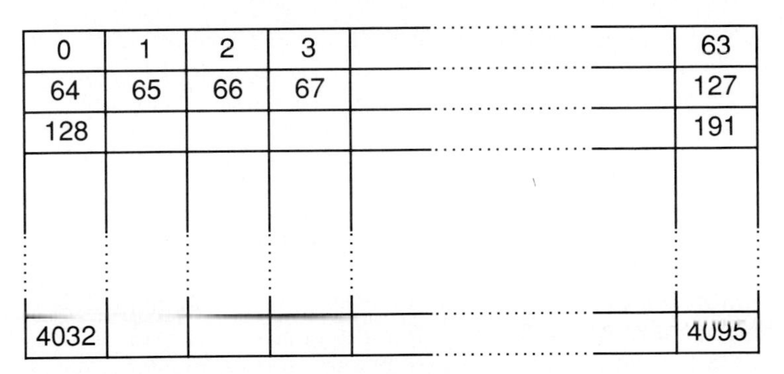

Schematic Representation of realdata

4096	4097				...	4159
4160					...	
					...	
⋮	⋮	⋮	⋮	⋮		⋮
8128					...	8191

Schematic Representation of imaginarydata

Figure 7.1 In-Place Buffers

Twiddle factors, sine values and cosine values used for evaluation of complex exponentials, are stored in two tables, *twid_real* and *twid_imag*. These twiddle factors are multiplied with the FFT data values. To exploit the Harvard architecture of the ADSP-2100, the twiddle factor tables are placed in program memory. A dual data fetch occurs in a single processor cycle, retrieving a twiddle factor fron program memory and a data value from data memory. The indexing of the twiddle factors is maintained with the index registers I4 and I5.

7 Two-Dimensional FFTs

Because row-column decomposition divides the 2D FFT into many 1D FFTs, a mechanism is needed to keep track of the current FFT. This program uses five variables for this purpose: *offset*, *rrowbase*, *irowbase*, *rcolbase*, and *icolbase*. The *rowbase* variables keep track of the real and imaginary row FFT starting points, and the *colbase* variables keep track of the real and imaginary column FFT starting points. *Offset* is used to update these variables as each FFT is completed. Three other variables—*groups*, *bflys_per_group*, and *node_space*—are used in each FFT.

Listing 7.1 contains the source code for the main module. The main module calls eight subroutines. The first, *initialize*, performs the once-only initialization of data pointers, length registers and modify registers. The second, *fft_start*, performs DIF FFTs on the row data. For an N-by-N input signal, this routine is called N times. After all of the row FFTs have been performed, the routine *unscr_start* unscrambles the row output data by bit-reversing the addresses. Then the *row_final_adj* routine adjusts the outputs of the row FFTs for bit growth.

For the column FFTs, *col_init*, an initialization routine, is called. Because sequential points in the column FFTs do not reside in sequential addresses, the *col_init* routine is needed to initialize the parameters that define the FFT bounds. The column FFT subroutine, *col_fft_strt*, is called N times to transform each column of data in the buffers *realdata* and *tempdata*. Once the column FFTs are done, the subroutine *col_unscr_start* bit-reverses the column outputs' addressing. Finally, the subroutine *col_final_adj* adjusts the column output data for bit growth.

The *initialize* and *col_init* routines are part of the main modules, but are described later in Listings 7.5 and 7.6. The other subroutines are in other modules.

```
.MODULE      fft_2d;

{     Performs a two dimensional FFT using the row column
      decomposition method. One dimensional FFTs are performed
      first on each row. The pointers rrowbase and irowbase keep
      track of row bounds. One dimensional FFTs are then
      performed on each column. The pointers rcolbase and
      icolbase keep track of the column data. The row FFTs and
      column FFTs are decimation in frequency: input in normal
      order and output in bit-reversed order.
```

```
     Variables
          realdata          Array (N X N) of real input data
          imaginarydata     Array (N X N) of imaginary input data
          twid_real         Real part of twiddle factors
          twid_imag         Imaginary part of twiddles
          groups            # of groups in current stage
          node_space        spacing between dual node points
          bflys_per_group   # of butterflys per group
          rrowbase          real data row pointer
          irowbase          imaginary data row pointer
          rcolbase          real data column pointer
          icolbase          imaginary data column pointer
          fft_start         entry point row FFTs
          col_fft_strt      entry point column FFTs
          shift_strt        entry point normalizer
}

.CONST                  N = 64, N_X_2 = 128;
.CONST                  N_div_2 = 32, log2N = 6;

.VAR/DM/RAM             realdata[4096];
.VAR/DM/RAM/ABS=h#1000  imaginarydata[4096];
.VAR/DM/RAM/ABS=h#2000  tempdata[4096];

.VAR/PM/RAM/CIRC   twid_real[N_div_2];
.VAR/PM/RAM/CIRC   twid_imag[N_div_2];
.VAR/DM            row_exponents[N];
.VAR/DM            col_exponents[N];
.VAR/PM/RAM        padding[8];
.VAR/DM/RAM        groups, node_space, bflys_per_group;
.VAR/DM/RAM        offset, rrowbase, irowbase, rcolbase, icolbase;
.VAR/DM/RAM        blk_exponent;
.VAR/DM/RAM        real_br_pointers[N];
.VAR/DM/RAM        imag_br_pointers[N];
.VAR/DM/RAM        c_real_br_pointers[N];
.VAR/DM/RAM        c_imag_br_pointers[N];
.VAR/DM/RAM        current_rrow;
.VAR/DM/RAM        current_irow;
.VAR/DM/RAM        current_rcol;
.VAR/DM/RAM        current_icol;
```

(listing continues on next page)

```
.INIT       realdata: <realdata.dat>;
.INIT       imaginarydata: <imagdata.dat>;
.INIT       twid_real:  <twid_real.dat>;
.INIT       twid_imag:  <twid_imag.dat>;
.INIT       real_br_pointers: <real_ptr.dat>;
.INIT       imag_br_pointers: <imag_ptr.dat>;
.INIT       c_real_br_pointers: <c_real_p.dat>;
.INIT       c_imag_br_pointers: <c_imag_p.dat>;
.INIT       padding: 0,0,0,0,0,0,0,0;
.INIT       groups: 1;
.INIT       blk_exponent: 0;
.INIT       node_space: N_div_2;
.INIT       bflys_per_group: N_div_2;

.GLOBAL     realdata, imaginarydata, twid_real, twid_imag;
.GLOBAL     groups, bflys_per_group, node_space, offset;
.GLOBAL     rrowbase, irowbase, icolbase, rcolbase, blk_exponent;
.GLOBAL     real_br_pointers, imag_br_pointers;
.GLOBAL     c_real_br_pointers, c_imag_br_pointers;
.GLOBAL     current_rrow, current_irow, tempdata;
.GLOBAL     current_rcol, current_icol;
.GLOBAL     row_exponents, col_exponents;

.EXTERNAL   fft_start, col_fft_strt, unscr_start;
.EXTERNAL   col_unscr_start, row_final_adj, col_final_adj;

            NOP;          NOP;
            NOP;          NOP;

            CALL initialize;
            CNTR = N;                 {row counter}
            DO rowloop UNTIL CE;      {do row FFTs}
               CALL fft_start;
rowloop:       NOP;

            CNTR = N;                 {bit reverse rows}
            I6 = ^real_br_pointers;
            I7 = ^imag_br_pointers;
            DO unscramble UNTIL CE;  {real data -> tempdata}
               CALL unscr_start;      {imaginary -> realdata}
unscramble:    NOP;
```

```
            CALL row_final_adj;
            I7 = ^col_exponents;
            CNTR = N;                   {column counter}
            DO colloop UNTIL CE;        {do column FFTs}
               CALL col_init;
               CALL col_fft_strt;
colloop:       NOP;

            CNTR = N;
            I6 = ^c_imag_br_pointers;
            I7 = ^c_real_br_pointers;
            DO col_unscramble UNTIL CE;
                CALL col_unscr_start;
col_unscramble: NOP;

            CALL col_final_adj;         {final BFP adjust}

            TRAP;
```

(initialize and col_init routines shown in Listings 7.5 and 7.6)

```
.ENDMOD;
```

Listing 7.1 Main Module

7.1.3.2 Row DIF Module

The *dif_fft* module operates on row data from the data buffers *realdata* and *imaginarydata*. Rows consist of 64 sequential locations and have start addresses that are multiples of 64. Figure 7.2, on the following page, shows row data from the data buffer *realdata*.

The *dif_fft* routine (entry point at *fft_start*) is called N times for an N-by-N-point image. Each time, the routine performs an FFT on a single row of complex data, actually two rows of data, one representing the real part and one representing the imaginary part of the data. Three data memory variables keep track of the current row of data: *rrowbase* contains the start address of the current real data row, *irowbase* contains the start address of the current imaginary data row, and *offset* calculates *rrowbase* and *irowbase* for the next data row.

0	1	2	3	......	63
64	65	66	67	......	127
128				......	191
⋮	⋮	⋮	⋮		⋮
4032				......	4095

Figure 7.2 Row Data

The FFT program is a looped program consisting of three nested loops. They are the *stage* loop, the *group* loop and the *butterfly* loop. This looped structure is identical to that of the DIF FFT programs found in Chapter 6. The innermost loop, the *butterfly* loop, performs the FFT calculations. A generalized DIF butterfly flow graph is shown in Figure 7.3.

Evaluation of the DFT requires multiplication by a complex exponential, W_N. This complex exponential can be divided into real and imaginary parts according to the following relationship:

$$W_N = e^{-j2\pi/N} = \cos(2\pi/N) - j\sin(2\pi/N)$$

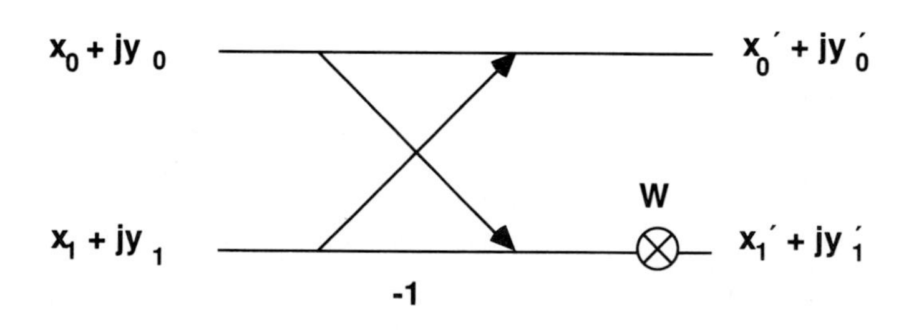

Figure 7.3 Radix-2 DIF Butterfly

The following equations calculate the real and imaginary parts of the DIF butterfly. The variables x_0 and y_0 are the real and imaginary parts of the primary node, and x_1 and y_1 are the real and imaginary parts of the dual node.

$$x'_0 = x_0 + x_1$$

$$y'_0 = y_0 + y_1$$

$$x'_1 = \mathbf{C}(x_0 - x_1) - (-\mathbf{S})\,(y_0 - y_1)$$

$$y'_1 = (-\mathbf{S})\,(x_0 - x_1) + \mathbf{C}(y_0 - y_1)$$

Listing 7.2 details the butterfly loop. The primary and dual nodes are separated in memory by the number of locations indicated by the program variable *node_space*. The primary node and dual node points are accessed throughout the program using index registers; I0 and I2 index the real and imaginary parts (respectively) of the primary node, and I1 and I3 index the real and imaginary parts of the dual node.

```
            DO bfly_loop UNTIL CE;
               AR=AX0+AY0, AX1=DM(I2,M0), MY0=PM(I4,M5);
                                          {AR=x0+x1,AX1=y0,MY0=C}
               DM(I0,M1)=AR, AR=AX1+AY1;  {x0=x0+x1,AR=y0+y1}
               DM(I2,M1)=AR, AR=AX0-AY0;  {y0=y0+y1,AR=x0-x1}
               MX0=AR, AR=AX1-AY1;        {MX0=x0-x1,AR=y0-y1}
               MR=MX0*MY0 (SS), AX0=DM(I0,M0), MY1=PM(I5,M5);
                        {MR=(x0-x1)C,AX0=next x0,MY1=(-S)}
               MR=MR-AR*MY1 (RND);
                        {MR=(x0-x1)C-(y0-y1)(-S),AY0=next x1}
               DM(I1,M1)=MR1, MR=AR*MY0 (SS);
                        {x1=(x0-x1)C-(y0-y1)(-S),MR=(y0-y1)C}
               MR=MR+MX0*MY1 (RND), AY0=DM(I1,M0);     {AY0=new x1}
                        {MR=(y0-y1)C+(x0-x1)(-S),AY1=next y1}
               DM(I3,M1)=MR1;                          {AY1=new y1}
                        {y1=(y0-y1)C+(x0-x1)(-S),check bit growth}
bfly_loop:     AY1=DM(I3,M0);
```

Listing 7.2 DIF Butterfly Loop

7 Two-Dimensional FFTs

The DIF algorithm accepts input in a normal order. Output of the DIF algorithm is scrambled and needs to be put through a bit-reverse algorithm to achieve normal ordering of output. For the 2D FFT, both the row and column FFTs use the DIF algorithm and the output of each requires a bit-reverse subroutine. These routines are described later under *Bit Reverse Modules*.

A complete description of the DIF FFT algorithm and its implementation on the ADSP-2100 can be found in Chapter 6. Listing 7.3 contains the complete row FFT module.

```
.MODULE   dif_fft;

{       DIF section for Row FFTs

        Calling Parameters
           realdata = Real input data normal order
           imaginarydata = Imaginary data normal order
           twid_real = Twiddle factor cosine values
           twid_imag = Twiddle factor sine values
           groups = 1
           bflys_per_group = N/2
           node_space = N/2
           rrowbase = 0
           irowbase = 4096

        Return Values
           realdata = row FFT results in bit-reversed order
           imaginarydata = column FFT results bit-reversed

        Altered Registers
           I0,I1,I2,I3,I4,I5,L0,L1,L2,L3,L4,L5
           M0,M1,M2,M3,M4,M5
           AX0,AX1,AY0,AY1,AR,AF
           MX0,MX1,MY0,MY1,MR,SB,SE,SR,SI

        Altered Memory
           realdata,imaginarydata, groups, node_space,
           bflys_per_group, irowbase, icolbase, offset
}

.CONST        N=64, N_div_2=32, log2N=6;

.EXTERNAL     realdata,imaginarydata, twid_real, twid_imag;
.EXTERNAL     offset, irowbase, rrowbase;
```

```
.EXTERNAL     groups,bflys_per_group,node_space;
.EXTERNAL     row_bfp, blk_exponent;

.ENTRY        fft_start;

fft_start:    AX0 = N;
              DM(offset) = AX0;
              CNTR=log2N;                {Initialize stage counter}
              DO stage_loop UNTIL CE;
                 SB = -2;                {block exponent}
                 I0 = DM(rrowbase);
                 I2 = DM(irowbase);
                 M2 = DM(node_space);
                 I1=I0;
                 MODIFY(I1,M2);          {I1 -> x1}
                 I3=I2;
                 MODIFY(I3,M2);          {I3 -> y1}

                 CNTR=DM(groups);        {Initialize group counter}
                 M5=CNTR;                {Init. twid factor modifier}

                 DO group_loop UNTIL CE;
                    CNTR=DM(bflys_per_group);  {Init. bfly counter}
                    AX0=DM(I0,M0);             {AX0=x0}
                    AY0=DM(I1,M0);             {AY0=x1}
                    AY1=DM(I3,M0);             {AY1=y1}
                    DO bfly_loop UNTIL CE;
                       AR=AX0+AY0, AX1=DM(I2,M0), MY0=PM(I4,M5);
                               {AR=x0+x1,AX1=y0,MY0=C}
                       SB = EXPADJ AR;
                       DM(I0,M1)=AR, AR=AX1+AY1;
                               {x0=x0+x1,AR=y0+y1}
                       SB = EXPADJ AR;
                       DM(I2,M1)=AR, AR=AX0-AY0;
                               {y0=y0+y1,AR=x0-x1}
                       MX0=AR, AR=AX1-AY1;
                               {MX0=x0-x1,AR=y0-y1}

                       MR=MX0*MY0(SS),AX0=DM(I0,M0),MY1=PM(I5,M5);
                               {MR=(x0-x1)C,AX0=next x0,MY1=(-S)}
                       MR=MR-AR*MY1 (SS);  MR=MR(RND);
                               {MR=(x0-x1)C-(y0-y1)(-S),AY0=next x1}
                       SB = EXPADJ MR1;
                       DM(I1,M1)=MR1, MR=AR*MY0 (SS);
                               {x1=(x0-x1)C-(y0-y1)(-S),MR=(y0-y1)C}
                       MR=MR+MX0*MY1 (RND), AY0=DM(I1,M0);
```

(listing continues on next page)

```
                              { AY0 = new x1 }
                              {MR=(y0-y1)C+(x0-x1)(-S),AY1=next y1}
                         DM(I3,M1)=MR1, SB = EXPADJ MR1;
                              { AY1 = new y1 }
                              {y1=(y0-y1)C+(x0-x1)(-S),check bit growth}
bfly_loop:               AY1=DM(I3,M0);

                       MODIFY(I0,M2); {I0->x0 of 1st bfly next group}
                       MODIFY(I1,M2); {I1->x1 of 1st bfly next group}
                       MODIFY(I2,M2); {I2->y0 of 1st bfly next group}
group_loop:            MODIFY(I3,M2); {I3->y1 of 1st bfly next group}

                   CALL row_bfp;

                   SI=DM(groups);
                   SR=LSHIFT SI BY 1 (LO);
                   DM(groups)=SR0;     {groups=groups X 2}
                   SI=DM(node_space);
                   SR=LSHIFT SI BY -1 (LO);
                   DM(node_space)=SR0; {node_space=node_space / 2}
stage_loop:        DM(bflys_per_group)=SR0;
                                       {bflys_per_group=bflys_per_group / 2}

               AX0 = DM(offset);       {calculate next fft base}
               AY0 = DM(rrowbase);
               AR = AX0 + AY0;
               I0 = AR;
               DM(rrowbase) = AR;
               AY0 = DM(irowbase);
               AR = AX0 + AY0;
               DM(irowbase) = AR;
               I2 = AR;
               AX1 = n_div_2;
               DM(node_space) = AX1;  {reset node_space}
               DM(bflys_per_group) = AX1;
               AR = 1;
               DM(groups) = AR;

               AX0 = DM(blk_exponent);
               DM(I6,M7) = AX0;
               AX0 = 0;
               DM(blk_exponent)=AX0;

               RTS;
.ENDMOD;
```

Listing 7.3 Row DIF Module

7.1.3.3 Column DIF Module

The *col_dif_fft* module operates on column data in the data buffers *realdata* and *tempdata*. Sequential points in a data column are not at sequential addresses. Figure 7.4 illustrates the arrangement of columns in the data buffer *tempdata*.

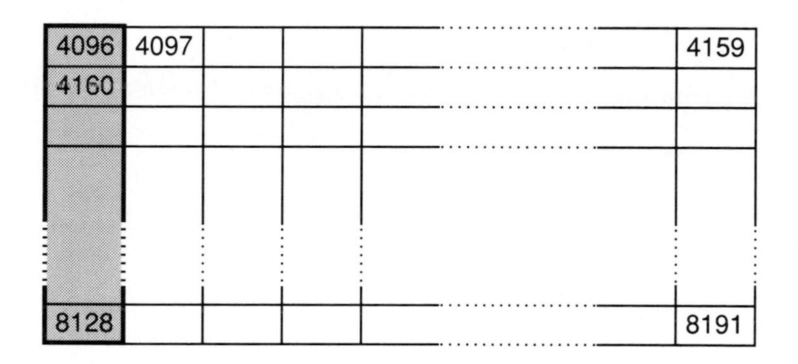

Figure 7.4 Column Data

Accessing data in a column format is easy with the ADSP-2100. In the first stage of a typical DIF FFT, the primary and dual nodes are separated by a node space of N/2. Because this implementation of the DIF algorithm relies on the data memory variable *node_space* to reference the spacing between primary and dual node, however, they need not be sequential. Before the start of each column FFT, *node_space* is initialized to the value 64 x N/2. As the algorithm progresses, the node spacing decrements by a factor of 2 with each stage.

Listing 7.4 contains the complete column FFT module.

```
.MODULE   col_dif_fft;

{       DIF section for Column FFTs

        Calling Parameters
           realdata = Real input data normal order
           imaginarydata = Imaginary data normal order
           twid_real = Twiddle factor cosine values
           twid_imag = Twiddle factor sine values
           groups = 1
           bflys_per_group = N/2
           node_space = N/2
           rrowbase = 0
           irowbase = 4096

        Return Values
           realdata = row FFT results in bit reversed order
           imaginARydata = column FFT results bit reversed

        Altered Registers
           I0,I1,I2,I3,I4,I5,L0,L1,L2,L3,L4,L5
           M0,M1,M2,M4,M5
           AX0,AX1,AY0,AY1,AR,AF
           MX0,MX1,MY0,MY1,MR,SB,SE,SR,SI

        Altered Memory
           realdta,imaginarydata, groups, node_space
           bflys_per_group, irowbase, icolbase, offset
}

.CONST       N=64, N_div_2=32, log2N=6;

.EXTERNAL    realdata,imaginarydata, twid_real, twid_imag;
.EXTERNAL    offset, icolbase, rcolbase;
.EXTERNAL    groups,bflys_per_group,node_space;
.EXTERNAL    col_bfp, blk_exponent;

.ENTRY       col_fft_strt;

col_fft_strt: CNTR=log2N;              {Initialize stage counter}
              DO stage_loop UNTIL CE;
                 SB = -2;              {block exponent}
                 I0 = DM(rcolbase);
                 I2 = DM(icolbase);
```

```
              M2 - DM(node_space);
              I1=I0;
              MODIFY(I1,M2);          {I1 -> x1}
              I3=I2;
              MODIFY(I3,M2);          {I3 -> y1}

              CNTR=DM(groups);        {Initialize group counter}
              M5=CNTR;                {Init twid factor modifier}

              DO group_loop UNTIL CE;
                 CNTR=DM(bflys_per_group); {Init bfly counter}
                 AX0=DM(I0,M0);             {AX0=x0}
                 AY0=DM(I1,M0);             {AY0=x1}
                 AY1=DM(I3,M0);             {AY1=y1}
                 DO bfly_loop UNTIL CE;
                    AR=AX0+AY0, AX1=DM(I2,M0), MY0=PM(I4,M5);
                                       {AR=x0+x1,AX1=y0,MY0=C}
                    SB = EXPADJ AR;
                    DM(I0,M1)=AR, AR=AX1+AY1;
                                       {x0=x0+x1,AR=y0+y1}
                    SB = EXPADJ AR;
                    DM(I2,M1)=AR, AR-AX0-AY0;
                                       {y0=y0+y1,AR=x0-x1}
                    MX0=AR, AR=AX1-AY1;
                                       {MX0=x0-x1,AR=y0-y1}
                    MR=MX0*MY0(SS),AX0=DM(I0,M0),MY1=PM(I5,M5);
                                       {MR=(x0-x1)C,AX0=next x0,MY1=(-S)}
                    MR=MR-AR*MY1 (SS);  MR = MR(RND);
                                       {MR=(x0-x1)C-(y0-y1)(-S),AY0=next x1}
                    SB = EXPADJ MR1;
                    DM(I1,M1)=MR1, MR=AR*MY0 (SS);
                                       {x1=(x0-x1)C-(y0-y1)(-S),MR=(y0-y1)C}
                    MR=MR+MX0*MY1 (RND), AY0=DM(I1,M0);
                                       {AY0 = new x1}
                     {MR=(y0-y1)C+(x0-x1)(-S),AY1=next y1}
                    DM(I3,M1)=MR1, SB = EXPADJ MR1;
                                       {AY1 = new y1}
                                       {y1=(y0-y1)C+(x0-x1)(-S),check bit growth}
bfly_loop:          AY1=DM(I3,M0);

                 MODIFY(I0,M2);  {I0->x0 of 1st bfly next grp}
                 MODIFY(I1,M2);  {I1->x1 of 1st bfly next grp}
                 MODIFY(I2,M2);  {I2->y0 of 1st bfly next grp}
group_loop:      MODIFY(I3,M2);  {I3->y1 of 1st bfly next grp}
```

(lisitng continues on next page)

```
                        CALL col_bfp;
                        SI=DM(groups);
                        SR=LSHIFT SI BY 1 (LO);
                        DM(groups)=SR0;         {groups=groups X 2}
                        SI=DM(node_space);
                        SR=LSHIFT SI BY -1 (LO);
                        DM(node_space)=SR0;    {node_space=node_space / 2}
                        SI=DM(bflys_per_group);
                        SR=LSHIFT SI BY -1 (LO);
stage_loop:             DM(bflys_per_group)=SR0;

                     AY0 = DM(rcolbase);
                     AR = AY0 + 1;
                     DM(rcolbase) = AR;
                     AY0 = DM(icolbase);
                     AR = AY0 + 1;
                     DM(icolbase) = AR;

                     AX0 = DM(blk_exponent); {take blk exponent for this}
                     DM(I7,M7) = AX0;        {fft and store in array}
                     AX0 = 0;
                     DM(blk_exponent) = AX0;

                     RTS;
.ENDMOD;
```

Listing 7.4 Column DIF Module

7.1.3.4 Initialization

There are two initialization subroutines in this implementation of the two-dimensional FFT. Both of the routines are part of the main module. The first, *initialize*, performs once-only initialization of index registers, modify registers and length registers. Pointers are used for data access as well as access to twiddle factors. This routine is listed below in Listing 7.5. The twiddle factors are stored circular buffers of length 32 (for a 64-point FFT) in program memory and use the index registers I4 and I5. Index registers I0 and I2 set up pointer access to the real and imaginary parts of the data buffers *realdata* and *imaginarydata*. In addition, the *initialize* routine declares the variables *rrowbase*, *irowbase*, and *offset* to their initial values.

```
{One-time only initialization. Pointers and modifiers}
{and length registers.}

initialize: L0 = 0;
            L1 = 0;
            L2 = 0;
            L3 = 0;
            L4 = N_div_2;
            L5 = N_div_2;
            L6 = 0;
            L7 - 0,

            I0 = ^realdata;
            I2 = ^imaginarydata;
            I3 = ^tempdata;
            I4 = ^twid_real;
            I5 = ^twid_imag;
            I6 = ^row_exponents;

            DM(current_rrow) = I0;
            DM(current_irow) = I2;

            DM(current_icol) = I0;
            DM(current_rcol)=  I3;

            M0 = 0;
            M1 = 1;
            M3 = 0;
            M7 = 1;

            AF = PASS 0;
            AX0 = 0;
            AY0 = 0;
            AY1 = 4096;
            DM(offset) = AY0;       {base address for rows}
            DM(rrowbase) = AY0;
            DM(irowbase) = AY1;
            AY1 = ^tempdata;
            DM(rcolbase) = AY1;
            DM(icolbase) = AY0;
            RTS;
```

Listing 7.5 Initialize Routine

The second initialization routine is *col_init*, shown in Listing 7.6. This routine initializes variables before the start of each column FFT. This initialization is necessary because the data points in each column FFT are not in sequential order.

```
{       Initialization for column DIF FFTs}

col_init:    AY1 = N_DIV_2;
             DM(bflys_per_group) = AY1; {bflys = 1}
             AY1 = 1;
             DM(groups) = AY1;           {groups = 32}
             AX1 = 2048;
             DM(node_space) = AX1;       {node_space = 64}
             M1 = 64;                    {modifier for col. points}
             L4 = N_DIV_2;
             I4 = ^twid_real;
             I5 = ^twid_imag;

             RTS;
```

Listing 7.6 Column Initialization Routine

7.1.3.5 Bit Reverse Modules

The 1D FFTs used to perform the 2D FFT use the DIF algorithm and produce frequency output points in bit-reversed addressing (scrambled) order. For a complete explanation of bit-reversing in FFTs, see Chapter 6. Bit-reversing the addresses of the output data puts the data back into sequential order (unscrambles it). Because the output for each 1D FFT needs to be bit-reversed, the bit-reverse routine is called 2N times. The bit-reverse subroutines for the row and column FFTs are nearly identical.

The ADSP-2100 has a bit-reversed addressing capability. Unscrambling a row or column requires a seed value (the bit-reversed address of the first location in the row or column) and a modify value. The modify value is the value of 2 raised to the difference between 14 (the number of address bits) and the number of bits to be reversed, i.e., 2^{14-N}. In this example of a 64x64-point FFT, the number of bits to be reversed is six, and so the modify value is 2^8 or 256.

With the bit-reverse capability of the ADSP-2100 enabled, adding the modify value to a current address provides the correct address for the next bit-reversed sample. Because the 2D FFT needs to perform bit-reversal on 2N 1D FFTs, the program needs the seed values (first location addresses, bit-reversed) for each row and column. The seed values for the rows and columns for a 64x64-point 2D FFT are stored in buffers called *real_br_pointers, imag_br_pointers, c_real_br_pointers* and *c_imag_br_pointers* (*c_* means "column").

Listings 7.7 and 7.8 contain the row bit-reverse routine and the column bit-reverse routine, respectively.

```
.MODULE dif_unscramble;

{
      Calling Parameters
         Real and imaginary scrambled output data in inplacedata

      Return Values
         Normal ordered real output data in real_results
         Normal ordered imag. output data in imaginary_results

      Altered Registers
         I0,I1,I4,M1,M4, AY1,CNTR

      Altered Memory
         real_results, imaginary_results
}

.VAR/DM     rownum;
.CONST      N=64, mod_value=256;         {Initialize constants}
.CONST      N_x_2 = 128, N_DIV_2 = 32;

.EXTERNAL   current_rrow, current_irow;
.EXTERNAL   real_br_pointers, imag_br_pointers;
.ENTRY      unscr_start;        {Declare entry point into module}

unscr_start:  I4=DM(current_rrow);
                        {I4->real part of 1st data point}
              M4=1;     {Modify by 2 to fetch only real data}
              L0=0;
              L4=0;
              SI = DM(I6,M4);
              I1 = SI;
              M1=mod_value;          {Modifier for FFT size}
              CNTR=N;                {N=number of real data points}
              ENA BIT_REV;           {Enable bit-reverse}
```

(listing continues on next page)

```
                DO bit_rev_real UNTIL CE;
                   AY1=DM(I4,M4);      {Read real data}
bit_rev_real:      DM(I1,M1)=AY1;      {Place in sequential order}
                DM(current_rrow) = I4;
                CNTR = N;
                I4 = DM(current_irow);
                SI = DM(I7,M4);
                I1 = SI;

                DO bit_rev_imag UNTIL CE;
                   AY1=DM(I4,M4);      {Read imag data}
bit_rev_imag:      DM(I1,M1)=AY1;      {Place in sequential order}
                DM(current_irow) = I4;
                DIS BIT_REV;           {Disable bit-reverse}
                RTS;
.ENDMOD;
```

Listing 7.7 Row Bit-Reverse Routine

```
.MODULE col_dif_unscramble;

{
     Calling Parameters
        Real and imaginary scrambled output data in inplacedata

     Return Values
        Normal ordered real output data in real_results
        Normal ordered imag output data in imaginary_results

     Altered Registers
        I0, I1, I4, M1, M4, AY1, CNTR

     Altered Memory
        real_results, imaginary_results
}

.VAR/DM      rownum;
.CONST       N=64, mod_value=4;        {Initialize constants}
.CONST       N_x_2 = 128, N_DIV_2 = 32;

.EXTERNAL    current_rcol, current_icol;
.ENTRY       col_unscr_start;  {Declare entry point into module}
```

```
col_unscr_start: I4 = DM(current_icol);
                                {I4->real part of 1st data point}
                 M4 = 64;       {Modify by 64 to fetch next col val}
                 M5 = 1;
                 L0 = 0;
                 L4 = 0;
                 SI = DM(I6,M5);
                 I1 = SI;
                 M1=mod_value; {Modifier for FFT size}
                 CNTR=N;        {N=number of real data points}
                 ENA BIT_REV;   {Enable bit-reverse}

                 DO bit_rev_real UNTIL CE;
                    AY1=DM(I4,M4);     {Read real data}
bit_rev_real:       DM(I1,M1)=AY1;     {Place in sequential order}
                 AY0 = DM(current_icol);
                                 {increment pointer to current}
                 AR = AY0 + 1; {imaginary column data}
                 DM(current_icol) = AR;

                 CNTR = N;
                 I4 = DM(current_rcol);
                 SI  = DM(I7,M5);
                 I1 = SI;

                 DO bit_rev_imag UNTIL CE;
                    AY1=DM(I4,M4);  {Read imag data}
bit_rev_imag:       DM(I1,M1)=AY1;  {Place in sequential order}
                 AY0 = DM(current_rcol);
                                 {increment pointer to current}
                 AR = AY0 + 1; {real column data}
                 DM(current_rcol) = AR;
                 DIS BIT_REV;  {Disable bit-reverse}
                 RTS;
.ENDMOD;
```

Listing 7.8 Column Bit-Reverse Routine

7.1.3.6 Block Floating-Point Adjustment

Block floating-point format provides extended dynamic range of fixed-point arithmetic without the computational burdens of full floating-point arithmetic. In a block floating-point implementation, there is a single exponent for a group of mantissas. In the implementation of the 2D FFT explained in this chapter, there is an associated exponent for each of the 2N 1D FFTs performed.

There are two block floating-point adjustment routines for each 1D FFT. The first block floating-point routine normalizes 1D FFT values. The 64 real and imaginary output values are normalized to a single block exponent. At the completion of all of the rows or all of the columns the second block floating-point routine is called. This routine normalizes the output of the entire output array to a single exponent.

The first adjustment routine is called from within the stage loop of each 1D FFT. If there is any bit growth, then the stage output is adjusted according to a shift value stored in the SB register. For an N-stage FFT, there can be at most N bits of growth. For every bit of growth, this routine shifts the results of the FFT one bit and stores the amount of shift as an exponent in an appropriate buffer. The buffers used are *row_exponents* for the row FFTs and *col_exponents* for the column FFTs.

The second block floating-point adjustment routine is called after all of the row FFTs or column FFTs are complete. These routines *row_final_adj* and *col_final_adj* (for the row FFTs and column FFTs, respectively) search the buffers *col_exponents* and *row_exponents* for the largest exponent. Each row or column value is shifted by the difference between the largest exponent and the exponent associated with the particular row or column. This has the effect of scaling the entire output array to one exponent.

The block floating-point adjustment routines are shown in Listings 7.9 through 7.12. Listings 7.9 and 7.10 show the routines for the stage outputs of the row FFTs and column FFTs, respectively. Listings 7.11 and 7.12 show the routines that adjust the entire output array for the row FFTs and column FFTs, respectively.

```
.MODULE     each_row_bfp_adjust;
{
        Calling Parameters
           FFT stage results in realdata & imaginarydata

        Return Parameters
           realdata & imaginarydata adjusted for bit growth

        Altered Registers
           I0,I1,AX0,AY0,AR,MX0,MY0,MR,CNTR

        Altered Memory
           inplacereal, inplaceimag, blk_exponent
}

.EXTERNAL   realdata, blk_exponent;    {Begin declaration section}
.EXTERNAL   imaginarydata, rrowbase, irowbase;
.CONST      buffer_size = 64;
.ENTRY      row_bfp;

row_bfp:       AY0=CNTR;          {Check for last stage}
               AR=AY0-1;
               IF EQ RTS;         {If last stage, return}
               AY0=-2;
               AX0=SB;
               AR=AX0-AY0;        {Check for SB=-2}
               IF EQ RTS;         {If SB=-2, no bit growth, return}
               I0=DM(rrowbase); {I0=read pointer}
               I1=DM(rrowbase); {I1=write pointer}
               AY0=-1;
               MY0=H#4000;        {Set MY0 to shift 1 bit right}
               AR=AX0-AY0,MX0=DM(I0,M1);
                   {Check if SB=-1; Get first sample}
               IF EQ JUMP strt_shift;
                   {If SB=-1, shift block data 1 bit}
               AY0=-2;            {Set AY0 for block exponent update}
               MY0=H#2000;        {Set MY0 to shift 2 bits right}
strt_shift:    CNTR=buffer_size - 1; {initialize loop counter}
               DO shift_loop UNTIL CE;  {Shift block of data}
                  MR=MX0*MY0(RND),MX0=DM(I0,M1);
                         {MR=shifted data,MX0=next value}
shift_loop:       DM(I1,M1)=MR1;    {Unshifted data=shifted data}
               MR=MX0*MY0(RND);     {Shift last data word}
```

(listing continues on next page)

```
            I0=DM(irowbase);        {I0=read pointer}
            I1=DM(irowbase);        {I1=write pointer}
            AY0=-1;
            MY0=H#4000;             {Set MY0 to shift 1 bit right}
            AR=AX0-AY0,MX0=DM(I0,M1);
                                    {Check if SB=-1; Get first sample}
            IF EQ JUMP i_strt_shift;
                                    {If SB=-1, shift block data 1 bit}
            AY0=-2;                 {Set AY0 for block exponent update}
            MY0=H#2000;             {Set MY0 to shift 2 bits right}
i_strt_shift: CNTR=buffer_size - 1;  {initialize loop counter}
            DO i_shift_loop UNTIL CE;  {Shift block of data}
               MR=MX0*MY0(RND),MX0=DM(I0,M1);
                                    {MR=shifted data,MX0=next value}
i_shift_loop:    DM(I1,M1)=MR1;     {Unshifted data=shifted data}
            MR=MX0*MY0(RND);        {Shift last data word}

            AY0 = DM(blk_exponent); {Update block exponent and}
            DM(I1,M1)=MR1,AR=AY0-AX0;
                                     {store last shifted sample}
            DM(blk_exponent)=AR;
            RTS;

.ENDMOD;
```

Listing 7.9 Row BFP Adjustment Routine

```
.MODULE      column_bfp_adjust;

.EXTERNAL    realdata, blk_exponent;   {Begin declaration section}
.EXTERNAL    tempdata, rcolbase, icolbase;
.CONST       buffer_size = 64;
.ENTRY       col_bfp;

col_bfp:      AY0=CNTR;         {Check for last stage}
              AR=AY0-1;
              IF EQ RTS;        {If last stage, return}
              AY0=-2;
              AX0=SB;
              AR=AX0-AY0;       {Check for SB=-2}
              IF EQ RTS;        {If SB=-2, no bit growth, return}
```

```
            I0=DM(rcolbase);               {I0=read pointer}
            I1=DM(rcolbase);               {I1=write pointer}
            M1=buffer_size;
            AY0=-1;
            MY0=H#4000;
                                           {Set MY0 to shift 1 bit right}
            AR=AX0-AY0,MX0=DM(I0,M1);
                                           {Check if SB=-1; Get first sample}
            IF EQ JUMP strt_shift;
                                           {If SB=-1, shift block data 1 bit}
            AY0=-2;                        {Set AY0 for block exponent update}
            MY0=H#2000;                    {Set MY0 to shift 2 bits right}
strt_shift: CNTR=buffer_size - 1;          {initialize loop counter}
            DO shift_loop UNTIL CE;        {Shift block of data}
               MR=MX0*MY0(RND),MX0=DM(I0,M1);
                                           {MR=shifted data,MX0=next value}
shift_loop:    DM(I1,M1)=MR1;              {Unshifted data=shifted data}
            MR=MX0*MY0(RND);
            DM(I1,M1)=MR1;                 {Shift last data word}
                                           {store last shifted sample}

            I0=DM(icolbase);               {I0=read pointer}
            I1=DM(icolbase);               {I1=write pointer}
            AY0=-1;
            MY0=H#4000;                    {Set MY0 to shift 1 bit right}
            AR=AX0-AY0,MX0=DM(I0,M1);
                                           {Check if SB=-1; Get first sample}
            IF EQ JUMP shift_start;
                                           {If SB=-1, shift block data 1 bit}
            AY0=-2;                        {Set AY0 for block exponent update}
            MY0=H#2000;                    {Set MY0 to shift 2 bits right}
shift_start: CNTR=buffer_size - 1;         {initialize loop counter}
            DO shft_loop UNTIL CE;         {Shift block of data}
               MR=MX0*MY0(RND),MX0=DM(I0,M1);
                                           {MR=shifted data,MX0=next value}
shft_loop:     DM(I1,M1)=MR1;              {Unshifted data=shifted data}
            MR=MX0*MY0(RND);               {Shift last data word}
            AY0=DM(blk_exponent);          {Update block exponent and}
            DM(I1,M1)=MR1,AR=AY0-AX0;      {store last shifted sample}
            DM(blk_exponent)=AR;

            RTS;

.ENDMOD;
```

Listing 7.10 Column BFP Adjustment Routine

```
.MODULE         all_row_bfp;

{
        This module does the final adjusting of the row
        block floating point exponents, to normalize the
        entire array. Each row has an associated BFP exponent.
        This module finds the greatest magnitude exponent,
        then shifts each row by the difference of the individual
        BFP exponent and the greatest magnitude exponent.
}

.VAR/DM   largest_re;
.CONST    N=64;
.EXTERNAL row_exponents, rrowbase, irowbase, realdata, tempdata;
.ENTRY    row_final_adj;

{find the greatest of the row exponents}

row_final_adj: I0 = ^row_exponents;
               M0 = 0;
               M1 = 1;
               AX1 = 0;
               AF = PASS AX1;

               AY0 = DM(I0,M1);     {find max of 1st two values}
               AX0 = DM(I0,M1);
               AR = AX0 - AY0;      {see which is greater}

               IF GE AF = PASS AX0; {AF gets greatest of 1st two}
               IF LT AF = PASS AY0;
               AX0 = DM(I0,M1);

               CNTR = N-2;          {buffer size less first two values}
               DO findmax UNTIL CE; {find greatest in the buffer}
                  AR = AX0 - AF;
                  IF GE AF = PASS AX0;
findmax:          AX0 = DM(I0,M1);
               AR = PASS AF;        {put the largest row exponent}
               DM(largest_re) = AR; {in memory}

{shift each row by the difference between the greatest
and the individual row exponent}

               I1=^tempdata;        {address of post scrambled real data}
               I2=^realdata;        {address of post scrambled imag data}
               I0=^row_exponents;
```

```
                AY0 = DM(largest_re);
                CNTR = N;
                AX0=DM(I0,M1);              {get individual row BFP}

                DO row UNTIL CE;
                   AR = AY0 - AX0;
                                            {diff between greatest and row BFP}
                   SE = AR;
                   AF = PASS 0;             {clear AC bit}
                   CNTR = N;
                   DO row_shift UNTIL CE;   {shift the row}
                      SI = DM(I1,M0);
                                            {get next value leave pointer}
                      SR = ASHIFT SI (HI);
                      DM(I1,M1) = SR1;
                                            {put shifted value back, inc pointer}
                      SI = DM(I2,M0);       {do same for imag values}
                      SR = ASHIFT SI (HI);
row_shift:            DM(I2,M1) = SR1;
row:               AX0=DM(I0,M1);           {get indiv. row BFP}

                RTS;

.ENDMOD;
```

Listing 7.11 Row Final Exponent Adjustment Routine

```
.MODULE         all_col_bfp;

{
     This module does the final adjusting of the column
     block floating point exponents, to normalize the
     entire array. Each column has an associated BFP exponent.
     This module finds the greatest magnitude exponent,
     then shifts each column by the difference of the individual
     BFP exponent and the greatest magnitude exponent.
}
.VAR/DM         largest_ce;
.CONST          N=64;
.EXTERNAL       col_exponents, realdata, imaginarydata;
.EXTERNAL       current_rcol, current_icol;
.ENTRY          col_final_adj;

{find the greatest of the column exponents}

col_final_adj: I0 = ^col_exponents;
               M0 = 0;
               M1 = 1;
               M2 = N;             {increment by N for columns}
               AX1 = 0;
               AY1 = h#1000;
               AF = PASS AX1;
               DM(current_rcol) = AX1;
               DM(current_icol) = AY1;

               AY0 = DM(I0,M1);    {find max of 1st two values}
               AX0 = DM(I0,M1);
               AR = AX0 - AY0;     {see which is greater}

               IF GE AF = PASS AX0; {AF gets greatest of 1st two}
               IF LT AF = PASS AY0;
               AX0 = DM(I0,M1);

               CNTR = N-2;         {buffer size less first two values}
               DO findcolmax UNTIL CE;
                                   {find greatest in the buffer}
                  AR = AX0 - AF;
                  IF GE AF = PASS AX0;
findcolmax:       AX0 = DM(I0,M1);
               AR = PASS AF;       {put the largest row exponent}
               DM(largest_ce) = AR; {in memory}
```

```
{shift each row by the difference between the greatest and the individual row exponent}

              I1=DM(current_rcol);
                                           {address of post scrambled real data}
              I2=DM(current_icol);
                                           {address of post scrambled imag data}
              I0 = ^col_exponents;

              AY0 = DM(largest_ce);
              CNTR = N;
              AX0=DM(I0,M1);               {get indiv. row BFP}

              DO col UNTIL CE;
                 AR = AY0 - AX0;
                                           {diff between greatest and row BFP}
                 SE = AR;
                 AF = PASS 0;              {clear AC bit}
                 CNTR = N;
                 DO col_shift UNTIL CE;    {shift the row}
                    SI = DM(I1,M0);
                                           {get next value leave pointer}
                    SR = ASHIFT SI (HI);
                    DM(I1,M2) = SR1;
                                           {put shifted value back, incr pointer}
                    SI = DM(I2,M0);        {do same for imag values}
                    SR = ASHIFT SI (HI);
col_shift:          DM(I2,M2) = SR1;
                 AY1=DM(current_rcol);     {incr col pointer real}
                 AR = AY1 + 1;
                 I1 = AR;
                 DM(current_rcol) = AR;
                 AY1=DM(current_icol);     {incr col pointer imag}
                 AR = AY1 + 1;
                 I2 = AR;
                 DM(current_icol) = AR;
col:             AX0=DM(I0,M1);            {get next indiv. row BFP}

              RTS;

.ENDMOD;
```

Listing 7.12 Column Final Exponent Adjustment Routine

7 Two-Dimensional FFTs

7.2 BENCHMARKS

Benchmarks for the 2D FFT programs are given below. All are for a 64-by-64-point 2D FFT.

Routine	*Number of Cycles*	*Execution Time* (12.5MHz ADSP-2100A)
fft_2d (main)	597	47.8 µs
initialize	33	2.64 µs
col_init	11 (called 64 times)	0.88 µs
dif_fft	3457 (called 64 times)	277.00 µs
col_dif	3396 (called 64 times)	272.00 µs
row_bfp	287 (multiple calls)*	23.00 µs
col_bfp	288 (multiple calls)*	23.00 µs
all_row_bfp	25105	2.01 ms
all_col_bfp	25118	2.01 ms
unscr	279 (called 64 times)	22.3 µs
col_unscr	280 (called 64 times)	22.4 µs

* This routine is data-dependent; the cycle count shown is worst case.

Experimentally, the 2D FFTs of complex signals ranged from a minimum of 543001 cycles (43.4ms) to a maximum 633701 cycles (50.7ms).

Image Processing ■ 8

8.1 OVERVIEW

Image processing often involves computation on large matrices (of data values) that represent digitized images. Each element of the array represents a pixel of the image; its location in the array corresponds to its location in the image, and its value determines the color or shading of the pixel.

The largest two-dimensional matrix that a full complement of ADSP-2100 data memory will hold contains 16384 elements (128 rows by 128 columns). If your application requires a larger array, you must use extended memory addressing. For more information on a hardware implementation of this type of memory architecture, contact Analog Devices' DSP Applications Group.

The CNTR register of the ADSP-2100 is a 14-bit unsigned register that can track loop iteration; a loop can be set to execute up to 16383 times. If more loop iterations are needed, they can be obtained by nesting loops. If the number of iterations needed can not be achieved easily through loop nesting, the loop can be programmed explicitly. In this method, the AF register is preloaded with the number of loops, and at the end of the loop it is decremented. A conditional jump tests the AF value after the decrement; if it is not zero, a jump to the top of the loop is executed, and if it is zero, no jump occurs, and the loop is exited. A loop using this method can be iterated up to 65535 times. Each loop execution incurs a two-cycle overhead penalty, however.

8.2 TWO-DIMENSIONAL CONVOLUTION

Two-dimensional convolution has a variety of applications in image processing. One common application is finding an object in an image using two-dimensional edge detection. To determine the orientation of the edge, an input window is convolved with several different templates. A direction is chosen based on the template that yields the maximum convolution value. Another application is pixel smoothing, which can be accomplished with two-dimensional convolution by varying the convolution coefficients. A smoothed pixel value is based on the weighted average of the unsmoothed value and those of the eight neighboring pixels.

When implementing the convolution, instead of giving the address of the pixel being convolved, the address of the upper left hand corner of the convolving window is given and the convolution window will not extend outside the input window. Due to this fact, the output window will be smaller (by two pixels per row and column) than the input window. Input and coefficient matrices stored by rows allow you to take advantage of the one-cycle address modification capability of the ADSP-2100 using modify (M) registers. When the pointer to the input window reaches the end of each coefficient row, it is updated with a different modify register that causes the pointer to move to the beginning of the next row of the convolution window. When the multiplication is complete, still another modify register moves the index register to the point at which the next convolution will begin.

The routine in Listing 8.1 uses two loops to allow for large input windows. Because the CNTR register is 14-bits wide, a single loop would restrict the input matrix to 16383 elements. The two loops also may be easily modified for extended memory addressing.

Before calling the routine you must store the address of the input matrix in I0, the coefficient matrix in I4, and the output matrix in I1. The length registers L0 and L1 should be set to zero. L4 should be set to nine to use the circular buffer modulo addressing with the pointer to the convolution matrix, I4, because the matrix is used repeatedly. When performing the convolution on an MxN matrix, CNTR should be set to M–2, the number of output rows, and M1 should be set to N–2, the number of output columns. M1 is also used to move the I0 pointer from one row to the next, which is necessary because the nine values of the input matrix used in the convolution are not contiguous in memory. M2 is set to –(2N +1) to move the I0 pointer back to the beginning of the convolution window, and M3 is set to two to move the I0 pointer to the next input row. M0 and M4 are set to one for sequential fetches from data memory and program memory.

The routine begins by reading the first data and coefficient values. The *in_row* loop is executed once for each row of the output matrix. The *in_col* loop executes once for each column of the output matrix. Inside this loop, one multiply/data fetch multifunction instruction is executed for each element of the convolution window.

The last instruction of the *in_col* loop saves the output data point. A MODIFY instruction that moves the input matrix pointer, I0, to the beginning of the next input row finishes the *in_row* loop.

```
.MODULE Two_Dimensional_Convolution;

{              3   3
   G(x,y) = Σ  Σ [H(i,j) × F(x+i, y+j)]
            i=1 j=1

   Calling Parameters
        I0 -> F(x,y), MxN Input Matrix                          L0 = 0
        I1 -> G(x,y), (M-2)x(N-2) Output Matrix                 L1 = 0
        I4 -> H(i,j), 3x3 Coefficient Matrix (circular)         L4 = 9
        CNTR = M-2        M3 = 2
        M0 = 1            M4 = 1
        M1 = N-2          M2 = -(2 × N + 1)

   Return Values
        G(x,y) Filled

   Altered Registers
        MX0,MY0,MR,I0,I1,I4

   Computation Time
        ((10 × (N-2) + 4) × (M-2)) + 3 + 10 cycles
}

.ENTRY  conv;

conv:   MX0=DM(I0,M0), MY0=PM(I4,M4);           {Get first data and coeff}
        DO in_row UNTIL CE;                     {Loop M-2 times}
           CNTR=M1;
           DO in_col UNTIL CE;                  {Loop N-2 times}
              MR=MX0*MY0 (SS), MX0=DM(I0,M0), MY0=PM(I4,M4);
              MR=MR+MX0*MY0 (SS), MX0=DM(I0,M1), MY0=PM(I4,M4);
              MR=MR+MX0*MY0 (SS), MX0=DM(I0,M0), MY0=PM(I4,M4);
              MR=MR+MX0*MY0 (SS), MX0=DM(I0,M0), MY0=PM(I4,M4);
              MR=MR+MX0*MY0 (SS), MX0=DM(I0,M1), MY0=PM(I4,M4);
              MR=MR+MX0*MY0 (SS), MX0=DM(I0,M0), MY0=PM(I4,M4);
              MR=MR+MX0*MY0 (SS), MX0=DM(I0,M0), MY0=PM(I4,M4);
              MR=MR+MX0*MY0 (SS), MX0=DM(I0,M2), MY0=PM(I4,M4);
              MR=MR+MX0*MY0 (RND), MX0=DM(I0,M0), MY0=PM(I4,M4);
in_col:       DM(I1,M0)=MR1;             {Save convolution value}
in_row:    MODIFY(I0,M0);                {Point to next input row}

       RTS;
.ENDMOD;
```

Listing 8.1 Two-Dimensional Convolution

8.3 SINGLE-PRECISION MATRIX MULTIPLY

Matrix multiplication is commonly used to translate or rotate an image. The routine presented in this section multiplies two input matrices: X, an RxS (R rows, S columns) matrix stored in data memory, and Y, an SxT (S rows, T columns) matrix stored in program memory. The output Z, an RxT (R rows, T columns) matrix, is written to data memory.

Before calling the routine, which is shown in Listing 8.2, the values must be set up as follows. The starting address of X must be in I2 and the starting address of Y in I5. The starting address of the result buffer Z must be in I1. All matrices are stored by rows; that is, each successive memory location contains the next sequential row element, and the last element of a row is followed by the first element of the next row. M0 and M4 must be set to one. M5 must contain the value of T (number of columns in Y), and M1 the value of S (number of columns in X). The CNTR register must contain the value of R (number of rows in X), and SE must contain the value necessary to shift the result of each multiplication into the desired format. For example, SE would be set to zero to obtain a matrix of 1.31 values from the multiplication of two matrices of 1.15 values.

The *row_loop* loop is executed once for each row of the result matrix (R times). Before the *column_loop* loop is entered, CNTR is set to the value of T and I5 is reset to point to the beginning of the Y matrix. The *column_loop* loop is executed once for each column of the result matrix (T times). The *element_loop* loop is executed to compute one element of the output matrix.

Before the element of the output matrix is computed, CNTR is set to S for the number of multiplies necessary, I0 is set to the first element of the current X matrix row, and I4 is set to the first element of the current Y matrix column. The MR register is cleared, the first element of the current X row is loaded into MX0, and the first element of the current Y column is loaded into MY0.

After an element of the output matrix is computed, it is adjusted in the shifter to maintain data integrity. For example, if each matrix is composed of 4.12 elements, each output element must be shifted to the left three bits to form a 4.12 value before being stored, if the output matrix is also to be composed of 4.12 elements. The magnitude of the shift is controlled by the value that was preloaded in the SE register. During the first shift operation, the multiword instruction also updates the pointer to the next column of the Y matrix.

The last instruction in the *column_loop* loop stores the value of the output element in memory. The *row_loop* loop finishes by modifying the pointer to the current row of the X matrix.

The time required to complete the multiplication depends on the size of the input matrices. The routine requires one cycle for the return instruction and one cycle to start the *row_loop* loop, which is executed R times. The *row_loop* loop requires four cycles in addition to the cycles used in *column_loop*, which is executed T times. The *column_loop* loop contains eight cycles overhead plus the *element_loop* loop, which executes in S cycles. Two more cycles are required for data reads from program memory before the code is completely contained in cache memory. The total execution time is therefore $((S + 8) \times T + 4) \times R + 4$ cycles.

```
.MODULE matmul;

{
   Single-Precision Matrix Multiplication

                    S
        Z(i,j) = Σ [X(i,k) × Y(k,j)]    i = 0 to R; j = 0 to T
                  k=0

        X is an RxS matrix
        Y is an SxT matrix
        Z is an RxT matrix

   Calling Parameters
        I1 -> Z buffer in data memory                L1 = 0
        I2 -> X, stored by rows in data memory       L2 = 0
        I6 -> Y, stored by rows in program memory    L6 = 0
        M0 = 1       M1 = S
        M4 = 1       M5 = T
        L0,L4,L5 = 0
        SE = Appropriate scale value
        CNTR = R

   Return Values
        Z Buffer filled by rows

   Altered Registers
        I0,I1,I2,I4,I5,MR,MX0,MY0,SR

   Computation Time
        ((S + 8) × T + 4) × R + 2 + 2 cycles
}
```

(listing continues on next page)

```
.ENTRY   spmm;

spmm:    DO row_loop UNTIL CE;
            I5=I6;                                    {I5 = start of Y}
            CNTR=M5;
            DO column_loop UNTIL CE;
               I0=I2;                                 {Set I0 to current X row}
               I4=I5;                                 {Set I4 to current Y col}
               CNTR=M1;
               MR=0, MX0=DM(I0,M0), MY0=PM(I4,M5);    {Get 1st data}
               DO element_loop UNTIL CE;
element_loop:     MR=MR+MX0*MY0 (SS), MX0=DM(I0,M0), MY0=PM(I4,M5);
               SR=ASHIFT MR1 (HI), MY0=DM(I5,M4);     {Update I5}
               SR=SR OR LSHIFT MR0 (LO);              {Finish shift}
column_loop:   DM(I1,M0)=SR1;                         {Save output}
row_loop:   MODIFY(I2,M1);                            {Update I2 to next X row}
         RTS;
.ENDMOD;
```

Listing 8.2 Single-Precision Matrix Multiply

8.4 HISTOGRAM

A histogram describes the frequency of occurrences of a particular value in a matrix (or, more generally, any set of data). A common image-processing application is determining the number of times a particular gray scale (pixel value) occurs in a digitized two-dimensional image.

To perform a histogram on a range of data, you must know the number of unique values each datum can have. The histogram contains one location for each value in which it records the occurrences of that value. For example, if each pixel in an image is represented by an 8-bit value, each pixel can take on 256 (28) different values. Its histogram has 256 different locations, one for each value.

The histogram routine is shown in Listing 8.3. The address of the input array is read from the address stored in I0; L0 and L4 must both be set to zero. The modify register M1 must be set to one, and M5 and M0 must be set to zero. All locations of the output array, whose address is stored in I4, must be initialized to zero before the routine is called.

The routine checks the value of each element in the input data array and increments the appropriate location for the value in the output array. The routine begins by reading in the first value of the input data array. This value is used to modify the index register, which initially points to the beginning of the output data array. The second input value is then read, and the CNTR is set to the number of remaining input points (the total number less two).

Each pass through the *histo_loop* loop calculates the modify value for the output buffer pointer (I4), reads in and increments the current counter, inputs the next value, and modifies the I4 register to point to the next location as it saves the new counter value. The instructions following the *histo_loop* loop read in and increment the counters for the last two data values.

No boundary checking is performed on the input data, so the calling routine must ensure that the output histogram has enough locations to accommodate the maximum possible range of the input data.

```
.MODULE Histogram_subroutine;

.CONST  N_less_2=2046;          {number of pixels - 2}

{
   Calculates histogram of input data

   Calling Parameters
        I0 -> Input buffer in data memory        L0 = 0
        I4 -> Histogram buffer in program memory L4 = 0
             (I4 initialized to first location;
              All buffer locations initialized to 0)
        M0 = 0
        M1 = 1
        M5 = 0

   Return Values
        Histogram buffer filled with results

   Altered Registers
        AX0,AX1,AY0,AF,AR,I0,I4,M4

   Computation Time
        7 + (4 × (Input buffer length - 2) + 2) + 12 cycles
}
```

(listing continues on next page)

```
.ENTRY  histo;

histo:  AX1=DM(I0,M1);                          {Get 1st data value}
        M4=AX1;
        MODIFY(I4,M4);                          {Point to its counter}
        AX0=DM(I0,M0);                          {Get 2nd data value}
        AF=PASS AX1;                            {Pass to y operand}

        CNTR=N_less_2;

        DO histo_loop UNTIL CE;
             AR=AX0-AF, AX1=DM(I0,M1), AY0=PM(I4,M5);
                                {Calc index, reread data, locate cntr}
             M4=AR;             {Transfer index to modify register}
             AR=AY0+1, AX0=DM(I0,M0);           {Incr cntr, load next data}
histo_loop:  PM(I4,M4)=AR, AF=PASS AX1;
             {Save updated cntr, point to next, pass to y}

   AR=AX0-AF, AY0=PM(I4,M5);                    {Calc index, locate cntr}
   M4=AR;                                       {Transfer index to modify register}
   AR=AY0+1;                                    {Increment cntr}
   PM(I4,M4)=AR;                                {Save updated cntr, point to next}
   AY0=PM(I4,M5);                               {Locate cntr}
   AR=AY0+1;                                    {Incr cntr}
   PM(I4,M5)=AR;                                {Store updated cntr}
   RTS;

.ENDMOD;
```

Listing 8.3 Histogram

8.5 REFERENCES

Offen, R.J. 1985. VLSI Image Processing. New York: McGraw-Hill Book Company.

Oppenheim, A.V. ed. 1978. Applications of Digital Signal Processing. Englewood Cliffs, N.J.: Prentice-Hall, Inc.

Graphics ■ 9

9.1 OVERVIEW

In graphics processing, geometric and topological information constitute the essence of the image. By contrast, in the more familiar image processing, an image consists of pixels. Graphics programs operate on data and data structures such as vector arrays, line lists, and polygons rather than pixels. Graphics processors use rotation, projection, ray tracing, and other techniques to *synthesize original* images, whereas image processors *enhance existing* image aspects such as contrast, definition, and edge delineation. Graphics applications include operations such as geometric modeling and drafting, solids and surface modeling, ray tracing, hidden line removal, shadow casting, texture mapping, perspective views, image synthesis, three-dimensional imagery, and animation.

Generally, high-end graphics applications are limited to 32-bit machines because of the greater resolution necessary in recursive transformations to avoid accumulating observable error. In fact, a floating-point format is often necessary to provide sufficient dynamic range to accommodate zooming and scaling operations. However, a low-end graphics engine that uses the 16-bit fixed-point format of the ADSP-2100 is more than adequate for applications such as video games and small computer graphics packages. To illustrate the graphics processing capabilities of the ADSP-2100, we present such an application in this chapter.

The complete graphics processor solution presented in this chapter consists solely of the ADSP-2100 single-chip microprocessor and some simple analog interface components. The ADSP-2100 performs all aspects of spatial rotation and display of a three-dimensional object in real time, demonstrating the principles of basic graphics operations discussed in this chapter. The example is expressly for demonstration and implements only a subset of the graphics processing capabilities available with the ADSP-2100.

9.2 GRAPHICS PROCESSING SYSTEM

Figure 9.1 shows a block diagram of the graphics processing system based on the ADSP-2100. An analog-to-digital converter (ADC) takes samples of a joystick's position as inputs. The graphics processor uses this data to control the amount of rotation of an object displayed on an oscilloscope. A digital-to-analog converter (DAC) generates beam deflection voltages for the oscilloscope from the output of the graphics processor to draw the object. The four-channel 8-bit ADC is memory-mapped into the ADSP-2100 data memory space and joystick input samples are obtained from this memory space. A quad 8-bit DAC is also mapped into the ADSP-2100 memory space; the ADSP-2100 writes data to this memory space to control the oscilloscope beam.

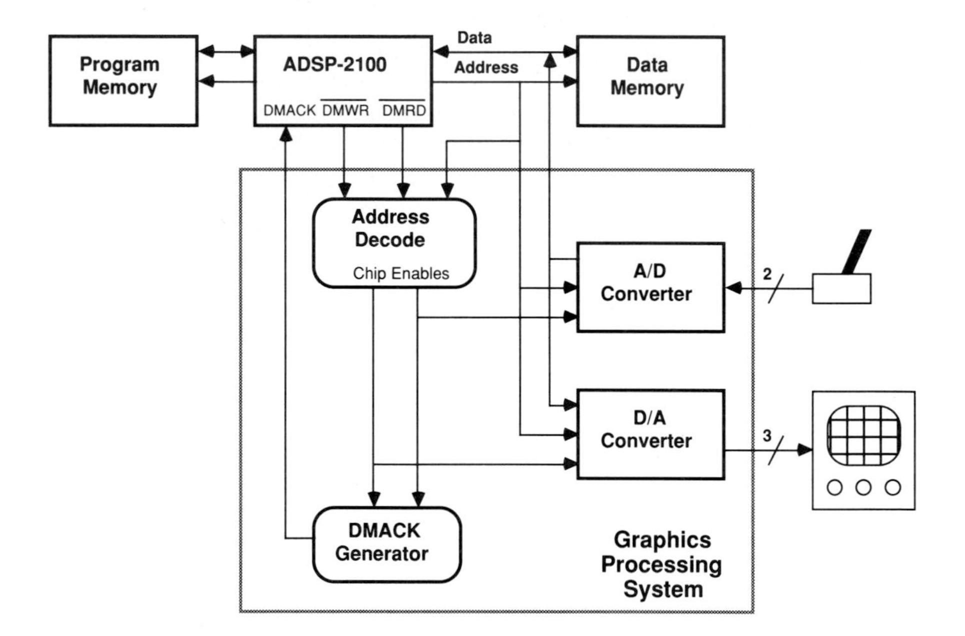

Figure 9.1 Graphics System Block Diagram

The reference object is stored in data memory as a series of (x, y, z) coordinate sets (vectors). Each vector represents a point or vertex of the object and all vertices are numbered. A line list (also stored in data memory) describes where (between which points) lines are to be drawn.

The software provides for rotating the object in four modes. The object can be rotated in each of three dimensions sequentially or in all three dimensions simultaneously. Rotation in these two modes is continuous and requires no joystick control. The object can also be rotated in the direction indicated by the joystick. The joystick position can be sampled and processed in two different ways, each of which produces a different effect on the motion of the displayed object. A pushbutton in hardware switches from one rotation mode to the next.

In the two joystick-controlled modes, each new set of joystick input samples starts the process of rotating the displayed object. A rotational transform is generated from the joystick data. In the other modes, the rotational transform is generated automatically in software. Matrix multiplication (described in Chapter 8) of the current position of the reference object by the rotational transform calculates the new position of the reference object, point by point. The rotated object is then projected from the three-dimensional spatial coordinate system onto a two-dimensional screen coordinate system to enable it to be displayed; this process is similar to casting the shadow of the object.

The wire-frame drawing of the object is done using the Bressenham line segment drawing algorithm (Foley and Van Dam, 1983). The line list tells the Bressenham algorithm where to draw lines. The actual line drawing is done by moving the oscilloscope beam along the path between the two endpoints of the line. Each line is drawn, a pixel at a time, until the entire object has been completed. The drawing sequence is then repeated ad infinitum.

Other topics covered in this chapter necessarily include data normalization and scaling, finite precision arithmetic, numerical overflow, and saturating arithmetic. Performance measures, data structures, schematics, and program listings pertaining to the example are also presented.

9.3 SETTING THE STAGE

Three-dimensional scenes use a four-dimensional transform space, just as two-dimensional scenes use a three-dimensional transform space, because the (x, y) and (x, y, z) coordinates of two-dimensional and three-dimensional vectors need an additional scale factor, generally referred to as W. In two-dimensional notation, the point P(x, y) is represented as P(Wx, Wy, W), with the scale factor $W \neq 0$. The coordinates for the point P(X, Y, W) are then $x=X/W$ and $y=Y/W$. The scale factor W preserves vector scaling through any transformation. In three-dimensional notation, the point P(x, y, z) is represented as P(Wx, Wy, Wz, W), and the coordinates are recovered similarly.

The ability to scale vectors on a pointwise basis is important because it allows equal resolution of coarse-grained and fine-grained features. For example, one display of a data base may be a view of a space shuttle from a distance of 50 meters, while a second view of the same data base may detail the 1/4-20 bolt positions of the gibulator inside the pod bay door control assembly.

Large values of W allow fine-grained coordinates, which would otherwise underflow an integer format, to be represented with resolution comparable to that of coarse-grained coordinates (which necessarily have smaller values of W). In essence, W can be considered as an exponent associated with each fixed-point coordinate set that is much the same as the exponent a full floating-point hardware implementation would provide.

For the sake of simplicity, we set W=1 in this example so the three-dimensional point P(x, y, z) is represented as P(X, Y, Z, 1), in which x=X, etc. All points are represented as row vectors with normal scaling and (x, y, z) components:

$$P(X, Y, Z, 1) = [x\ y\ z\ 1]$$

The left-handed coordinate system shown in Figure 9.2 is used because this system provides for larger z values to be displayed as being further from the viewer: a more intuitive convention than the familiar right-

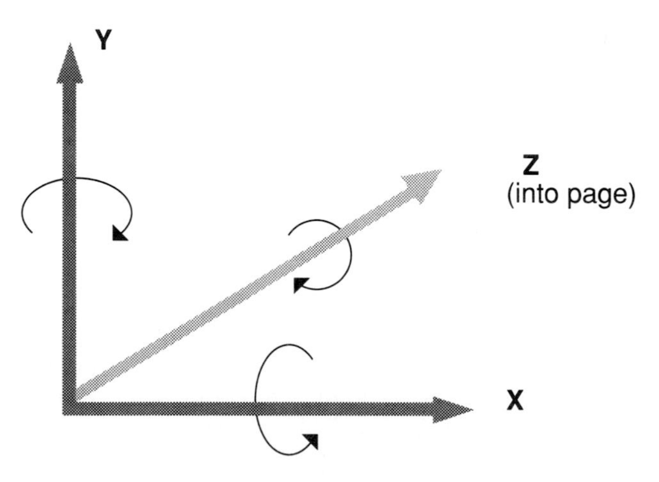

Figure 9.2 Left-Handed Coordinate System

handed system in which the z-axis comes out of the page. Positive rotations for the left-handed system are always clockwise when viewing the origin from a positive axis.

Individual transforms can be concatenated by matrix multiplication to form a single complex transform. The complex transform has the same total effect as each simple transform applied sequentially. Thus, multiple operations can be performed simultaneously, rather that sequentially, saving valuable processor time.

In general, a four-dimensional transform matrix is comprised of various submatrixes corresponding to different operations. Rotational operators comprise a 3x3 submatrix justified to the upper left corner of the 4x4 matrix, translation operators constitute a 1x3 submatrix in the lower left corner, perspective operators constitute a 3x1 in the upper right corner, and zooming (the simultaneous scaling of all three components) uses only a 1x1 element in the lower right corner, as shown in Figure 9.3.

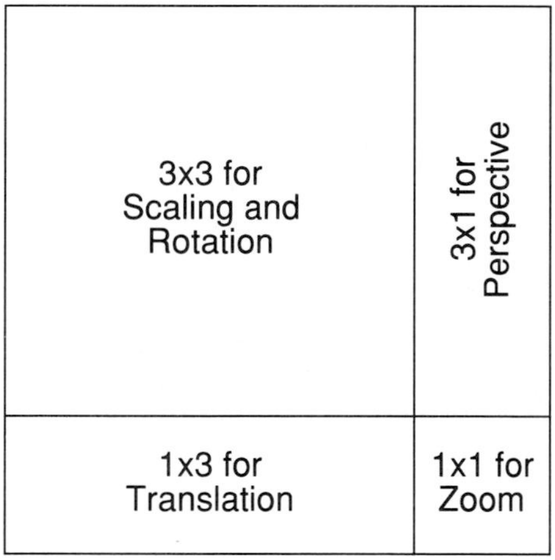

3x3 for Scaling and Rotation	3x1 for Perspective
1x3 for Translation	1x1 for Zoom

Figure 9.3 Components of the 4x4 Transformation Matrix

The conventional geometric operations that can be performed on three-dimensional coordinates (in a four-dimensional space) are rotation (see Figures 9.4 through 9.6), translation (Figure 9.7), and scaling (Figure 9.8). In these figures, Cx and Sy in the example matrixes denote the trigonometric functions cos(x) and sin(y), in which x and y are angles of rotation. Perspective transformations and zooming are neglected for the moment.

$$\begin{bmatrix} 1 & 0 & 0 & 0 \\ 0 & Cx & Sx & 0 \\ 0 & -Sx & Cx & 0 \\ 0 & 0 & 0 & 1 \end{bmatrix}$$

Figure 9.4 Rotation About the X Axis

$$\begin{bmatrix} Cy & 0 & -Sy & 0 \\ 0 & 1 & 0 & 0 \\ Sy & 0 & Cy & 0 \\ 0 & 0 & 0 & 1 \end{bmatrix}$$

Figure 9.5 Rotation About the Y Axis

$$\begin{bmatrix} Cz & Sz & 0 & 0 \\ -Sz & Cz & 0 & 0 \\ 0 & 0 & 1 & 0 \\ 0 & 0 & 0 & 1 \end{bmatrix}$$

Figure 9.6 Rotation About the Z Axis

$$\begin{bmatrix} 1 & 0 & 0 & 0 \\ 0 & 1 & 0 & 0 \\ 0 & 0 & 1 & 0 \\ \Delta x & \Delta y & \Delta z & 1 \end{bmatrix}$$

Figure 9.7 Translation by (Δx, Δy, Δz)

$$\begin{bmatrix} Sx & 0 & 0 & 0 \\ 0 & Sy & 0 & 0 \\ 0 & 0 & Sz & 0 \\ 0 & 0 & 0 & 1 \end{bmatrix}$$

Figure 9.8 Scaling by (Sx, Sy, Sz)

9.4 COMPUTATIONAL REDUCTIONS IN TRANSFORMATIONS

The transformation matrix can be simplified to reduce computational requirements. There is a tradeoff in the complexity of the graphic display, but we will show that the tradeoff is not significant for this application.

Any number of rotation, scaling, and translation matrixes can be multiplied together before being applied to the object. The result is always a single matrix, **M**, of the form shown in Figure 9.9.

$$\mathbf{M} = \begin{bmatrix} r_{11} & r_{12} & r_{13} & 0 \\ r_{21} & r_{22} & r_{23} & 0 \\ r_{31} & r_{32} & r_{33} & 0 \\ t_x & t_y & t_z & 1 \end{bmatrix}$$

Figure 9.9 Combined Rotation, Scaling and Translation Matrix

As shown in Figure 9.9, the upper-left 3x3 submatrix, R, gives the aggregate rotation and scaling of all the premultiplied matrixes, while the lower-left 1x3 submatrix **T** gives the aggregate translation. A reduction in the amount of numerical processing to evaluate the overall transform is obtained by the simplification:

$$[x'\ y'\ z'] = [x\ y\ z] \bullet \mathbf{R} + \mathbf{T}$$

rather than by implementing the full 1x4•4x4 multiplication directly:

$$[x'\ y'\ z'\ 1] = [x\ y\ z\ 1] \bullet \mathbf{M}$$

The 3x3 matrix provides a much simpler and faster implementation because only 9 multiplications and 6 additions are needed to transform each vector, as opposed to 16 multiplications and 12 additions for the 4x4 matrix: a 56% savings in multiplications alone!

The 3x3 matrix structure preserves both rotation and translation, although the zoom and perspective functions are lost. Applications needing zoom must either preserve the 4x4 transform structure and sustain increased computational load or use the 3x3 structure and apply any zoom operations as a postprocess to the rotation transform. The decision depends on how many vectors there are and what the throughput requirements are. Because we have no great dynamics in this example (W=1 for all points), the loss of the zoom function is of no consequence.

Perspective transformations introduce realism by use of one or more vanishing points. Without perspective, parallel lines converge at a point located at infinity. Vanishing points are imaginary points usually set at some finite distance from the object along a major axis. They move the convergence point of parallel lines in from infinity, introducing foreshortening in which foreground objects appear larger and background objects appear smaller, creating an illusion of realism (see Figure 9.10).

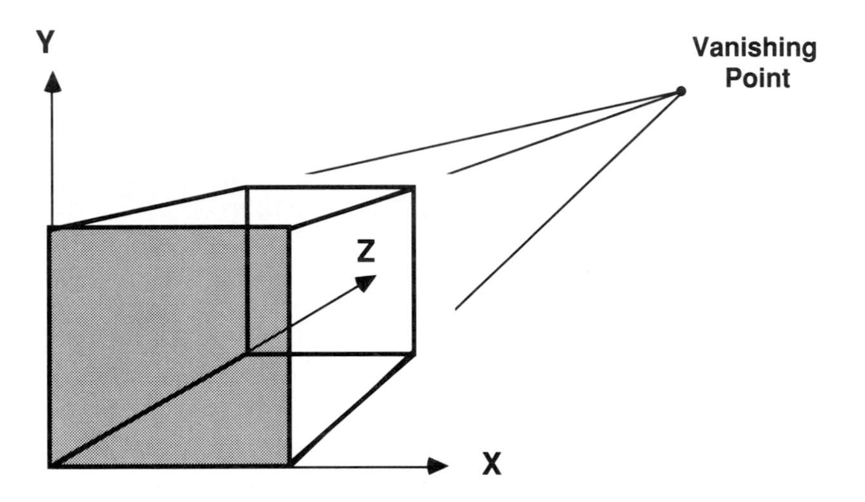

Figure 9.10 Perspective Projection Using One Vanishing Point

Nonzero elements in the 3x1 perspective submatrix migrate the vanishing point associated with the corresponding axis away from infinity. Perspective transformations are lost by the simplification to the more efficient 3x3 matrix structure because all perspective elements are assumed to be zero.

The example shown here uses a simple parallel projection technique and does not need perspective projection. Therefore, the loss of perspective transformations is of no consequence. If perspective transformations are needed, then the transform matrix size must be increased and the efficiency of the whole computational process suffers.

The forfeit of the perspective and zoom transformations for improved efficiency are somewhat subjective decisions. Only a viewing of the final result will determine whether the correct cost/performance tradeoff has been made. The criteria for making these decisions consist of aesthetic and performance considerations; if more complex (and realistic) visual effects are needed, then use the perspective and zoom transformations.

The combined rotational transform matrix, **R**, used in the example is shown in Figure 9.11. Any translation (matrix **T**) would be applied after calculating **R** using simple addition as shown above. Translation, however, is not demonstrated in this example.

$$\mathbf{R} = \begin{bmatrix} CyCz & CySz & -Sy \\ SxSyCz - CxSz & SxSySz + CxCz & SxCy \\ CxSyCz + SxSz & CxSySz - SxCz & CxCy \end{bmatrix}$$

Figure 9.11 Concatenated Rotation Matrix

9.5 PROJECTION TECHNIQUES

Once the scene has been transformed in three dimensions, it must be projected onto a two-dimensional screen to be viewed. This operation is like casting a shadow onto the sidewalk; a three-dimensional object is projected onto the two-dimensional sidewalk by the sun. Two types of projection techniques exist: the perspective projection and the parallel projection.

Perspective projections are visually more pleasing and intuitive. These projections use vanishing points to which parallel lines (other than those parallel to the projection plane) converge. As a result of this convergence, distant objects seem to be further away because they are smaller than closer objects.

One or two vanishing points are generally used, depending upon the degree of realism desired. The two-vanishing-point scheme (see Figure 9.12, on the next page) is commonly used in engineering sketches, the graphic arts, and architectural and industrial design. Two-point renderings usually preserve vertical parallelism while "parallel" lines in the other two dimensions actually converge to their respective vanishing points.

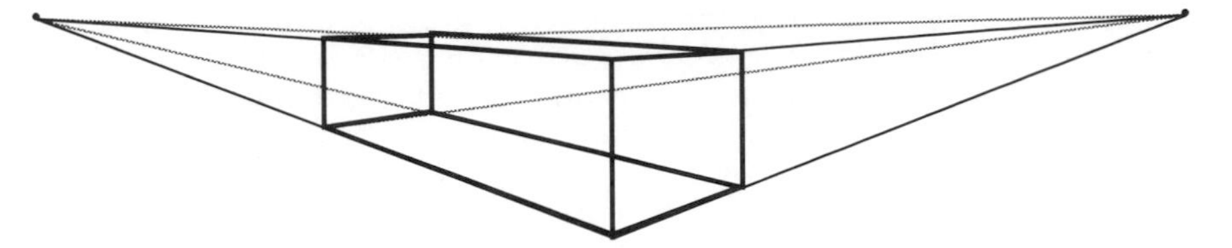

Figure 9.12 Prospective Projection Using Two Vanishing Points

Parallel projections do not use vanishing points at all; parallel lines in three-dimensional space remain parallel in two-dimensional space. The most common parallel projection is the orthographic projection, shown in Figure 9.13, in which top, end, and side views convey the essence of an object. Orthographic projections are so named because the normal of the projection plane is parallel to the projection direction.

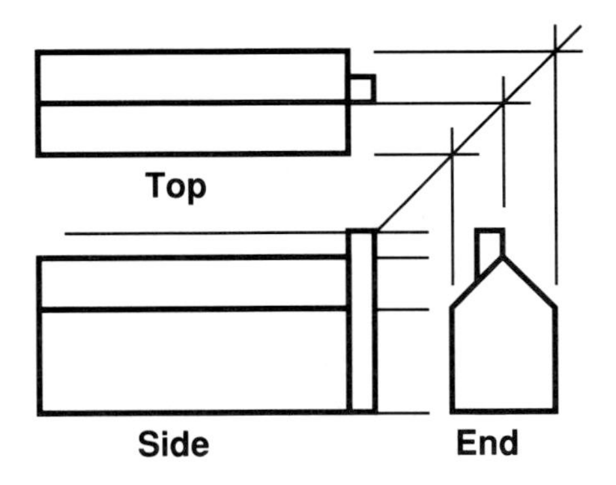

Figure 9.13 Three Orthographic Views of a House

The orthographic projection is used to depict machine parts and building structures because length and angular measures are preserved. Orthographic projections are difficult to visualize because they show only head-on projections of the different sides of objects, and the viewer must conceptualize the image.

A second class of parallel projections is the oblique projection. Oblique projections share the general orthographic property of the projection plane being normal to a principal axis, but differ slightly in that the projection (or viewing) direction is not. Oblique projections preserve linear and angular measure for faces which are parallel to the projection plane. Faces which are not parallel to the projection plane preserve only linear measures, whereas angles are distorted.

Oblique parallel projections, as shown in Figure 9.14, are used extensively because they are easy to draw. Everything remains parallel, yet objects look realistic. Two common oblique projections are the cavalier and the cabinet projections. Each makes a specific angle with the projection plane, cavalier being 45° and cabinet being the arccotangent of 1/2. Generally, cabinet projections look more realistic, but cavalier projections preserve uniform linear measure.

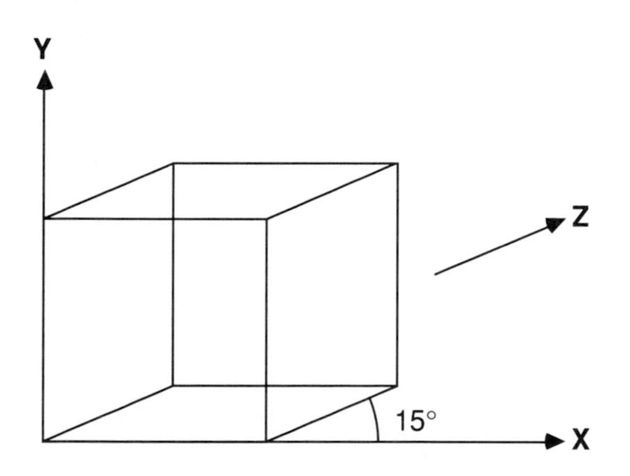

Figure 9.14 Oblique Parallel Projection of a Cube

We use the oblique parallel projection to generate the display data in this example because of its relative simplicity and effective realism. The two-dimensional screen coordinate (x_s, y_s) display data is derived from the transformed three-dimensional coordinates (x, y, z) using simple trigonometry:

$x_s = x + z \cos(15°)$
$y_s = y + z \sin(15°)$

Lines parallel to the z-axis appear to make a 15° angle with the x-axis (as projected on the screen) due to the angle which the projection screen normal and viewpoint (which share a common direction) make with the xy plane. Any angle may be used with varying manifestations in the appearance of the projection.

Note that if the full 4x4 matrix structure is utilized, both the transformation and projection operations may be combined into a single matrix. These operations are distinct here for simplicity and illustration. Foley and Van Dam (see *References* at the end of this chapter) discuss this issue more fully in section 8.2 of their book.

9.6 DATA FORMAT

Hardware multipliers don't know the difference between 101.01012 and 1010.101_2; the placement of the binary point is purely arbitrary as far as the hardware is concerned. However, ADSP-2100 users are strongly encouraged to use the 1.15 format (shown in Figure 9.15) because the ADSP-2100 multiplier is optimized for the 1.15 format.

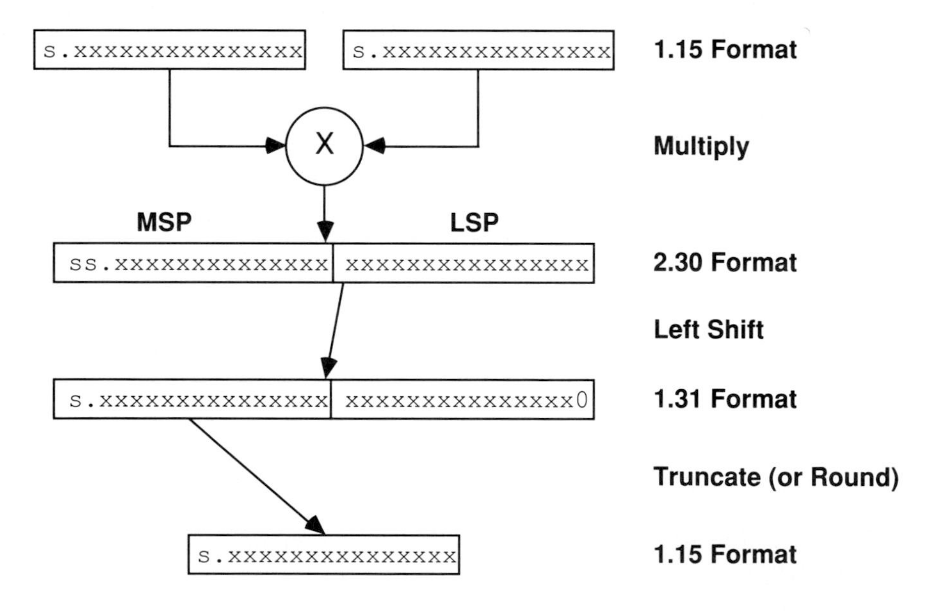

Figure 9.15 The 1.15 Data Format

The ADSP-2100 left-shifts products to renormalize them automatically to a destination format similar to that of the input operands, with the result that there is no binary point migration as long as both input operands are in 1.15 format. Two multiplicands in the 1.15 format produce a 32-bit product in the 2.30 format which is left-shifted to 1.31 format and then and rounded (or truncated) to the MSP (Most Significant 16-bit half of the 32-bit Product) format of 1.15. Automatic normalization works only with the 1.15 format; other formats will manifest binary point migration. Hence, in this example, all data is normalized to the 1.15 format prior to processing.

Note that it is the 16-bit MSP that contains the result in 1.15 format, although product summing is performed in the full 40-bit resolution of the accumulator (in 1.31 format) before rounding (or truncating) to the MSP. (See the complete block diagram and functional description of the MAC in the *ADSP-2100 User's Manual.*)

9.7 NORMALIZATION AND SCALING

Two operations must be performed on the input data before the program will run without data overflows. All input data must be normalized to the largest value, then all normalized data must be adjusted (by upshifting) to the 1.15 format.

Normalization is the division of a set of numbers by their largest member so that the largest number is normalized to unity and the rest of the numbers are all guaranteed to be less than or equal to one. Data normalization is necessary to guarantee that products get smaller after multiplication instead of larger and therefore do not overflow.

Normalized data is upshifted to the 1.15 format so that the automatic renormalization by left-shift works as described above. This upshift is accomplished by multiplying the normalized data with the 16-bit twos-complement positive full scale value ($7FFF_{16} = 2^{15}-1 = 32767$).

In the source data of this example, the largest component of any point vector is 21, so normalization entails the division of all vector components by 21. However, normalization to unity yields a few numbers which are still large enough to cause intermediate results to overflow during the transformation process due to addition operations. Therefore, we increase the normalization factor to guarantee that all overflows are eliminated. By trial and error, we determine that a normalization factor of 30 is sufficient.

Normalization of the source data therefore entails division by 30. For example, the normalized value of 21 is $21 \div 30 = 7/10$. The normalized data is then multiplied by positive full scale to produce the source data used in the transformation process: for example, 7/10 (32767) = 5999_{16}.

The finite precision of hardware processors' numerical formats and the selection of a data normalization factor (resolution) play crucial roles in the successful development of any numerical processing application. Too much resolution in the data (a small normalization constant) results in less headroom (allowance for overflow of intermediate results) within the fixed word size, while too little resolution (a large normalization constant) distorts the data. The key to success is to balance the normalization with the word size to maintain sufficient headroom throughout the process without compromising resolution to the point of introducing too much distortion.

An example of the problems that arise when too little headroom is provided is illustrated in the two photographs shown in Figure 9.16. Both examples show a slight overflow of the screen coordinate system resulting in points wrapping around the screen edges. Wraparound is due to insufficient normalization scaling (not enough headroom) which produces arithmetic overflows.

The first photograph illustrates that such wrapped points produce lines which must cross the screen to make their connections. The second photograph illustrates saturating arithmetic (an optional mode of operation on the ADSP-2100 ALU and MAC) in which any overflows are automatically saturated, or set to full scale. Points which would otherwise wrap around the screen are constrained to the edge (clipped). The effect of saturation arithmetic is an appreciable reduction in the severity of overflow distortion.

The upshifting and normalization of input data are necessary to ensure data integrity through transformation and projection. Before displaying the data, however, the output data must be further scaled to adapt to the display driver.

A simple example of a vector graphic display is the oscilloscope. The hardware used in this example employs a straight binary-coded quad 8-bit (not twos-complement) DAC to drive the x and y deflection inputs of an oscilloscope (see the Joystick and Scope Interface schematic at the end of this chapter). The 8-bit resolution of the DAC provides a screen resolution of 256x256 pixels upon which to display the rotating object.

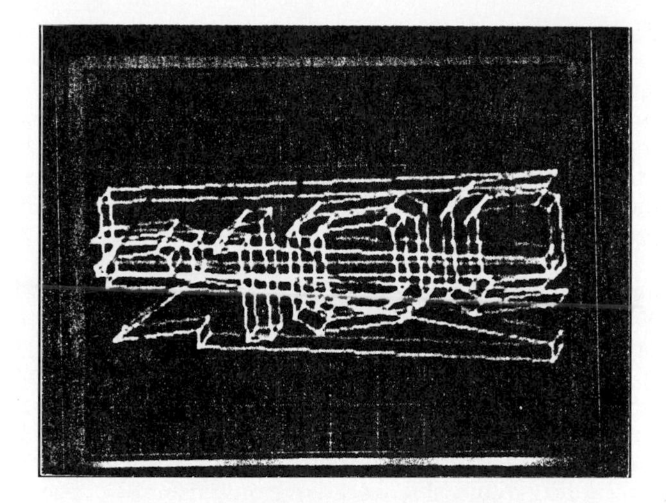

Overflow without Saturation Logic

Overflow with Saturation Logic

Figure 9.16 Overflows With and Without Saturation Logic

The three-dimensional coordinate system of the source data has its origin located in the center of the object with points (vertices of the object) assuming ± twos-complement values (corresponding to the format of the ADSP-2100) in three dimensions. All two-dimensional display data must therefore be converted to the unsigned 8-bit binary format used by the DAC prior to display. This is done by multiplying each screen coordinate (x_s, y_s) by the DAC's half-scale value (80_{16}) and then adding an offset of half-scale to shift the center of the object to the DAC's half-scale point.

In the example, the maximum value of all source coordinates is 21, which when normalized and converted to 1.15 format, becomes 5999_{16}. Assuming that the worst case gain through rotation and projection is unity, the maximum display value is 5999_{16}. Prior to being written to the DAC, this value is multiplied by the DAC's half-scale value, 80_{16}, which translates the normalized value to a corresponding voltage of the DAC's output range. The left-shifted resultant product is $(5999 \times 0080 = 0059\ 9900)_{16}$, which after rounding to the MSP (recall, we only use MR1), becomes $5A_{16}$, a worst case screen coordinate value.

Adding 80_{16} to $5A_{16}$ yields DA_{16}. This addition simply moves the object to center screen and has no scaling effect. The final value which is written to the DAC for display is DA_{16}. Note that the ratio of the worst case screen coordinate value to the positive full scale DAC value, $(5A{:}80)_{16}$, is the same as the original source coordinate to the normalization factor (21:30).

Figure 9.17 uses a number line analogy to summarize all the data format transitions and dynamics during operations, and available headroom for each of six stages described above. In summary, these stages are:

Stage 1: The actual source data consisting of manually quantized (x,y,z) coordinates of the object is edited into a data file.

Stage 2: The quantized data is normalized by a Pascal program.

Stage 3: The same Pascal program formats the normalized data producing a hexadecimal data file. This data is ultimately loaded into the allocated area of data RAM on the target system by the ADSP-2100 assembler INIT directive in the main program.

Stage 4: After the ADSP-2100 has performed rotation and projection transformations, the same limits and headroom are present as in the previous stage, but during the processing between these two stages, the computational dynamics of the operations require the headroom to avoid data overflow.

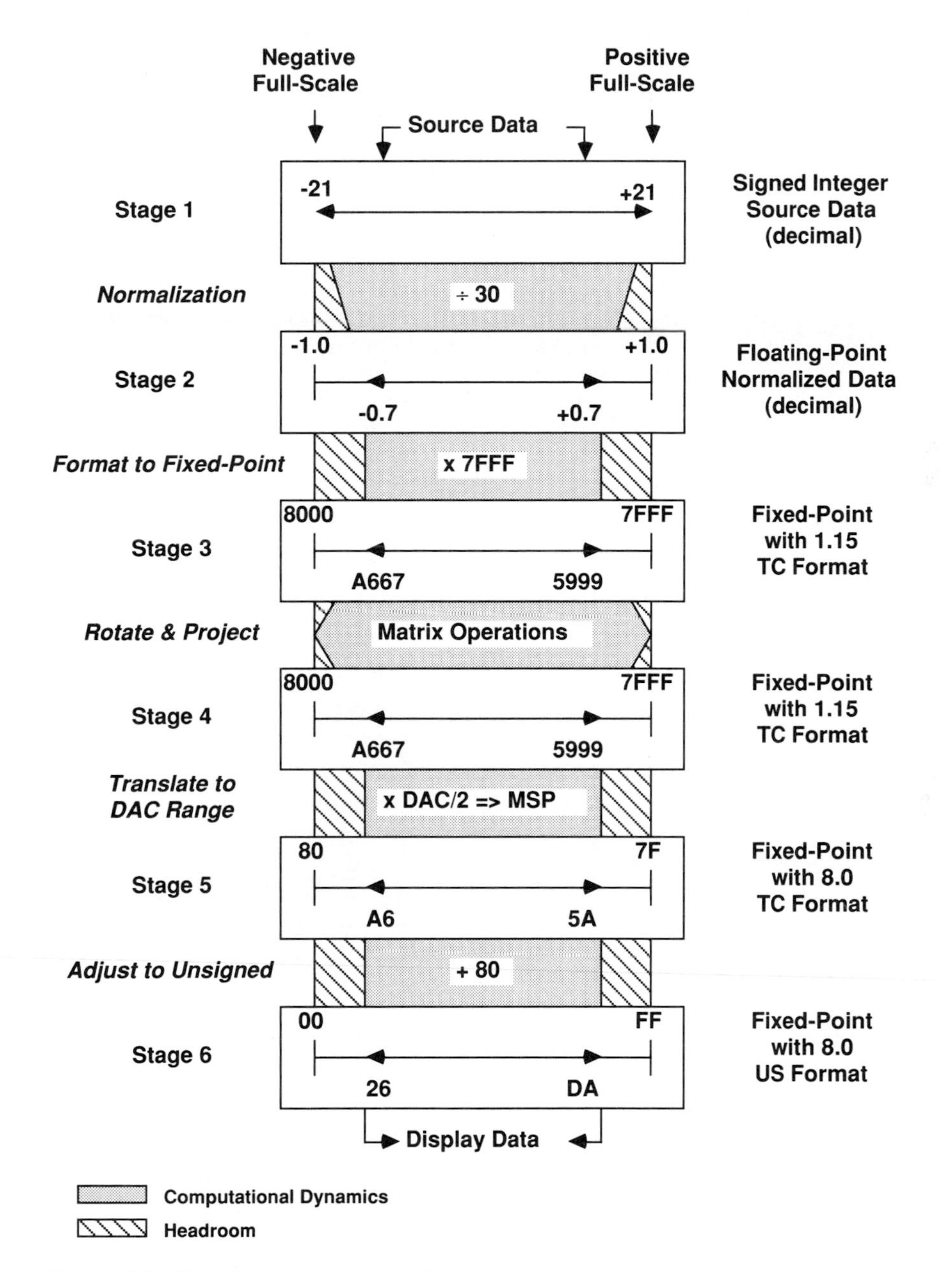

Figure 9.17 Data Format Transition Summary

Stage 5: The data has been multiplied by the half-scale DAC value (the MSP of the MAC contains the result) to translate the twos-complement data range to a corresponding full-scale range for the DAC (remember that the twos-complement format provides for only half the actual range as the unsigned format does).

Stage 6: The last step is to compensate the data for the unsigned format of the DAC by adding the half-scale DAC value to all data. This operation moves the twos-complement negative full-scale value to zero, zero to mid-scale, and positive full-scale to positive full-scale. The resulting data is what actually defines the vertices of the two-dimensional object between which the line segment drawing routines (see *Display Driver*, section 9.9) draw lines.

9.8 PROGRAM AND FILE DESCRIPTIONS

This section presents the files and programs used in the example graphics application. The flowchart in Figure 9.18 illustrates the various operations and how they interrelate; files are shown as ovals, and operations are shown as rectangles. The brief descriptions in this section give general explanations of each file.

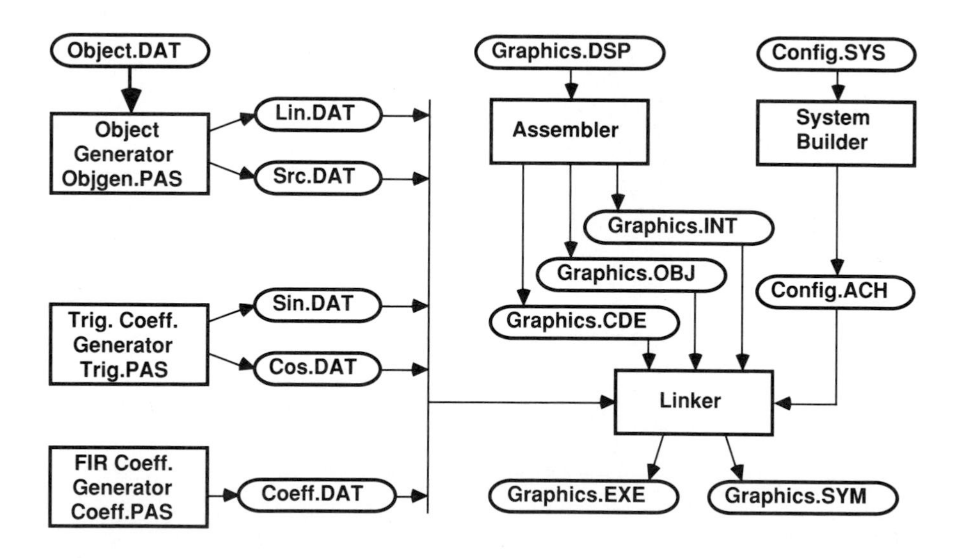

Figure 9.18 Program and File Flowchart

9.8.1 Object Generation

The OBJGEN.PAS program is a Pascal program that translates textual information representing vector coordinates, connectivity, scaling, and number of vectors (all from OBJECT.DAT) to produce hexadecimal versions of the source vector coordinates, fully normalized and formatted, and the line list (SRC.DAT and LIN.DAT, respectively). These files are used as resources in the main GRAPHICS.DSP file, through INIT directives that load the data into the arrays allocated in RAM during the linking process.

```
{Contents of the file OBJGEN.PAS}
program objgen(input,output,object,src,lin);
const
        fs=                                         32767;
var
        object, src, lin:                           text;
        scale,numpoints,points:                     integer;
        x,y,z,s:                                    integer;

begin
reset(object);
rewrite(src);
read(object,scale);
read(object,numpoints);
for points:= 1 to numpoints do begin
        read(object,x,y,z,s);
        x:= trunc(x/scale*fs);
        y:= trunc(y/scale*fs);
        z:= trunc(z/scale*fs);
        writeln(src,hex(x,4,4));
        writeln(src,hex(y,4,4));
        writeln(src,hex(z,4,4));
        end;
writeln(src,'');
close(src);

rewrite(lin);
repeat
        read(object,x);
        writeln(lin,hex(x,4,4));
        until x= -1;
writeln(lin,'');
close(lin);
end.
```

Listing 9.1 Object Generation Program

```
{Contents of the file OBJECT.DAT}
30                        {Normalization Constant}
112                       {Number of Vectors comprosing object}

   -21    3   -2    1          {Vector list for foreground "2"}
   -21    5   -2    1
   -19    7   -2    1
   -13    7   -2    1
   -11    5   -2    1
   -11    1   -2    1
   -18   -5   -2    1
   -11   -5   -2    1
   -11   -7   -2    1
   -21   -7   -2    1
   -21   -5   -2    1
   -13    2   -2    1
   -13    4   -2    1
   -14    5   -2    1
   -18    5   -2    1
   -19    4   -2    1
   -19    3   -2    1

    -9    4   -2    1          {Vector list for foreground "1"}
    -9    5   -2    1
    -7    7   -2    1
    -5    7   -2    1
    -5   -7   -2    1
    -7   -7   -2    1
    -7    4   -2    1

    -3    5   -2    1          {Vector list for 1st foreground "0"}
    -1    7   -2    1
     6    7   -2    1
     8    5   -2    1
     8   -5   -2    1
     6   -7   -2    1
    -1   -7   -2    1
    -3   -5   -2    1
    -1    4   -2    1
     0    5   -2    1
     5    5   -2    1
     6    4   -2    1
     6   -4   -2    1
```

```
   5   -5   -2    1
   0   -5   -2    1
  -1   -4   -2    1

  10    5   -2    1        {Vector list for 2nd foreground "0"}
  12    7   -2    1
  19    7   -2    1
  21    5   -2    1
  21   -5   -2    1
  19   -7   -2    1
  12   -7   -2    1
  10   -5   -2    1
  12    4   -2    1
  13    5   -2    1
  18    5   -2    1
  19    4   -2    1
  19   -4   -2    1
  18   -5   -2    1
  13   -5   -2    1
  12   -4   -2    1

 -21    3    2    1        {Vector list for background "2"}
 -21    5    2    1
 -19    7    2    1
 -13    7    2    1
 -11    5    2    1
 -11    1    2    1
 -18   -5    2    1
 -11   -5    2    1
 -11   -7    2    1
 -21   -7    2    1
 -21   -5    2    1
 -13    2    2    1
 -13    4    2    1
 -14    5    2    1
 -18    5    2    1
 -19    4    2    1
 -19    3    2    1

  -9    4    2    1        {Vector list for background "1"}
  -9    5    2    1
  -7    7    2    1
  -5    7    2    1
```

(listing continues on next page)

```
    -5   -7    2    1
    -7   -7    2    1
    -7    4    2    1

    -3    5    2    1          {Vector list for 1st background "0"}
    -1    7    2    1
     6    7    2    1
     8    5    2    1
     8   -5    2    1
     6   -7    2    1
    -1   -7    2    1
    -3   -5    2    1
    -1    4    2    1
     0    5    2    1
     5    5    2    1
     6    4    2    1
     6   -4    2    1
     5   -5    2    1
     0   -5    2    1
    -1   -4    2    1

    10    5    2    1          {Vector list for 2nd background "0"}
    12    7    2    1
    19    7    2    1
    21    5    2    1
    21   -5    2    1
    19   -7    2    1
    12   -7    2    1
    10   -5    2    1
    12    4    2    1
    13    5    2    1
    18    5    2    1
    19    4    2    1
    19   -4    2    1
    18   -5    2    1
    13   -5    2    1
    12   -4    2    1

{Connection list using vector numbers, 0=penup mode}
0 1 2 3 4 5 6 7 8 9 10 11 12 13 14 15 16 17 1 0 {foreground "2"}
18 19 20 21 22 23 24 18 0                      {foreground "1"}
25 26 27 28 29 30 31 32 25 0                   {foreground 1st "0"}
33 34 35 36 37 38 39 40 33 0                   {foreground 1st "0"}
```

```
41 42 43 44 45 46 47 48 41 0                    {foreground 2nd "0"}
49 50 51 52 53 54 55 56 49 0                    {foreground 2nd "0"}
57 58 59 60 61 62 63 64 65 66 67 68 69 70 71 72 73 57 0
                                                {background "2"}
74 75 76 77 78 79 80 74 0                       {background "1"}
81 82 83 84 85 86 87 88 81 0                    {background 1st "0"}
89 90 91 92 93 94 95 96 89 0                    {background 1st "0"}
97 98 99 100 101 102 103 104 97 0               {background 2nd "0"}
105 106 107 108 109 110 111 112 105 0           {background 2nd "0"}

{connect foreground to background}
1 57 0
2 58 0
3 59 0
4 60 0
5 61 0
6 62 0
7 63 0
8 64 0
9 65 0
10 66 0
11 67 0
12 68 0
13 69 0
14 70 0
15 71 0
16 72 0
17 73 0
18 74 0
19 75 0
20 76 0
21 77 0
22 78 0
23 79 0
24 80 0
25 81 0
26 82 0
27 83 0
28 84 0
29 85 0
30 86 0
31 87 0
32 88 0
```

(listing continues on next page)

```
33 89 0
34 90 0
35 91 0
36 92 0
37 93 0
38 94 0
39 95 0
40 96 0
41 97 0
42 98 0
43 99 0
44 100 0
45 101 0
46 102 0
47 103 0
48 104 0
49 105 0
50 106 0
51 107 0
52 108 0
53 109 0
54 110 0
55 111 0
56 112 -1
```

Listing 9.2 Object Data File

9.8.2 Trigonometric Coefficient Generation

The TRIG.PAS program is a Pascal program that generates 256 uniformly spaced samples of the sine and cosine functions corresponding to the 256 possible positions (using 8-bit quantization) which the joystick may assume. These hexadecimal data files (SIN.DAT and COS.DAT) are fully normalized and formatted. The lookup tables used during the generation of the transformation matrix are initialized with data from these files. Note that zero is positioned in the middle of the arrays to correspond with a zero rotation at the center joystick position. The data in these files is loaded during the linking process through INIT directives.

```
{Contents of the file TRIG.PAS}
program trig(input,output,sin_table,cos_table);

const
        pi=3.141592654;

var
        sin_table:                                  text;
        cos_table:                                  text;
        degree:                                     integer;
        arg:                                        real;
        result:                                     integer;

begin
rewrite(sin_table);
rewrite(cos_table);
for degree:= -128 to 127 do begin
        arg:= 360 * degree / 256 * pi / 180;
        result:= trunc(sin(arg) * 32767);
        writeln(sin_table, hex(result,4,4));
        result:= trunc(cos(arg) * 32767);
        writeln(cos_table, hex(result,4,4));
        end;
writeln(sin_table,'');
writeln(cos_table,'');
close(cos_table);
close(sin_table);
end.
```

Listing 9.3 Sine and Cosine Table Generation Program

9.8.3 FIR Filter Coefficient Generation

The FIR coefficient generator program COEFF.PAS generates underdamped FIR filter coefficients which are used in the filtered joystick display mode (described in *Display Driver*, section 9.9). The response of the filter was derived experimentally by plotting various exponentially damped sine waves as a function of ringing and settling time. The best response (a subjective determination) was taken as the impulse response of the desired filter; quantizing this response into 128 samples produced the FIR coefficients. The actual settling time of the filter is the number of taps divided by the frame rate, or $128 \div 90 \approx 1.5$ seconds. Coefficients were normalized (uniformly scaled so that their sum was approximately equal to one) to produce a unity gain filter and then converted to 1.15 format. The hexadecimal values were stored in the COEFF.DAT file, allowing the INIT directive to load them during the linking process.

```
{Contents of the file COEFF.PAS}
PROGRAM the_function (input, output, coeff);

CONST
        pi=         3.141592654;
        cycles=     19.84;
        scale=      0.362951735;
        tc=         -15;

VAR
        x, y:       real;
        i, ypt:     integer;
        coeff:      text;

BEGIN
        rewrite(coeff);
        FOR i := 127 DOWNTO 0 DO BEGIN
          x := i * cycles * pi / 180;
          y := scale * exp(i/tc) * sin(x);
          ypt:= trunc(y * 32767 + 0.5);
          writeln(coeff, hex(ypt,4,4));
        END;
        close(coeff);
END.
```

Listing 9.4 FIR Filter Coefficient Generation Program

9.8.4 System Configuration

System configuration is mandatory for all ADSP-2100 applications. The GRAPHICS.SYS file is used by the System Builder to specify the target system configuration. The memory and peripheral mapping defined in GRAPHICS.SYS must correspond to the target system memory configuration and peripheral address decoding. This file defines RAM and ROM segments and their locations. Device interfaces are also declared using the PORT directive. Notice that data memory can be interleaved with peripheral devices, so long as contiguous arrays are kept smaller than the allocation block size. The System Builder produces the GRAPHICS.ACH file required by the Linker.

```
{Contents of the file GRAPHICS.SYS}
.SYSTEM   graphics_config;

{allocate a 2K (0-07FF) block of PMC}
.SEG/RAM/ABS=h#0000/PM/CODE     pmc[h#0800];

{allocate a 2K (0-07FF) block of PMD}
.SEG/RAM/ABS=h#4000/PM/DATA     pmd[h#0800];

{allocate a 4K (0000-0FFF) block of DMD}
.SEG/RAM/ABS=h#0000/DM/DATA     dmd1[h#0800];

{allocate I/O map starting at DMD location h#1000}
.CONST  ioblk=h#1000;

{define x and y joystick inputs on the 4-channel adc}
.PORT/ABS=ioblk+h#0             adx;
.PORT/ABS=ioblk+h#1             ady;

{define deflection and intensity scope channels on the quad dac}
.PORT/ABS=ioblk+h#8             dax;
.PORT/ABS=ioblk+h#9             day;
.PORT/ABS=ioblk+h#A             daz;

.ENDSYS;
```

Listing 9.5 System Configuration File

9.8.5 Main Source Program

The GRAPHICS.DSP file contains the actual ADSP-2100 source code. The ADSP-2100 Assembler assembles the source code and allocates variable storage. The Assembler produces the three files (GRAPHICS.INT, GRAPHICS.OBJ, and GRAPHICS.CDE) used by the Linker. The Linker accepts the various data files mentioned above as INIT directive arguments to initialize the various RAM arrays. It produces the GRAPHICS.EXE file (executable image) and GRAPHICS.SYM file (symbol table) which can both be downloaded to a RAM-based system. If ROMs are to be burned, then an additional formatting step is required.

Only 950 lines of source code are used for this example (approximately 2000 lines of executable code), although as the performance benchmarks indicate (see *Performance*, section 9.10), much of the code consists of loops. Note that the main loop takes about 94,000 instruction cycles to complete one iteration (this includes all iterations of inner loops), which corresponds to roughly a 100:1 cycles-to-line-of-code ratio. Various allocation and initialization steps are performed at startup before the program enters the main loop. The main loop consists of building a new transform, applying the transform to the object, projecting the object, and displaying the object; this loop is repeated over and over again. A manual interrupt button on the target board generates IRQ2, whose service routine sequences between the four display modes.

An interesting technique is used in the display routine: an indexed indirect jump. A jump table consists of different JUMP LABEL instructions. An index into the jump table is created and added to the base address of the jump table. Then, an indirect jump into the table is performed. The index determines which jump instruction in the jump table gets executed.

The code in Listing 9.6 below is part of the display routine. The AF register stores the index value. In this case, the index determines in which of eight possible octants the point is located. The signs (positive or negative) of the Δx and Δy values select a quadrant, and the difference in magnitude between the values ($|\Delta x| - |\Delta y|$) selects one of the two octants in that quadrant. The index picks out the jump instruction to the correct octant routine to draw the line to the point.

```
AF=PASS 0;          {init for indirect jump offset}
AX1=DM(newx);       {compute delta x}
AY1=DM(oldx);
AR=AX1-AY1;
DM(dx)=AR;
AX1=4;
IF GT AF=AX1 OR AF;   {set bit 2 if delta x is
                       positive}
AX1=DM(newy);       {compute delta y}
AY1=DM(oldy);
AR=AX1-AY1;
DM(dy)=AR;
AX1=2;              {set bit 1 if delta y is positive}
IF GT AF=AX1 OR AF;
AX1=DM(dy);         {compute |dx|-|dy|}
AR=ABS AX1;
AY1=AR;
AX1=DM(dx);
AR=ABS AX1;
```

```
                AX1=AR;
                AR=AX1-AY1;
                AX1=1;            {set bit 0 if delta x is greater}
                IF GT AF=AX1 OR AF;
                AX1=^jump_table; {add jump table base address}
                AR=AX1+AF;
                I4=AR;
                JUMP (I4);        {do the indirect jump}
jump_table:     JUMP octant6;
                JUMP octant5;
                JUMP octant3;
                JUMP octant4;
                JUMP octant7;
                JUMP octant8;
                JUMP octant2;
                JUMP octant1;
```

Listing 9.6 Jump Table

9.8.6 Data Structures

Some of the arrays and variables in the GRAPHICS.DSP source code (see the program listing at the end of the chapter) are explained in this section.

xfm_array
The nine coefficients of the 3x3 transformation are stored in this circular buffer. The circular buffer organization eliminates the need to reinitialize the transform coefficient address pointer after each vector has been transformed.

coeff
The 128 FIR coefficients associated with the filtered display mode (see next section) are stored in this array. These coefficients are applied to the joystick input samples to introduce a little ringing to the joystick response. The effect is as though the display object were resistant to changes in position.

xbuff and *ybuff*
These arrays are also used in the filtered display mode (see next section) to store the previous joystick input samples (delay line). The FIR routine can then perform the convolution of the delayed samples with the coefficients to produce the filtered version of the joystick control signal.

sin_array and *cos_array*
These arrays hold the sine wave and cosine wave values generated by the TRIG.PAS program. During the generation of the transformation array, the various sine and cosine values dictated by the joystick inputs are fetched from these arrays.

src_array
This array stores the actual reference data describing the source object. Each new transformation always uses this source data as a starting point to avoid introducing the recursive errors that are found in systems that transform previous transforms.

line_list
The connection information describing which vectors have lines connected to them and where those lines go is stored in this array. As described below, a 0 in the line list means that no line should be drawn to the next point, nonzero values denote point numbers to which lines should be drawn (all points are numbered), and a –1 means that all lines have been drawn and thus another transform can start.

wcs_array
The "World Coordinate System" (WCS) is used in this context to refer to the transformed source data which is still in three-dimensional coordinates.

ecs_array
The "Eye Coordinate System" (ECS) is used in this context to refer to the WCS data that has been projected to a two-dimensional space and is ready for display.

xpntr and *ypntr*
These two data RAM pointers are used to keep track of the current starting position of the *xbuff* and *ybuff* arrays during the FIR filtering of the joystick input samples. These arrays are circular buffers, and therefore the starting position circulates through the buffer as new samples are brought in after each filter pass; the pointers keep track of the changing starting position within each buffer.

oldx and *oldy*
These two pointers into the *ecs_array* locate the last point which the line drawing routine processed. This data is necessary because the line list structure only indicates the next point to which a line should be drawn, not the starting point.

newx and *newy*
These temporary variables hold the location of the current x and y coordinates in the *ecs_array* during the line drawing routine for repeated access, so as to avoid having to calculate these locations more than once.

dx and *dy*
These variables hold the differential change in x and y between the last point and the current point. They are used to determine the octant in which the new point resides.

mode
This variable keeps track of the display mode that is currently activated (see next section).

rotation
This variable, used in the *autorotate* mode (see next section), tracks the amount of rotation to apply to the object. The value of *rotation* is incremented by one after each iteration of the main loop.

xyorzflag
This variable holds a flag which is used in the *autoxyz* display mode (see next section) to indicate which axis is currently being rotated: x, y or z.

9.9 DISPLAY DRIVER

The display driver has the job of drawing the wire-frame object on the screen. The line list describes points from which lines are drawn and to which points the lines go to form the polygons that comprise the object. Zeros in the line list indicate to the line-drawing routine to jump to the next point without drawing a line (equivalent to a plotter "penup"), as in the start of a new polygon. Nonzero numbers in the line list mean to draw a line from the last point to the next point ("pendown"), which is identified by the number. A –1 value (8000_{16}) in the line list indicates that no more points remain and the drawing is complete.

Four display modes are demonstrated in this example: 1) automatic rotation about the x, y, and z axes sequentially; 2) automatic rotation about all three axes simultaneously; 3) averaged joystick control in x and y axes; and 4) filtered joystick control in x and y axes. The display is advanced from one mode to the next by a debounced pushbutton connected to the IRQ2 interrupt input of the ADSP-2100 (see the *ir2_serve* routine in the listing).

The autorotate modes (1 and 2) rotate the object about 1.5 degrees per frame, which corresponds to the resolution of the trigonometric function tables (360°/revolution ÷ 256 entries/revolution). The averaged joystick mode sums 128 samples and then downshifts the result by seven bits to produce an average reading for each direction (x axis and y axis). Averaging reduces the potentiometer jitter associated with the joystick wiper action. The filtered mode applies an FIR filter to both x axis and y axis readings. Two 128-tap delay lines are used to track historic samples of x and y. These samples are convolved with the FIR coefficients of an exponentially underdamped sine wave. The filter parameters were selected to introduce a sense of inertial mass, complete with overshoot and ringing (see FIR filters in Chapter 5).

Before drawing each line, the program determines in which of eight octants (see Figure 9.19) the end point is relative to the first. Eight octants are used because certain aspects of the line segment routines vary uniquely for each octant. The determination of the octant in which the new point resides is made by calculating the three parameters: Δx, Δy, and $|\Delta x - \Delta y|$. The first two determine in which of four quadrants the second point resides, while the last test essentially checks whether the slope of the line is greater than or less than one, and thus determines which half of the quadrant (which octant) the second point is in.

The actual line is drawn pixel-by-pixel using an optimized Bressenham's algorithm (see references) for generating line segments between endpoints

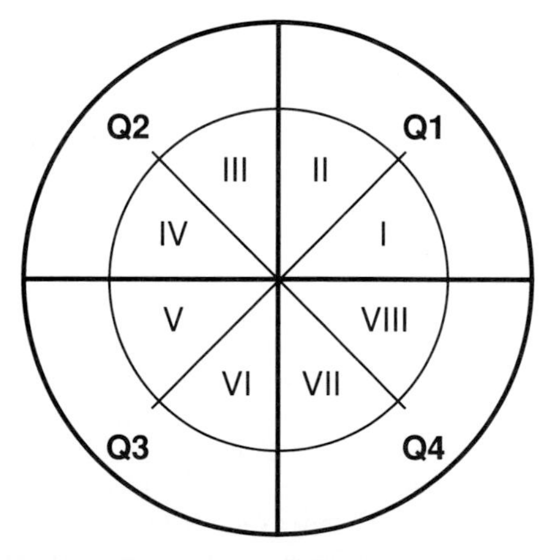

Figure 9.19 Quadrants and Octants

quickly. Bressenham's algorithm is particularly attractive for this hardware implementation because it requires no division or multiplication, only simple integer arithmetic (see program listings). Depending on which octant the new point resides in, either the x axis or y axis (whichever has the faster rate of change) is either incremented or decremented (depending on the direction) a pixel at a time while the other axis is conditionally incremented or decremented. An error term that is tracked with each iteration determines whether the conditional increment or decrement is made (see program listings). Typical object images are shown in Figure 9.20, found on the following page.

Moves to new points ("penup" moves) on the display screen are made with the beam turned off and are accomplished by writing FF_{16} to the z-axis DAC. The output of the z-axis DAC is connected to the z-axis input (or beam intensity) control, found on the back of most scopes. After the DACs are updated with the (x, y) coordinates of each new pixel, a beam-on macro (see program listings at the end of this chapter) turns on the beam for about ten cycles to make the pixel visible. The beam-on macro ends by turning the beam off again. The z-axis modulation eliminates extraneous display artifacts such as retrace and DAC transitions.

The background register set of the ADSP-2100 is used during the Bressenham algorithm because the other operations (matrix generation, transformation and projection) have many constants already stored. Having a complete set of background registers which can be instantaneously activated makes the time-consuming process of context switching (push data - process new context - pop data) obsolete.

9.10 PERFORMANCE

The object in this example consists of 112 three-dimensional vectors and 170 line segments. The major program loop consists of the following functions, with execution times shown in cycles for each:

Program Phase	*Duration*
Generate or gather new position data*	
autoxyz mode:	26 cycles
autorotate mode:	20 cycles
averaged mode:	2,848 cycles
filtered mode:	330 cycles
Generate a new transform	133 cycles
Apply the transform	1,927 cycles
Project and scale the scene to two dimensions	1,595 cycles
Draw the scene	89,695 cycles

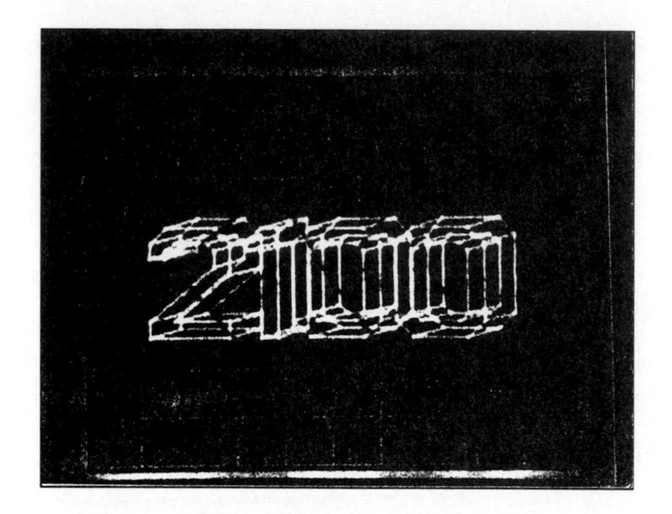

The Object

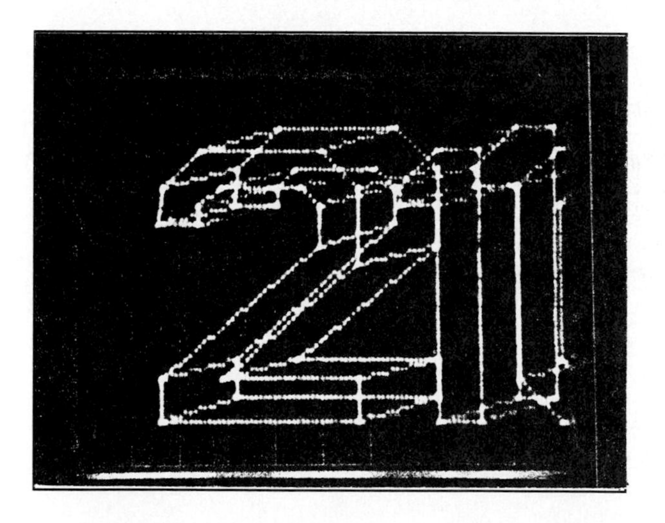

Closeup

Figure 9.20 Typical Displays

* Joystick input samples are either averaged or filtered over 128 samples, hence some modes require less time than others. See program listings for details.

The process is repeated almost 90 times a second, three times faster than necessary for a perception of continuous rotation. In other words, the ADSP-2100 could handle three times as complex an application and still convey the illusion of smooth rotation! In fact, the 90 frames/second display rate already includes performance enhancements (such as the joystick filtering and averaging modes) which are nice but unnecessary. A 33% processor utilization leaves ample processing power for extras such as hidden line removal, shading and texture mapping, shadow casting, etc. (see *Overview* at the beginning of this chapter for other ideas).

The key number in the benchmarks is the 1,927 cycles required for the entire transformation subroutine (see *doxfm* in the program listings). This measure is made from the subroutine call to the return and includes all the subroutine setup overhead instructions. A way to put this benchmark in perspective is to normalize it by the number of transforms which are actually performed: 112 1x3 vectors each multiplied by a 3x3 transform matrix produces a transform rate of 1927÷112=17.21 cycles/transform. (Although the loop is only 9 instructions long, the iterations require some overhead.) Within each 17-odd cycle transform, the following steps are performed:

- Fetch the instructions (the cache RAM is used after the first iteration),
- Fetch the nine coefficients and three vector components,
- Perform nine 16-bit multiply/accumulates,
- Store three results, and
- Maintain RAM pointers to both the transform and data arrays on each cycle.

The number of cycles in the transformation subroutine is

[(4 inner loop instructions x 3 columns) + 5 outer loop instructions] x 112 vectors + 23 overhead instructions = 1927 cycles

as shown above. If the transform matrix size were increased from 3x3 to 4x4, the number of cycles would be

[(5 inner loop instructions x 4 columns) + 5 outer loop instructions] x 112 vectors + 23 overhead instructions = 2823 cycles

or a 46% increase.

However, the impact of this increase in the overall context is relatively insignificant. Tallying the above benchmarks for the 3x3 structure, we have about 93,373 cycles per frame (using an auto display mode), which corresponds to an 85.7 frame rate if an 8MHz processor is used. A similar tally for the 4x4 structure gives about 94,313 cycles (allowing for the increased transform time and an estimated increase for the transform build function), corresponding to an 84.8 frame rate. We may conclude that going to the full 4x4 structure (which includes the translation, zoom and perspective operations), would cost a mere 1% decrease in the frame rate.

A more significant factor affecting the overall performance is the beam dwell time (see the beam-on macro in the program listings). The beam dwell is used to saturate the screen phosphor of the oscilloscope at each pixel long enough to leave a nice bright trace, but not longer. The value for the beam dwell used in the benchmark measures is 10 cycles per pixel. Because the vast majority of time is spent drawing lines, variations in the beam dwell time produce large changes in the overall frame rate. In fact, cutting the dwell time in half increases the frame rate from 85 to 106, decreasing the processor utilization from 35% to 28% while still producing an acceptable display.

The reason for the discrepancy in computation time between the averaged and filtered display modes is that filtering takes one sample from the joystick and 127 samples from the delay line maintained in data RAM, while averaging takes 128 new joystick samples of both x and y fore each frame. The filtered mode is faster because most of the data is already available in the data buffer. Resampling each value takes about 20 cycles per sample due to the relatively long ADC conversion time.

9.11 SCHEMATICS

The schematics for the graphics processor are shown in Figures 9.21 through 9.23. Figure 9.21 shows the ADC and DAC connections. The AD7824 is a four-channel, 8-bit ADC with a 2.4μs (20 cycles of the ADSP-2100) conversion time. The scope inputs are driven by an AD7226 quad 8-bit DAC. The $\overline{\text{IOSEL}}$ signal is a predecoded bank select into which both the ADC and the DAC are mapped. Reads from the $\overline{\text{IOSEL}}$ memory region come from the ADC, whereas writes to the same region go to the DAC (see the GRAPHICS.SYS system configuration file listing).

Read/write decoding and DMACK generation are provided by the circuit in Figure 9.22. Control signals from the ADSP-2100 are decoded to provide the WR and $\overline{\text{CNVT}}$ control signals for the DAC and ADC, respectively. The $\overline{\text{INT}}$ output of the ADC, which goes active LO when the conversion is complete, is used to produce the data memory acknowledge signal, DMACK, for the ADSP-2100. DMACK generates wait states during ADC and DAC conversion times by delaying the ADSP-2100 the appropriate amount of time.

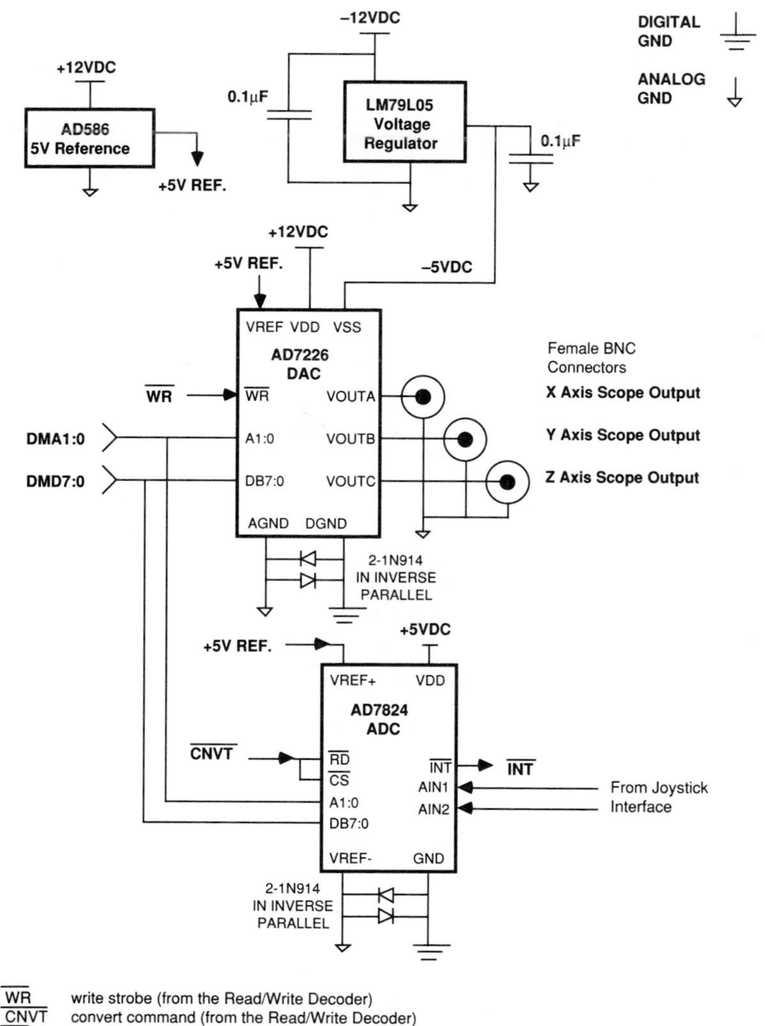

Figure 9.21 ADC and DAC Connections

The A/D conversion is started by the assertion of $\overline{\text{IOSEL}}$ and $\overline{\text{DMRD}}$, which issues the $\overline{\text{CNVT}}$ signal to the ADC. The ADC converts within 2.4µs, during which time the ADSP-2100 is held in a "slow peripheral read" mode by DMACK; wait states (nops) are executed until the conversion is complete. DAC writes also hold off DMACK to expand the write pulse width of the ADSP-2100 to meet the longer requirement of the AD7226.

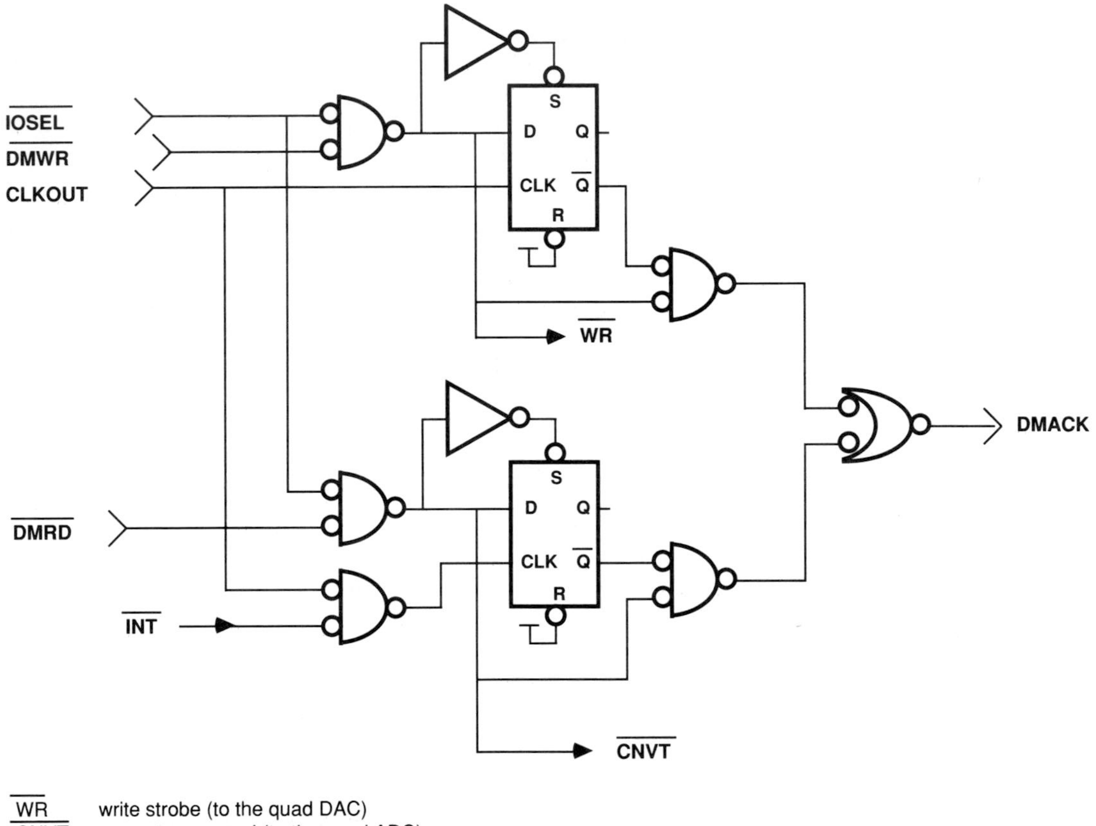

$\overline{\text{WR}}$ write strobe (to the quad DAC)
$\overline{\text{CNVT}}$ convert command (to the quad ADC)
$\overline{\text{INT}}$ conversion complete handshake (from the ADC)

(writes delay one cycle, reads delay until $\overline{\text{INT}}$)

Figure 9.22 Read/Write Decoder and DMACK Logic

The joystick interface circuit is shown in Figure 9.23. The RC4558 dual op amp buffers the joystick x and y inputs to the ADC. The op amp also low-pass filters some of the joystick potentiometer noise to stabilize the display.

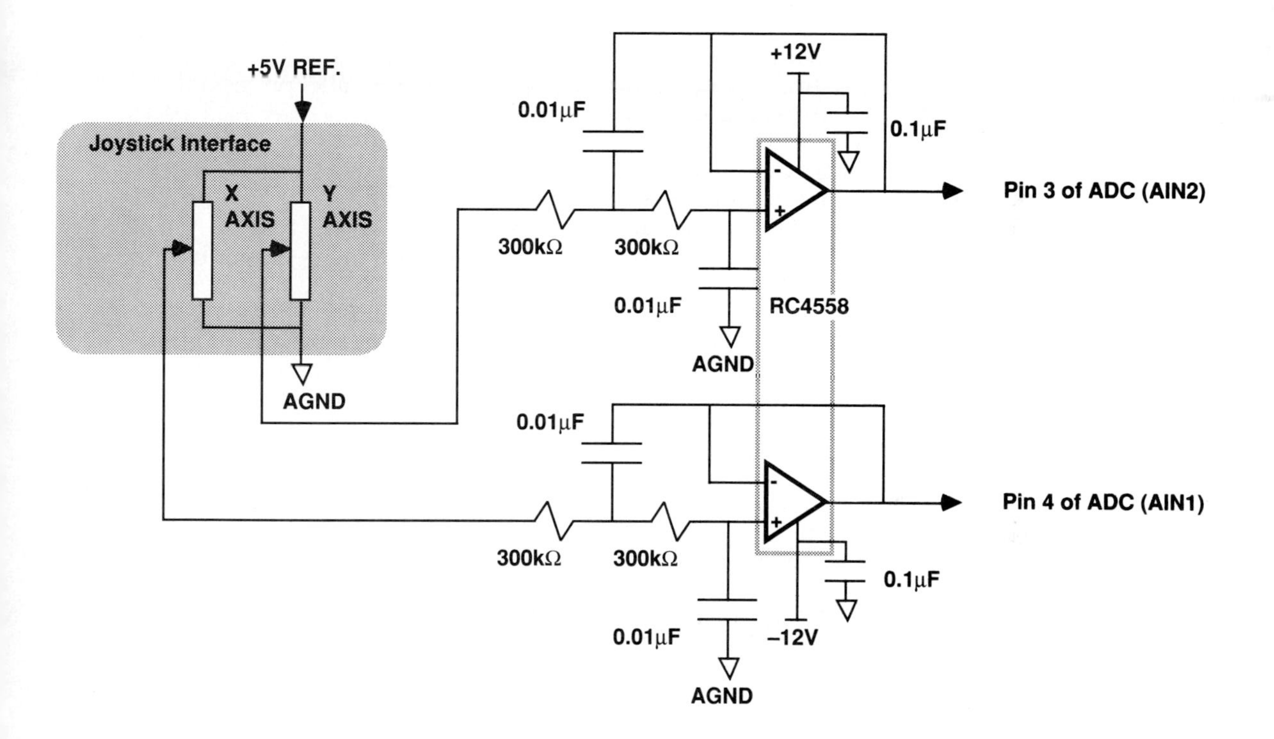

Figure 9.23 Joystick Interface

9.12 SUMMARY

The ADSP-2100 can be the basis of a complete, hardware-oriented application for performing graphics operations on a three-dimensional database. The example application presented in this chapter performs normalization and formatting to avoid overflow and preserve data formats through the transformation operation. It uses data structures that facilitate the object rendering by the Bressenham line segment drawing algorithm. A 3x3 rotation matrix has been derived for this application; the means for implementing translation, scaling, perspective, and zoom are also described in this chapter. Both perspective and parallel projection techniques have been discussed as well.

Software and the accompanying benchmarks show that a three-dimensional object can be rotated smoothly in a real-time display on an oscilloscope. Miscellaneous support software illustrates the basic techniques of generating source data and coefficients and getting them into the program.

The ADSP-2100 proves to be more than adequate in graphics-oriented applications. In fact, the complete application presented in this chapter uses less than a third of the available processing power of the ADSP-2100; three times as complex an application as is shown could be implemented on the ADSP-2100 while maintaining the 30Hz frame rate needed for smooth display.

9.13 REFERENCES

Foley, J. D. and Van Dam, A. 1983. *Fundamentals of Interactive Computer Graphics*. Reading, MA: Addison-Wesley Publishing Company.

Newman, William M. and Sproull, Robert F. 1983. *Principals of Interactive Computer Graphics*. New York, NY: McGraw-Hill Book Company.

Stone, Maureen C., ed. 1987. *SIGGRAPH '87 Conference Proceedings: Computer Graphics, Vol. 21, No. 4.* Anaheim, CA: Association for Computing Machinery's Special Interest Group on Computer Graphics.

9.14 PROGRAM LISTING

This section contains the complete program listing for the graphics application described in this chapter.

Graphics 9

```
{Contents of the file GRAPHICS.DSP}
.MODULE/RAM/ABS=h#0000   graphics;

.VAR/PM/RAM/CIRC     xfm_array[h#0009];
.VAR/PM/RAM          coeff[128];

.VAR/DM/RAM/CIRC     xbuff[128];
.VAR/DM/RAM/CIRC     ybuff[128];
.VAR/DM/RAM          sin_table[h#0100];
.VAR/DM/RAM          cos_table[h#0100];
.VAR/DM/RAM          src_array[h#0150];
.VAR/DM/RAM          line_list[h#0132];
.VAR/DM/RAM          wcs_array[h#0150];
.VAR/DM/RAM          ecs_array[h#00E0];
.VAR/DM/RAM          xpntr;
.VAR/DM/RAM          ypntr;
.VAR/DM/RAM          oldx;
.VAR/DM/RAM          oldy;
.VAR/DM/RAM          newx;
.VAR/DM/RAM          newy;
.VAR/DM/RAM          dx;
.VAR/DM/RAM          dy;
.VAR/DM/RAM          mode;
.VAR/DM/RAM          rotation;
.VAR/DM/RAM          xyorzflag;

.CONST          numpoints=112;
.CONST          numpoints_2=224;
.CONST          half_scale=128;
.CONST          sin_angle=h#2121;          {sin(15 deg)*32767}
.CONST          cos_angle=h#7BA2;          {cos(15 deg)*32767}

.PORT           adx, ady, adz, dax, day, daz;

.INIT           sin_table: <sin.dat>;
.INIT           cos_table: <cos.dat>;
.INIT           src_array: <src.dat>;
.INIT           line_list: <lin.dat>;
.INIT           coeff:     <coeff.dat>;

.MACRO          nops;
                NOP; NOP; NOP; NOP; NOP; NOP;
.ENDMACRO;

.MACRO          beam_on;
.LOCAL          dwell;
```

(listing continues on next page)

```
                CNTR=10;
                DM(daz)=AX1;  {turn beam on}
                DO dwell UNTIL CE;
dwell:             NOP; {wait}
                DM(daz)=AX0;  {turn beam off}
.ENDMACRO;

{initializations...}
                RTI; RTI; JUMP ir2_serve; RTI;
{default to linear addressing}
                L0=0; L1=0; L2=0; L3=0; L4=0; L5=0; L6=0; L7=0;
                PX=0;   {clear bus exchange register}
                ENA SEC_REG;  {init secondary registers}
                MY0=1;  {used in line segment drawing}
                MY1=-1;
                AX0=h#00FF;   {to turn off the beam}
                AX1=h#0000;   {to turn it on}
                DIS SEC_REG;

                AY0=0;                {initialization value}
                DM(rotation)=AY0;     {init autorotation counter}
                DM(mode)=AY0;         {init display mode}
                DM(xyorzflag)=AY0;    {init x y or z flag}

                I0=^xbuff;    {init xbuff and ybuff delay lines}
                I1=^ybuff;
                M0=1;
                CNTR=128;     {clear samples of 128 tap filter}
                DO initloop UNTIL CE;
                   DM(I0,M0)=AY0;
initloop:          DM(I1,M0)=AY0;

                AY0=^xbuff;
                DM(xpntr)=AY0;        {init new x sample pointer}
                AY0=^ybuff;
                DM(ypntr)=AY0;        {init new y sample pointer}

                ICNTL=h#0004;         {make IRQ2 edge-sensitive}
                IMASK=h#0004;         {enable IRQ2}

{begin actual code}
mainloop:       CALL bldxfm;  {read adcs, build new transform matrix}
                CALL doxfm;   {transform source by new matrix}
                CALL doproj;  {calculate 2D projection of 3D object}
                CALL display; {drive xyz axes on scope using quad dac}
                JUMP mainloop;{display until next rotation}
```

```
{interrupt routine to sequence the display mode through
autorotate, unfiltered joystick and filtered joystick control}
ir2_serve:      AX0=DM(mode);
                AY0=0;
                AR=AX0 XOR AY0;
                IF EQ JUMP make1;
                AY0=1;
                AR=AX0 XOR AY0;
                IF EQ JUMP make2;
                AY0=2;
                AR=AX0 XOR AY0;
                IF EQ JUMP make3;
                AY0=3;
                AR=AX0 XOR AY0;
                IF EQ JUMP makez;
make1:          AX0=1;
                DM(mode)=AX0;
                RTI;
make2:          AX0=2;
                DM(mode)=AX0;
                RTI;
make3:          AX0=3;
                DM(mode)=AX0;
                RTI;
makez:          AX0=0;
                DM(mode)=AX0;
                RTI;


{BLDXFM
Module to build the master transformation matrix from the three
rotational axes components as sampled by the quad ADC.

The following registers may be overwritten by this routine,
depending upon the display mode:
    I0, I4
    L0
    M0-M3, M4
    AX0, AY0, AR
    MX0, MY0, MF, MR
}
```

(listing continues on next page)

```
bldxfm:     AX0=DM(mode);           {check out display mode}
            AY0=0;
            AR=AX0 XOR AY0;
            IF EQ JUMP autoxyz;
            AY0=1;
            AR=AX0 XOR AY0;
            IF EQ JUMP autorotate;
            AY0=2;
            AR=AX0 XOR AY0;
            IF EQ JUMP averaged;
            AY0=3;
            AR=AX0 XOR AY0;
            IF EQ JUMP filtered;
            TRAP;                   {should never get here}

autoxyz:    AY0=DM(rotation);
            AR=AY0+1;
            AY0=255;
            AR=AR AND AY0;
            DM(rotation)=AR;
            M1=AR;
            M2=AR;
            M3=AR;

            AX0=DM(xyorzflag);      {get xy or z flag}
            AY0=0;
            AR=AX0 XOR AY0;         {check for zero ==> rotate only x}
            IF EQ JUMP xonly;
            AY0=1;
            AR=AX0 XOR AY0;         {check for zero ==> rotate only y}
            IF EQ JUMP yonly;
            AY0=2;
            AR=AX0 XOR AY0;         {check for zero ==> rotate only z}
            IF EQ JUMP zonly;

xonly:      AX0=M1;                 {get current x rotation}
            M2=128;                 {zero out y}
            M3=128;                 {zero out z}
            AY0=128;                {check current rotation}
            AR=AX0 XOR AY0;         {against zero}
            IF NE JUMP calculate;   {if not zero, keep going with x}
            AR=1;                   {otherwise, change axis of rotation}
            DM(xyorzflag)=AR;       {to y before going on}
            JUMP calculate;
```

```
yonly:        AX0=M2;                    {get current y rotation}
              M1=128;                    {zero out x}
              M3=128;                    {zero out z}
              AY0=128;                   {check current rotation}
              AR=AX0 XOR AY0;            {against zero}
              IF NE JUMP calculate;      {if not zero, keep going with y}
              AR=2;                      {otherwise, change axis of rotation}
              DM(xyorzflag)=AR;          {to z before going on}
              JUMP calculate;

zonly:        AX0=M3;                    {get current z rotation}
              M1=128;                    {zero out x}
              M2=128;                    {zero out y}
              AY0=128;                   {check current rotation}
              AR=AX0 XOR AY0;            {against zero}
              IF NE JUMP calculate;      {if not zero, keep going with z}
              AR=0;                      {otherwise, change axis of rotation}
              DM(xyorzflag)=AR;          {to x before going on}
              JUMP calculate;

autorotate:   AY0=DM(rotation);
              AR=AY0+1;
              AY0=255;
              AR=AR AND AY0;
              DM(rotation)=AR;
              M1=AR;
              M2=AR;
              M3=AR;
              JUMP calculate;

{average both x and y axis joysticks over 256 samples}
averaged:     AX1=h#00FF;                {mask bits for a/d samples}
              AY1=h#00FF;
              nops;
              AX0=DM(adx);               {get 1st sample}
              NOP;
              AR=AX0 AND AY1;
              AX0=AR;
              nops;
              AY0=DM(adx);               {get 2nd sample}
              NOP;
              AR=AX1 AND AY0;
              AY0=AR;
              AF=AX0+AY0;                {add 1st and 2nd samples}
              CNTR=126;
```

(listing continues on next page)

```
            DO xaverage UNTIL CE;
               nops;
               AX0=DM(adx);
               NOP;
               AR=AX0 AND AY1;
               AX0=AR;
xaverage:      AF=AX0+AF;          {add in 126 more samples}
            AR=PASS AF;
            SI=AR;
            SR=LSHIFT SI BY -7 (HI);         {divide by 128 to get average}
            AX0=SR1;
            AR=AX0 AND AY1;
            M1=AR;

            nops;
            AX0=DM(ady);
            NOP;
            AR=AX0 AND AY1;
            AX0=AR;
            nops;
            AY0=DM(ady);
            NOP;
            AR=AX1 AND AY0;
            AY0=AR;
            AF=AX0+AY0;
            CNTR=126;
            DO yaverage UNTIL CE;
               nops;
               AX0=DM(ady);
               NOP;
               AR=AX0 AND AY1;
               AX0=AR;
yaverage:      AF=AX0+AF;
            AR=PASS AF;
            SI=AR;
            SR=LSHIFT SI BY -7 (HI);
            AX0=SR1;
            AR=AX0 AND AY1;
            M2=AR;

            M3=128;                {no z on manual rotation}
            JUMP calculate;

{filter x axis}
filtered:   L0=%xbuff;             {setup circular buffer for sample buffer}
            M0=1;                  {init buffer increment}
```

```
            M4=1;                       {init coeff increment}
            I0=DM(xpntr);               {get current buffer pointer}
            I4=^coeff;                  {init coeff pointer}
            nops;
            AX0=DM(adx);                {get a sample}
            NOP;
            AY0=h#00FF;
            AR=AX0 AND AY0;             {mask out upper bits}
            MX0=AR;                     {load it into multiplier}
            DM(I0,M0)=MX0;              {add it to the delay line}
            MY0=PM(I4,M4);              {load 1st coeff}
            MR=0;
            CNTR=127;
            DO xfilter UNTIL CE;
xfilter:       MR=MR+MX0*MY0(SS), MX0=DM(I0,M0), MY0=PM(I4,M4);
            MR=MR+MX0*MY0(SS);
            AX0=MR1;
            AR=AX0 AND AY0;             {mask out any wraparound}
            MODIFY(I0,M0);              {increment sample buffer once more}
            DM(xpntr)=I0;               {save buffer pointer}
            DM(newx)=AR;                {save filtered x}

{filter y axis}
            I0=DM(ypntr);               {get current buffer pointer}
            I4=^coeff;                  {init coeff pointer}
            nops;
            AX0=DM(ady);                {get a sample}
            NOP;
            AY0=h#00FF;
            AR=AX0 AND AY0;             {mask out upper bits}
            MX0=AR;                     {load it into multiplier}
            DM(I0,M0)=MX0;              {add it to the delay line}
            MY0=PM(I4,M4);              {load 1st coeff}
            MR=0;
            CNTR=127;
            DO yfilter UNTIL CE;
yfilter:       MR=MR+MX0*MY0(SS), MX0=DM(I0,M0), MY0=PM(I4,M4);
            MR=MR+MX0*MY0(SS);
            AX0=MR1;
            AR=AX0 AND AY0;             {mask out any wraparound}
            MODIFY(I0,M0);              {increment sample buffer once more}
            DM(ypntr)=I0;               {save buffer pointer}
            DM(newy)=AR;                {save filtered x}

            L0=0;                       {restore linear addressing}
            M1=DM(newx);                {load filtered x value}
```

(listing continues on next page)

```
            M2=DM(newy);              {load filtered y value}
            M3=128;                   {no z on manual rotation}

calculate:  I4=^xfm_array;            {reset xfm pointer}
            M4=1;                     {to walk through xfm array}
            M0=0;                     {no modify after trig table lookup}

{calculate element xfm(11)...}
            I0=^cos_table;
            MODIFY(I0,M2);
            MX0=DM(I0,M0);            {cy}
            I0=^cos_table;
            MODIFY(I0,M3);
            MY0=DM(I0,M0);            {cz}
            MR=MX0*MY0(RND);          {cy*cz}
            PM(I4,M4)=MR1;

{calculate element xfm(21)...}
            I0=^sin_table;
            MODIFY(I0,M1);
            MX0=DM(I0,M0);            {sx}
            I0=^sin_table;
            MODIFY(I0,M2);
            MY0=DM(I0,M0);            {sy}
            MF=MX0*MY0(RND);          {sx*sy}
            I0=^cos_table;
            MODIFY(I0,M3);
            MX0=DM(I0,M0);            {cz}
            MR=MX0*MF(SS);            {sx*sy*cz}
            I0=^cos_table;
            MODIFY(I0,M1);
            MX0=DM(I0,M0);            {cx}
            I0=^sin_table;
            MODIFY(I0,M3);
            MY0=DM(I0,M0);            {sz}
            MR=MR-MX0*MY0(SS);        {sx*sy*cz-cx*sz}
            MR=MR(RND);               {loose round bug by not rounding above}
            PM(I4,M4)=MR1;

{calculate element xfm(31)...}
            I0=^cos_table;
            MODIFY(I0,M1);
            MX0=DM(I0,M0);            {cx}
            I0=^sin_table;
            MODIFY(I0,M2);
            MY0=DM(I0,M0);            {sy}
```

```
            MF=MX0*MY0(RND);          {cx*sy}
            I0=^cos_table;
            MODIFY(I0,M3);
            MX0=DM(I0,M0);            {cz}
            MR=MX0*MF(SS);            {sx*sy*cz}
            I0=^sin_table;
            MODIFY(I0,M1);
            MX0=DM(I0,M0);            {sx}
            I0=^sin_table;
            MODIFY(I0,M3);
            MY0=DM(I0,M0);            {sz}
            MR=MR+MX0*MY0(RND);       {cx*sy*cz+sx*sz}
            PM(I4,M4)=MR1;

{calculate element xfm(12)...}
            I0=^cos_table;
            MODIFY(I0,M2);
            MX0=DM(I0,M0);            {cy}
            I0=^sin_table;
            MODIFY(I0,M3);
            MY0=DM(I0,M0);            {sz}
            MR=MX0*MY0(RND);          {cy*sz}
            PM(I4,M4)=MR1;

{calculate element xfm(22)...}
            I0=^sin_table;
            MODIFY(I0,M1);
            MX0=DM(I0,M0);            {sx}
            I0=^sin_table;
            MODIFY(I0,M2);
            MY0=DM(I0,M0);            {sy}
            MF=MX0*MY0(RND);          {sx*sy}
            I0=^sin_table;
            MODIFY(I0,M3);
            MX0=DM(I0,M0);            {sz}
            MR=MX0*MF(SS);            {sx*sy*sz}
            I0=^cos_table;
            MODIFY(I0,M1);
            MX0=DM(I0,M0);            {cx}
            I0=^cos_table;
            MODIFY(I0,M3);
            MY0=DM(I0,M0);            {cz}
            MR=MR+MX0*MY0(RND);       {sx*sy*sz+cx*cz}
            PM(I4,M4)=MR1;
```

(listing continues on next page)

```
{calculate element xfm(32)...}
            I0=^cos_table;
            MODIFY(I0,M1);
            MX0=DM(I0,M0);          {cx}
            I0=^sin_table;
            MODIFY(I0,M2);
            MY0=DM(I0,M0);          {sy}
            MF=MX0*MY0(RND);        {cx*sy}
            I0=^sin_table;
            MODIFY(I0,M3);
            MX0=DM(I0,M0);          {sz}
            MR=MX0*MF(SS);          {cx*sy*sz}
            I0=^sin_table;
            MODIFY(I0,M1);
            MX0=DM(I0,M0);          {sx}
            I0=^cos_table;
            MODIFY(I0,M3);
            MY0=DM(I0,M0);          {cz}
            MR=MR-MX0*MY0(SS);      {cx*sy*sz-sx*cz}
            MR=MR(RND);             {loose round bug by not rounding above}
            PM(I4,M4)=MR1;

{calculate element xfm(13)...}
            I0=^sin_table;
            MODIFY(I0,M2);
            MR1=DM(I0,M0);          {sy}
            AR=-MR1;                {-sy}
            PM(I4,M4)=AR;

{calculate element xfm(23)...}
            I0=^sin_table;
            MODIFY(I0,M1);
            MX0=DM(I0,M0);          {sx}
            I0=^cos_table;
            MODIFY(I0,M2);
            MY0=DM(I0,M0);          {cy}
            MR=MX0*MY0(RND);        {sx*cy}
            PM(I4,M4)=MR1;

{calculate element xfm(33)...}
            I0=^cos_table;
            MODIFY(I0,M1);
            MX0=DM(I0,M0);          {cx}
            I0=^cos_table;
            MODIFY(I0,M2);
```

```
            MY0=DM(I0,M0);          {cy}
            MR=MX0*MY0(RND);        {cx*cy}
            PM(I4,M4)=MR1;

            RTS;

{DOXFM
Module to perform the actual transformation of the raw source data
by the master transformation matrix.

The transformation, xfm_array, is organized sequentially, first by columns,
then by rows:

                        | 11 12 13 |
        xfm_array =     | 21 22 23 |
                        | 31 32 33 |

i.e., xfm_array is stored in a nine location buffer in PMD as follows:

        xfm_array[1..9] = (11, 21, 31, 12, 22, 32, 13, 23, 33)

The following registers are blown away by this routine:
        I0, I1, I4
        L4
        M0, M1, M5
        MX0, MY0, MR
}

doxfm:
        I0=^src_array;              {get source dm array pointer}
        I1=^wcs_array;              {get wcs dm array pointer}
        I4=^xfm_array;              {get transform pm array pointer}

        L4=%xfm_array;    {to run modulo9 through the transform array}

        M0=2;             {to get the next xyz for each new transform}
        M2=-2;            {retard src_array pointer by 2 for each column}
        M1=1;             {general purpose for simple incrementing}
        M5=-1;            {special decrement for xfm at end of point loop}

        CNTR=numpoints;                           {transform all point vectors}
        DO points UNTIL CE;
           MX0=DM(I0,M1), MY0=PM(I4,M4);          {load 1st set of operands}
           CNTR=3;                                {do three columns}
              DO columns UNTIL CE;
                 MR=MX0*MY0(SS), MX0=DM(I0,M1), MY0=PM(I4,M4);
                  {1st multiply clears}
```

(listing continues on next page)

```
                MR=MR+MX0*MY0(SS), MX0=DM(I0,M2), MY0=PM(I4,M4);
                MR=MR+MX0*MY0(SS), MX0=DM(I0,M1), MY0=PM(I4,M4);
columns:        DM(I1,M1)=MR1;     {store the transformed component}
          MODIFY(I4,M5);           {retard xfm pointer by one}
points:   MODIFY(I0,M0);           {pick up next set of vectors}

        L4=0;

        RTS;                       {jump back}

{DOPROJ
Module to do the 3D to 2D object projection

The following registers are overwritten by this routine:
     I0-I3, M0, M1
     AX0, AY0, MX0, MX1, MY0
     MR, AR
}

doproj:
               I0=^wcs_array;      {x-component pointer}
               I1=^wcs_array+1;    {y-component pointer}
               I2=^wcs_array+2;    {z-component pointer}
               I3=^ecs_array;      {2D result pointer}

               M0=0;               {for no-modify access}
               M1=1;               {simple increment for ecs}
               M3=3;               {skip thru wcs by 3s}

               MX0=sin_angle;      {load constants}
               MX1=cos_angle;

               MY0=DM(I2,M3);      {preload z to start pipeline}

               CNTR=numpoints;
               ENA AR_SAT;         {saturate over/underflows}
          DO project UNTIL CE;
               MR=MX1*MY0(RND), AY0=DM(I0,M3);  {z*cos(angle), get x}
               AR=MR1+AY0;                      {x+z*cos(angle)}
               DM(I3,M1)=AR;                    {store x projection}

               MR=MX0*MY0(RND), AY0=DM(I1,M3);  {z*sin(angle), get y}
               AR=MR1+AY0, MY0=DM(I2,M3);       {y+z*sin(angle), get next z}
project:       DM(I3,M1)=AR;                    {store y projection}
               DIS AR_SAT;                    {restore normal ALU operation}
```

```
{in-place adjust ecs data for fullscale dac range}
                MX0=half_scale;      {load scale factor}
                AY0=half_scale;      {load axis offset}
                I3=^ecs_array;       {init 2D result pointer}
                CNTR=numpoints_2;    {twice numpoints for x&y}
                DO scale UNTIL CE;
                   MY0=DM(I3,M0);
                   MR=MX0*MY0(RND);  {applly scaling}
                   AR=MR1+AY0;       {shift axis}
scale:             DM(I3,M1)=AR;     {save scaled x&y}

                RTS;

{DISPLAY
Module to display the 2D image in ecs_array on scope by writing to the dacs the
point vectors and z data (for beam on and beam off).

The line segment drawing routines for each octant are derived from the
Bressenham Algorithm which may be found in any computer graphics text.

The backround registers are used here during the actual line segment drawing
routines.

The following primary registers are blown away by this routine:
     AX0, AX1, AY0, AR, AF
     I0, I1, I4
}

display:        IMASK=0;             {disable interrupts}
                AX0=^ecs_array-2;    {-2 to adjust for 0/1 starting}
                I1=^line_list;

repeat:
                AX1=DM(I1,M1);       {load next linelist value}
                AR=PASS AX1;
                IF LT JUMP done;     {watch for -1 flag to get out}
                IF EQ JUMP moveto;   {move to new line with beam off}
                JUMP lineto;         {draw line from old to new vector}

moveto:         SI=DM(I1,M1);        {do a left shift by one to account}
                SR=LSHIFT SI BY 1 (HI);    {for xy interleave of ecs array}
                AY0=SR1;
                AR=AX0+AY0;          {add in base address of ecs-2}
                I0=AR;               {transfer to ag1}
                AR=DM(I0,M1);        {get new x-coordinate}
                DM(dax)=AR;          {write to dac with beam off}
```

(listing continues on next page)

```
                DM(oldx)=AR;         {update old x-coordinate}
                AR=DM(I0,M1);        {get new y-coordinate}
                DM(day)=AR;          {write to dac with beam off}
                DM(oldy)=AR;         {update old y-coordinate}
                JUMP repeat;

lineto:                              {get 1st point and}
                SI=AX1;              {do a left shift by one to account}
                SR=LSHIFT SI BY 1 (HI);   {for xy interleave of ecs array}
                AY0=SR1;             {with respect to line list}
                AR=AX0+AY0;          {add in base address of ecs-2}
                I0=AR;               {transfer x-addr of ecs to ag1}
                AR=DM(I0,M1);
                DM(newx)=AR;         {update new x-coordinate}
                AR=DM(I0,M1);
                DM(newy)=AR;         {update new y-coordinate}
                AF=PASS 0;           {init for indirect jump offset}
                AX1=DM(newx);        {do delta x}
                AY1=DM(oldx);
                AR=AX1-AY1;
                DM(dx)=AR;
                AX1=4;
                IF GT AF=AX1 OR AF;
                AX1=DM(newy);        {do delta y}
                AY1=DM(oldy);
                AR=AX1-AY1;
                DM(dy)=AR;
                AX1=2;
                IF GT AF=AX1 OR AF;
                AX1=DM(dy);          {do |dx|-|dy|}
                AR=ABS AX1;
                AY1=AR;
                AX1=DM(dx);
                AR=ABS AX1;
                AX1=AR;
                AR=AX1-AY1;
                AX1=1;
                IF GT AF=AX1 OR AF;
                AX1=^jump_table;     {add in jump table base address}
                AR=AX1+AF;
                I4=AR;
                JUMP (I4);           {do the indirect jump}

jump_table:     JUMP octant6;
                JUMP octant5;
                JUMP octant3;
```

```
                JUMP octant4;
                JUMP octant7;
                JUMP octant8;
                JUMP octant2;
                JUMP octant1;
{
In the following code segments:
     MY0' holds +2 (really +1, but shift makes it +2)
     MY1' holds -2 (really -1, but shift makes it -2)
     SR0' holds incr1
     SR1' holds incr2
     AY0' holds current x pixel
     AY1' holds current y pixel
     AX0' holds h#00FF to turn off the beam with
     AX1' holds h#0000 to turn on the beam
     AF   tracks the error term, d
     MX0' holds intermediate stuff
     MR gets hosed
     AR' holds intermediate stuff
}
octant1:        ENA SEC_REG;
                MX0=DM(dy);
                MR=MX0*MY0(SS);
                SR0=MR0;                {incr1 = 2dy}
                SR1=DM(dx);
                AY1=DM(dy);
                AR=SR1-AY1;
                MR=AR*MY1(SS);
                SR1=MR0;                {incr2 = -2(dx-dy)}
                AR=DM(dx);
                AF=ABS AR;
                AR=AF+1;
                CNTR=AR;                {draw line with |dx|+1 pixels}
                AY0=DM(dx);             {init AF with d = incr1-dx}
                AF=PASS AY0;
                AF=SR0-AF;
                AY0=DM(oldx);           {start at last point}
                AY1=DM(oldy);
                DO octant1loop UNTIL CE;
                   DM(dax)=AY0;        {move beam to new x along line segment}
                   DM(day)=AY1;        {move beam to new y along line segment}
                   beam_on;
                   AR=AY0+1;            {increment x}
                   AY0=AR;
                   AR=PASS AF;          {check sign of error term, d}
                   PUSH STS;            {save status for 'else' test}
```

(listing continues on next page)

```
                        IF LT AF=SR0+AF;      {'if then' clause: d = d+incr1}
                        POP STS;
                        IF LT JUMP octant1loop;
                        AR=AY1+1;             {'else' clause...}
                        AY1=AR;               {increment y}
                        AF=SR1+AF;            {d = d+incr2}
octant1loop:            NOP;
                     JUMP update;

octant2:             ENA SEC_REG;
                     MX0=DM(dx);
                     MR=MX0*MY0(SS);
                     SR0=MR0;                 {incr1 = 2dx}
                     SR1=DM(dx);
                     AY1=DM(dy);
                     AR=SR1-AY1;
                     MR=AR*MY0(SS);
                     SR1=MR0;                 {incr2 = 2(dx-dy)}
                     AR=DM(dy);
                     AF=ABS AR;
                     AR=AF+1;
                     CNTR=AR;                 {draw line with |dy|+1 pixels}
                     AY0=DM(dy);              {init AF with d = incr1-dy}
                     AF=PASS AY0;
                     AF=SR0-AF;
                     AY0=DM(oldx);            {start at last point}
                     AY1=DM(oldy);
                     DO octant2loop UNTIL CE;
                        DM(dax)=AY0;         {move beam to new x along line segment}
                        DM(day)=AY1;         {move beam to new y along line segment}
                        beam_on;
                        AR=AY1+1;             {increment y}
                        AY1=AR;
                        AR=PASS AF;           {check sign of error term, d}
                        PUSH STS;             {save status for 'else' test}
                        IF LT AF=SR0+AF;      {'if then' clause: d = d+incr1}
                        POP STS;
                        IF LT JUMP octant2loop;
                        AR=AY0+1;             {'else' clause...}
                        AY0=AR;               {increment x}
                        AF=SR1+AF;            {d = d+incr2}
octant2loop:            NOP;
                     JUMP update;

octant3:             ENA SEC_REG;
                     MX0=DM(dx);
```

```
                MR=MX0*MY1(SS);
                SR0=MR0;                {incr1 = -2dx}
                SR1=DM(dx);
                AY1=DM(dy);
                AR=SR1+AY1;
                MR=AR*MY1(SS);
                SR1=MR0;                {incr2 = -2(dx+dy)}
                AR=DM(dy);
                AF=ABS AR;
                AR=AF+1;
                CNTR=AR;                {draw line with |dy|+1 pixels}
                AY0=DM(dy);             {init AF with d = incr1-dy}
                AF=PASS AY0;
                AF=SR0-AF;
                AY0=DM(oldx);           {start at last point}
                AY1=DM(oldy);
                DO octant3loop UNTIL CE;
                   DM(dax)=AY0;        {move beam to new x along line segment}
                   DM(day)=AY1;        {move beam to new y along line segment}
                   beam_on;
                   AR=AY1+1;            {increment y}
                   AY1=AR;
                   AR=PASS AF;          {check sign of error term, d}
                   PUSH STS;            {save status for 'else' test}
                   IF LT AF=SR0+AF;     {'if then' clause: d = d+incr1}
                   POP STS;
                   IF LT JUMP octant3loop;
                   AR=AY0-1;            {'else' clause...}
                   AY0=AR;              {decrement x}
                   AF=SR1+AF;           {d = d+incr2}
octant3loop:       NOP;
                JUMP update;

octant4:        ENA SEC_REG;
                MX0=DM(dy);
                MR=MX0*MY0(SS);
                SR0=MR0;                {incr1 = 2dy}
                SR1=DM(dx);
                AY1=DM(dy);
                AR=SR1+AY1;
                MR=AR*MY0(SS);
                SR1=MR0;                {incr2 = 2(dx+dy)}
                AR=DM(dx);
                AF=ABS AR;
                AR=AF+1;
                CNTR=AR;                {draw line with |dx|+1 pixels}
```

(listing continues on next page)

```
                  AY0=DM(dx);                {init AF with d = incr1+dx}
                  AF=PASS AY0;
                  AF=SR0+AF;
                  AY0=DM(oldx);              {start at last point}
                  AY1=DM(oldy);
                  DO octant4loop UNTIL CE;
                     DM(dax)=AY0;            {move beam to new x along line segment}
                     DM(day)=AY1;            {move beam to new y along line segment}
                     beam_on;
                     AR=AY0-1;               {decrement x}
                     AY0=AR;
                     AR=PASS AF;             {check sign of error term, d}
                     PUSH STS;               {save status for 'else' test}
                     IF LT AF=SR0+AF;        {'if then' clause: d = d+incr1}
                     POP STS;
                     IF LT JUMP octant4loop;
                     AR=AY1+1;               {'else' clause...}
                     AY1=AR;                 {increment y}
                     AF=SR1+AF;              {d = d+incr2}
octant4loop:         NOP;
                  JUMP update;

octant5:            ENA SEC_REG;
                  MX0=DM(dy);
                  MR=MX0*MY1(SS);
                  SR0=MR0;                   {incr1 = -2dy}
                  SR1=DM(dx);
                  AY1=DM(dy);
                  AR=SR1-AY1;
                  MR=AR*MY0(SS);
                  SR1=MR0;                   {incr2 = 2(dx-dy)}
                  AR=DM(dx);
                  AF=ABS AR;
                  AR=AF+1;
                  CNTR=AR;                   {draw line with |dx|+1 pixels}
                  AY0=DM(dx);                {init AF with d = incr1+dx}
                  AF=PASS AY0;
                  AF=SR0+AF;
                  AY0=DM(oldx);              {start at last point}
                  AY1=DM(oldy);
                  DO octant5loop UNTIL CE;
                     DM(dax)=AY0;            {move beam to new x along line segment}
                     DM(day)=AY1;            {move beam to new y along line segment}
                     beam_on;
                     AR=AY0-1;               {decrement x}
                     AY0=AR;
```

```
                   AR=PASS AF;          {check sign of error term, d}
                   PUSH STS;            {save status for 'else' test}
                   IF LT AF=SR0+AF;     {'if then' clause: d = d+incr1}
                   POP STS;
                   IF LT JUMP octant5loop;
                   AR=AY1-1;            {'else' clause...}
                   AY1=AR;              {decrement y}
                   AF=SR1+AF;           {d = d+incr2}
octant5loop:       NOP;
                JUMP update;

octant6:        ENA SEC_REG;
                MX0=DM(dx);
                MR=MX0*MY1(SS);
                SR0=MR0;                {incr1 = -2dx}
                SR1=DM(dx);
                AY1=DM(dy);
                AR=SR1-AY1;
                MR=AR*MY1(SS);
                SR1=MR0;                {incr2 = -2(dx-dy)}
                AR=DM(dy);
                AF=ABS AR;
                AR=AF+1;
                CNTR=AR;                {draw line with |dy|+1 pixels}
                AY0=DM(dy);             {init AF with d = incr1+dy}
                AF=PASS AY0;
                AF=SR0+AF;
                AY0=DM(oldx);           {start at last point}
                AY1=DM(oldy);
                DO octant6loop UNTIL CE;
                   DM(dax)=AY0;        {move beam to new x along line segment}
                   DM(day)=AY1;        {move beam to new y along line segment}
                   beam_on;
                   AR=AY1-1;            {decrement y}
                   AY1=AR;
                   AR=PASS AF;          {check sign of error term, d}
                   PUSH STS;            {save status for 'else' test}
                   IF LT AF=SR0+AF;     {'if then' clause: d = d+incr1}
                   POP STS;
                   IF LT JUMP octant6loop;
                   AR=AY0-1;            {'else' clause...}
                   AY0=AR;              {decrement x}
                   AF=SR1+AF;           {d = d+incr2}
octant6loop:       NOP;
                JUMP update;
```

(listing continues on next page)

```
octant7:        ENA SEC_REG;
                MX0=DM(dx);
                MR=MX0*MY0(SS);
                SR0=MR0;                {incr1 = 2dx}
                SR1=DM(dx);
                AY1=DM(dy);
                AR=SR1+AY1;
                MR=AR*MY0(SS);
                SR1=MR0;                {incr2 = 2(dx+dy)}
                AR=DM(dy);
                AF=ABS AR;
                AR=AF+1;
                CNTR=AR;                {draw line with |dy|+1 pixels}
                AY0=DM(dy);             {init AF with d = incr1+dy}
                AF=PASS AY0;
                AF=SR0+AF;
                AY0=DM(oldx);           {start at last point}
                AY1=DM(oldy);
                DO octant7loop UNTIL CE;
                   DM(dax)=AY0;        {move beam to new x along line segment}
                   DM(day)=AY1;        {move beam to new y along line segment}
                   beam_on;
                   AR=AY1-1;            {decrement y}
                   AY1=AR;
                   AR=PASS AF;          {check sign of error term, d}
                   PUSH STS;            {save status for 'else' test}
                   IF LT AF=SR0+AF;     {'if then' clause: d = d+incr1}
                   POP STS;
                   IF LT JUMP octant7loop;
                   AR=AY0+1;            {'else' clause...}
                   AY0=AR;              {increment x}
                   AF=SR1+AF;           {d = d+incr2}
octant7loop:       NOP;
                JUMP update;

octant8:        ENA SEC_REG;
                MX0=DM(dy);
                MR=MX0*MY1(SS);
                SR0=MR0;                {incr1 = -2dy}
                SR1=DM(dx);
                AY1=DM(dy);
                AR=SR1+AY1;
                MR=AR*MY1(SS);
                SR1=MR0;                {incr2 = -2(dx+dy)}
                AR=DM(dx);
                AF=ABS AR;
```

```
                  AR=AF+1;
                  CNTR=AR;               {draw line with |dx|+1 pixels}
                  AY0=DM(dx);            {init AF with d = incr1-dx}
                  AF=PASS AY0;
                  AF=SR0-AF;
                  AY0=DM(oldx);          {start at last point}
                  AY1=DM(oldy);
                  DO octant8loop UNTIL CE;
                     DM(dax)=AY0;       {move beam to new x along line segment}
                     DM(day)=AY1;       {move beam to new y along line segment}
                     beam_on;
                     AR=AY0+1;           {increment x}
                     AY0=AR;
                     AR=PASS AF;         {check sign of error term, d}
                     PUSH STS;           {save status for 'else' test}
                     IF LT AF=SR0+AF;    {'if then' clause: d = d+incr1}
                     POP STS;
                     IF LT JUMP octant8loop;
                     AR=AY1-1;           {'else' clause...}
                     AY1=AR;             {decrement y}
                     AF=SR1+AF;          {d = d+incr2}
octant8loop:           NOP;
                  JUMP update;

update:           AY0=DM(newx);          {update old with last new point}
                  DM(oldx)=AY0;
                  AY1=DM(newy);
                  DM(oldy)=AY1;
                  DIS SEC_REG;
                  JUMP repeat;

done:             IMASK=h#0004;          {re-enable irq2}
                  RTS;

.ENDMOD;
```

Linear Predictive Speech Coding 10

10.1 OVERVIEW

The linear predictive method of speech analysis approximates the basic parameters of speech. This method is based on the assumption that a speech sample can be approximated as a linear combination of previous speech samples. The application of linear predictive analysis to estimate speech parameters is often called linear predictive coding (LPC).

The LPC method models the production of speech as shown in Figure 10.1. The time-varying digital filter has coefficients that represent the vocal tract parameters. This filter is driven by a function e(t). For voiced speech (sounds created by the vibration of the vocal folds), e(t) is a train of unit impulses at the pitch (fundamental) frequency. For unvoiced speech (sounds generated by the lips, tongue, etc., without vocal-fold vibration), e(t) is random noise with a flat spectrum.

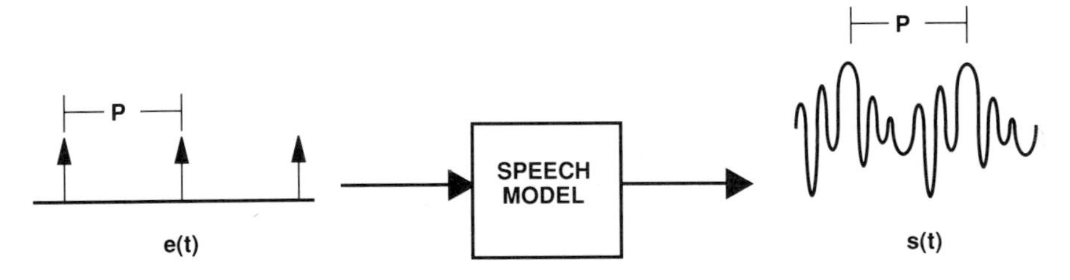

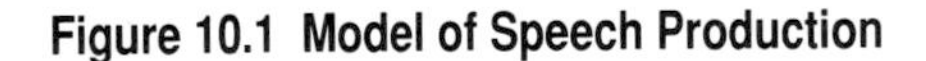

Figure 10.1 Model of Speech Production

The LPC method can be used to create a voice coding system for low-bit-rate transmission. Figure 10.2, on the next page, shows a block diagram of this system. The speech signal is input to the coding system, which derives a set of filter coefficients for the signal and determines whether the signal is voiced. For a voiced signal, the pitch is also calculated. The pitch and filter coefficients are transmitted to a receiving system. This system synthesizes the voice signal by creating a digital filter with the given coefficients and driving the filter with either a train of impulses at the given pitch (for voiced sounds) or a random noise sequence (for unvoiced sounds). The driving function is multiplied by a gain factor, G. For simplicity, in this example we assume unity gain.

10 Linear Predictive Coding

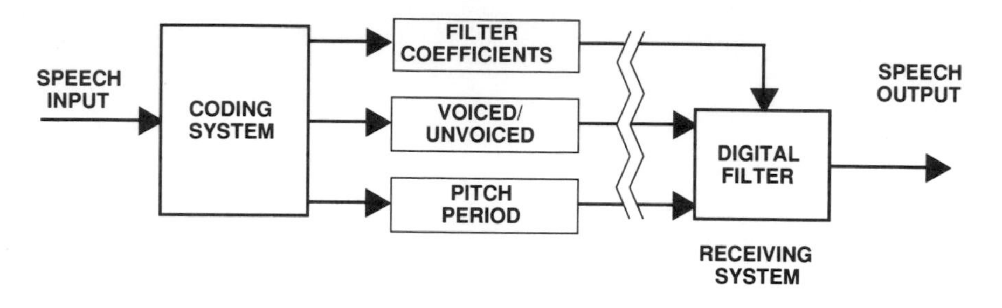

Figure 10.2 Simplified System For Speech Analysis and Synthesis

The coding system predicts the value of an input signal based on the weighted values of the previous P input samples; P is the number of filter coefficients (the order of the synthesis filter). The difference between the actual input and the predicted value is the prediction error. The problem of linear prediction is to determine a set of P coefficients that minimizes the average squared prediction error (E) over a short segment (window) of the input signal. The routines in this chapter are based on a 20-millisecond window, which at a 12-KHz sampling rate yields 240 input samples.

The most efficient method for finding the coefficients is the Levinson-Durbin recursion, which is described by the four equations below. Note that the prime symbol (´) indicates the value to be used in the next pass of the recursion; e.g., E´ is the next value of E.

$$k_i' = [r_s(i) - \sum_{j=1}^{i-1} a_j r_s(i-j)] \div E$$

$$a_i' = k_i'$$

$$a_j' = a_j - k_i' a_{i-j} \qquad j = 1 \text{ to } i-1$$

$$E' = (1-(k_i')^2)\, E$$

where

k_i The negatives of the reflection coefficients used in the LPC synthesis filter (an all-pole lattice filter)

a_i The coefficients used to predict the value of the next input sample

$r_s(i)$ The autocorrelation function of the input signal

E The average squared error between the actual input and predicted input

The autocorrelation function, which is explained further in the next section, is defined as

$$r(k) = \sum_{m=-\infty}^{+\infty} s(m)\, s(m+k)$$

The autocorrelation function of the input signal s(n) is therefore

$$r_s(k) = \sum_{m=0}^{N-1-k} s(m)\, s(m+k)$$

To find the LPC coefficients (k-values), E is initialized to $r_s(0)$. Then the four equations are solved recursively for i = 1 to P. On each pass of the recursion, the first two equations yield another k-value and a-value. In the third equation, all previously calculated a-values are recalculated using the new k-value. The last equation produces a new value for E.

Once all of the k-values have been determined, we can determine whether the input sample is voiced, and if so, what the pitch is. We use the modified autocorrelation analysis algorithm to calculate the pitch using the autocorrelation sequence of the predicted input signal, $r_e(k)$, which can be expressed in terms of the autocorrelation sequence of the actual input and the autocorrelation sequence of the prediction coefficients, a_i.

$$r_e(k) = \sum_{j=1}^{P} r_a(j)\, r_s(k-j)$$

10 Linear Predictive Coding

The autocorrelation function for a_i is defined as

$$r_a(j) = \sum_{i=1}^{P} a_i a_{i+j}$$

The pitch is detected by finding the peak of the normalized autocorrelation sequence ($r_e(n)/r_e(0)$) in the time interval that corresponds to 3 to 15 milliseconds inside the 20-millisecond sampling window. If the value of this peak is at least 0.25, the window is considered voiced with a pitch equal to the value of n at the peak (the pitch period, n_p) divided by the sampling frequency (f_s). If the peak value is less than 0.25, the frame is considered unvoiced and the pitch is zero.

The values of the LPC coefficients (k-values) and the pitch period are transmitted from the coding system to the receiving system. The synthesizer is a lattice filter with coefficients that are the negatives of the calculated k-values. This filter is excited by a signal that is a train of impulses at the pitch frequency. If the pitch is zero, the excitation signal is random noise with a flat spectrum. The excitation function is scaled by the gain value, which is assumed to be one in this example.

The subroutines in this chapter implement linear predictive speech coding. We first present the *correlate* subroutine, which we use whenever a correlation operation (described in the next section) is needed. The *l_p_analysis* subroutine calculates the k-values and the pitch in three main steps. First, the *correlation* subroutine autocorrelates the input signal using the *correlate* subroutine. Next, the *levinson* subroutine finds the k-values using Levinson-Durbin recursion. Last, the *pitch_decision* subroutine determines whether the frame is voiced and computes the pitch period if it is. The *l_p_synthesis* routine, presented at the end of this chapter, generates speech output using the parameters calculated by the *l_p_analysis* routine and the all-pole lattice filter routine presented in Chapter 5.

10.2 CORRELATION

Correlation is an operation performed on two functions of the same variable that are both measures of the same property (voltage, for example). There are two types of correlation functions: cross-correlation and autocorrelation. The cross-correlation of two signals is the sum of the scalar products of the signals in which the signals are displaced in time with respect to one another. The displacement is the independent variable of the cross-correlation function. Cross-correlation is a measure of the similarity between two signals; it is used to detect time-shifted or periodic similarities. Autocorrelation is the cross-correlation of a signal with a copy of the same signal. It compares the signal with itself, providing information about the time variation of the signal.

The cross-correlation of x(n) with y(n) is described by the equation below. L is the number of samples used for both inputs and L–k–1 is number of "overlapping" samples at the displacement k.

$$R(k) = \sum_{n=0}^{L-k-1} (x(n) \times y(n+k))$$

In autocorrelation, x(n) and y(n) are the same signal.

Correlation is required three times in the computation of the linear prediction coefficients and pitch. First, the input signal must be autocorrelated to determine $r_s(k)$. The first P (= number of k-values) values of $r_s(k)$ are used in the Levinson-Durbin recursion and the rest are used in the pitch determination. Next, the a-values calculated in the Levinson-Durbin recursion are autocorrelated to yield $r_a(k)$. The $r_a(k)$ sequence is then cross-correlated with the autocorrelation sequence of the original input signal, $r_s(k)$, to yield $r_e(k)$, which is used to determine the pitch.

Listing 10.1 shows a correlation routine developed for the ADSP-2100. Before the routine is called, one of the input sequences must be stored in a program memory buffer whose starting address is in register I5. The other input sequence must be stored in a data memory buffer whose starting address is in I1. I6 should point to the start of the result buffer in program memory. I2, which is used as a down counter, must be initialized to the length of the input data buffer (both buffers have the same length), and M2 must be initialized to –1, to allow efficient counter manipulation. The CNTR register should be set with the number of correlation samples desired (N). The SE register, which controls output data scaling, must be set to an appropriate value to shift the products, if necessary, into the desired output format. (For example, if two 4.12 numbers are multiplied, the product is a 7.23 number. To obtain a product in 9.21 format, the SE register must be set to –2.) The modify registers M0, M4, M5, and M6 should all be set to one, and the circular buffer length registers must be set to zero.

The routine executes the *corr_loop* loop to produce the number of correlation samples specified by the CNTR register. Address registers I0 and I4 are set to the starting values of the input data buffers. Each time the loop is executed, I0 fetches the same input data, but the value of I4 is moved forward to fetch the next data sample in the program memory buffer. The CNTR register is then loaded with the length of the multiply/accumulate operation required to produce the current term of the correlation sequence; this length decreases each time the *corr_loop* loop is executed because N–k–1 decreases as k increases. The *data_loop* loop performs the multiply/accumulate operation. The result is then scaled to maintain a valid format. During the scaling operation, the routine takes advantage of multifunction instructions to update various pointers. I5, which points to the start of the program memory buffer, is incremented, and I2, which holds the length of the multiply/accumulate operation for the next loop, is decremented. The values in MX0 and MY0 are extraneous and are overwritten.

```
.MODULE         Correlation;

{               Correlate Routine

                Calling Parameters
                   I1 -> Data Memory Buffer                L1 = 0
                   I2 -> Length of Data Buffer             L2 = 0
                   I5 -> Program Memory Buffer             L5 = 0
                   I6 -> Program Memory Result Buffer      L6 = 0
                   M0,M4,M5,M6 = 1       M2 = -1
                   L0,L4 = 0
                   SE = scale value
                   CNTR = output buffer length

                Return Values
                   Result Buffer Filled

                Altered Registers
                   I0,I1,I2,I4,I5,I6,MX0,MY0,MR,SR

                Computation Time
                   Output Length × (Input Length + 8 - ((Output Length - 1) ÷ 2)) + 2
}
```

```
.ENTRY          correlate;

correlate:      DO corr_loop UNTIL CE;
                   I0=I1;
                   I4=I5;
                   CNTR=I2;
                   MR=0, MY0=PM(I4,M4), MX0=DM(I0,M0);
                   DO data_loop UNTIL CE;
data_loop:            MR=MR+MX0*MY0(SS),MY0=PM(I4,M4),MX0=DM(I0,M0);
                   MY0=PM(I5,M5), SR=LSHIFT MR1 (HI);
                   MX0=DM(I2,M2), SR=SR OR LSHIFT MR0 (LO);
corr_loop:         PM(I6,M6)=SR1;
                   RTS;
.ENDMOD;
```

Listing 10.1 Correlation

10.3 LEVINSON-DURBIN RECURSION

The *l_p_analysis* routine, shown in Listing 10.2, calculates the coefficients of the LPC synthesis filter and determines the pitch in approximately 30,000 cycles. This routine calls three other subroutines. First, the *correlation* subroutine autocorrelates the input signal. The *levinson* subroutine uses this autocorrelation sequence to find the LPC coefficients. The *pitch_decision* routine determines the pitch by calling the *pitch_detect* routine, which is presented in the next section.

The subroutines presented in this section calculate the LPC coefficients using the Levinson-Durbin recursion equations:

$$k_i' = [r_s(i) - \sum_{j=1}^{i-1} a_j r_s(i-j)] \div E$$

$$a_i' = k_i'$$

$$a_j' = a_j - k_i' a_{i-j} \qquad j = 1 \text{ to } i-1$$

$$E' = (1-(k_i')^2)\, E$$

The LPC coefficients are the negatives of the k-values.

The *correlation* routine, shown in Listing 10.2, calls the *correlate* routine presented in the previous section to compute the autocorrelation sequence of the input data. The length of the sequence (N) is given by the parameter *wndolength* in the constant file, *lpcconst.h*. The autocorrelation of the original input signal takes up the vast majority of the computation time, almost 25,000 clock cycles.

The *levinson* subroutine, also shown in Listing 10.2, uses two data buffers (*a_ping*, *a_pong*) to store the a-values from the previous iteration of the recursion and the new a-values being computed in the current iteration, as necessitated by the third equation in the Levinson-Durbin recursion. The *in_a* pointer points to the start of the input buffer (old a-values) and the *out_a* pointer points to the start of the output buffer (new a-values). The locations of these pointers are swapped at the end of each iteration of the recursion. On the next pass, new values (a_j') become old values (a_j) and the previous old values are overwritten by the newly calculated values.

The *levinson* subroutine calls the *initialize* routine to set up various parameters for the recursion algorithm. The pointers to the a-value buffers (*in_a, out_a*) are initialized. The SE register is set to an appropriate scaling value. The last location of the output buffer is found by adding one less than the number of LPC coefficients that will be produced (P–1) to the starting location of the output data buffer; the resulting value is stored in I1. This location is needed because the LPC synthesis routine, presented later in this chapter, uses the coefficients in reverse order, and thus the routine stores the coefficients beginning with the last location of the output buffer. The location at which to store the pitch value is determined by adding P to the starting location of the buffer; this value is stored in SI. The 4.12 fixed-point representation for a one is stored in MF and AR; these values are used to adjust the result in some multiplications. The first term of the autocorrelation sequence of the input signal ($r_s(0)$) is stored as the startup value for the error (E).

The first pass of the recursion algorithm is executed outside of the *recursion* subroutine in the *pass_1* subroutine. Although it performs the same operations as other passes, this pass requires much less processing, since it computes only the first value. The *pass_1* subroutine uses an external *divide* routine (see Chapter 2) to divide the second term of the autocorrelation sequence of the input signal, $r_s(1)$, by the initial value of E, shifted left by three, to yield k_1 in 4.12 format. By Levinson-Durbin recursion, this is also a_1. In this pass, there are no old a-values to recalculate, so all that remains is to determine the new value of E. First, the MR register is loaded with a 7.25 representation of a one (by multiplying registers AR and MF, which were initialized to the necessary values). The squared k_1 value just calculated is subtracted from the MR register value. This difference ($1–k_i^2$) is then multiplied by the old E value at the same time that a_1 is stored at the location in I0 and the first filter coefficient (negated k_1) is stored at the location in I1. The subroutine finishes by storing the new E value and swapping the *in_a* and *out_a* pointers to the a-value buffers.

The *recursion* subroutine produces the remaining P–1 filter coefficients, one for each iteration of the *durbin* loop. To keep track of loop iterations, the CNTR register is loaded with P–1 and the counting variable *i*, in DM(I), is initialized to a one. Inside the *durbin* loop, pointers to the buffers that contain the a-values and the r_s values are set up. The *loop_1* loop finds the product of a previous a-value and the corresponding $r_s(i–j)$ value and subtracts this product from $r_s(i)$; this operation is performed until all previous i–1 a-values have been used. The result is divided by E to produce the new k_i. The *loop_2* loop recalculates i–1 a-values, one per iteration. It multiplies the old a_{i-j} value from the DM(*in_a*) buffer (pointer in I6) by k_i and subtracts this product from the old a_i value (pointer in I4). The resulting new a-value is stored in the DM(*out_a*) buffer (pointer in I5). The new value of E is calculated in the same way as in the *pass_1* routine. The quantity $(1–k_i^2)$ in MR is multiplied by the old value of E fetched from data memory to produce the new E while a_i and the filter coefficient are being stored. The loop finishes by incrementing the counting variable *i* and swapping the *in_a* and *out_a* pointers to the a-value buffers.

```
.MODULE Predictor;

{   This routine computes the LPC coefficients for the input data.

    Calling Parameters
         I0 —> Input Buffer                                    L0 = 0
         I1 —> Output Buffer                                   L1 = 0
         L2,L3,L4,L5,L6,L7 = 0

    Return Values
         Output buffer filled
         k[10].....k[1], PITCH

    Altered Registers
         I0,I1,I2,I4,I5,I6,M0,M1,M2,M4,M5,M6,M7
         AX0,AY0,AX1,AY1,AR,AF
         MX0,MY0,MX1,MY1,MR,MF
         SI,SE,SR

    Computation Time
         34,000 cycles (approximately)
}
```

(listing continues on next page)

10 Linear Predictive Coding

```
.INCLUDE          <divide.mac>;
.INCLUDE          <lpcconst.h>;

.VAR/DM/RAM       a_ping[p], a_pong[p], dmhold[p];
.VAR/DM/RAM       e, i, in_a, out_a;

.VAR/PM/RAM       hold[length], zeroed[p], r[ptchlength];
.VAR/PM/RAM       ra[p], re[ptchlength];

.EXTERNAL         correlate, pitch_detect;

.INIT             zeroed : <zero.dat>;

.ENTRY            l_p_analysis;

l_p_analysis:   CALL correlation;
                CALL levinson;
                CALL pitch_decision;
                RTS;

correlation:    AY0=I1;I1=I0;M1=1;
                I4=^hold;M5=1;
                CNTR=length;
                DO trans UNTIL CE;          {copy signal into PM}
                   AX0=DM(I0,M1);
trans:             PM(I4,M5)=AX0;
                SE=5;CNTR=ptchlength;       {set parameters for correlate}
                M6=1;M2=-1;M4=1;M0=1;
                I5=^hold;I2=length;I6=^r;
                CALL correlate;
                RTS;

levinson:       CALL initialize;
                CALL pass_1;
                CALL recursion;
                RTS;

pitch_decision: AY0=^r; I0=^a_ping;
                CALL pitch_detect;
                RTS;

initialize:     M0=0;M4=0;M6=-1;            {set up pointers for recursion}
                AX0=^a_ping;DM(in_a)=AX0;
                AX0=^a_pong;DM(out_a)=AX0;
                SE=3;
                AX0=p-1;
                AR=AX0+AY0;
                I1=AR;                      {point to k buffer}
                AX0=p;
                AR=AX0+AY0;
                SI=AR;                      {save pitch pointer}
```

```
               AR=H#1000;MY1=H#8000;
               MF=AR*MY1 (SU);           {MF = formatted one}
               I4=^r;AX0=PM(I4,M5);
               DM(e)=AX0;                {E = r(0)}
               RTS;

pass_1:        AY1=PM(I4,M6);            {compute k}
               SR1=DM(e);
               SR=LSHIFT SR1 (HI);
               AX0=SR1;AY0=SR0;
               divide(AX0,AY1);
               MY0=AY0;
               I0=DM(out_a);
               MX0=AY0, MR=AR*MF (SS);
               MY1=AX0, MR=MR-MX0*MY0 (SS);
               AR=-AY0;
               DM(I0,M0)=AY0, SR=LSHIFT MR1 (HI);    {compute next E}
               DM(I1,M2)=AR, MR=SR1*MY1 (SS);
               DM(e)=MR1;                            {store next E}
               SR1=DM(in_a);
               SR0=DM(out_a);
               DM(out_a)=SR1;
               DM(in_a)=SR0;
               RTS;

recursion:     CNTR=p-1;
               AX1=1;
               DM(i)=AX1;
               AX0=H#1000;
               DO durbin UNTIL CE;
                  I2=DM(in_a);I6=DM(in_a);
                  I4=^r+1;I5=DM(out_a);M7=DM(i);
                  MX0=PM(I4,M7), AR=PASS AX0;
                  MY1=PM(I4,M6);
                  MR=AR*MY1 (SS), MY0=PM(I4,M6), MX0=DM(I2,M1);
                  CNTR=DM(i);
                  DO loop_1 UNTIL CE;          {compute k values}
loop_1:              MR=MR-MX0*MY0 (SS),MX0=DM(I2,M1),MY0=PM(I4,M6);
                  SR=LSHIFT MR1 (HI);
                  AY1=SR1;
                  SR1=DM(e);
                  AY0=SR0, SR=LSHIFT SR1 (HI);
                  AX1=SR1;
                  divide(AX1,AY1);             {divide by E}
                  I4=DM(in_a);
                  MX1=DM(I4,M5);
                  MODIFY(I6,M7);
                  MODIFY(I6,M6);
                  CNTR=DM(i);
                  MY0=AY0;
                  DO loop_2 UNTIL CE;          {compute new a values}
```

(listing continues on next page)

```
                        MR=MX1*MF(SS), MX0=DM(I6,M6);
                        MR=MR-MX0*MY0 (SS);
                        SR=LSHIFT MR1 (HI), MX1=DM(I4,M5);
loop_2:                 DM(I5,M5)=SR1;
                     MY1=DM(e);
                     I6=DM(out_a);
                     MX0=MY0, MR=AR*MF (SS);             {MR = 1}
                     SR0=DM(I6,M7), MR=MR-MX0*MY0 (SS); {MR = 1-k^2}
                     AR=-AY0;
                     DM(I1,M2)=AR, SR=LSHIFT MR1 (HI);
                     DM(I6,M6)=AY0, MR=SR1*MY1 (SS);
                     SR=LSHIFT MR1 (HI);
                     DM(e)=SR1;
                     AY0=DM(i);
                     AR=AY0+1;
                     DM(i)=AR;
                     SR1=DM(in_a);
                     SR0=DM(out_a);
                     DM(in_a)=SR0;
durbin:              DM(out_a)=SR1;
                  RTS;
.ENDMOD;
```

Listing 10.2 LPC Coefficient Calculation

10.4 PITCH DETECTION

The pitch detection routine is shown in Listing 10.3. Two separate correlation operations (calls to the *correlate* subroutine shown earlier in this chapter) are performed. The first call occurs in the *coeff_corr* routine to autocorrelate the sequence of a-values, which were computed by the Levinson-Durbin recursion routine in the previous section. The second call occurs in the *error_corr* routine which cross-correlates the autocorrelation sequence of the a-values with the autocorrelation of the original input data to calculate the value of $r_e(k)$, as given by the equation:

$$r_e(k) = \sum_{j=1}^{P} r_a(j)\, r_s(j-k) \qquad k = 0 \text{ to } \textit{wndolength}$$

Because this equation is not a true cross-correlation, the calculation requires a few variations from the normal execution of the *correlate* routine. M4 is set to –1, not the usual 1, to scan the sequence $r_s(n)$ backward instead of forward. To eliminate the possibility of generating errors by using values of $r_s(n)$ in which n is less than zero, P zeros are appended to the beginning of the $r_s(n)$ data buffer. M2 is set to zero, so that the number of multiplies remains the same instead of decreasing for each execution of the loop.

In the *pitch_period* routine, the sequence $r_e(n)$ is searched over the interval from 3 to 15 milliseconds for the peak value. The starting point of the search is given by *ptchstrt*, which has been set to the sample number that corresponds to 3 milliseconds. The routine first determines whether $r_e(0)$ is positive or negative. If it is negative, the routine jumps to the *nomaxabs* loop, which finds the negative value with the greatest magnitude. If it is positive, the routine jumps to the *nomax* loop, which finds the positive value with the greatest magnitude. After the peak value is found, if its magnitude is greater than $r_e(0)$, then we know that $r_e(peak)/r_e(0)$ must be greater than one; thus, the routine jumps to the *compute* label to compute the pitch. Otherwise, $r_e(peak)$ is divided by $r_e(0)$ to determine if this value is greater than 0.25. Only the first three bits are calculated because more precision is not required. If $r_e(peak)/r_e(0) > 0.25$, the window is considered voiced and the pitch is calculated by multiplying the peak position time value by the sampling period (in *iperiodh* and *iperiodl*), which requires fewer cycles than dividing by the sampling frequency. If $r_e(peak)/r_e(0) \leq 0.25$, the routine returns with a pitch of zero to indicated an unvoiced window.

```
.MODULE       pitching;

{             This routine computes the pitch period for a speech sample.
              It is used in conjunction with a Linear Predictive Coder.

              Calling Parameters
                 AY0 -> Autocorrelation buffer
                 I0 -> LPC Coefficient buffer       L0 = 0
                 SI -> Pitch buffer
                 L1,L2,L3,L4,L5,L6,L7 = 0

              Return Values
                 Pitch buffer filled

              Altered Registers
                 I0,I1,I2,I4,I5,I6,M1,M2,M4,M5,M6,AX0,AX1,AY0,AY1,AR,AF
                 MX0,MX1,MY0,MY1,MR,MF,SI,SE,SR

              Computation Time
                 approximately 1800 cycles
}
```

(listing continues on next page)

10 Linear Predictive Coding

```
.INCLUDE        <lpcconst.h>;
.VAR/DM/RAM     rahold[p];
.VAR/PM/RAM     hold[p], ra[p], re[ptchlength];
.GLOBAL         pitch_detect;
.EXTERNAL       correlate;

pitch_detect:   CALL coeff_corr;
                CALL error_corr;
                CALL pitch_period;
                RTS;

coeff_corr:     M1=1;M5=1;M2=-1;
                I1=I0;I4=^hold;
                CNTR=p;
                DO move_coeff UNTIL CE;          {copy coeff. to PM}
                  AX0=DM(I0,M1);
move_coeff:       PM(I4,M5)=AX0;
                I5=^hold;CNTR=p;
                M0=1;M4=1;M6=1;I2=p;I6=^ra;
                CALL correlate;
                RTS;

error_corr:     I0=^rahold;I4=^ra;
                CNTR=p;
                DO move_ra UNTIL CE;             {copy r_a to DM}
                  AX0=PM(I4,M5);
move_ra:          DM(I0,M1)=AX0;
                SE=5;I1=^rahold;I5=AY0;CNTR=ptchlength;
                M4=-1;M2=0;I2=P;I6=^re;
                CALL correlate;
                RTS;
pitch_period:   I5=^re;                          {point to r_e}
                M7=ptchstrt;
                MX0=M7;
                AX0=PM(I5,M7);
                AX1=PM(I5,M5), AR=ABS AX0;
                CNTR=wndolength;
                AY0=PM(I5,M5);
                IF POS JUMP max;
                DO nomaxabs UNTIL CE;            {find largest neg. number}
                  AR=AX1-AY0;
                  IF LT JUMP nomaxabs;
                  AX1=AY0;                       {find peak value}
                  AX0=I5;
                  AY0=^re;
                  AF=AX0-AY0;
                  AR=AF-1;
                  MX0=AR;                        {MX0 = peak value}
```

```
nomaxabs:         AY0=PM(I5,M5);
                JUMP pitch_compute;
max:            DO nomax UNTIL CE;
                  AR=AX1-AY0;                    {find peak value}
                  IF GT JUMP nomax;
                  AX1=AY0;
                  AX0=I5;
                  AY0=^re;
                  AF=AX0-AY0;
                  AR=AF-1;
                  MX0=AR;                        {MX0 = peak value}
nomax:            AY0=PM(I5,M5);
pitch_compute:  M0=0;
                I5=^re;
                I1=SI;
                AX0=PM(I5,M5), AR=PASS 0;
                DM(I1,M0)=AR;
                AY0=AR, AR=ABS AX0;
                SR0=AR, AF=ABS AX1;
                AR=SR0-AF;                       {r(0) < r(peak)?}
                IF LE JUMP compute;              {yes, compute pitch}
                AF=PASS AX1;                     {no, find r(peak)÷r(0)}
                AX1=3;
                DIVS AF,AX0;
                DIVQ AX0;
                DIVQ AX0;
                AR=AX1 AND AY0;                  {r(peak)÷r(0) < 0.25?}
                IF EQ JUMP done_compute;         {yes, pitch = 0}
compute:        MY0=iperiodh;                    {no, compute pitch}
                MR=MX0*MY0 (ss);
                MR1=MR0;
                MR0=0;
                MY0=iperiodl;
                MR=MR+MX0*MY0 (su);
                SR=LSHIFT MR1 BY -1 (LO);
                DM(I1,M0)=SR0;
done_compute:   RTS;

.ENDMOD;
```

Listing 10.3 Pitch Detection

10.5 LINEAR PREDICTIVE CODING SYNTHESIZER

The receiving end of the LPC system uses the recursive IIR lattice filter routine presented in Chapter 5 to synthesize an approximation of the original voice signal according to the the LPC voice model. The coefficients of this filter are the negatives of the reflection coefficients (k-values); the k-values were negated and stored in memory in the *levinson* routine. The filter is driven by an excitation function based on the pitch calculated by the *pitch_detect* routine. The gain factor is assumed to be one in this example.

The *l_p_synthesis* routine shown in Listing 10.4 multiplies the pitch by the sampling frequency to yield the pitch period, n_p. It fills the driving function buffer with impulses at the pitch period, and zeros elsewhere. If the window is unvoiced (pitch is zero), the synthesizer uses a driving function of random data in a buffer called *white_noise*. You must initialize this buffer before executing the routine. This can be done using the uniform random number generator presented in Chapter 4.

Various parameters are then set to call the *lattice_filter* subroutine. Note that the length registers L1 and L4 must be set to P, the number of k-values; they are set to zeros after the routine has been executed to ensure that their values do not interfere with any subsequent routines.

```
.MODULE      Synthesizer;

{            Lattice Filter LPC synthesizer

             Calling Parameters
                I1 -> Coefficient Buffer    L0,L1,L2,L3=0
                I2 -> Output Buffer         L4,L5,L6,L7=0

             Return Values
                Output Buffer Filled

             Altered Registers
                I0,I1,I2,I4,M0,M1,M4,M5,M6,M7,AX0,AY0,AX1,AY1,AR,AF
                MX0,MY0,MY1,MR,L1,L4,SE

             Computation Time
                16,000 cycles (approximately)
}

.INCLUDE          <lpcconst.h>;

.VAR/DM           white_noise[length];
.VAR/DM/RAM       pitch_driver[length];
.VAR/PM/RAM/CIRC  delay[p];
```

```
.EXTERNAL         p_latt;
.INIT             delay: <zero.dat>;
.INIT             white_noise: <random.dat>;

.ENTRY            l_p_synthesis;

l_p_synthesis: CALL set_driving;
               CALL synthesis;
               RTS;

set_driving:   M0=1;
               AX0=I1;
               AY0=p;
               AR=AX0+AY0;
               I0=AR;
               AX0=DM(I0,M0);
               I0=^white_noise;            {Point to random data buffer}
               MX0=AX0, AR=PASS AX0;
               IF EQ RTS;                  {If pitch = 0, return}
               I0=^pitch_driver;           {Compute pitch period}
               MY0=samplefreq;
               MR=MX0*MY0 (SS);
               CNTR=length;
               AY1=0;
               AY0=MR1, AF=AY1+1;
               AX1=impulse;
               DO fill_buffer UNTIL CE;    {Fill buffer with impulses at}
                  AX0=AY1, AF=AF-1;        {the pitch period}
                  IF NE JUMP fill_buffer;
                  AX0=AX1, AF=AY0+1;
fill_buffer:      DM(I0,M0)=AX0;
               I0=^pitch_driver;
               RTS;

synthesis:     CNTR=length;I4=^delay;            {Set parameters for lattice}
               L1=p;L4=p;M1=-1;M4=1;             {filter routine}
               M5=-1;M6=3;M7=-2;AR=H#1000;
               AX0=p-1;SE=3;
               CAll p_latt;
               L1=0;L4=0;
               RTS;
.ENDMOD;
```

Listing 10.4 LPC Synthesizer

10.6 REFERENCES

Levinson-Durbin Recursion:
Rabiner, L. R. and Schafer, R. W. 1978. *Digital Processing of Speech Signals*. Englewood Cliffs, N.J.: Prentice-Hall, Inc.

Pitch Detection:
Markel, J. D., and Gray, A. H., Jr. 1980. *Linear Prediction of Speech*. New York: Springer-Verlag.

Pulse Code Modulation 11

11.1 OVERVIEW

Pulse code modulation (PCM) is a method of digitizing or quantizing an analog waveform that is used primarily in the transmission of speech signals, for example, in telephone communication. As in any analog-to-digital (A/D) conversion, the quantization process produces an estimate of the signal sample, possibly introducing an error into the digital representation because of the finite number of bits available to represent the value. In theory, this error can be made insignificant by representing the estimate with a large number of bits (high precision). In practice, however, there must be tradeoff between the amount of error and the size of the data representation. The goal is to quantize the data in the smallest number of bits that results in a tolerable error. In the case of speech signals, a linear quantization with 13 or 14 bits is the minimum required to produce a digital representation of the full range of speech signals accurately.

The number of bits required is reduced to eight in the CCITT recommendation G.711 by exploiting a nonlinear characteristic of human hearing. The human ear is more sensitive to quantization noise in small signals than to noise in large signals. G.711 applies a non-uniform (logarithmic) quantization function to adjust the data size in proportion to the input signal. Thus, smaller signals are approximated with greater precision.

Two quantization functions, or encoding laws, are defined by G.711: μ-law and A-law. In most cases, the United States and Japan use μ-law, whereas Europe uses A-law. The ADSP-2100 implementations of PCM encoder and decoder for both laws are provided in this chapter. In each case, algorithmic versions of the encoder and decoder are provided. In theory, the decoder could also be implemented by table lookup, which provides faster conversion at the cost of additional memory. This implementation is straightforward and is not described here.

11 Pulse Code Modulation

In practice, an inexpensive codec is often used to perform the A/D conversion. The following routines accept encoded speech samples from a codec and generate a linear sample (decoding), and prepare a linear sample for output to a codec (encoding).

11.2 PULSE CODE MODULATION USING μ-LAW

Rather that taking the logarithm of the linear input directly, which can be difficult, μ-law PCM matches a logarithmic curve with a piecewise linear approximation. Eight straight line segments along the curve produce a close approximation to the logarithmic function. Each of these lines is called a segment. A sample value is represented by its segment and its position within the segment.

The CCITT recommendation provides a μ-law conversion table. This table has several regular characteristics, however, so that the conversion can be implemented without storing the entire table. The PCM value is in signed-magnitude format, so the conversion table for negative numbers is the same as for positive numbers except for the sign. In addition, adding 33 to the segment endpoints produces boundaries at even powers of two.

The format of the μ-law PCM 8-bit word consists of three parts. The most significant bit (MSB) is the sign bit, the next three bits contain the segment number, and the last four bits indicate the position within the segment. All bits of the number are inverted from their actual values to increase the density of 1s, (because speech is typically low-energy) a property that can be used by error-correcting circuitry on transmission lines.

11.2.1 μ-Law PCM Encoder

The ADSP-2100 μ-law encoder subroutine, shown in Listing 11.1, is based on the CCITT recommendation G.711. The only deviation from G.711 is the input format. Because the ADSP-2100 is optimized for full fractional numbers (1.15 format), the subroutine accepts input in 1.15 format, instead of the integer format specified in G.711. To use this routine with integer input values, you would shift the input two bits to the left, maintaining the sign, before calling the routine.

The *u_compress* routine adds 132 (33 shifted left two bits) to the absolute value of the input, then normalizes the result. The most significant non-sign bit is zeroed by exclusive-ORing it with H#4000. The position within the segment is equal to the six MSBs of this number (the two MSBs are zero, giving the four bits). A 10-bit right shift moves these bits to the proper position.

The sign of the output word is generated by testing the sign of the original input, which is stored in the AS flag of the ASTAT register by the absolute value instruction at the start of the routine. The AR register is cleared to all zeros if the sign is positive or set to H#4000 if the sign is negative. This value is shifted to the right seven bits and ORed with the position bits.

The segment number is generated by adding seven to the number of bits the input value was shifted for normalization (stored in the SE register). This number is shifted to the left four bits to move it into the proper position to be ORed with the sign and position bits.

The last step in the encoding process is to invert all bits. This is accomplished with the NOT instruction. The eight LSBs of the AR register hold the encoded PCM value.

```
.MODULE/ROM  u_compression;
{
             Linear to u-law Compression Subroutine

             Calling parameters:
                AR = Input linear value

             Returns with:
                AR = 8-bit u-law value (with all bits inverted)

             Altered Registers:
                AX0,AY0,AR,SE,SR

             Computation Time:
                18 cycles
}

.ENTRY     u_compress;

u_compress:  AR=ABS AR;                         { Absolute value }
             AY0=132;
             AR=AR+AY0;
             SR1=32636;                         {This instruction and the next can}
             IF AV AR=PASS SR1;                 {be removed if input ≤ 32636}
             SE=EXP AR (HI);
             SR=NORM AR (LO);                   { Normalize input }
             AY0=H#4000;
             AR=SR0 XOR AY0;                    { clear NMSB }
             SR=LSHIFT AR BY -10 (LO);          { position bits }
             AX0=SE, AR=PASS AY0;
```

(listing continues on next page)

```
            IF POS AR=PASS 0;
            SR=SR OR LSHIFT AR BY -7 (LO);    { sign bit }
            AY0=7;
            AR=AX0+AY0;
            SR=SR OR LSHIFT AR BY 4 (LO);     { segment bits }
            AR=NOT SR0;                       { Invert all bits }
            RTS;
.ENDMOD;
```

Listing 11.1 μ-Law Encoder

11.2.2 μ-Law PCM Decoder

The μ-law PCM decoder expands data received from a transmission line or a codec to the linear domain. The *u_expand* routine decodes a PCM value, requiring 18 cycles to produce a result.

As in the encoder, the only deviation from G.711 is the format of the linear output data. Because the ADSP-2100 is optimized for full fractional data, the output is in 1.15 format, rather than the integer format described in G.711. If integer values are required, you should arithmetically shift the decoder output two bits to the right or remove the shift-by-2 instruction at the end of each routine.

The *u_expand* routine, shown in Listing 11.2, masks out the upper eight bits of the PCM input value and inverts all of the remaining lower eight bits. The segment number is moved, in integer format, from the input word to the SE register. To determine the sign of the input number, H#FF80 is added to the input value. If the sign bit is set, a number greater than zero results; otherwise, the result is a number less than zero.

Control passes to the *negval* block if the input number was negative or *posval* block if it was positive. In the *negval* block the input number (position bits only, no sign or segment bits) is added to itself, effectively shifting the value one bit to the left, then added to 33. The value is shifted by the segment number and stored in the SE register. The shifted value is subtracted from 33 (the same as subtracting 33 from the shifted value and negating the result). Then the number is shifted two bits to the left to place it in 1.15 format.

In the *posval* block, the input number (position bits only) is added to itself, effecting a one-bit left shift. After 33 is added to this value, the result is shifted by the segment number, in the SE register. Then 33 is subtracted from the value, and the result is shifted into 1.15 format, producing a linear value.

```
.MODULE/ROM  u_law_expansion;
{
    This routine determines the 14-bit linear PCM value (right-
    justified) from the 8-bit (right-justified) log (u-law) value.

            Calling parameters:
               AR = 8-bit u-law value

            Return values:
               AR = 16-bit linear value (right-justified)

            Altered Registers:
               AR,AF,AX0,AY0,SR,SE

            Computation Time:
               17 Cycles

}

.ENTRY     u_expand;

u_expand: AY0=H#FF;                  { mask unwanted bits }
          AF=AR AND AY0, AX0=AY0;
          AF=AX0 XOR AF;             { invert bits }
          AX0=H#70;
          AR=AX0 AND AF;             { isolate segment bits }
          SR=LSHIFT AR BY -4 (LO);   { shift to LSBs }
          SE=SR0, AR=AR XOR AF;      { remove segment bits }
          AY0=H#FF80;
          AF=AR+AY0;
          IF LT JUMP posval;         { determine sign }

negval:   AR=PASS AF;
          AR=AR+AF;                  { shift left one bit }
          AY0=33;
          AR=AR+AY0;                 { add segment offset }
          SR=ASHIFT AR (LO);         { position bits }
          AR=AY0-SR0;                { remove segment offset }
          RTS;
```

(listing continues on next page)

```
posval:     AF=PASS AR;
            AR=AR+AF;                { shift left one bit }
            AY0=33;
            AR=AR+AY0;               { add segment offset }
            SR=ASHIFT AR (LO);
            AR=SR0-AY0;              { remove segment offset }
            RTS;

.ENDMOD;
```

Listing 11.2 μ-Law Decoder

11.3 PULSE CODE MODULATION USING A-LAW

A-law PCM uses the same approach as μ-law PCM in approximating the logarithmic curve using eight line segments. In A-law conversion, however, the segment endpoints are at even powers of two, rather than offset by 33. On output, only the even bits of the encoded number are inverted in A-law. Otherwise, the conversion is the same. The format of the 8-bit A-law PCM word is the same as the μ-law format; the most significant bit (MSB) is the sign bit, the next three bits contain the segment number, and the last four bits indicate the position within the segment.

11.3.1 A-Law PCM Encoder

The ADSP-2100 A-law encoder, shown in Listing 11.3, is based on the CCITT recommendation G.711. The only deviation from G.711 is the input format. The encoder accepts its input in full fractional format (1.15), instead of the integer format specified G.711, because the ADSP-2100 is optimized for this numeric format. To use this routine with integer input values, you would shift the input three bits to the left (A-law requires a 3-bit shift because it is normalized to 13 bits), maintaining the sign, before calling the routine.

The zero segment values in A-law companding are computed in a slightly different fashion than those of other segments. If the input is less than 128, the segment value is simply the input downshifted by four bits. A test at the beginning of the *a_compress* routine determines whether the input is less than 128 and the routine branches accordingly.

For values in all other segments, the routine determines the exponent of the absolute value of the input using the EXP instruction. The segment value is the value in SE plus seven. After normalization, the most significant non-sign bit is removed by exclusive-ORing with H#4000. The four bits of the segment are positioned by shifting to the right ten bits. The sign and segment bits are determined in the same way as with μ-law encoding.

The final step of the compression process is to invert the even bits of the output. This is accomplished by exclusive-ORing the output with H#55.

```
.MODULE     a_law_compression;

{           This routine determines the 8-bit A-law value from the
            16-bit (left-justified) linear input.

               Calling Parameters
                  AR=16-bit (left-justified) linear input

               Return Values
                  AR=8-bit log (A-law) value with even bits inverted

               Altered Registers
                  AR, AF, AXO, AY0, SR, SE

               Computation Time
                  19 cycles

}

.ENTRY         a_compress;

a_compress:    AR=ABS AR;                 {Take absolute value}
               AY0=127;                   {Check for zero segment}
               AF=AR-AY0;
               IF GT JUMP upper_seg;
               SR=LSHIFT AR BY -4 (LO);
               AR=H#4000;
               IF NEG AR=PASS 0;
               SR=SR OR LSHIFT AR BY -7 (LO);
               AY0=H#55;
               AR=SR0 XOR AY0;
               RTS;
upper_seg:     SE=EXP AR (HI);              {Find exponent adjustment}
```

(listing continues on next page)

```
            AX0=SE, SR=NORM AR (LO);     {Normalize input}
            AY0=H#4000;
            AR=SR0 XOR AY0;          {Remove first significant bit}
            SR=LSHIFT AR BY -10 (LO);   {Shift position bits}
            AR=PASS AY0;
            IF NEG AR=PASS 0;           {Create sign bit}
            SR=SR OR LSHIFT AR BY -7 (LO); {Position sign bit}
            AY0=7;
            AR=AX0+AY0;                 {Compute segment}
            IF LT AR=PASS 0;
            SR=SR OR LSHIFT AR BY 4 (LO);{Position segment bits}
            AY0=H#55;
            AR=SR0 XOR AY0;             {Invert bits}
            RTS;
.ENDMOD;
```

Listing 11.3 A-Law Encoder

11.3.2 A-Law PCM Decoder

The *a_expand* routine decodes an A-law PCM value, requiring 18 cycles to produce a result. As in the encoder, the only deviation from the standard is the format of the linear output data, which is 1.15 format rather than the integer format as described in G.711. If integer outputs are required, arithmetically shift the decoder output two bits to the right or remove the shift-by-3 instruction at the end of each routine.

The *a_expand* routine, shown in Listing 11.4, ORs the shifted input value with H#00080800 in the SR register (32-bit register). This simultaneously sets the LSB of the interval and sets the sign bit (takes the absolute value of) the input code word. Also, the 12-bit shift of the input places the segment value in SR1, and the interval in SR0. This value is used to shift the position bits to their proper location.

The sign of the input is determined by adding H#FF80 to it. If the input is negative, the result is greater than zero, and the linear value is negated. In some cases it is necessary to add the interval MSB (32) that was removed during compression. Either 32 or 0 is stored in AF and ORed with the interval bits.

```
.MODULE    A_Law_Expansion;
{
            This routine determines the 16-bit (left-justified)
            linear value from an 8-bit log (a-law) input


            Calling Parameters
                   AR = 8-bit log (a-law) value

            Return Values
                   AR = 16-bit linear output

            Altered Registers
                   AR, AF, AX0, AY0,
                   SR, SE

            Cycle Count
                   19 Cycles

}

.ENTRY                  a_expand;

a_expand:               AY0=H#0055;          {Set mask for inversion}
                        AR=AR XOR AY0;       {Even bit inversion}
                        SR1=H#0008;          {Always set sign bit}
                        SR0=H#0800;          {Set LSB of interval}
                        SR=SR OR LSHIFT AR
                        BY 12 (LO);          {Isolate segment,
                                             interval}
                        AY0=32;
                        AX0=AR, AF=PASS AY0;
                        AY0=9;               {Segment bias}
                        AR=SR1-AY0;          {Determine shift value}
                        IF LT AF=PASS 0;     {No extra MSB bit}
                        IF EQ AR=PASS 0;     {No less then zero bits}
                        SR=LSHIFT SR0 BY -11 (LO);{Isolate Interval}
                        SE=AR, AR=SR0 OR AF; {Add bit if necessary}
                        SR=LSHIFT AR (LO);   {Position output}
                        SR=LSHIFT SR0 BY 3 (LO);
```

(listing continues on next page)

```
                    AY0=H#FF80;
                    AR=SR0, AF=AX0+AY0;          {Is sign bit set?}
                    IF LT AR=-SR0;               {Yes, invert word}
                    RTS;

.ENDMOD;
```

Listing 11.4 A-Law Decoder

Adaptive Differential Pulse Code Modulation 12

12.1 OVERVIEW

Pulse code modulation (PCM) samples an input signal using a fixed quantizer to produce a digital representation. This technique, although simple to implement, does not take advantage of any of the redundancies in speech signals. The value of the current input sample does not have an effect on the coding of future samples. Adaptive differential PCM (ADPCM), on the other hand, uses an adaptive predictor, one that adjusts according to the value of each input sample, and thereby reduces the number of bits required to represent the data sample from eight (non-adaptive PCM) to four.

ADPCM does not transmit the value of the speech sample, but rather the difference between a predicted value and the actual sample value. Typically, an ADPCM transcoder is inserted into a PCM system to increase its voice channel capacity. Therefore, the ADPCM encoder accepts PCM values as input, and the ADPCM decoder outputs PCM values. For a complete description of PCM implementation on the ADSP-2100, see Chapter 11, *Pulse Code Modulation*.

12.2 ADPCM ALGORITHM

The ADSP-2100 implementation of ADPCM presented in this chapter is based on the CCITT recommendation G.721 and the identical ANSI recommendation T1.301-1987. The routines have been checked with the digital test sequences provided for the standards and are fully compatible with the recommendation. The terms and equations used in the recommendation are too numerous to be explained fully in this chapter. You should refer to the complete text of the recommendation, which can be obtained from the CCITT or ANSI (see *References* at the end of this chapter).

Figure 12.1, on the following page, shows a block diagram of the ADPCM algorithm. An 8-bit PCM value is input and converted to a 14-bit linear format. The predicted value is subtracted from this linear value to generate a difference signal. Adaptive quantization is performed on this difference, producing the 4-bit ADPCM value to be transmitted.

12 ADPCM

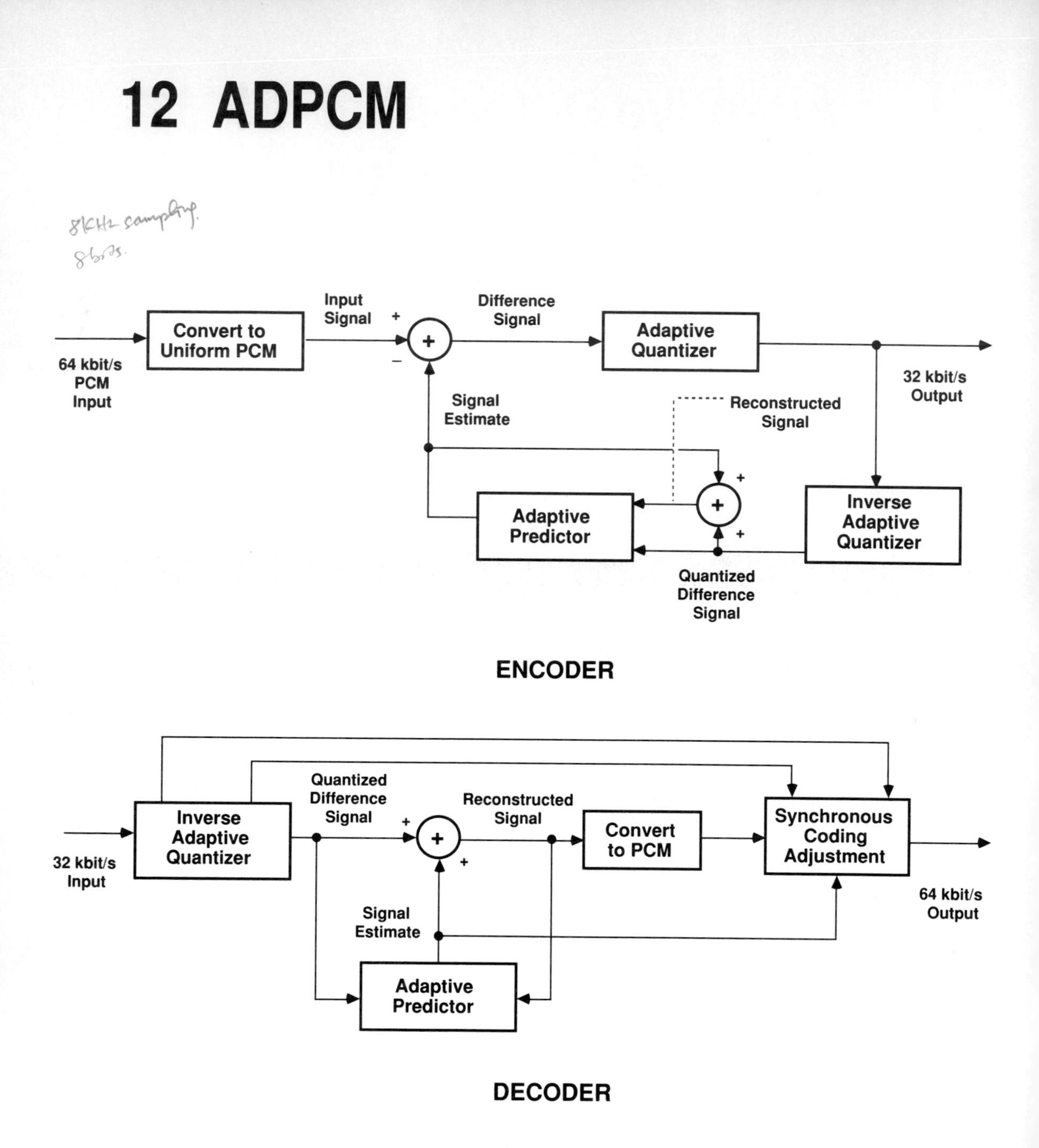

Figure 12.1 ADPCM Block Diagram

Both the encoder and decoder update their internal variables based on only the generated ADPCM value. This ensures that the encoder and decoder operate in synchronization without the need to send any additional or sideband data. A full decoder is embedded within the encoder to ensure that all variables are updated based on the same data.

In the receiving decoder as well as the decoder embedded in the encoder, the transmitted ADPCM value is used to update the inverse adaptive quantizer, which produces a dequantized version of the difference signal. This dequantized value is added to the value generated by the adaptive predictor to produce the reconstructed speech sample. This value is the output of the decoder.

The adaptive predictor computes a weighted average of the last six dequantized difference values and the last two predicted values. The coefficients of the filter are updated based on their previous values, the current difference value, and other derived values.

The ADPCM transcoder program is presented in the listings at the end of this chapter. Two versions are provided: one that conforms fully to the recommendation and a faster version that does not conform fully. This program has two sections, *adpcm_encode* and *adpcm_decode*. Both routines update and maintain the required variables and can be called independently. The program listings indicate the registers that must be initialized before calling each routine.

The code presented in this chapter executes the ADPCM algorithm on both the ADSP-2100 and the ADSP-2101. The routines duplicate the variable and function names indicated in G.721 whenever possible, making it easy to locate the code that implements each functional block.

The format of many of the variables specified in the standard are sign-extended to the full 16-bit word size of the ADSP-2100. This data size works efficiently with the ADSP-2100 and does not affect the ADPCM algorithm. In all cases, this implementation provides at least the smallest data format required.

12.3 ADPCM ENCODER

The ADPCM encoder is shown in Figure 12.2, on the next page. The 8-bit PCM input value is converted to a 14-bit linear representation in the *expand* routine using PCM decoder routines described in Chapter 11, *Pulse Code Modulation*. This linear value is stored in the data memory location *sl*.

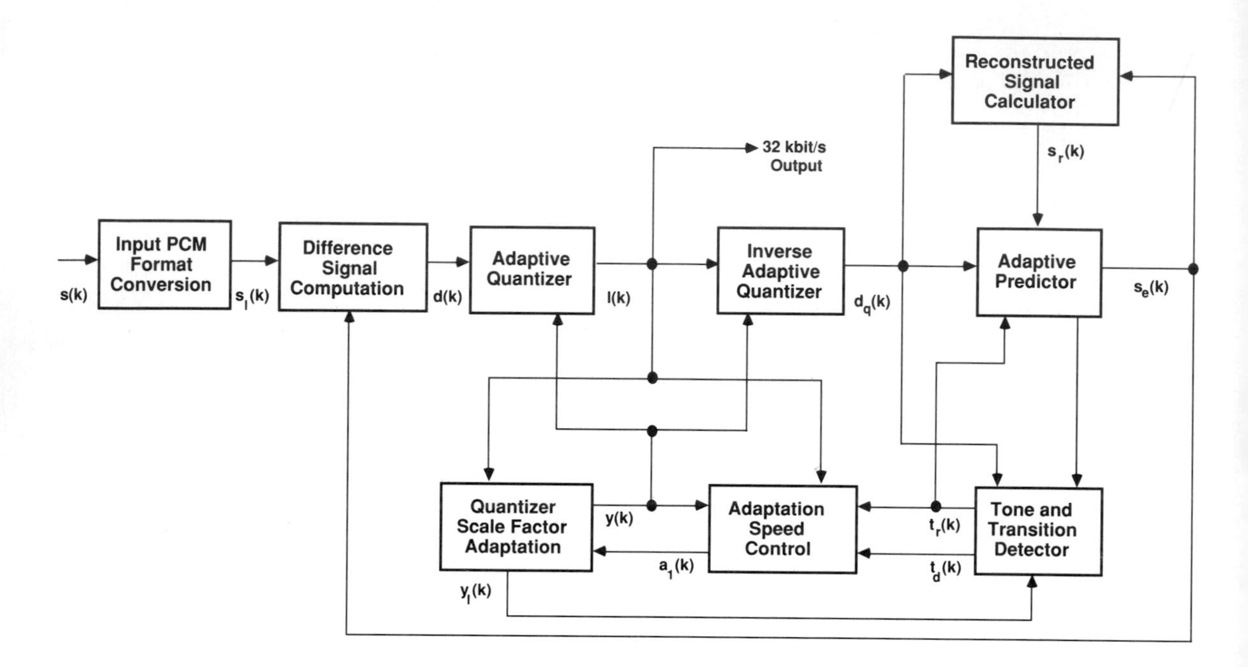

Figure 12.2 Encoder (Transmitter) Block Diagram

12.3.1 Adaptive Predictor

The predicted value of the linear sample is computed by the following equation

$$s_e(k) = \sum_{i=1}^{2} a_i(k-1)\, s_r(k-i) + s_{ez}(k)$$

where

$$s_{ez}(k) = \sum_{i=1}^{6} b_i(k-1)\, d_q(k-i)$$

The *predict* routine produces the predicted value $s_e(k)$ in the variable *s_e* and its component $s_{ez}(k)$ in the variable *sez*, implementing the FMULT and ACCUM functional blocks of the ADPCM algorithm. The running total the WAn and WBn variables is stored in the AF register during this routine. The sum of the WBns are computed first in the *sez_cmp* loop to produce *sez*, the zero component of the predicted value. The second loop, *s_e_cmp*, is identical to the first except for the address registers used; it computes the two WAns and adds them to the sum to produce the predicted sample value, *s_e*.

The implementation of the *s_r* and *dq* delay lines in the *predict* routine is slightly more efficient than the method used in the G.721 recommendation. The variables of the algorithm are in floating-point format. The three parts (sign, exponent, mantissa) of each delay line variable are not packed in a single memory location (as specified in the recommendation), but occupy three separate locations. This is a more efficient method because packing and unpacking the data is not necessary. Because the ADSP-2100 can perform a data fetch in parallel with other operations, using this method does not affect performance.

Once the floating-point components have been determined, the mantissa of the result (W-value) must be generated. During this operation, a fixed offset is added to the product before shifting. In this implementation the MR registers are loaded with the properly shifted product, then the offset is added. The MX1 and MY1 registers are used to generate the properly oriented offset value. The output is in the MR1 register, ready to be shifted by the exponent value.

12.3.2 Adaptive Quantizer

Two routines update the parameters of the adaptive quantizer. The first, *lima*, computes the new value for the speed control variable, *al*, which is used to weight the two components of the quantizer scale factor. The speed control variable varies between 0 and 1, tending towards 1 for speech signals. It is computed using a predicted value, *ap*, which can vary between 0 and 2. The relationship between these two variables is defined by:

$$a_l(k) = \begin{cases} 1, & a_p(k-1) \geq 1 \\ a_p(k-1), & a_p(k-1) \leq 1 \end{cases}$$

Because *ap* is stored in 8.8 format, it is compared to 256 (1 in 8.8 format) and capped at that value. It is shifted down twice to remove the two LSBs, then shifted up nine bits to its proper position for the *mix* routine.

The *mix* routine, using the speed control variable *al*, adds two factors, *yu*, the unlocked factor, and *yl*, the locked factor, to produce the quantizer scale factor variable *y*. The equation for *y* is

$$y(k) = a_l(k)\, y_u(k-1) + [1 - a_l(k)]\, y_l(k-1)$$

The scale factor *y* adapts in a different fashion for speech signals and voice band data.

The *mix* routine computes the difference between *yu* and the upper bits of *yl*. The absolute value of this difference is multiplied by *al*. The product is then added (or subtracted, based on the sign of the difference) to *yl* to produce *y*. The value *y* downshifted twice is also calculated, because this value is needed by other routines.

The difference between the linear input signal and the predicted signal is computed. This value is used by the quantizer to compute the 4-bit ADPCM value. The quantizer computes the base 2 logarithm of the difference value, then subtracts the scale factor *y*. The result is used to index a table that contains the ADPCM quantizer values. This table contains the quantizer segment endpoints and is used to determine the final ADPCM value.

The *log* routine that produces the quantized value takes the absolute value of the difference value. The exponent detector determines the number of redundant sign bits, then 14 is added to generate the exponent value. The $\log_2$ approximation is generated by normalizing the difference and masking out the two MSBs (which is equivalent to subtracting 1).

The exponent value is shifted up seven bits and ORed with the $\log_2$ approximation of the difference shifted down seven bits. The value *y* is subtracted, then a table lookup is performed to retrieve the ADPCM value. The lower end point of each quantizer range is stored in memory, and the AF register is initialized to 7. As each segment end point is checked, AF is decremented if the value is less than the lower end point. When the lower end point is less than or equal to the quantized value, AF holds the ADPCM value.

12.3.3 Inverse Adaptive Quantizer

The calculated 4-bit ADPCM value (I value) is used by the *reconst* routine. The magnitude of the I value is mapped to a data memory location that contains the value (DQLN in the recommendation) to be added to the scale factor. The inverse $\log_2$ of this sum is the quantized difference value, *dq*, which is used by the prediction and update routines.

The *reconst* routine takes the absolute value of the ADPCM input. This value is placed in the M3 register to be used in this and other routines for table lookup. The dequantized value is read from memory, and *y* is added to this value. Log-to-linear conversion is done by first isolating the exponent in SR1. The mantissa is then isolated in AR, and 1 (H#80) is added to it. The magnitude of the shift is determined by subtracting 7 from the exponent in SR1. In the last step, the sign of the dequantized difference signal is determined from the sign of the ADPCM input (I) value.

12.3.4 Adaptation Speed Control

Several of the internal variables of the algorithm are updated in one large subroutine, beginning with the scale factor components. The unlocked component, *yu*, able to adapt to a quickly changing signal, is based on the scale factor and the recently produced ADPCM value. This factor is explicitly limited to the range from 1.06 to 10.00. Other factors derived from *yu* (*y* and *yl*) are therefore implicitly limited to this same range.

The locked factor, *yl*, which adapts more slowly than *yu*, is based on the previous locked factor value as well as the current unlocked factor. The unlocked and locked factors are updated by *filtd* and *filte*, respectively. The unlocked scale factor *yu* is computed as follows:

$$y_u(k) = (1-2^{-5})\ y(k) + 2^{-5}W[I(k)]$$

The function W(I) is determined using a table lookup in the *functw* routine. The code block labeled *filtd* shifts W[I(k)] into its proper format and subtracts *y*. This double-precision remainder is downshifted five bits to accommodate the time factor. This downshifted value (gain) is added to *y* to produce *yu*.

The code block labeled *limb* limits *yu* to the range $1.06 \le yu(k) \le 10.00$. This explicit limitation on *yu* implicitly limits both *yl* and *y* to the same range. The locked factor is determined as follows:

$$y_l(k) = (1-2^{-6})\ y_l(k-1) + 2^{-6}\, y_u(k)$$

The update of *yl* is accomplished by first negating *yl* (in double precision) and adding the MSW of the remainder to the new *yu*. This sum is downshifted six bits and added to the original value of *yl* in double precision to produce the updated value.

The long- and short-term averages of the ADPCM value must also be updated. These values are used to compute the predicted speed control factor, *ap*, used in the next cycle of the loop. The code blocks at the *filta* and *filtb* labels update the averages, while the *filtc* code computes the new predicted weighting factor.

The short-term average (*dms*) is updated by the *filta* routine. The required F-value is determined by a table lookup. The old short-term average is subtracted from the F-value and downshifted five bits. This gain is added to the previous short-term value to generate the updated value.

$$d_{ms}(k) = (1-2^{-5})\, d_{ms}(k-1) + 2^{-5}\mathbf{F}[I(k)]$$

The long-term average is updated by the *filtb* routine.

$$d_{ml}(k) = (1-2^{-7})\, d_{ml}(k-1) + 2^{-7}\mathbf{F}[I(k)]$$

12.3.5 Predictor Coefficient Update

The *update_filter* routine has several parts, all of which are used to compute the new b and a filter coefficients. The first step is to update the b filter coefficients, whose new values are based their previous values plus a weighted product of the signs of the current quantized difference value, and the associated *dq* value contained in the filter delay line. The equation implicitly limits the coefficients to a maximum absolute value of 2.

To update the a-filter coefficients, the routine computes the sum of the quantized difference signal and the portion of the estimated signal computed by the b-coefficients. The sign of this value and the previous values are used when computing the new coefficients. Each a-value computed is also explicitly limited, to increase the stability of the filter.

The *update_filter* routine first computes the magnitude of the gain for the b-coefficients. If the current *dq* value is 0, then the gain is also 0; otherwise the magnitude of the gain is 128 (2^{-7} in 2.14). The sign of the gain is the exclusive-OR of the sign of the current *dq* and the delayed *dq* associated with each coefficient. The gain is added to the old coefficient and the leak factor ($b(k) \times 2^{-8}$) is subtracted for that sum. The new coefficient is stored in program memory and the loop is re-executed for the remaining coefficients. The relationship between the old and new b-coefficient values is defined by:

$$b_i(k) = (1-2^{-8})\, b_i(k-1) + 2^{-7}\mathrm{sgn}[d_q(k)]\ \mathrm{sgn}[d_q(k-i)]$$

After the b-coefficients are updated, the current dequantized difference value (*dq*) is placed in the delay line. This requires a fixed-to-floating-point conversion. Remember that the delay line variable used in this implementation is three separate words instead of one packed word.

$$a_2(k) = (1-2^{-7})\, a_2(k-1) + 2^{-7}\,\{\mathrm{sgn}[p(k)]\ \mathrm{sgn}[p(k-2)] - f[a_1(k-1)]\ \mathrm{sgn}[p(k)]\ \mathrm{sgn}[p(k-1)]\}$$

The update of the *pk* variables, which are the current and delayed sign values of the partial signal estimate, occurs in the code labeled *update_p*. Two of the MR registers are loaded with update values based on the *pk* values. This are used later in the update sections for the a-values.

$$a_1(k) = (1-2^{-8})\, a_1(k-1) + (3 \times 2^{-8})\ \mathrm{sgn}[p(k)]\ \mathrm{sgn}[p(k-1)]$$

The variable a_2 is updated first, because its value is used to limit the new value of a_1. The first step is to generate the function f, which is based on the previous value of a_1. The value for $f(a_1)$ is added to (or subtracted from) the gain, in double precision. The new value for a_2 is generated at the code labeled *upa2*. After being limited, the value of a_2 is stored and used to check for a partial band signal. The tone detector uses the a_2 value as described later in this document.

The variable a_1 is updated by the code labeled *upa1*. This final step of the update process limits a_1 based on the new value of a_2. The code labeled *limd* performs this operation.

12.3.6 Tone and Transition Detector

The last step of the algorithm checks for a tone on the input. If the tone is present, the prediction coefficients are set to 0 and the predicted weighting factor to 256 (1 in 8.8). This configuration allows rapid adaptation and improved performance with FSK modems. The *trigger_true* routine is called if the tone is detected (*tr* = 1) to set the variables to the appropriate values.

The code labeled *tone* checks whether a_2 is less than –0.71875. If it is, the variable *tdp* (tone detect) is set to 1. The *tdp* variable is used in both the *subtc* and *trans* routines.

The *trans* routine implements the transition detector. If *tdp* is a 1 and the absolute value of *dq* exceeds a predetermined threshold, the *tr* (trigger) variable is set to 1. After the filter coefficients are updated, *tr* is checked; if it is a 1, the prediction coefficients are set to zero.

12.4 ADPCM DECODER

The ADPCM decoder is shown in Figure 12.3. The core of the decoder is the same as the decoder embedded within the encoder, so most of the decoder is described in the previous encoder sections. The major difference is that the decoder routines are called with different variables (see *Program Listings* at the end of this chapter). In the decoder, all unique variables and code have an *_r* appended to their names.

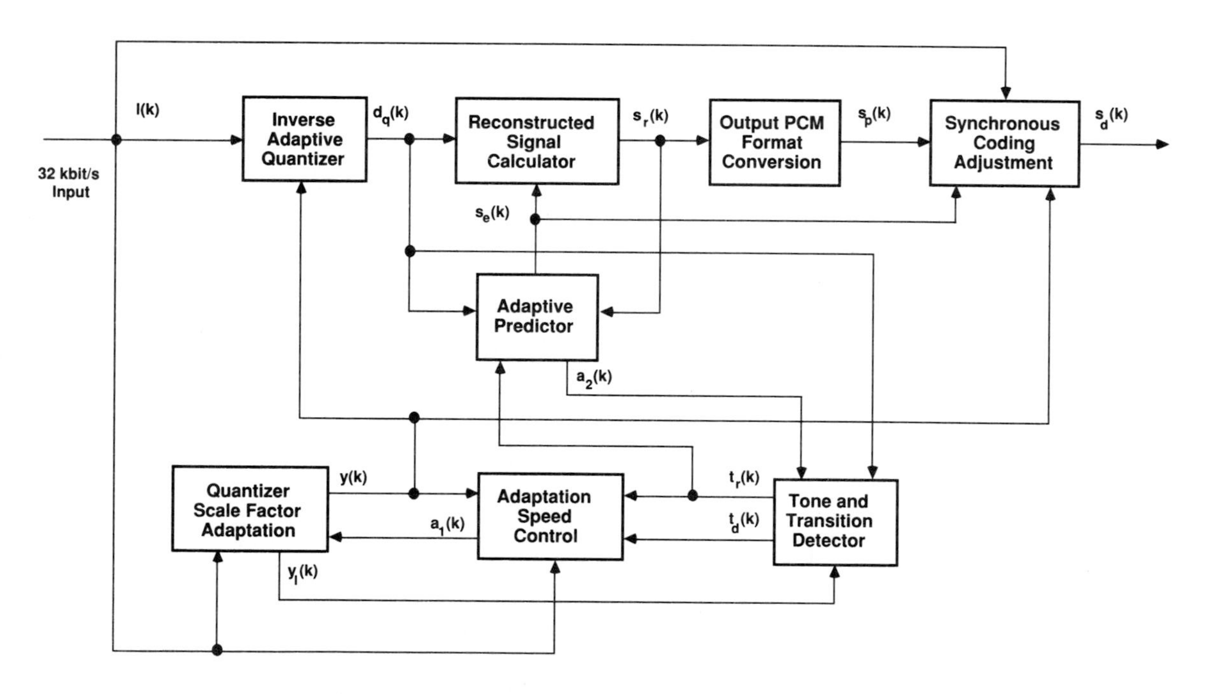

Figure 12.3 Decoder (Receiver) Block Diagram

The decoder filter update and trigger routines are the same as for the encoder, but because they use variable names within this code, it is more efficient to have separate routines rather than have the encoder and decoder call the same routines.

The decoder contains two additional routines. The *compress* routine converts a linear-PCM value to a logarithmic-PCM value. This routine operates on the reconstructed value (*s_r*).

The *sync* routine adjusts the final logarithmic-PCM output. This synchronous coding adjustment ensures no cumulative distortion of the speech signal on tandem (ADPCM-PCM-ADPCM) codings. It performs this function by determining whether another downstream ADPCM encoder would produce I values different from those of the decoder; if so, it increments or decrements the logarithmic-PCM output value by an LSB to prevent this distortion.

12.5 NONSTANDARD ADPCM TRANSCODER

In some applications (voice storage, for example) not all properties of the standard ADPCM algorithms are needed. An application that codes only speech data and is not used with voice band data does not require the tone and transition detectors. The synchronous coding adjustment can be removed if the transcoder is not used in a telecommunications network in which multiple tandem codings can occur.

A version of the ADPCM code with these three sections (*trigger, trans,* and *sync*) removed is shown in *Program Listings*, at the end of this chapter. The nonstandard encoder and decoder routines are called *ns_adpcm_encode* and *ns_adpcm_decode,* respectively. There is no noticeable difference in the quality of the speech, and it is possible to operate two full-duplex channels on a 12.5MHz ADSP-2101.

12.6 COMPANDING TECHNIQUES

The CCITT version of ADPCM works with both the A-law and μ-law PCM companding techniques. The ADPCM algorithm itself does not change, only the PCM input and output are different. The appropriate *expand* and *compress* routines must be in the code to ensure proper operation. Both programs in this chapter are based on μ-law companding.

Changes to three routines are needed to adapt the programs in this chapter to A-law companding: *expand, compress* and *sync*. Chapter 11, *Pulse*

12 ADPCM

Code Modulation, contains the A-law companding routines for *expand* and *compress*; the *sync* routine for A-law values is described below. The ADSP-2101 performs both A-law and μ-law companding in hardware, so for ADSP-2101 operation, only the *sync* routine needs to be changed.

The synchronous coding adjustment (*sync* routine) reduces errors on tandem ADPCM-PCM-ADPCM codings by adjusting the logarithmic-PCM output by (at most) one LSB on output of the decoder. The A-law *sync* routine is shown in Listing 12.1.

```
sync:           AX0=DM(a_ik_r);             {Get input value of I}
                AY1=AR, AF=ABS AR;
                IF NEG AR=AY1+1;            {Convert 1s comp to 2s comp}
                AY1=AX0, AF=ABS AX0;
                IF NEG AF=AY1+1;            {Same for new I value }
                AR=AR-AF;
                AR=DM(sp_r);
                IF GT JUMP decrement_sp;    {Next most negative value}
                IF LT JUMP increment_sp;    {Next most positive value}
                AF=PASS AX0;                {Check for invalid 0 input}
                IF NE RTS;

increment_sp: SR=LSHIFT AR BY 8 (HI);       {Get sign of PCM value}
                AY0=H#AA;
                AF=AR-AY0;                  {Check for maximum value}
                IF EQ RTS;                  {Already maximum value}
                AY0=H#55;                   {Check for sign change}
                AF=AR-AY0;
                IF NE JUMP no_pos_sgn;      {Jump if no sign change}
                AR=H#D5;
                RTS;
```

```
no_pos_sgn:     AF=ABS SR1;
                AF=AR XOR AY0;
                AR=AF-1;                    {Compute adjusted PCM value}
                IF NEG AR=AF+1;
                AR=AR XOR AY0;
                RTS;

decrement_sp:   SR=LSHIFT AR BY 8 (HI);     {Get sign of PCM value}
                AY0=H#2A;
                AF=AR-AY0;                  {Check for minimum value}
                IF EQ RTS;                  {Already minimum value}
                AY0=H#D5;
                AF=AR-AY0;
                IF NE JUMP no_sign_chn;     {If input is H#D5}
                AR=H#55;                    {New output will be H#55}
                RTS;

no_sign_chn:    AF=ABS SR1;                 {Otherwise adjust by 1}
                AY0=H#55;
                AF=AR XOR AY0;
                AR=AF+1;                    {Compute adjusted PCM value}
                IF NEG AR=AF-1;
                AR=AR XOR AY0;
                RTS;
```

Listing 12.1 A-Law Synchronous Coding Adjustment Routine

12.7 BENCHMARKS AND MEMORY REQUIREMENTS

The following tables indicate the number of cycles and execution time of both the ADPCM algorithms on each of the ADSP-210X processors. The memory requirements for the standard ADPCM algorithm are 628 words of program memory and 137 words of data memory. The nonstandard algorithm uses slightly less program memory space.

	Standard ADPCM			*Nonstandard ADPCM*		
	Cycle Count	*Time (μs)*	*Loading (%)*	*Cycle Count*	*Time (μs)*	*Loading (%)*
ADSP-2100 (8 MHz)						
Encode	467	58.3	46.7	437	54.63	43.7
Decode	514	64.2	51.4	409	51.13	40.9
Full Transcode	981	122	98.1	846	105.75	84.6
ADSP-2100A (12.5 MHz)						
Encode	467	37.3	29.8	437	34.96	27.97
Decode	514	41.1	32.8	409	32.72	26.18
Full Transcode	981	78.4	62.7	846	67.68	54.16
ADSP-2101 (12.5 MHz)						
Encode	433	34.6	27.7	403	32.24	25.80
Decode	459	36.7	29.3	375	30.00	24.00
Full Transcode	892	71.3	57.1	778	62.24	49.80

12.8 REFERENCES

American National Standards Institute, Inc. 1987. *American National Standard for Telecommunications: Digital Processing of Voice-Band Signals—Algorithm and Line Format for 32 kbit/s Adaptive Differential Pulse-Code Modulation (ADPCM)*. New York: ANSI, Inc.

International Telegraph and Telephone Consultative Committee. 1986. Study Group XVIII—Report R26(C), Recommendation G.721. *32 kbit/s Adaptive Differential Pulse-Code Modulation (ADPCM).*

12.9 PROGRAM LISTINGS

The complete listings for both the standard and nonstandard ADPCM transcoders are presented in this section.

12.9.1 Standard ADPCM Transcoder Listing

The code below represents a full-duplex ADPCM transcoder. This program has been developed in accordance with ANSI specification T1.301-1987 and CCITT G.721 (bis). It is fully bit-compatible with the test vectors supplied by both of these organizations.

```
.MODULE        Adaptive_Differential_PCM;

{       Calling Parameters
                AR =    Companded PCM value (encoder)
                        or ADPCM I value (decoder)

                M0=3;   L0=18;
                M1=1;   L1=6;
                M2=-1
                        L3=0;
                M4=0    L4=6;
                M5=1    L5=2
                M6=-1   L6=5

        Return Values
                AR =    ADPCM I value (encoder)
                        or Companded PCM value (decoder)

        Altered Registers
                AX0, AX1, AY0, AY1, AF, AR,
                MX0, MX1, MY0, MY1, MR,
                I0, I1, I3, I4, I5, I6
                SI, SR
                M3

        Cycle Count
                467 cycles for encode
                514 cycles for decode

}
```

(listing continues on next page)

12 ADPCM

```
.ENTRY adpcm_encode, adpcm_decode;

.VAR/PM/CIRC   b_buf[6];                {b coefficients for encode}
.VAR/PM/CIRC   a_buf[2];                {a coefficients for encode}
.VAR/PM/CIRC   b_buf_r[6];              {b coefficients for decode}
.VAR/PM/CIRC   a_buf_r[2];              {a coefficients for decode}

.VAR/DM/CIRC   b_delay_r[18];           {dq delay for decode}
.VAR/DM/CIRC   a_delay_r[6];            {sr delay for decode}
.VAR/DM/CIRC   b_delay[18];             {dq delay for encode}
.VAR/DM/CIRC   a_delay[6];              {sr delay for encode}

.VAR/DM/CIRC   mult_data[5];            {predictor immediate data}

.VAR/DM        qn_values[10],dq_values[8]; {quantizer & dequantizer data}
.VAR/DM        f_values[12], w_values[8];  {update coefficient data}
.VAR/DM        a_data[10];

.VAR/DM        s_e,s_r,a_ik,dq,p;
.VAR/DM        sez,sl,yu,yl_h,yl_l,y,y_2,ap,p_o,p_o_o,dms,dml,tdp,tr;
.VAR/DM        a_ik_r,dq_r,p_r;
.VAR/DM        yu_r,yl_h_r,yl_l_r,ap_r,p_o_r;
.VAR/DM        p_o_o_r,dms_r,dml_r,tdp_r;

.VAR/DM        sp_r;                    {PCM code word for synchronous adj}

.VAR/DM        hld_a_t, hld_b_t, hld_a_r, hld_b_r;
```

```
.INIT  qn_values:   7, 14, H#3F80, 400, 349, 300, 246, 178, 80, H#FF84;
.INIT  dq_values :  h#F800, 4, 135, 213, 273, 323, 373, 425;
.INIT  f_values : -5, 0, 5120, 544, 0, 0, 0, 512, 512, 512, 1536, 3584;
.INIT  w_values:    65344, 288, 656, 1024, 1792, 3168, 5680, 17952;
.INIT  mult_data :  H#1FFF, H#4000, h#7E00, H#7FFF, H#FFFE;
.INIT  a_data :     H#1FFF, 2, 16384, 0, -7, 192, H#3000, H#D000,
                    H#D200,  H#3C00;

.INIT  hld_a_t : ^a_delay;
.INIT  hld_b_t : ^b_delay;
.INIT  hld_a_r : ^a_delay_r;
.INIT  hld_b_r : ^b_delay_r;

.INIT  b_buf : 0,0,0,0,0,0;                          {2.14}
.INIT  a_buf : 0,0;                                  {2.14}
.INIT  b_delay : 0,0,0,0,0,0,0,0,0,0,0,0,0,0,0,0,0,0; {16.0, 16.0, 0.16}
.INIT  a_delay : 0,0,0,0,0,0;                        {16.0, 16.0, 0.16}
.INIT  p : 0;                                        {16.0}
.INIT  yu :0;                                        {7.9}
.INIT  yl_h : 0;                                     {7.9}
.INIT  yl_l : 0;                                     {0.16}
.INIT  ap : 0;                                       {8.8}
.INIT  p_o : 0;                                      {16.0}
.INIT  p_o_o : 0;                                    {16.0}
.INIT  dms : 0;                                      {7.9}
.INIT  dml : 0;                                      {5.11}
.INIT  tdp : 0;                                      {16.0}

.INIT  b_buf_r : 0,0,0,0,0,0;                        {2.14}
.INIT  a_buf_r : 0,0;                                {2.14}
.INIT  b_delay_r:0,0,0,0,0,0,0,0,0,0,0,0,0,0,0,0,0,0; {16.0, 16.0, 0.16}
.INIT  a_delay_r : 0,0,0,0,0,0;                      {16.0, 16.0, 0.16}
.INIT  p_r : 0;                                      {16.0}
.INIT  yu_r :0;                                      {7.9}
.INIT  yl_h_r : 0;                                   {7.9}
.INIT  yl_l_r : 0;                                   {0.16}
.INIT  ap_r : 0;                                     {8.8}
.INIT  p_o_r : 0;                                    {16.0}
.INIT  p_o_o_r : 0;                                  {16.0}
.INIT  dms_r : 0;                                    {7.9}
.INIT  dml_r : 0;                                    {5.11}
.INIT  tdp_r : 0;                                    {16.0}
```

(listing continues on next page)

```
adpcm_encode: I4=^b_buf;              {Set pointer to b-coefficients}
              I5=^a_buf;              {Set pointer to a-coefficients}
              I6=^mult_data;          {Set pointer to predictor data}
              I1=DM(hld_a_t);         {Restore pointer to s_r delay}
              I0=DM(hld_b_t);         {Restore pointer to dq delay}

              CALL expand;            {Expand 8-bit log-PCM to 12 bits}
              DM(sl)=AR;              {Store linear PCM value in sl}
              CALL predict;           {Call s_e and sez predictors}
              AX1=DM(ap);
              AY0=DM(yl_h);
              AX0=DM(yu);
              CALL lima;              {Limit ap and compute y}
              DM(y)=AR;               {Save y for later updates}
              DM(y_2)=SR1;            {Save y>>2 for log and reconst}
              AX0=DM(sl);
              AY0=DM(s_e);
              AY1=SR1, AR=AX0-AY0;    {Compute difference signal, d}
              CALL log;               {Determine I value from d}
              DM(a_ik)=AR;
              CALL reconst;           {Compute dq based ONLY on }
              DM(dq)=AR;
              AY0=DM(s_e);
              AR=AR+AY0;              {Compute reconstructed signal}
              DM(s_r)=AR;

              DM(I1,M1)=AR, AR=ABS AR; {Convert s_r to floating point}
              SR1=H#4000;             {Set SR1 to minimum value}
              SE=EXP AR (HI);                   {Determine exponent adjust}
              AX0=SE, SR=SR OR NORM AR (HI);    {Normalize into SR}
              SR=LSHIFT SR1 BY -9 (HI); {Delete lower bits}
              AY0=11;                 {Base exponent}
              SR=LSHIFT SR1 BY 2 (HI); {Adjust for ADSP-210x version}
              AR=AX0+AY0;             {Compute exponent}
              DM(I1,M1)=AR;           {Save exponent}
              DM(I1,M1)=SR1;          {Save mantissa}

              MR1=DM(tdp);
              SI=DM(yl_h);
              AX1=DM(dq);
              CALL trans;             {Compute new trigger }
              DM(tr)=AR;
```

```
AR=PASS AR;                  {Check state of trigger}
IF EQ CALL update_filter; {Update filter if trigger false}

MR0=DM(ap);                  {Load variables for updating}
MR1=DM(y);
MR2=DM(tdp);                 {Always load MR2 after MR1!}
MY0=DM(yl_h);
MY1=DM(yl_l);
AY0=DM(y);
MX0=DM(dms);
MX1=DM(dml);
CALL functw;                 {Update variables}
DM(ap)=AR;                   {Store updated variables}
DM(yu)=AX1;
DM(yl_l)=MY1;
DM(yl_h)=MY0;
DM(dms)=MX0;
DM(dml)=MX1;

AY1=DM(tr);                  {Load trigger }
AF=PASS AY1;                 {Check state of trigger}
IF NE CALL trigger_true;     {Call only if trigger true}

AX0=DM(a_ik);                {Get I value for return}
AY0=H#F;                     {Only 4 LSBs are used}
AR=AX0 AND AY0;              {So mask redundant sign bits}

DM(hld_a_t)=I1;              {Save s_r delay pointer}
DM(hld_b_t)=I0;              {Save dq delay pointer}

RTS;                         {Return to caller}
```

(listing continues on next page)

```
adpcm_decode: I1=DM(hld_a_r);             {Restore s_r delay pointer}
              I0=DM(hld_b_r);             {Restore dq delay pointer}
              I4=^b_buf_r;                {Set pointer to b-coefficients}
              I5=^a_buf_r;                {Set pointer to a-coefficients}
              I6=^mult_data;              {Set pointer to predictor data}

              SR=LSHIFT AR BY 12 (HI);  {Get sign of ADPCM I value here}
              SR=ASHIFT SR1 BY -12 (HI);{Sign extend ADPCM value to 16}
              DM(a_ik_r)=SR1;             {Save I value}
              CALL predict;               {Call s_e and sez predictor}
              AX1=DM(ap_r);
              AY0=DM(yl_h_r);
              AX0=DM(yu_r);
              CALL lima;                  {Limit ap and compute y}
              DM(y)=AR;
              DM(y_2)=SR1;
              AY1=DM(y_2);
              AR=DM(a_ik_r);
              CALL reconst;               {Compute dq from received I}
              DM(dq_r)=AR;
              AY0=DM(s_e);
              AR=AR+AY0;                  {Compute reconstructed signal}
              DM(s_r)=AR;

              DM(I1,M1)=AR, AR=ABS AR;  {Make s_r floating point}
              SR1=H#4000;                 {Set SR1 to minimum value}
              SE=EXP AR (HI);                     {Determine exponent adjust}
              AX0=SE, SR=SR OR NORM AR (HI);      {Normalize value}
              SR=LSHIFT SR1 BY -9 (HI); {Remove LSBs per spec}
              AY0=11;                     {Base exponent}
              SR=LSHIFT SR1 BY 2 (HI);  {Adjust for ADSP-210x version}
              AR=AX0+AY0;                 {Compute exponent}
              DM(I1,M1)=AR;               {Store exponent}
              DM(I1,M1)=SR1;              {Store mantissa}

              MR1=DM(tdp_r);
              SI=DM(yl_h_r);
              AX1=DM(dq_r);
              CALL trans;                 {Compute new trigger}
              DM(tr)=AR;
```

```
AR=PASS AR;                  {Check state of trigger}
IF EQ CALL update_filter_r;{Update filter if trigger false}

AY0=DM(y);                   {Load variables for updating}
MY1=DM(yl_l_r);
MY0=DM(yl_h_r);
MR0=DM(ap_r);
MR1=DM(y);
MR2=DM(tdp_r);               {Always load MR2 after MR1}
MX0=DM(dms_r);
MX1=DM(dml_r);
CALL functw;                 {Update variables}
DM(yu_r)=AX1;                {Stored updated variables}
DM(yl_l_r)=MY1;
DM(yl_h_r)=MY0;
DM(ap_r)=AR;
DM(dms_r)=MX0;
DM(dml_r)=MX1;

AY1=DM(tr);                  {Load current trigger}
AF=PASS AY1;                 {Check state of trigger}
IF NE CALL trigger_true_r;{Call only if trigger true}

CALL compress;               {Compress PCM value}
DM(sp_r)=AR;                 {Save original value for sync}
CALL expand;                 {Expand for sync coding adj}
AY0=DM(s_e);
AR=AR-AY0;                   {Compute dx for sync coding}
AY1=DM(y_2);
CALL log;                    {Compute new dqx value}
CALL sync;                   {Adjust PCM value by +1,-1, 0}

DM(hld_a_r)=I1;              {Save s_r delay pointer}
DM(hld_b_r)=I0;              {Save dq delay pointer}

RTS;
```

(listing continues on next page)

```
compress:    AR=DM(s_r);                      {Get reconstructed signal}
             AR=ABS AR;                       {Take absolute value}
             AY0=33;                          {Add offset of boundries}
             AR=AR+AY0;
             AY0=8191;                        {Maximum PCM value}
             AF=AR-AY0;                       {Cap input}
             IF GT AR=PASS AY0;               {If in excess}
             SE=EXP AR (HI);                  {Find exponent adjustmet}
             AX0=SE, SR=NORM AR (LO);         {Normalize input}
             AY0=H#4000;
             AR=SR0 XOR AY0;                  {Remove first significant bit}
             SR=LSHIFT AR BY -10 (LO);        {Shift position bits}
             AR=PASS AY0;
             IF POS AR=PASS 0;                {Create sign bit}
             SR=SR OR LSHIFT AR BY -7 (LO);        {Position sign bit}
             AY0=9;
             AR=AX0+AY0;                      {Compute segment}
             IF LT AR=PASS 0;
             SR=SR OR LSHIFT AR BY 4 (LO);         {Position segment bits}
             AY0=H#FF;
             AR=SR0 XOR AY0;                  {Invert bits}
             RTS;

expand:      AY0=H#FF;                        {Mask unwanted bits}
             AF=AR AND AY0, AX0=AY0;
             AF=AX0 XOR AF;                   {Invert bits}
             AX0=H#70;
             AR=AX0 AND AF;                   {Isolate segment bits}
             SR=LSHIFT AR BY -4 (LO);         {Shift to LSBs}
             SE=SR0, AR=AR XOR AF;            {Remove segment bits}
             AY0=H#FF80;
             AF=AR+AY0;
             IF LT JUMP posval;               {Detemine sign}
             AR=PASS AF;
             AR=AR+AF;                        {Shift left by 1 bit}
             AY0=33;
             AR=AR+AY0;                       {Add segment offset}
             SR=ASHIFT AR (LO);               {Position bits}
             AR=AY0-SR0;                      {Remove segment offset}
             RTS;
```

```
posval:         AF=PASS AR;
                AR=AR+AF;                        {Shift left by 1}
                AY0=33;
                AR=AR+AY0;                       {Add segment offset}
                SR=ASHIFT AR (LO);
                AR=SR0-AY0;                      {Remove segment offset}
                RTS;

predict:        AX1=DM(I0,M2), AY1=PM(I4,M6);        {Point to dq6 and b6}
                AF=PASS 0, SI=PM(I4,M4);         {AF hold partial sum}
                AY0=DM(I6,M5);
                MX1=3;                           {This multiply will give the}
                MY1=32768;                       {+48>>4 term}
                SR=ASHIFT SI BY -2 (HI);         {Downshift b6 per spec}
                CNTR=6;                          {Loop once for each b}
                DO sez_cmp UNTIL CE;
                   AR=ABS SR1, SR1=DM(I6,M5);        {Get absolute value of b}
                   AR=AR AND AY0, AY0=DM(I6,M5);     {Mask bits per spec}
                   SE=EXP AR (HI), MY0=DM(I0,M2);    {Find exponent adjust}
                   AX0=SE, SR=SR OR NORM AR (HI);    {Compute bnMANT}
                   AR=SR1 AND AY0, AY0=DM(I0,M2);    {Mask bits per spec}
                   MR=AR*MY0 (SS), AX1=DM(I0,M2), AY1=PM(I4,M6);
                   AR=AX0+AY0, AY0=DM(I6,M5);        {Compute WbnEXP}
                   SE=AR, MR=MR+MX1*MY1 (SU);        {Compute WbnMANT}
                   SR=LSHIFT MR1 (HI), SE=DM(I6,M5);{Compute Wbn}
                   AR=SR1 AND AY0, SI=PM(I4,M4);     {Mask Wbn per spec}
                   AX0=AR, AR=AX1 XOR AY1;           {Determine sign of Wbn}
                   AR=AX0, SR=ASHIFT SI (HI);        {Downshift b(n-1) per spec}
                   IF LT AR=-AX0;                    {Negate Wbn if necessary}
sez_cmp:           AF=AR+AF, AY0=DM(I6,M5);          {Add Wbn to partial sum}
                AR=PASS AF, AX1=DM(I0,M1), AY1=PM(I5,M6);   {Get sezi}
                SR=ASHIFT AR BY -1 (HI);             {Downshift to produce sez}
                DM(sez)=SR1;
                SI=PM(I5,M4);                    {Get a2}
                SR=ASHIFT SI (HI);               {Downshift a2 per spec}
                AX1=DM(I1,M2), AY1=PM(I4,M5);        {Restore bn and dqn pointers}
                CNTR=2;                          {Loop once for each a}
```

(listing continues on next page)

```
            DO s_e_cmp UNTIL CE;
               AR=ABS SR1, SR1=DM(I6,M5);        {Get absolute value of a}
               AR=AR AND AY0, AY0=DM(I6,M5);     {Mask bits per spec}
               SE=EXP AR (HI), MY0=DM(I1,M2);    {Get exponent adjust for a}
               AX0=SE, SR=SR OR NORM AR (HI);    {Compute WanMANT}
               AR=SR1 AND AY0, AY0=DM(I1,M2);    {Mask bits per spec}
               MR=AR*MY0(SS), AX1=DM(I1,M2), AY1=PM(I5,M6);
               AR=AX0+AY0, AY0=DM(I6,M5);        {Compute WanEXP}
               SE=AR, MR=MR+MX1*MY1 (SU);        {Complete WanMANT computation}
               SR=LSHIFT MR1 (HI), SE=DM(I6,M5);         {Compute Wan}
               AR=SR1 AND AY0, SI=PM(I5,M4);     {Mask Wan per spec}
               AX0=AR, AR=AX1 XOR AY1;           {Determine sign of Wan}
               AR=AX0, SR=ASHIFT SI (HI);        {Downshift a1 per spec}
               IF LT AR=-AX0;                    {Negate Wan if necessary}
s_e_cmp:       AF=AR+AF, AY0=DM(I6,M5);          {Add Wan to partial sum}
            AR=PASS AF, AX1=DM(I1,M1), AY1=PM(I5,M5);   {Get sei}
            SR=ASHIFT AR BY -1 (HI);             {Compute se}
            DM(s_e)=SR1;
            RTS;

lima:       AY1=256;                     {Maximum value for ap}
            AR=AX1, AF=AX1-AY1;          {Cap if it exceeds}
            IF GE AR=PASS AY1;
            SR=ASHIFT AR BY -2 (HI);     {>>2 to produce al}
            SR=LSHIFT SR1 BY 9 (HI);     {Adjust for ADSP-210x version}

mix:        MY0=SR1, AR=AX0-AY0;         {MY0=al, AR=diff}
            AR=ABS AR;                   {Take absolute value of diff}
            MR=AR*MY0 (SU);              {Generate prod}
            AR=MR1+AY0;                  {Add to yu}
            IF NEG AR=AY0-MR1;           {Subtract if diff < 0}
            SR=ASHIFT AR BY -2 (HI);     {Generate y>>2}
            RTS;
```

```
log:            I3=^qn_values;                {Point to data array}
                AR=ABS AR, AX1=DM(I3,M1);     {Take absolute of d}
                SE=EXP AR (HI), AX0=DM(I3,M1);      {Determine exponent adjust}
                AY0=SE, SR=NORM AR (HI);      {Normalize}
                AR=AX0+AY0, AY0=DM(I3,M1);    {Compute exponent}
                IF LT AR=PASS 0;              {Check for exponent -1}
                SI=AR, AR=SR1 AND AY0;        {Mask mantissa bits}
                SR=LSHIFT AR BY -7 (HI);      {Position mantissa}
                SR=SR OR LSHIFT SI BY 7 (HI);       {Position exponent}

subtb:          AR=SR1-AY1, AY0=DM(I3,M1);    {Subtract y>>2 for log}
                AX0=AR, AF=PASS AX1;          {Setup for quantizing}

quan:           AR=AX0-AY0, AY0=DM(I3,M1);    {Is dl less then upper limit?}
                IF LT AF=AF-1;
                AR=AX0-AY0, AY0=DM(I3,M1);    {Continue to check for }
                IF LT AF=AF-1;
                AR=AX0-AY0, AY0=DM(I3,M1);    {where dl fits in quantizer}
                IF LT AF=AF-1;
                AR=AX0-AY0, AY0=DM(I3,M1);
                IF LT AF=AF-1;
                AR=AX0-AY0, AY0=DM(I3,M1);
                IF LT AF=AF-1;
                AR=AX0-AY0, AY0=DM(I3,M1);
                IF LT AF=AF-1;
                AR=AX0-AY0;
                IF LT AF=AF-1;
                AR=PASS AF;
                IF NEG AR=NOT AF;             {Negate value if ds negative}
                IF EQ AR=NOT AR;              {Send 15 for 0}
                RTS;
```

(listing continues on next page)

```
reconst:        AF=ABS AR;
                IF NEG AR=NOT AR;          {Find absolute value}
                M3=AR;                     {Use this for table lookup}
                I3=^dq_values;             {Point to dq table}
                MODIFY(I3,M3);             {Set pointer to proper spot}
                AX1=DM(I3,M1);             {Read dq from table}

adda:           AR=AX1+AY1;                {Add y>>2 to dq}

antilog:        SR=ASHIFT AR BY 9 (LO);    {Get antilog of dq}
                AY1=127;                   {Mask mantisa}
                AX0=SR1, AR=AR AND AY1;    {Save sign of DQ+Y in AX0}
                AY1=128;                   {Add 1 to mantissa}
                AR=AR+AY1;
                AY0=-7;                    {Compute magnitude of shift}
                SI=AR, AR=SR1+AY0;
                SE=AR;
                SR=ASHIFT SI (HI);         {Shift mantissa }
                AR=SR1, AF=PASS AX0;
                IF LT AR=PASS 0;           {If DQ+Y <0, set to zero}
                IF NEG AR=-SR1;            {Negate DQ if I value negative}
                RTS;

trans:          SR=ASHIFT SI BY -9 (HI);   {Get integer of yl}
                SE=SR1;                    {Save for shift}
                SR=LSHIFT SR0 BY -11 (HI); {Get 5 MSBs of fraction of yl}
                AY0=32;
                AR=SR1+AY0;                {Add one to fractional part}
                AX0=SE, SR=LSHIFT AR (HI); {Shift into proper format}
                AY0=8;
                AR=H#3E00;                 {Maximum value}
                AF=AX0-AY0;
                IF LE AR=PASS SR1;         {Cap at maximum value}
                AF=ABS AX1, AY0=AR;        {Get absolute value of dq}
                SR=LSHIFT AR BY -1 (HI);
                AR=SR1+AY0;
                SR=LSHIFT AR BY -1 (HI);
                AF=SR1-AF, AR=MR1;         {tdp must be set for tr true}
                IF GE AR=PASS 0;           {If dq exceeds threshold no tr}
                RTS;
```

```
functw:     I3=^w_values;               {Get scale factor multiplier}
            MODIFY(I3,M3);              {Based on I value}
            AF=PASS 0, SI=DM(I3,M1);
            I3=^f_values;

filtd:      SR=ASHIFT SI BY 1 (LO);     {Update fast quantizer factor}
            AR=SR0-AY0, SE=DM(I3,M1);   {Compute difference}
            SI=AR, AR=SR1-AF+C-1;       {in double precision}
            SR=ASHIFT AR (HI), AX0=DM(I3,M1); {Time constant is 1/32}
            SR=SR OR LSHIFT SI (LO), AY1=DM(I3,M1);
            AR=SR0+AY0, AY0=DM(I3,M1);          {Add gain}

limb:       AF=AR-AY1, SI=DM(I3,M3);    {Limit fast scale factor}
            IF GT AR=PASS AY1;          {Upper limit 10}
            AF=AR-AY0, AY1=MY1;
            IF LT AR=PASS AY0;          {Lower limit 1.06}

filte:      AF=AX0-AY1, AY0=MY0;        {Update quantizer slow factor}
            AF=AX0-AY0+C-1, AX0=DM(I3,M1);      {Compute difference}
            AX1=AR, AR=AR+AF;
            SR=ASHIFT AR BY -6 (HI);    {Time constant is 1/64}
            AR=SR0+AY1, AY1=MX0;        {Add gain}
            MY1=AR, AR=SR1+AY0+C;       {in double precision}

filta:      MY0=AR, AR=AX0-AY1;         {Update short term I average}
            SR=ASHIFT AR (HI), SI=AX0;  {Time constant is 1/32}
            AR=SR1+AY1, AY0=MX1;        {Add gain}

filtb:      SR=LSHIFT SI BY 2 (HI);     {Update long term I average}
            MX0=AR, AR=SR1-AY0;
            SR=ASHIFT AR BY -7 (HI);    {Time constant is 1/128}
            AR=SR1+AY0, SI=MX0;         {Add gain}

subtc:      SR=ASHIFT AR BY -3 (HI);    {Compute difference of long}
            AF=PASS AR, AX0=SR1;        {and short term I averages}
            SR=ASHIFT SI BY 2 (HI);
            MX1=AR, AR=SR1-AF;
            AF=ABS AR;
            AR=MR2, AF=AX0-AF;          {tdp must be true for ax 0}
            IF LE AR=PASS 1;
            AY0=1536;
            AF=MR1-AY0, AY0=MR0;
            IF LT AR=PASS 1;            {Y>3 for ax to be 0}
```

(listing continues on next page)

```
filtc:          SR=ASHIFT AR BY 9 (HI);      {Update speed control}
                AR=SR1-AY0;                  {Compute difference}
                SR=ASHIFT AR BY -4 (HI);     {Time constant is 1/16}
                AR=SR1+AY0;                  {Add gain}
                RTS;

trigger_true:   CNTR=6;                      {Only called when trigger true}
                AX0=0;
                DO trigger UNTIL CE;
trigger:           PM(I4,M5)=AX0;            {Set all b-coefficients to 0}
                AX1=DM(dq);

                DM(tdp)=AX0;                 {Set tdp to 0}

                PM(I5,M5)=AX0;               {Set a2 to 0}
                PM(I5,M5)=AX0;               {Set a1 to 0}

                AR=ABS AX1;                  {Add dq to delay line}
                SE=EXP AR (HI);
                AX0=SE, SR=NORM AR (HI);
                SR=LSHIFT SR1 BY -9 (HI);
                AY0=11;
                AY1=32;
                AR=SR1 OR AY1;
                AY1=DM(a_ik);
                SR=LSHIFT AR BY 2 (HI);
                AR=AX0+AY0, DM(I0,M1)=AY1;
                DM(I0,M1)=AR;
                DM(I0,M1)=SR1;

                AY0=DM(sez);                 {Compute new p values}
                AR=AX1+AY0;
                AX0=DM(p);
                AY0=DM(p_o);
                DM(p)=AR;
                DM(p_o)=AX0;
                DM(p_o_o)=AY0;

                AR=256;
                DM(ap)=AR;                   {Set ap to triggered value}

                RTS;
```

```
update_filter: AX0=DM(dq);                    {Get value of current dq}
               AR=128;
               AF=PASS AX0, AY1=DM(I0,M0); {Read sign of dq(6)}
               IF EQ AR=PASS 0;             {If dq 0 then gain 0}
               SE=-8;                       {Time constant is 1/256}
               AX1=AR;
               CNTR=6;
               DO update_b UNTIL CE;        {Update all b-coefficients}
                  AF=AX0 XOR AY1, AY0=PM(I4,M4);  {Get sign of update}
                  IF LT AR=-AX1;
                  AF=AR+AY0, SI=AY0;        {Add update to original b}
                  SR=ASHIFT SI (HI), AY1=DM(I0,M0); {Get next dq(k)}
                  AR=AF-SR1;                       {Subtract leak factor}
update_b:         PM(I4,M5)=AR, AR=PASS AX1;       {Write out new b-coefficient}

place_dq:      AR=ABS AX0, AY0=DM(I0,M2); {Take absolute value of dq}
               SE=EXP AR (HI);              {Determine exponent adjust}
               SR1=H#4000;                  {Set minimum value into SR1}
               AX1=SE, SR=SR OR NORM AR (HI);     {Normalize dq}
               AY0=11;                      {Used for exponent adjustment}
               SR=LSHIFT SR1 BY -9 (HI);    {Remove lower bits}
               SR=LSHIFT SR1 BY 2 (HI);     {Adjust for ADSP-210x version}
               DM(I0,M2)=SR1, AR=AX1+AY0;   {Save mantisa, compute exp.}
               DM(I0,M2)=AR;                {Save exponent}
               AX1=DM(a_ik);                {Use sign of I, not dq}
               DM(I0,M0)=AX1;               {Save sign}

update_p:      AY0=DM(sez);                 {Get result of predictor}
               AR=AX0+AY0;                  {Use dq from above}
               AY1=DM(p);                   {Delay all old p's by 1}
               AY0=DM(p_o);
               DM(p)=AR;
               DM(p_o)=AY1;
               DM(p_o_o)=AY0;
               AX1=AR, AR=AR XOR AY0;       {Compute p xor poo}
               MR1=AR, AR=AX1 XOR AY1;      {Compute p xor po}
               MR0=AR;
```

(listing continues on next page)

```
upa2:       I3=^a_data;
            SI=PM(I5,M5);                  {Hold a2 for later}
            AR=PM(I5,M5);                  {Get a1 for computation of f}
            AR=ABS AR, AY0=DM(I3,M1);      {Cap magnitude of a1 at 1/2}
            AF=AR-AY0, SE=DM(I3,M1);
            IF GT AR=PASS AY0;
            IF NEG AR=-AR;                 {Restore sign}
            SR=ASHIFT AR (LO), AY0=DM(I3,M1);
            AF=ABS MR0, AY1=DM(I3,M1);     {If p xor po = 0 negate f}
            AR=SR0, AF=PASS SR1;
            IF POS AR=AY1-SR0;             {Double precision}
            IF POS AF=AY1-SR1+C-1;
            SR0=AR, AR=PASS AF;
            SR1=AR, AF=ABS MR1;            {If p xor poo = 1 subtract}
            AR=SR0+AY0, SE=DM(I3,M1);
            AF=SR1+AY1+C, AX0=DM(I3,M1);
            IF NEG AR=SR0-AY0;
            IF NEG AF=SR1-AY1+C-1;
            SR=LSHIFT AR (LO);
            AR=PASS AF;
            SR=SR OR ASHIFT AR (HI), AY0=SI;
            AY1=SR0, SR=ASHIFT SI (HI);    {Downshift a2 for adjustment}
            AR=AY0-SR1, AY0=DM(I3,M1);
            AF=PASS AX1;
            IF NE AR=AR+AY1;               {If sigpk = 1, no gain}

limc:       AF=AR-AY0, AY1=DM(I3,M1);      {Limit a2 to .75 max}
            IF GT AR=PASS AY0;
            AF=AR-AY1, AY0=DM(I3,M1);      {Limit a2 to -.75 min}
            IF LT AR=PASS AY1;
            PM(I5,M5)=AR;                  {Store new a2}

tone:       AF=AR-AY0, AY1=AR;             {If a2 < .71, tone = 1}
            AR=0;
            IF LT AR=PASS 1;
            DM(tdp)=AR;                    {Store new tdp value (for ap)}
```

```
upa1:           AR=AX0, AF=PASS MR0;
                IF LT AR=-AX0;
                AF=PASS AX1, SI=PM(I5,M4);
                IF EQ AR=PASS 0;
                SR=ASHIFT SI BY -8 (HI);     {Leak Factor = 1/256}
                AF=PASS AR, AR=SI;
                AF=AF-SR1;
                AR=AR+AF, AX1=DM(I3,M1);

limd:           AX0=AR, AR=AX1-AY1;          {Limit a1 based on a2}
                AY0=AR, AR=AY1-AX1;
                AY1=AR, AR=PASS AX0;
                AF=AR-AY0;
                IF GT AR=PASS AY0;           {Upper limit 1 - 2^-4 - a2}
                AF=AR-AY1;
                IF LT AR=PASS AY1;           {Lower limit a2 - 1 + 2^-4}
                PM(I5,M5)=AR;                {Store new a1}

                RTS;

trigger_true_r: CNTR=6;                      {Here only if trigger true}
                AX0=0;
                DO trigger_r UNTIL CE;
trigger_r:      PM(I4,M5)=AX0;               {Set all b-coefficients to 0}
                AX1=DM(dq_r);

                DM(tdp_r)=AX0;               {Set tdp to 0}

                PM(I5,M5)=AX0;               {Set a2 to 0}
                PM(I5,M5)=AX0;               {Set a1 to 0}

                AR=ABS AX1;                  {Add dq_r to delay line}
                SE=EXP AR (HI);
                AX0=SE, SR=NORM AR (HI);
                SR=LSHIFT SR1 BY -9 (HI);
                AY0=11;
                AY1=32;
                AR=SR1 OR AY1;
                AY1=DM(a_ik_r);
                SR=LSHIFT AR BY 2 (HI);
                AR=AX0+AY0, DM(I0,M1)=AY1;
                DM(I0,M1)=AR;
                DM(I0,M1)=SR1;
```

(listing continues on next page)

```
            AY0=DM(sez);                   {Compute new p_r's}
            AR=AX1+AY0;
            AX0=DM(p_r);
            AY0=DM(p_o_r);
            DM(p_r)=AR;
            DM(p_o_r)=AX0;
            DM(p_o_o_r)=AY0;

            AR=256;
            DM(ap_r)=AR;                   {Set ap_r to triggered value}

            RTS;

update_filter_r: AX0=DM(dq_r);             {Get dq_r}
            AR=128;                        {Set possible gain}
            AF=PASS AX0, AY1=DM(I0,M0);    {Get sign of dq(6)}
            IF EQ AR=PASS 0;               {If dq_r 0, gain 0}
            SE=-8;                         {Leak factor 1/256}
            AX1=AR;
            CNTR=6;
            DO update_b_r UNTIL CE;        {Update all b-coefficients}
               AF=AX0 XOR AY1, AY0=PM(I4,M4);   {Get sign of gain}
               IF LT AR=-AX1;
               AF=AR+AY0, SI=AY0;               {Add gain to original b}
               SR=ASHIFT SI (HI);               {Time constant is 1/256}
               AR=AF-SR1, AY1=DM(I0,M0);        {Compute new b-value}
update_b_r:    PM(I4,M5)=AR, AR=PASS AX1;       {Store new b-value}

place_dq_r: AR=ABS AX0, AY0=DM(I0,M2);          {Get absolute value fo dq_r}
            SE=EXP AR (HI);                {Determine exponent adjustment}
            SR1=H#4000;                    {Set SR to minimum value}
            AX1=SE, SR=SR OR NORM AR (HI);      {Normalize dq_r}
            AY0=11;                        {Used for exponent adjust}
            SR=LSHIFT SR1 BY -9 (HI);      {Remove lower bits}
            SR=LSHIFT SR1 BY 2 (HI);       {Adjust for ADSP-210x version}
            DM(I0,M2)=SR1, AR=AX1+AY0;     {Store mantissa, compute exp}
            AX1=DM(a_ik_r);                {Use sign of I, not dq}
            DM(I0,M2)=AR;                  {Store exponent}
            DM(I0,M0)=AX1;                 {Store sign}
```

```
update_p_r:     AY0=DM(sez);                    {Compute new p}
                AR=AX0+AY0;                     {Use dq_r from above}
                AY1=DM(p_r);                    {Delay old p's by 1}
                AY0=DM(p_o_r);
                DM(p_r)=AR;
                DM(p_o_r)=AY1;
                DM(p_o_o_r)=AY0;
                AX1=AR, AR=AR XOR AY0;          {Compute p and poo}
                MR1=AR, AR=AX1 XOR AY1;         {Compute p and po}
                MR0=AR;

upa2_r:         I3=^a_data;
                SI=PM(I5,M5);                   {Hold a2 for later}
                AR=PM(I5,M5);                   {Get a1 for computation of f}
                AR=ABS AR, AY0=DM(I3,M1);       {Cap magnitude of a1 to 1/2}
                AF=AR-AY0, SE=DM(I3,M1);
                IF GT AR=PASS AY0;
                IF NEG AR=-AR;                  {Restore sign of f}
                SR=ASHIFT AR (LO), AY0=DM(I3,M1);
                AF=ABS MR0, AY1=DM(I3,M1);      {If p_r xor poo_r =1 subtract}
                AR=SR0, AF=PASS SR1;
                IF POS AR=AY1-SR0;
                IF POS AF=AY1-SR1+C-1;
                SR0=AR, AR=PASS AF;
                SR1=AR, AF=ABS MR1;
                AR=SR0+AY0, SE=DM(I3,M1);
                AF=SR1+AY1+C, AX0=DM(I3,M1);
                IF NEG AR=SR0-AY0;
                IF NEG AF=SR1-AY1+C-1;
                SR=LSHIFT AR (LO);
                AR=PASS AF;
                SR=SR OR ASHIFT AR (HI), AY0=SI;
                AY1=SR0, SR=ASHIFT SI (HI);     {Leak factor of 1/128}
                AR=AY0-SR1, AY0=DM(I3,M1);
                AF=PASS AX1;
                IF NE AR=AR+AY1;                {If sigpk = 1 , no gain}

limc_r:         AF=AR-AY0, AY1=DM(I3,M1);       {Limit a2 to .75 max}
                IF GT AR=PASS AY0;
                AF=AR-AY1, AY0=DM(I3,M1);       {Limit a2 to -.75 min}
                IF LT AR=PASS AY1;
                PM(I5,M5)=AR;                   {Store new a2}
```

(listing continues on next page)

```
tone_r:       AF=AR-AY0, AY1=AR;
              AR=0;
              IF LT AR=PASS 1;            {If a2 < .71, tdp = 1}
              DM(tdp_r)=AR;

upa1_r:       AR=AX0, AF=PASS MR0;
              IF LT AR=-AX0;
              AF=PASS AX1, SI=PM(I5,M4);
              IF EQ AR=PASS 0;
              SR=ASHIFT SI BY -8 (HI);    {Leak Factor = 1/256}
              AF=PASS AR, AR=SI;
              AF=AF-SR1;
              AR=AR+AF, AX1=DM(I3,M1);

limd_r:       AX0=AR, AR=AX1-AY1;         {Limit a1 based on a2}
              AY0=AR, AR=AY1-AX1;
              AY1=AR, AR=PASS AX0;
              AF=AR-AY0;
              IF GT AR=PASS AY0;          {Upper limit 1 - 2^-4 -a2}
              AF=AR-AY1;
              IF LT AR=PASS AY1;          {Lower limit a2 - 1 + 2^-4}
              PM(I5,M5)=AR;               {Store new a1}

              RTS;

sync:         AX0=DM(a_ik_r);             {Get input value of I}
              AY1=AR, AF=ABS AR;
              IF NEG AR=AY1+1;            {Convert 1's comp to 2's comp}
              AY1=AX0, AF=ABS AX0;
              IF NEG AF=AY1+1;            {Same for new I value }
              AR=AR-AF;
              AR=DM(sp_r);
              IF GT JUMP decrement_sp;    {Next most negative value}
              IF LT JUMP increment_sp;    {Next most positive value}
              AF=PASS AX0;                {Check for invalid 0 input}
              IF NE RTS;
```

```
increment_sp: SR=LSHIFT AR BY 8 (HI);     {Get sign of PCM value}
              AY0=H#80;
              AF=AR-AY0;
              IF EQ RTS;                  {Already maximum value}
              AF=ABS SR1;
              AF=PASS 1;
              IF NEG AF=PASS -1;          {If negative, subtract 1}
              AR=AR+AF;                   {Compute adjusted PCM value}
              RTS;

decrement_sp: SR=LSHIFT AR BY 8 (HI);     {Get sign of PCM value}
              AR=PASS AR;
              IF EQ RTS;                  {Already minimum value}
              AY0=H#FF;
              AF=AR-AY0;
              IF NE JUMP no_sign_chn;     {If input is H#FF}
              AR=H#7E;                    {New output will be h#7E}
              RTS;

no_sign_chn:  AF=ABS SR1;                 {Otherwise adjust by 1}
              AF=PASS -1;
              IF NEG AF=PASS 1;           {Add 1 for negative values}
              AR=AR+AF;                   {Compute adjusted PCM value}
              RTS;

.ENDMOD;
```

Listing 12.2 Standard ADPCM Transcoder Routine

12.9.2 Nonstandard ADPCM Transcoder Listing

The code below represents a full-duplex ADPCM transcoder. Although developed in accordance with ANSI specification T1.301-1987 and CCITT G.721 (bis), it has been modified to improve its speed. The modifications include the removal of the synchronous coding adjustment and the tone and transition detectors. These deletions do not noticeably affect speech-only coding.

```
.MODULE         Adaptive_Differential_PCM;

{       Calling Parameters
                AR = Companded PCM value (encoder)
                       or ADPCM I value (decoder)

                M0=3;  L0=18;
                M1=1;  L1=6;
                M2=-1
                       L3=0;
                M4=0   L4=6;
                M5=1   L5=2
                M6=-1  L6=5

        Return Values
                AR = ADPCM I value (encoder)
                       or Companded PCM value (decoder)

        Altered Registers
                AX0, AX1, AY0, AY1, AF, AR,
                MX0, MX1, MY0, MY1, MR,
                I0, I1, I3, I4, I5, I6
                SI, SR
                M3

        Cycle Count
                437 cycles for encode
                409 cycles for decode

}
```

ADPCM 12

```
.ENTRY          ns_adpcm_encode, ns_adpcm_decode;

.VAR/PM/CIRC    b_buf[6];                      {b coefficients for encode}
.VAR/PM/CIRC    a_buf[2];                      {a coefficients for encode}
.VAR/PM/CIRC    b_buf_r[6];                    {b coefficients for decode}
.VAR/PM/CIRC    a_buf_r[2];                    {a coefficients for decode}

.VAR/DM/CIRC    b_delay_r[18];                 {dq delay for decode}
.VAR/DM/CIRC    a_delay_r[6];                  {sr delay for decode}
.VAR/DM/CIRC    b delay[18];                   {dq delay for encode}
.VAR/DM/CIRC    a_delay[6];                    {sr delay for encode}

.VAR/DM/CIRC    mult_data[5];                  {Predictor immediate data}

.VAR/DM         qn_values[10],dq_values[8]; {quantizer & dequantizer data}
.VAR/DM         f_values[12], w_values[8];  {Update coefficient data}
.VAR/DM         a_data[10];

.VAR/DM         s_e,s_r,a_ik,dq,p;
.VAR/DM         sez,sl,yu,yl_h,yl_l,y,y_2,ap,p_o,p_o_o,dms,dml,tdp,tr;
.VAR/DM         a_ik_r,dq_r,p_r;
.VAR/DM         yu_r,yl_h_r,yl_l_r,ap_r,p_o_r;
.VAR/DM         p_o_o_r,dms_r,dml_r,tdp_r;

.VAR/DM         sp_r;                          {PCM code word for synchronous adj}

.VAR/DM         hld_a_t, hld_b_t, hld_a_r, hld_b_r;

.INIT  qn_values:    7, 14, H#3F80, 400, 349, 300, 246, 178, 80, H#FF84;
.INIT  dq_values :   h#F800, 4, 135, 213, 273, 323, 373, 425;
.INIT  f_values :    -5, 0, 5120, 544, 0, 0, 0, 512, 512, 512, 1536, 3584;
.INIT  w_values:     65344, 288, 656, 1024, 1792, 3168, 5680, 17952;
.INIT  mult_data :   H#1FFF, H#4000, h#7E00, H#7FFF, H#FFFE;
.INIT  a_data :      H#1FFF, 2, 16384, 0, -7, 192, H#3000, H#D000,
                     H#D200, H#3C00;

.INIT           hld_a_t : ^a_delay;
.INIT           hld_b_t : ^b_delay;
.INIT           hld_a_r : ^a_delay_r;
.INIT           hld_b_r : ^b_delay_r;
```

(listing continues on next page)

```
.INIT          b_buf : 0,0,0,0,0,0;                  {2.14}
.INIT          a_buf : 0,0;                          {2.14}
.INIT          b_delay : 0,0,0,0,0,0,0,0,0,0,0,0,0,0,0,0,0,0;
                                                     {16.0, 16.0, 0.16}
.INIT          a_delay : 0,0,0,0,0,0;                {16.0, 16.0, 0.16}
.INIT          p : 0;                                {16.0}
.INIT          yu :0;                                {7.9}
.INIT          yl_h : 0;                             {7.9}
.INIT          yl_l : 0;                             {0.16}
.INIT          ap : 0;                               {8.8}
.INIT          p_o : 0;                              {16.0}
.INIT          p_o_o : 0;                            {16.0}
.INIT          dms : 0;                              {7.9}
.INIT          dml : 0;                              {5.11}
.INIT          tdp : 0;                              {16.0}

.INIT          b_buf_r : 0,0,0,0,0,0;                {2.14}
.INIT          a_buf_r : 0,0;                        {2.14}
.INIT          b_delay_r : 0,0,0,0,0,0,0,0,0,0,0,0,0,0,0,0,0,0;
                                                     {16.0, 16.0, 0.16}
.INIT          a_delay_r : 0,0,0,0,0,0;              {16.0, 16.0, 0.16}
.INIT          p_r : 0;                              {16.0}
.INIT          yu_r :0;                              {7.9}
.INIT          yl_h_r : 0;                           {7.9}
.INIT          yl_l_r : 0;                           {0.16}
.INIT          ap_r : 0;                             {8.8}
.INIT          p_o_r : 0;                            {16.0}
.INIT          p_o_o_r : 0;                          {16.0}
.INIT          dms_r : 0;                            {7.9}
.INIT          dml_r : 0;                            {5.11}
.INIT          tdp_r : 0;                            {16.0}

ns_adpcm_encode: I4=^b_buf;                          {Set pointer to b-coefficients}
               I5=^a_buf;                            {Set pointer to a-coefficients}
               I6=^mult_data;                        {Set pointer to predictor data}
               I1=DM(hld_a_t);                       {Restore pointer to s_r delay}
               I0=DM(hld_b_t);                       {Restore pointer to dq delay}

               CALL expand;                          {Expand 8-bit log-PCM to12 bits}
               DM(sl)=AR;                            {Store linear PCM value in sl}
               CALL predict;                         {Call s_e and sez predictors}
```

```
AX1=DM(ap);
AY0=DM(yl_h);
AX0=DM(yu);
CALL lima;                          {Limit ap and compute y}
DM(y)=AR;                           {Save y for later updates}
DM(y_2)=SR1;                        {Save y>>2 for log and reconst}
AX0=DM(sl);
AY0=DM(s_e);
AY1=SR1, AR=AX0-AY0;                {Compute difference signal, d}
CALL log;                           {Determine I value from d}
DM(a_ik)=AR;
CALL reconst;                       {Compute dq based ONLY on }
DM(dq)=AR;
AY0=DM(s_e);
AR=AR+AY0;                          {Compute reconstructed signal}
DM(s_r)=AR;

DM(I1,M1)=AR, AR=ABS AR;            {Convert s_r to floating point}
SR1=H#4000;                         {Set SR1 to minimum value}
SE=EXP AR (HI);                     {Determine exponent adjust}
AX0=SE, SR=SR OR NORM AR (HI);      {Normalize into SR}
SR=LSHIFT SR1 BY -9 (HI);           {Delete lower bits}
AY0=11;                             {Base exponent}
SR=LSHIFT SR1 BY 2 (HI);            {Adjust for ADSP-210x version}
AR=AX0+AY0;                         {Compute exponent}
DM(I1,M1)=AR;                       {Save exponent}
DM(I1,M1)=SR1;                      {Save mantissa}

CALL update_filter;                 {Update filter if trigger false}

MR0=DM(ap);                         {Load variables for updating}
MR1=DM(y);
MR2=DM(tdp);                        {Always load MR2 after MR1!}
MY0=DM(yl_h);
MY1=DM(yl_l);
AY0=DM(y);
MX0=DM(dms);
MX1=DM(dml);
CALL functw;                        {Update variables}
DM(ap)=AR;                          {Store updated variables}
DM(yu)=AX1;
DM(yl_l)=MY1;
DM(yl_h)=MY0;
```

(listing continues on next page)

```
                DM(dms)=MX0;
                DM(dml)=MX1;

                AX0=DM(a_ik);                   {Get I value for return}
                AY0=H#F;                        {Only 4 LSBs are used}
                AR=AX0 AND AY0;                 {So mask redundant sign bits}

                DM(hld_a_t)=I1;                 {Save s_r delay pointer}
                DM(hld_b_t)=I0;                 {Save dq delay pointer}
                RTS;                            {Return to caller}

ns_adpcm_decode: I1=DM(hld_a_r);                {Restore s_r delay pointer}
                I0=DM(hld_b_r);                 {Restore dq delay pointer}
                I4=^b_buf_r;                    {Set pointer to b-coefficients}
                I5=^a_buf_r;                    {Set pointer to a-coefficients}
                I6=^mult_data;                  {Set pointer to predictor data}

                SR=LSHIFT AR BY 12 (HI);        {Get sign of ADPCM I value here}
                SR=ASHIFT SR1 BY -12 (HI);
                                                {Sign extend ADPCM value to 16}
                DM(a_ik_r)=SR1;                 {Save I value}
                CALL predict;                   {Call s_e and sez predictor}
                AX1=DM(ap_r);
                AY0=DM(yl_h_r);
                AX0=DM(yu_r);
                CALL lima;                      {Limit ap and compute y}
                DM(y)=AR;
                DM(y_2)=SR1;
                AY1=DM(y_2);
                AR=DM(a_ik_r);
                CALL reconst;                   {Compute dq from received I}
                DM(dq_r)=AR;
                AY0=DM(s_e);
                AR=AR+AY0;                      {Compute reconstructed signal}
                DM(s_r)=AR;

                DM(I1,M1)=AR, AR=ABS AR;        {Make s_r floating point}
                SR1=H#4000;                     {Set SR1 to minimum value}
                SE=EXP AR (HI);                 {Determine exponent adjust}
                AX0=SE, SR=SR OR NORM AR (HI);  {Normalize value}
                SR=LSHIFT SR1 BY -9 (HI);       {Remove LSBs per spec}
                AY0=11;                         {Base exponent}
```

```
            SR=LSHIFT SR1 BY 2 (HI);          {Adjust for ADSP-210x version}
            AR=AX0+AY0;                       {Compute exponent}
            DM(I1,M1)=AR;                     {Store exponent}
            DM(I1,M1)=SR1;                    {Store mantissa}

            CALL update_filter_r;             {Update filter if trigger false}

            AY0=DM(y);                        {Load variables for updating}
            MY1=DM(yl_l_r);
            MY0=DM(yl_h_r);
            MR0=DM(ap_r);
            MR1=DM(y);
            MR2=DM(tdp_r);                    {Always load MR2 after MR1!}
            MX0=DM(dms_r);
            MX1=DM(dml_r);
            CALL functw;                      {Update variables}
            DM(yu_r)=AX1;                     {Stored updated variables}
            DM(yl_l_r)=MY1;
            DM(yl_h_r)=MY0;
            DM(ap_r)=AR;
            DM(dms_r)=MX0;
            DM(dml_r)=MX1;

            CALL compress;                    {Compress PCM value}

            DM(hld_a_r)=I1;                   {Save s_r delay pointer}
            DM(hld_b_r)=I0;                   {Save dq delay pointer}
            RTS;

compress:   AR=DM(s_r);                  {Get reconstructed signal}
            AR=ABS AR;                   {Take absolute value}
            AY0=33;                      {Add offset of boundries}
            AR=AR+AY0;
            AY0=8191;                    {Maximum PCM value}
            AF=AR-AY0;                   {Cap input}
            IF GT AR=PASS AY0;           {If in excess}
            SE=EXP AR (HI);              {Find exponent adjustmet}
            AX0=SE, SR=NORM AR (LO);     {Normalize input}
            AY0=H#4000;
            AR=SR0 XOR AY0;              {Remove first significant bit}
            SR=LSHIFT AR BY -10 (LO);    {Shift position bits}
            AR=PASS AY0;
            IF POS AR=PASS 0;            {Create sign bit}
```

(listing continues on next page)

```
                SR=SR OR LSHIFT AR BY -7 (LO);      {Position sign bit}
                AY0=9;
                AR=AX0+AY0;                 {Compute segment}
                IF LT AR=PASS 0;
                SR=SR OR LSHIFT AR BY 4 (LO);       {Position segment bits}
                AY0=H#FF;
                AR=SR0 XOR AY0;             {Invert bits}
                RTS;

expand:         AY0=H#FF;                   {Mask unwanted bits}
                AF=AR AND AY0, AX0=AY0;
                AF=AX0 XOR AF;              {Invert bits}
                AX0=H#70;
                AR=AX0 AND AF;              {Isolate segment bits}
                SR=LSHIFT AR BY -4 (LO);    {Shift to LSBs}
                SE=SR0, AR=AR XOR AF;       {Remove segment bits}
                AY0=H#FF80;
                AF=AR+AY0;
                IF LT JUMP posval;          {Detemine sign}
                AR=PASS AF;
                AR=AR+AF;                   {Shift left by 1 bit}
                AY0=33;
                AR=AR+AY0;                  {Add segment offset}
                SR=ASHIFT AR (LO);          {Position bits}
                AR=AY0-SR0;                 {Remove segment offset}
                RTS;

posval:         AF=PASS AR;
                AR=AR+AF;                   {Shift left by 1}
                AY0=33;
                AR=AR+AY0;                  {Add segment offset}
                SR=ASHIFT AR (LO);
                AR=SR0-AY0;                 {Remove segment offset}
                RTS;

predict:        AX1=DM(I0,M2), AY1=PM(I4,M6);       {Point to dq6 and b6}
                AF=PASS 0, SI=PM(I4,M4);    {AF hold partial sum}
                AY0=DM(I6,M5);
                MX1=3;                      {This multiply will give the}
                MY1=32768;                  {+48>>4 term}
                SR=ASHIFT SI BY -2 (HI);    {Downshift b6 per spec}
                CNTR=6;                     {Loop once for each b}
```

```
            DO sez_cmp UNTIL CE;
               AR=ABS SR1, SR1=DM(I6,M5);       {Get absolute value of b}
               AR=AR AND AY0, AY0=DM(I6,M5);    {Mask bits per spec}
               SE=EXP AR (HI), MY0=DM(I0,M2);   {Find exponent adjust}
               AX0=SE, SR=SR OR NORM AR (HI);   {Compute bnMANT}
               AR=SR1 AND AY0, AY0=DM(I0,M2);   {Mask bits per spec}
               MR=AR*MY0 (SS), AX1=DM(I0,M2), AY1=PM(I4,M6);
               AR=AX0+AY0, AY0=DM(I6,M5);       {Compute WbEXP}
               SE=AR, MR=MR+MX1*MY1 (SU);       {Compute WbnMANT}
               SR=LSHIFT MR1 (HI), SE=DM(I6,M5);        {Compute Wbn}
               AR=SR1 AND AY0, SI=PM(I4,M4);    {Mask Wbn per spec}
               AX0=AR, AR=AX1 XOR AY1;          {Determine sign of Wbn}
               AR=AX0, SR=ASHIFT SI (HI);       {Downshift b(n-1) per spec}
               IF LT AR=-AX0;                   {Negate Wbn if necessary}
sez_cmp:       AF=AR+AF, AY0=DM(I6,M5);         {Add Wbn to partial sum}
            AR=PASS AF, AX1=DM(I0,M1), AY1=PM(I5,M6);  {Get sezi}
            SR=ASHIFT AR BY -1 (HI);            {Downshift to produce sez}
            DM(sez)=SR1;
            SI=PM(I5,M4);                       {Get a2}
            SR=ASHIFT SI (HI);                  {Downshift a2 per spec}
            AX1=DM(I1,M2), AY1=PM(I4,M5);       {Restore bn and dqn pointers}
            CNTR=2;                             {Loop once for each a}
            DO s_e_cmp UNTIL CE;
               AR=ABS SR1, SR1=DM(I6,M5);       {Get absolute value of a}
               AR=AR AND AY0, AY0=DM(I6,M5);    {Mask bits per spec}
               SE=EXP AR (HI), MY0=DM(I1,M2);   {Get exponent adjust for a}
               AX0=SE, SR=SR OR NORM AR (HI);   {Compute anMANT}
               AR=SR1 AND AY0, AY0=DM(I1,M2);   {Mask bits per spec}
               MR=AR*MY0(SS), AX1=DM(I1,M2), AY1=PM(I5,M6);
               AR=AX0+AY0, AY0=DM(I6,M5);       {Compute WanEXP}
               SE=AR, MR=MR+MX1*MY1 (SU);       {Complete WanMANT computation}
               SR=LSHIFT MR1 (HI), SE=DM(I6,M5);        {Compute Wan}
               AR=SR1 AND AY0, SI=PM(I5,M4);    {Mask Wan per spec}
               AX0=AR, AR=AX1 XOR AY1;          {Determine sign of Wan}
               AR=AX0, SR=ASHIFT SI (HI);       {Downshift a1 per spec}
               IF LT AR=-AX0;                   {Negate Wan if necessary}
s_e_cmp:       AF=AR+AF, AY0=DM(I6,M5);         {Add Wan to partial sum}
            AR=PASS AF, AX1=DM(I1,M1), AY1=PM(I5,M5);  {Get sei}
            SR=ASHIFT AR BY -1 (HI);            {Compute se}
            DM(s_e)=SR1;
            RTS;
```

(listing continues on next page)

```
lima:       AY1=256;                      {Maximum value for ap}
            AR=AX1, AF=AX1-AY1;           {Cap if it exceeds}
            IF GE AR=PASS AY1;
            SR=ASHIFT AR BY -2 (HI);      {>>2 to produce al}
            SR=LSHIFT SR1 BY 9 (HI);      {Adjust for ADSP-210x version}

mix:        MY0=SR1, AR=AX0-AY0;          {MY0=al, AR=diff}
            AR=ABS AR;                    {Take absolute value of diff}
            MR=AR*MY0 (SU);               {Generate prod}
            AR=MR1+AY0;                   {Add to yu}
            IF NEG AR=AY0-MR1;            {Subtract if diff < 0}
            SR=ASHIFT AR BY -2 (HI);      {Generate y>>2}
            RTS;

log:        I3=^qn_values;                {Point to data array}
            AR=ABS AR, AX1=DM(I3,M1);     {Take absolute of d}
            SE=EXP AR (HI), AX0=DM(I3,M1);     {Determine exponent adjust}
            AY0=SE, SR=NORM AR (HI);      {Normalize}
            AR=AX0+AY0, AY0=DM(I3,M1);    {Compute exponent}
            IF LT AR=PASS 0;              {Check for exponent -1}
            SI=AR, AR=SR1 AND AY0;        {Mask mantissa bits}
            SR=LSHIFT AR BY -7 (HI);      {Position mantissa}
            SR=SR OR LSHIFT SI BY 7 (HI);      {Position exponent}

subtb:      AR=SR1-AY1, AY0=DM(I3,M1);    {Subtract y>>2 for log}
            AX0=AR, AF=PASS AX1;          {Setup for quantizing}

quan:       AR=AX0-AY0, AY0=DM(I3,M1);    {Is dl less then upper limit?}
            IF LT AF=AF-1;
            AR=AX0-AY0, AY0=DM(I3,M1);    {Continue to check for}
            IF LT AF=AF-1;
            AR=AX0-AY0, AY0=DM(I3,M1);    {where dl fits in quantizer}
            IF LT AF=AF-1;
            AR=AX0-AY0, AY0=DM(I3,M1);
            IF LT AF=AF-1;
            AR=AX0-AY0, AY0=DM(I3,M1);
            IF LT AF=AF-1;
            AR=AX0-AY0, AY0=DM(I3,M1);
            IF LT AF=AF-1;
            AR=AX0-AY0;
            IF LT AF=AF-1;
            AR=PASS AF;
            IF NEG AR=NOT AF;             {Negate value if ds negative}
            IF EQ AR=NOT AR;              {Send 15 for 0}
            RTS;
```

```
reconst:    AF=ABS AR;
            IF NEG AR=NOT AR;            {Find absolute value}
            M3=AR;                       {Use this for table lookup}
            I3=^dq_values;               {Point to dq table}
            MODIFY(I3,M3);               {Set pointer to proper spot}
            AX1=DM(I3,M1);               {Read dq from table}

adda:       AR=AX1+AY1;                  {Add y>>2 to dq}

antilog:    SR=ASHIFT AR BY 9 (LO);      {Get antilog of dq}
            AY1=127;                     {Mask mantisa}
            AX0=SR1, AR=AR AND AY1;      {Save sign of DQ+Y in AX0}
            AY1=128;                     {Add 1 to mantissa}
            AR=AR+AY1;
            AY0=-7;                      {Compute magnitude of shift}
            SI=AR, AR=SR1+AY0;
            SE=AR;
            SR=ASHIFT SI (HI);           {Shift mantissa}
            AR=SR1, AF=PASS AX0;
            IF LT AR=PASS 0;             {If DQ+Y <0, set to zero}
            IF NEG AR=-SR1;              {Negate DQ if I value negative}
            RTS;

functw:     I3=^w_values;                {Get scale factor multiplier}
            MODIFY(I3,M3);               {Based on I value}
            AF=PASS 0, SI=DM(I3,M1);
            I3=^f_values;

filtd:      SR=ASHIFT SI BY 1 (LO);      {Update fast quantizer factor}
            AR=SR0-AY0, SE=DM(I3,M1);    {Compute difference}
            SI=AR, AR=SR1-AF+C-1;        {in double precision}
            SR=ASHIFT AR (HI), AX0=DM(I3,M1); {Time constant is 1/32}
            SR=SR OR LSHIFT SI (LO), AY1=DM(I3,M1);
            AR=SR0+AY0, AY0=DM(I3,M1);   {Add gain}

limb:       AF=AR-AY1, SI=DM(I3,M3);     {Limit fast scale factor}
            IF GT AR=PASS AY1;           {Upper limit 10}
            AF=AR-AY0, AY1=MY1;
            IF LT AR=PASS AY0;           {Lower limit 1.06}

filte:      AF=AX0-AY1, AY0=MY0;         {Update quantizer slow factor}
            AF=AX0-AY0+C-1, AX0=DM(I3,M1);      {Compute difference}
            AX1=AR, AR=AR+AF;
            SR=ASHIFT AR BY -6 (HI);     {Time constant is 1/64}
            AR=SR0+AY1, AY1=MX0;         {Add gain}
            MY1=AR, AR=SR1+AY0+C;        {in double precision}
```

(listing continues on next page)

```
filta:         MY0=AR, AR=AX0-AY1;          {Update short term I average}
               SR=ASHIFT AR (HI), SI=AX0;   {Time constant is 1/32}
               AR=SR1+AY1, AY0=MX1;         {Add gain}

filtb:         SR=LSHIFT SI BY 2 (HI);      {Update long term I average}
               MX0=AR, AR=SR1-AY0;
               SR=ASHIFT AR BY -7 (HI);     {Time constant is 1/128}
               AR=SR1+AY0, SI=MX0;          {Add gain}

subtc:         SR=ASHIFT AR BY -3 (HI);     {Compute difference of long}
               AF=PASS AR, AX0=SR1;         {and short term I averages}
               SR=ASHIFT SI BY 2 (HI);
               MX1=AR, AR=SR1-AF;
               AF=ABS AR;
               AR=MR2, AF=AX0-AF;           {tdp must be true for ax 0}
               IF LE AR=PASS 1;
               AY0=1536;
               AF=MR1-AY0, AY0=MR0;
               IF LT AR=PASS 1;             {Y>3 for ax to be 0}
filtc:         SR=ASHIFT AR BY 9 (HI);      {Update speed control}
               AR=SR1-AY0;                  {Compute difference}
               SR=ASHIFT AR BY -4 (HI);     {Time constant is 1/16}
               AR=SR1+AY0;                  {Add gain}
               RTS;

update_filter: AX0=DM(dq);                  {Get value of current dq}
               AR=128;
               AF=PASS AX0, AY1=DM(I0,M0); {Read sign of dq(6)}
               IF EQ AR=PASS 0;             {If dq 0 then gain 0}
               SE=-8;                       {Time constand is 1/256}
               AX1=AR;
               CNTR=6;
               DO update_b UNTIL CE;        {Update all b-coefficients}
                  AF=AX0 XOR AY1, AY0=PM(I4,M4);      {Get sign of update}
                  IF LT AR=-AX1;
                  AF=AR+AY0, SI=AY0;        {Add update to original b}
                  SR=ASHIFT SI (HI), AY1=DM(I0,M0);  {Get next dq(k)}
                  AR=AF-SR1;                {Subtract leak factor}
update_b:         PM(I4,M5)=AR, AR=PASS AX1;          {Write out new b-coefficient}

place_dq:      AR=ABS AX0, AY0=DM(I0,M2);  {Take absolute value of dq}
               SE=EXP AR (HI);              {Determine exponent adjust}
```

```
               SR1=H#4000;                 {Set minimum value into SR1}
               AX1=SE, SR=SR OR NORM AR (HI);          {Normalize dq}
               AY0=11;                     {Used for exponent adjustment}
               SR=LSHIFT SR1 BY -9 (HI);   {Remove lower bits}
               SR=LSHIFT SR1 BY 2 (HI);    {Adjust for ADSP-210x version}
               DM(I0,M2)=SR1, AR=AX1+AY0;  {Save mantisa, compute exp.}
               DM(I0,M2)=AR;               {Save exponent}
               AX1=DM(a_ik);               {Use sign of I, not dq}
               DM(I0,M0)=AX1;              {Save sign}

update_p:      AY0=DM(sez);                {Get result of predictor}
               AR=AX0+AY0;                 {Use dq from above}
               AY1=DM(p);                  {Delay all old p's by 1}
               AY0=DM(p_o);
               DM(p)=AR;
               DM(p_o)=AY1;
               DM(p_o_o)=AY0;
               AX1=AR, AR=AR XOR AY0;      {Compute p xor poo}
               MR1=AR, AR=AX1 XOR AY1;     {Compute p xor po}
               MR0=AR;

upa2:          I3=^a_data;
               SI=PM(I5,M5);               {Hold a2 for later}
               AR=PM(I5,M5);               {Get a1 for computation of f}
               AR=ABS AR, AY0=DM(I3,M1);   {Cap magnitude of a1 at 1/2}
               AF=AR-AY0, SE=DM(I3,M1);
               IF GT AR=PASS AY0;
               IF NEG AR=-AR;              {Restore sign}
               SR=ASHIFT AR (LO), AY0=DM(I3,M1);
               AF=ABS MR0, AY1=DM(I3,M1);  {If p xor po = 0 negate f}
               AR=SR0, AF=PASS SR1;
               IF POS AR=AY1-SR0;          {Double precision}
               IF POS AF=AY1-SR1+C-1;
               SR0=AR, AR=PASS AF;
               SR1=AR, AF=ABS MR1;         {If p xor poo = 1 subtract}
               AR=SR0+AY0, SE=DM(I3,M1);
               AF=SR1+AY1+C, AX0=DM(I3,M1);
               IF NEG AR=SR0-AY0;
               IF NEG AF=SR1-AY1+C-1;
               SR=LSHIFT AR (LO);
               AR=PASS AF;
               SR=SR OR ASHIFT AR (HI), AY0=SI;
               AY1=SR0, SR=ASHIFT SI (HI); {Downshift a2 for adjustment}
```

(listing continues on next page)

```
              AR=AY0-SR1, AY0=DM(I3,M1);
              AF=PASS AX1;
              IF NE AR=AR+AY1;             {If sigpk = 1, no gain}

limc:         AF=AR-AY0, AY1=DM(I3,M1);    {Limit a2 to .75 max}
              IF GT AR=PASS AY0;
              AF=AR-AY1, AY0=DM(I3,M1);    {Limit a2 to -.75 min}
              IF LT AR=PASS AY1;
              PM(I5,M5)=AR;                {Store new a2}

upa1:         AR=AX0, AF=PASS MR0;
              IF LT AR=-AX0;
              AF=PASS AX1, SI=PM(I5,M4);
              IF EQ AR=PASS 0;
              SR=ASHIFT SI BY -8 (HI);     {Leak Factor = 1/256}
              AF=PASS AR, AR=SI;
              AF=AF-SR1;
              AR=AR+AF, AX1=DM(I3,M1);

limd:         AX0=AR, AR=AX1-AY1;          {Limit a1 based on a2}
              AY0=AR, AR=AY1-AX1;
              AY1=AR, AR=PASS AX0;
              AF=AR-AY0;
              IF GT AR=PASS AY0;           {Upper limit 1 - 2^-4 - a2}
              AF=AR-AY1;
              IF LT AR=PASS AY1;           {Lower limit a2 - 1 + 2^-4}
              PM(I5,M5)=AR;                {Store new a1}
              RTS;

update_filter_r:AX0=DM(dq_r);              {Get dq_r}
              AR=128;                      {Set possible gain}
              AF=PASS AX0, AY1=DM(I0,M0);  {Get sign of dq(6)}
              IF EQ AR=PASS 0;             {If dq_r 0, gain 0}
              SE=-8;                       {Leak factor 1/256}
              AX1=AR;
              CNTR=6;
              DO update_b_r UNTIL CE;      {Update all b-coefficients}
                 AF=AX0 XOR AY1, AY0=PM(I4,M4);   {Get sign of gain}
                 IF LT AR=-AX1;
                 AF=AR+AY0, SI=AY0;               {Add gain to original b}
                 SR=ASHIFT SI (HI);               {Time constant is 1/256}
                 AR=AF-SR1, AY1=DM(I0,M0);        {Compute new b-value}
update_b_r:      PM(I4,M5)=AR, AR=PASS AX1;       {Store new b-value}
```

```
place_dq_r:   AR=ABS AX0, AY0=DM(I0,M2);  {Get absolute value fo dq_r}
              SE=EXP AR (HI);             {Determine exponent adjustment}
              SR1=H#4000;                 {Set SR to minimum value}
              AX1=SE, SR=SR OR NORM AR (HI);      {Normalize dq_r}
              AY0=11;                     {Used for exponent adjust}
              SR=LSHIFT SR1 BY -9 (HI);   {Remove lower bits}
              SR=LSHIFT SR1 BY 2 (HI);    {Adjust for ADSP-210x version}
              DM(I0,M2)=SR1, AR=AX1+AY0;  {Store mantissa, compute exp}
              AX1=DM(a_ik_r);             {Use sign of I, not dq}
              DM(I0,M2)=AR;               {Store exponent}
              DM(I0,M0)=AX1;              {Store sign}

update_p_r:   AY0=DM(sez);                {Compute new p}
              AR=AX0+AY0;                 {Use dq_r from above}
              AY1=DM(p_r);                {Delay old p's by 1}
              AY0=DM(p_o_r);
              DM(p_r)=AR;
              DM(p_o_r)=AY1;
              DM(p_o_o_r)=AY0;
              AX1=AR, AR=AR XOR AY0;      {Compute p and poo}
              MR1=AR, AR=AX1 XOR AY1;     {Compute p and po}
              MR0=AR;

upa2_r:       I3=^a_data;
              SI=PM(I5,M5);               {Hold a2 for later}
              AR=PM(I5,M5);               {Get a1 for computation of f}
              AR=ABS AR, AY0=DM(I3,M1);   {Cap magnitude of a1 to 1/2}
              AF=AR-AY0, SE=DM(I3,M1);
              IF GT AR=PASS AY0;
              IF NEG AR=-AR;              {Restore sign of f}
              SR=ASHIFT AR (LO), AY0=DM(I3,M1);
              AF=ABS MR0, AY1=DM(I3,M1);  {If p_r xor poo_r =1 subtract}
              AR=SR0, AF=PASS SR1;
              IF POS AR=AY1-SR0;
              IF POS AF=AY1-SR1+C-1;
              SR0=AR, AR=PASS AF;
              SR1=AR, AF=ABS MR1;
              AR=SR0+AY0, SE=DM(I3,M1);
              AF=SR1+AY1+C, AX0=DM(I3,M1);
              IF NEG AR=SR0-AY0;
              IF NEG AF=SR1-AY1+C-1;
              SR=LSHIFT AR (LO);
              AR=PASS AF;
```

(listing continues on next page)

```
            SR=SR OR ASHIFT AR (HI), AY0=SI;
            AY1=SR0, SR=ASHIFT SI (HI); {Leak factor of 1/128}
            AR=AY0-SR1, AY0=DM(I3,M1);
            AF=PASS AX1;
            IF NE AR=AR+AY1;            {If sigpk = 1 , no gain}

limc_r:     AF=AR-AY0, AY1=DM(I3,M1);   {Limit a2 to .75 max}
            IF GT AR=PASS AY0;
            AF=AR-AY1, AY0=DM(I3,M1);   {Limit a2 to -.75 min}
            IF LT AR=PASS AY1;
            PM(I5,M5)=AR;               {Store new a2}

upa1_r:     AR=AX0, AF=PASS MR0;
            IF LT AR=-AX0;
            AF=PASS AX1, SI=PM(I5,M4);
            IF EQ AR=PASS 0;
            SR=ASHIFT SI BY -8 (HI);    {Leak Factor = 1/256}
            AF=PASS AR, AR=SI;
            AF=AF-SR1;
            AR=AR+AF, AX1=DM(I3,M1);

limd_r:     AX0=AR, AR=AX1-AY1;         {Limit a1 based on a2}
            AY0=AR, AR=AY1-AX1;
            AY1=AR, AR=PASS AX0;
            AF=AR-AY0;
            IF GT AR=PASS AY0;          {Upper limit 1 - 2^-4 -a2}
            AF=AR-AY1;
            IF LT AR=PASS AY1;          {Lower limit a2 - 1 + 2^-4}
            PM(I5,M5)=AR;               {Store new a1}
            RTS;

.ENDMOD;
```

Listing 12.3 Nonstandard ADPCM Transcoder Routine

High-Speed Modem Algorithms 13

13.1 OVERVIEW

In high-speed data communication systems, there often arises the need for digital signal processing techniques. In the implementation of medium-speed (up to 2400 bps) to high-speed (4800 bps and higher) modems, certain effects of the limited-bandwidth communications channel (typically a voice-band telephone line) present themselves as obstacles. The most notable of these effects is intersymbol interference, which is the "smearing together" of the transmitted symbols over a time-dispersive channel (Lucky, et al, 1968). This effect is a problem in virtually all pulse-modulation systems, including pulse-amplitude modulation (PAM), frequency-shift keying (FSK), phase-shift keying (PSK), and quadrature-amplitude modulation (QAM) systems.

The basic action in most methods of reducing the effects of intersymbol interference is to pass the received signal through a filter that approximates the inverse transfer function of the communications channel; this process is called equalization. The implementation of an equalizer usually depends upon the speed of the modem. For medium-speed modems (generally PSK) "compromise" equalization is often adequate. Compromise equalization is performed using a short transversal filter with fixed coefficients that compensate for a wide range of channel characteristics. High-speed modems (generally QAM) usually require adaptive equalization using an adaptive filter to compensate for the excessively wide range of channel characteristics encountered in the switched telephone network (Qureshi, 1982).

13.2 SP COMPLEX-VALUED TRANSVERSAL FILTER

In the implementation of PSK and QAM modems, two double-sideband suppressed-carrier AM signals are sent by the transmitter and separated at the receiver. Orthogonal (quadrature) carrier signals are used for modulation and demodulation. It is customary to represent the in-phase and quadrature components of the received signal as the real and imaginary parts of a complex signal. Thus, the equalizer will operate upon this complex signal in order to reduce the effects of intersymbol interference. In practice, the equalizer may be inserted either before (passband equalization) or after (baseband equalization) the demodulation of the received signal (Qureshi, 1982).

The subroutine shown in Listing 13.1 presents an FIR filter routine for complex-valued data and coefficients that could be used to implement an equalizer. This routine implements the same sum-of-products operation as the nonadaptive (fixed-coefficient) FIR filter presented in Chapter 5; it has been modified to operate upon complex values. The filter is described by the equation on the next page.

$$y(n) = \sum_{k=0}^{N-1} h_k \, x(n-k)$$

The first loop, *realloop*, computes the real output by computing the sum of products of the real data values and the real coefficients, and subtracting the sum of products of the imaginary data values and the imaginary coefficients. The second loop, *imagloop*, is similar in that it computes the imaginary output as the sum of products of the real data values and the imaginary coefficients, added to the sum of products of the imaginary data values and the real coefficients. The outputs in both cases are rounded and conditionally saturated.

```
.MODULE cfir_sub;

{  Single-Precision Complex FIR Filter Subroutine

   Calling Parameters
        I0 -> Oldest data value in real delay line (Xr's)
        L0 = filter length (N)
        I1 -> Oldest data value in imaginary delay line (Xi's)
        L1 = filter length (N)
        I4 -> Beginning of real coefficient table (Hr's)
        L4 = filter length (N)
        I5 -> Beginning of imaginary coefficient table (Hi's)
        L5 = filter length (N)
        M0,M4 = 1
        AX0 = filter length minus one (N-1)
        CNTR = filter length minus one (N-1)
```

```
    Return Values
         I0 -> Oldest data value in real delay line
         I1 -> Oldest data value in imaginary delay line
         I4 -> Beginning of real coefficient table
         I5 -> Beginning of imaginary coefficient table
         SR1 = real output (rounded and conditionally saturated)
         MR1 = imaginary output (rounded and conditionally saturated)

    Altered Registers
         MX0,MY0,MR,SR1

    Computation Time
         2 × (N-1) + 2 × (N-1) + 13 + 8 cycles

    All coefficients and data values are assumed to be in 1.15 format.
}

.ENTRY  cfir;

cfir:   MR=0, MX0=DM(I1,M0), MY0=PM(I5,M4);
        DO realloop UNTIL CE;
             MR=MR-MX0*MY0(SS), MX0=DM(I0,M0), MY0=PM(I4,M4);    {Xi × Hi}
realloop:    MR=MR+MX0*MY0(SS), MX0=DM(I1,M0), MY0=PM(I5,M4);    {Xr × Hr}
        MR=MR-MX0*MY0(SS), MX0=DM(I0,M0), MY0=PM(I4,M4);         {Last Xi × Hi}
        MR=MR+MX0*MY0(RND);                                      {Last Xr × Hr}
        IF MV SAT MR;
        SR1=MR1;                                                 {Store Yr}
        MR=0, MX0=DM(I0,M0), MY0=PM(I5,M4);
        CNTR=AX0;
        DO imagloop UNTIL CE;
             MR=MR+MX0*MY0(SS), MX0=DM(I1,M0), MY0=PM(I4,M4);    {Xr × Hi}
imagloop:    MR=MR+MX0*MY0(SS), MX0=DM(I0,M0), MY0=PM(I5,M4);    {Xi × Hr}
        MR=MR+MX0*MY0(SS), MX0=DM(I1,M0), MY0=PM(I4,M4);         {Xr × Hi}
        MR=MR+MX0*MY0(RND);                                      {Xi × Hr}
        IF MV SAT MR;                                            {MR1=Yi}
        RTS;
.ENDMOD;
```

Listing 13.1 Single-Precision Complex FIR Filter

13.3 COMPLEX-VALUED STOCHASTIC GRADIENT

As mentioned previously, non-adaptive or compromise equalization is usually only adequate in medium-speed modems. High-speed modems require the equalizer coefficients to be adapted because of changing channel characteristics. In fact, even many 2400-bps modems incorporate adaptive equalization.

Although many adaptive filtering algorithms exist, virtually all adaptive equalizers in high-speed modems utilize the stochastic gradient (SG) algorithm (described in Chapter 5). This is primarily because it generally provides adequate performance and requires the least computation for a given filter order as compared to the other adaptive algorithms. Using the SG algorithm, filter coefficients at time T, $c_j(T)$, are adapted through the following equation:

$$c_j(T + 1) = c_j(T) + ße_c(T)\, y^*(T - j + 1)$$

In this equation, $e_c(T)$ is the estimation error formed by the difference between the signal it is desired to estimate, d(T), and a weighted linear combination of the current and past input values y(T).

$$e_c(T) = d(T) - \sum_{j=1}^{n} c_j(T)\, y(T - j + 1)$$

The value $y(T - j + 1)$ represents the past value of the input signal "contained" in the jth tap of the transversal filter. For example, y(T), the present value of the input signal, corresponds to the first tap and $y(T - 42)$ corresponds to the forty-third filter tap. The step size ß controls the "gain" of the adaptation.

The coefficients are usually adapted during some training period after connection has been established. This involves the transmission of some known training sequence to the modem, during which time the equalizer adapts its coefficients according to a synchronized version of the received training sequence. Upon completion of the training period, slight variations in the channel characteristics may be tracked by performing the adaptation based on the estimate of the received symbol. This is referred to as decision-directed adaptation, and in some cases it is relied upon to perform the initial adaptation as well.

A subroutine for performing adaptation of complex FIR filter coefficients according to the stochastic gradient algorithm is given in Listing 13.2. In this subroutine, the cache memory is utilized very effectively, since four program memory accesses are made each time through the loop.

```
.MODULE csg_sub;

{  Single-Precision Complex SG Update Subroutine

   Calling Parameters
        I0 -> Oldest data value in real delay line                    L0 = N
        I1 -> Oldest data value in imag delay line                    L1 = N
        I4 -> Beginning of real coefficient table                     L4 = N
        I5 -> Beginning of imag coefficient table                     L5 = N
        MX0 = real part of Beta × Error
        MX1 = imag part of Beta × Error
        M0,M5 = 1
        M4=0
        M1= -1
        CNTR = Filter length (N)

   Return Values
        Coefficients updated
        I0 -> Oldest data value in real delay line
        I1 -> Oldest data value in imaginary delay line
        I4 -> Beginning of real coefficient table
        I5 -> Beginning of imaginary coefficient table

   Altered Registers
        MY0,MY1,MR,SR,AY0,AY1,AR

   Computation Time
        6 × N + 10 cycles

   All coefficients and data values are assumed to be in 1.15 format.
}
```

(listing continues on next page)

```
.ENTRY   csg;

csg:     MY0=DM(I0,M0);                          {Get Xr}
         MR=MX0*MY0(SS), MY1=DM(I1,M0);          {Er × Xr, get Xi}
         DO adaptc UNTIL CE;
            MR=MR+MX1*MY1(RND), AY0=PM(I4,M4);   {Ei × Xi, get Hr}
            AR=AY0-MR1, AY1=PM(I5,M4);           {Hr-Er × Xr+Ei × Xi, get Hi}
            PM(I4,M5)=AR, MR=MX1*MY0(SS);        {Store Hr, Er × Xi}
            MR=MR-MX0*MY1(RND), MY0=DM(I0,M0);   {Ei × Xr, get Xr}
            AR=AY1-MR1, MY1=DM(I1,M0);           {Hi-Er × Xi-Ei × Xr, get Xi}
adaptc:     PM(I5,M5)=AR, MR=MX0*MY0(SS);        {Store Hi, Er × Xr}
         MODIFY(I0,M1);
         MODIFY(I1,M1);
         RTS;
.ENDMOD;
```

Listing 13.2 Single-Precision Complex Stochastic Gradient

13.4 EUCLIDEAN DISTANCE

In the receiver of a high-speed modem, some method must be established for determining to which values from the space of possibilities the real and imaginary parts of the received sample correspond. In some QAM modems with large signal constellations, this can be a rather non-trivial process. For example, the CCITT V.29 standard calls for a 16-point signal constellation. One means of determining the value of samples is the Euclidean distance measure. This method involves computing the distance (error) between the received sample value and all possible candidates for the transmitted sample, given the signal constellation. The error is given by the following equation:

$$e(j) = ((x_r - c_r(j))^2 + (x_i - c_i(j))^2)^{1/2}$$

In this equation, the error e(j) is the distance between the received signal value, *x*, and the jth signal constellation value, *c*, in the real-imaginary plane. The value of *j* for which e(j) is minimum then selects the constellation point.

A subroutine for computing the Euclidean distance is shown in Listing 13.3. The *ptloop* loop is executed once for each point in the given signal constellation. The (squared) distance between *x* and each point is computed. AF is loaded with this value if it is less than the previous minimum, and the index corresponding to that constellation value (obtained from the current CNTR value) is stored in SI. After all distances have been computed, SI contains the index of the point that corresponds to the minimum e(j). This index can be used to select the constellation value.

```
.MODULE dist_sub;

{  Euclidean Distance Subroutine

   Calling Parameters
        I1 -> Start of constellation (C) table
        AX0 contains Xr
        AX1 contains Xi
        L1 = length of constellation table
        M0 = 1
        M1 = -1
        CNTR = length of constellation table

   Return Values
        SI contains the decision index j
        AF contains the minimum distance (squared)
        I1 -> Beginning of constellation table

   Altered Registers
        AY0,AY1,AF,AR,MX0,MY0,MY1,MR,SI

   Computation Time
        10 × N + 5 (maximum)
}

.ENTRY  dist;

dist:   AY0=32767;                          {Init min distance to largest possible value}
        AF=PASS AY0, AY0=DM(I1,M0);         {Get Cr}
        DO ptloop UNTIL CE;
           AR=AX0-AY0, AY1=DM(I1,M0);       {Xr-Cr, Get Ci}
           MY0=AR, AR=AX1-AY1;              {Copy Xr-Cr, Xi-Ci}
           MY1=AR;                          {Copy Xi-Ci}
           MR=AR*MY1(SS), MX0=MY0;          {(Xi-Ci)^2, Copy Xr-Cr}
           MR=MR+MX0*MY0 (RND);             {(Xr-Cr)^2}
           AR=MR1-AF;                       {Compare with previous minimum}
           IF GE JUMP ptloop;
           AF=PASS MR1;                     {New minimum if MR1<AF}
           SI-CNTR;                         {Record the constellation index}
ptloop:    AY0=DM(I1,M0);
        MODIFY(I1,M1);                      {Point back to beginning of table}
        RTS;
.ENDMOD;
```

Listing 13.3 Euclidean Distance

13.5 REFERENCES

Lucky, R. W.; Salz, J.; and Weldon, E. J., Jr. 1968. *Principles of Data Communication*. New York: McGraw-Hill.

Qureshi, S. U. H. 1982. Adaptive Equalization. *IEEE Communications*. March 1982. P. 9-16.

Dual-Tone Multi-Frequency Coding 14

14.1 INTRODUCTION

DTMF is the generic name for pushbutton telephone signaling equivalent to the Bell System's TouchTone®. Dual-Tone Multi-Frequency (DTMF) signaling is quickly replacing dial-pulse signaling in telephone networks worldwide. In addition to telephone call signaling, DTMF is becoming popular in interactive control applications, such as telephone banking or electronic mail systems, in which the user can select options from a menu by sending DTMF signals from a telephone.

To generate (encode) a DTMF signal, the ADSP-2100 adds together two sinusoids, each created by software. For DTMF decoding, the ADSP-2100 looks for the presence of two sinusoids in the frequency domain using modified Goertzel algorithms. This chapter shows how to generate and decode DTMF signals in both single channel and multi-channel environments. Realizable hardware is briefly mentioned.

DTMF signals are interfaced to the analog world via codec (coder/decoder) chips or linear analog-to-digital (A/D) converters and digital-to-analog (D/A) converters. Codec chips contain all the necessary A/D, D/A, sampling and filtering circuitry for a bidirectional analog/digital interface. These codecs with on-chip filtering are sometimes called codec/filter combo chips, or combo chips for short. They are referred to as codecs in this chapter.

The codec channel used in this example is bandlimited to pass only frequencies between 200Hz and 3400Hz. The codec also incorporates companding (audio compressing/expanding) circuitry for either of the two companding standards (A-law and μ-255 law). These two standards are explained in Chapter 11, *Pulse Code Modulation*. Companding is the process of logarithmically compressing a signal at the source and expanding it at the destination to maintain a high end-to-end dynamic range while reducing the dynamic range requirement within the communication channel.

In the example of DTMF signal generation shown in this chapter, the ADSP-2100 reads DTMF digits stored in data memory in a relocatable look-up list. Alternatively, a DTMF keypad could be used for digit entry. In either case, the resultant DTMF tones are generated mathematically and added together. The values are logarithmically compressed and passed to the codec chip for conversion to analog signals. Multi-channel DTMF signal generation is performed by simply time-multiplexing the processor among the channels.

On the receiving end, the ADSP-2100 reads the logarithmically compressed, digital data from the codec's 8-bit parallel data bus, logarithmically expands it to its 16-bit linear format, performs a Goertzel algorithm — a fast DFT (discrete Fourier transform) calculation — for each tone to detect, then passes the results through several tests to verify whether a valid DTMF digit was received. The result is coded and written to a memory-mapped I/O port. Multi-channel DTMF decoding is also performed by time-multiplexing the channels.

14.2 ADVANTAGES OF DIGITAL IMPLEMENTATION

Several chips are available which employ analog circuitry to generate and decode DTMF signals for a single channel. This function can be digitally implemented using the ADSP-2100. The advantages of a digital system include better accuracy, precision, stability, versatility, and reprogrammability as well as lower chip count, and thereby reduced board-space requirements.

Table 14.1 compares the tone accuracy expected of a DTMF tone dialer chip with the accuracy of tones generated by the ADSP-2100. Note that a tone dialer chip is an application-specific device with preprogrammed frequencies for use on a single channel, whereas the ADSP-2100 is a general purpose microprocessor which can be programmed to generate any frequency for many separate channels. When using the ADSP-2100, the frequency values can be specified to within 16 bits of resolution, implying an accuracy of 0.003%.

By simply changing the frequency values, the ADSP-2100 tone generator can be fine-tuned or reprogrammed for other tone standards, such as CCITT 2-of-6 Multi-Frequency (MF), call progress tones, US Air Force 412L, and US Army TA-341/PT. Since the numbers stored in data memory do not change in value over time or temperature, the precision and stability of a digital solution surpasses any analog equivalent. DTMF encoding and decoding can be written as a subfunction of a larger program, eliminating the need for separate components and specialized interface circuitry.

Dual-Tone Multi-Frequency 14

DTMF Standard Frequency	*Typical frequency from hybrid tone-dialer chip using 3.579545MHz crystal*	% deviation	*Frequency from ADSP-2100 program*	% deviation
697.0Hz	699.1Hz	+0.31%	697.0Hz	±0.003%
770.0Hz	766.2Hz	–0.49%	770.0Hz	±0.003%
852.0Hz	847.4Hz	–0.54%	852.0Hz	±0.003%
941.0Hz	948.0Hz	+0.74%	941.0Hz	±0.003%
1209.0Hz	1215.9Hz	+0.57%	1209.0Hz	±0.003%
1336.0Hz	1331.7Hz	–0.32%	1336.0Hz	±0.003%
1477.0Hz	1471.9Hz	–0.35%	1477.0Hz	±0.003%
1633.0Hz	1645.0Hz	+0.73%	1633.0Hz	±0.003%

Table 14.1 Precision of Analog and Digital Tone Generation

14.3 DTMF STANDARDS

The DTMF tone signaling standard is also known as TouchTone or MFPB (Multi-Frequency, Push Button). TouchTone was developed by Bell Labs for use by AT&T in the American telephone network as an in-band signaling system to supersede the dial-pulse signaling standard. Each administration has defined its own DTMF specifications. They are all very similar to the CCITT standard, varying by small amounts in the guardbands (tolerances) allowed in frequency, power, twist (power difference between the two tones) and talk-off (speech immunity). The CCITT standard appears as Recommendations Q.23 and Q.24 in Section 4.3 of the CCITT Red Book, Volume VI, Fascicle VI.1. Other standards (AT&T, CEPT, etc.) are listed in *References*, at the end of this chapter.

Two tones are used to generate a DTMF digit. One tone is chosen out of four row tones, and the other is chosen out of four column tones. Two of eight tones can be combined so as to generate sixteen different DTMF digits. Of the sixteen keys shown in Figure 14.1, on the next page, twelve are the familiar keys of a TouchTone keypad, and four (column 4) are reserved for future uses.

A 90-minute audio-cassette tape to test DTMF decoders is available from Mitel Semiconductor (part number CM7291). There also exists a standard describing requirements for systems which test DTMF systems. This standard is available from the IEEE as ANSI/IEEE Std. 752-1986.

		Column 1	Column 2	Column 3	Column 4
		1209Hz	1336Hz	1477Hz	1633Hz
Row 1	697Hz	1	2	3	A
Row 2	770Hz	4	5	6	B
Row 3	852Hz	7	8	9	C
Row 4	941Hz	*	0	#	D

DTMF digit = Row Tone + Column Tone

Figure 14.1 DTMF Digits

14.4 DTMF DIGIT GENERATION PROGRAM

Generation of DTMF digits is relatively straightforward. Digital samples of two sine waves are generated (mathematically or by look-up tables), scaled and then added together. The sum is logarithmically compressed and sent to the codec for conversion into the analog domain.

A sine look-up table is not used in this example because the sine can be computed quickly without using the large amount of data memory a look-up table would require. The sine computation routine is efficient, using only five data memory locations and executing in 25 instruction cycles. The routine used to compute the sine is from Chapter 4, *Function Approximation*. This routine evaluates the sine function to 16 significant bits using a fifth-order Taylor polynomial expansion. The sine computation routine is called from the tone generation program as an external routine; refer to that chapter for details on the sine computation routine.

Dual-Tone Multi-Frequency 14

To build a sine wave, the tone generation program utilizes two values. One, called *sum*, keeps track of where the current sample is along the time axis, and the other, called *advance*, increments that value for the next sample. Since the DTMF tone generation program generates two tones, there are two different *sum* values and two different frequency values (stored in the variables *hertz*). The value *sum* is stored in data memory in a variable called *sum*. *Sum* is modified every time a new sample is calculated. The value *advance* is calculated from a data memory variable called *hertz*. *Hertz* is a constant for a given tone frequency. The ADSP-2100 calculates the *advance* value from a stored *hertz* variable instead of storing *advance* as the data memory variable because this allows you to read the frequency being generated in Hz directly from the data memory display (in decimal mode) of the ADSP-2100 Simulator or Emulator.

The *sum* values are dm(sin1) and dm(sin2) in Listing 14.1 (see *Program Listings* at the end of this chapter). The *advance* values are derived from the variables dm(hertz1) and dm(hertz2) in Listing 14.1. For readable source code and easy debugging, Listing 14.1 uses two data memory variables dm(sin1) and dm(sin2) to store the value returned from each call to the sine subroutine rather than storing the value in a register. These variables are then added together, resulting in a DTMF output signal value.

The sampling frequency of telephone systems is 8kHz. Therefore, the ADSP-2100 must output samples every 125µs. An 8kHz TTL square wave is applied to an interrupt (IRQ3 in this case) pin of the ADSP-2100. The ADSP-2100 is initialized for edge-sensitive interrupts with interrupt-nesting mode disabled (see Listing 14.1, two lines immediately preceding the label *wait_int* near the top of executable code). The sampling frequency, in conjunction with the *advance* value, determines the frequency of the sine wave generated.

Circular movement around a unit circle is analogous to linear motion along the time axis of a sine wave, one revolution of the circle corresponding to one period of the wave. The range of inputs to the sine function approximation subroutine is $-\pi$ to 0 to $+\pi$ radians. This range maps to the 16-bit hexadecimal numbers H#8000 to H#FFFF and H#0000 to H#7FFF (see Figure 14.2 on the next page and Table 14.2 following it). All of the 16-bit numbers are equally spaced around the unit circle, dividing it into 65536 parts. The *advance* value is added to the *sum* value during each interrupt. To generate a 4kHz sine wave, the *advance* value would have to be 32768, equivalent to π radians, or a jump halfway around the unit circle. Because Nyquist theory dictates that 4kHz is the highest frequency that can be represented in an 8kHz sampling-frequency system, this is the maximum *advance* value that can be used. For a sine wave frequency of less than 4kHz, the *advance* value would be proportionally less (see Figure 14.3).

$$\textit{advance} \quad \text{value} = 65536\left(\frac{\text{tone frequency desired}}{\text{sampling frequency}}\right)$$

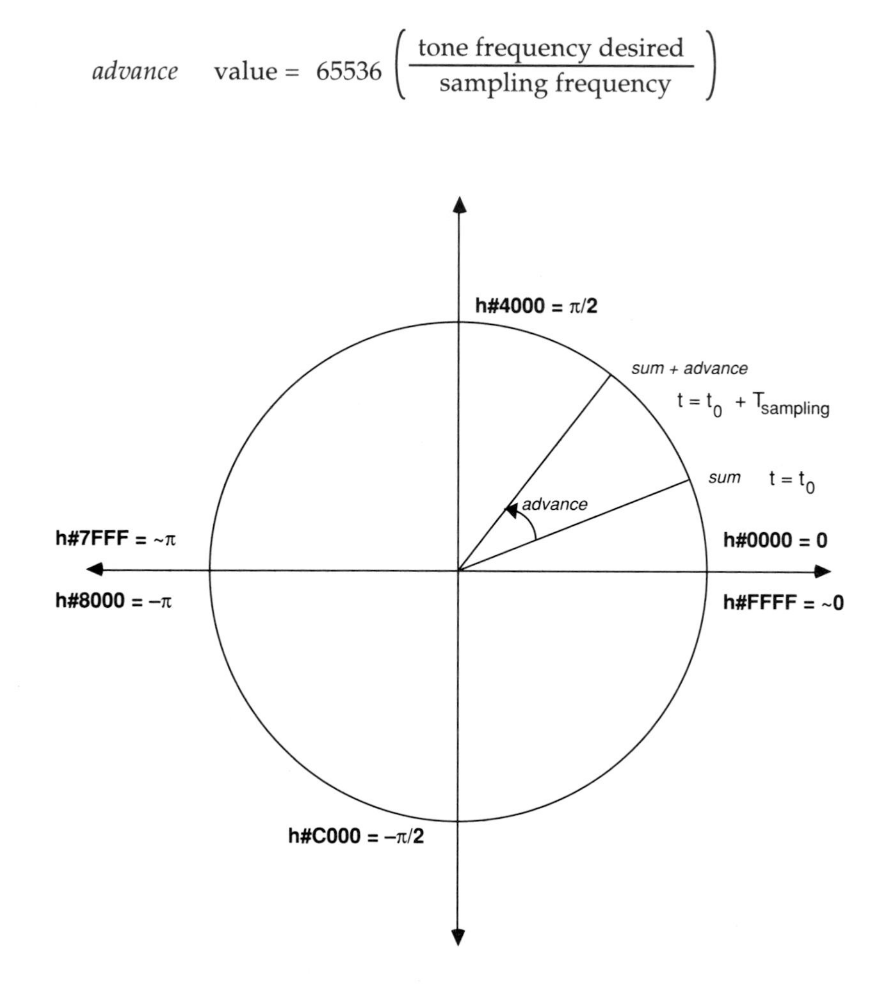

Figure 14.2 Sine Routine Input Angle Mapping

Input to Sine Approximation Routine	*Equivalent Input Angle (radians/degrees)*	
H#0000	0	0
H#2000	$\pi/4$	45
H#4000	$\pi/2$	90
H#6000	$3\pi/4$	135
H#7FFF	$\sim\pi$	~180
H#8000	$-\pi$	–180
H#A000	$-3\pi/4$	–135
H#C000	$-\pi/2$	–90
H#E000	$-\pi/4$	–45
H#FFFF	~0	~0

Table 14.2 Sine Routine Input Angle Mapping

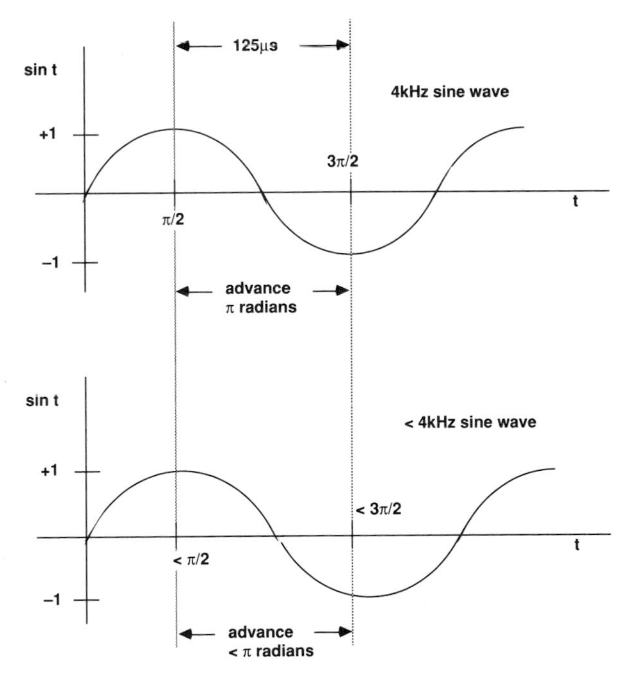

Figure 14.3 Sine Wave Frequency Determination

For examples of some tones and their *advance* values, see Table 14.3. Since telephone applications require an 8kHz sampling frequency, the formula above can be reduced to:

advance required = 8.192 (tone frequency desired in Hz)

Advance Value	*Tone Frequency (Hz)*
H#0000	0
H#2000	$(1/8)f_{sampling}$
H#4000	$(1/4)f_{sampling}$
H#6000	$(3/8)f_{sampling}$
H#7FFF	$\sim(1/2)f_{sampling}$

Table 14.3 Some Advance Values and Frequencies

The program shown in Listing 14.1 reads the frequency out of data memory at dm(hertz1) and dm(hertz2). The multiplication by 8.192 is implemented as a multiplication by 8, by 0.512, and then by 2. This approach ensures optimal precision. The multiplications by 8 and by 2 are done in the shifter, and the multiplication by 0.512 is done in the multiplier. The multiplications by 8 and by 2 do not cause any loss of precision. For the multiplication by 0.512, the value 0.512 is represented in 1.15 format, i.e., the full 15 bits of fractional precision. A multiplication by 8.192, which must be represented in at least 5.11 format, would leave only 11 bits of fractional precision. Although not explicitly shown here, this proves to be too little precision for high frequency tones.

After the *advance* value has been added to the *sum* value, the result is written back to the *sum* location in data memory, overwriting the past contents. The result is also passed in register AX0 to the sine function approximation subroutine. That subroutine calculates the sine in 25 cycles and returns the result in the AR register. The sine result is then scaled by downshifting (right arithmetic shift) by the amount specified in dm(scale). This scaling is to avoid overflow when adding the two sine values together later on. The *scale* amount is stored as a variable in data memory at dm(scale) so you can adjust the DTMF amplitude. The scaled result is stored in data memory in either the dm(sin1) or dm(sin2) locations, depending which sine is being evaluated.

When both sines have been calculated, the scaled sine values are recalled out of data memory and added together. That 16-bit, linear result is then passed via the AR register to the μ-255 law, logarithmic compression subroutine (called *u_compress* in Listing 14.1; this routine is listed in Chapter 11, *Pulse Code Modulation*), and the compressed result is finally written to the memory-mapped codec chip.

14.4.1 Digit Entry

There are two methods for entering DTMF digits into the ADSP-2100 program for conversion into DTMF signals. One method is to memory-map a keypad so that when you press a key, the resultant DTMF digit is generated. The other method is to have the ADSP-2100 read a data memory location which contains the DTMF digit in the four LSBs of the 16-bit number (see Figure 14.4). The program shown in Listing 14.1 uses the latter of the two methods. In either case, the row and column frequencies are determined by a look-up table. The keypad method is described in theory, then the data memory method is described, referring to Listing 14.1.

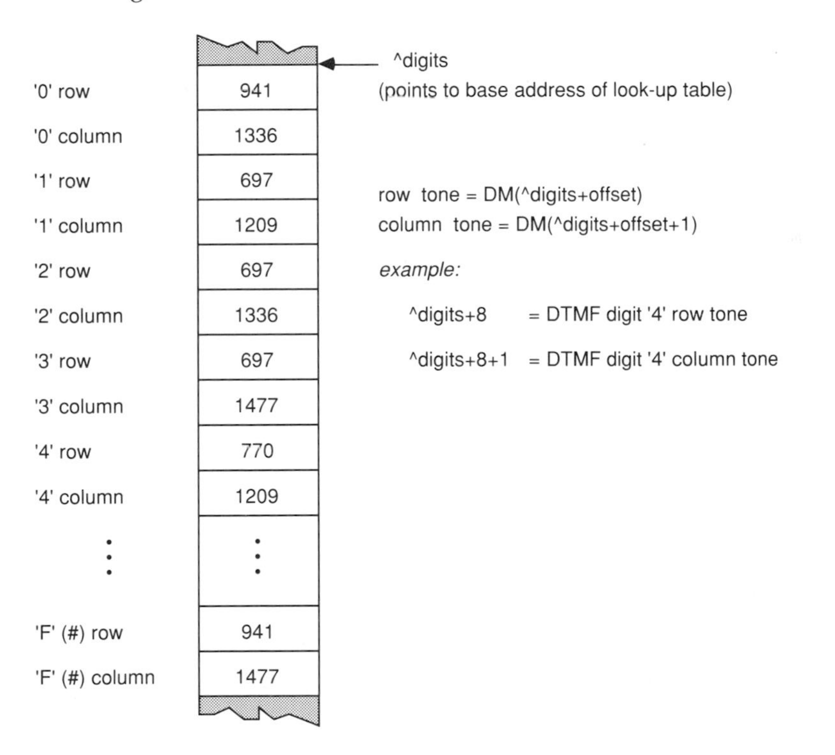

Figure 14.4 Tone Look-Up Table

14.4.1.1 Key Pad Entry

For DTMF digit entry using the keypad, a 74C922 16-key encoder chip is used in conjunction with a 16-key SPST switch matrix and address decoding circuitry. An example of this circuit is shown in Figure 14.5.

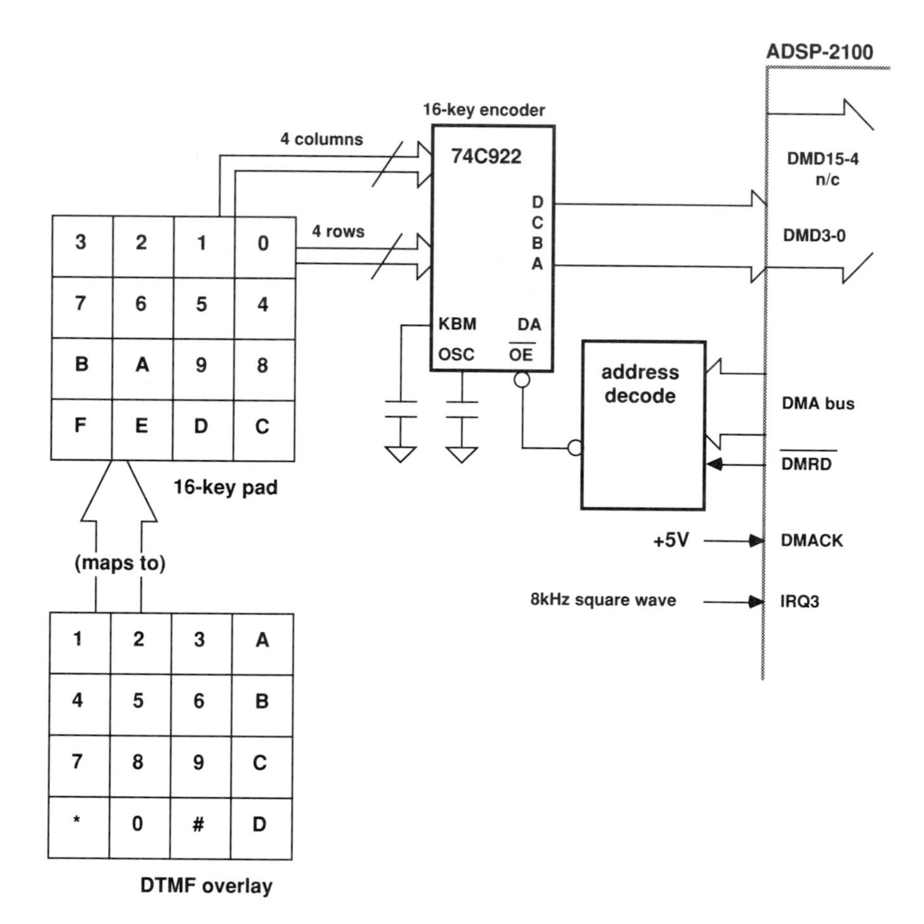

Figure 14.5 Keypad Entry Circuit

The CMOS key encoder provides all the necessary logic to fully encode an array of SPST switches. The keyboard scan is implemented by an external capacitor. An internal debounce circuit also needs a external capacitor as shown in Figure 14.5. The Data Available (DA) output goes high when a valid keyboard entry has been made. The Data Available output returns low when the entered key is released, even if another key is pressed. The Data Available will return high to indicate acceptance of the new key after a normal debounce period. To interrupt the ADSP-2100 when a key has been pressed, invert the Data Available signal and tie it directly to one of the four independent hardware interrupt pins. An internal register in the keypad encoder chip remembers the last key pressed even after the key is released.

Because a DTMF keypad maps to a hexadecimal keypad as shown in Figure 14.5 and Table 14.4, a program using the keypad for digit entry would use a row and column tone look-up list which is similar to that used in Listing 14.1, but its contents would be slightly different. The operation of the tone look-up list is the same as in Listing 14.1, although the values stored in data memory within the look-up list reflect the remapping of the keypad.

DTMF Key	*74C922 Output*	*DTMF Key*	*74C922 Output*
0	xxxE	8	xxxA
1	xxx3	9	xxx9
2	xxx2	A	xxx0
3	xxx1	B	xxx4
4	xxx7	C	xxx8
5	xxx6	D	xxxC
6	xxx5	*	xxxF
7	xxxB	#	xxxD

Table 14.4 DTMF to Keypad Encoder Conversion

14.4.1.2 Data Memory List

For DTMF digit entry by reading the digit in data memory (as in Listing 14.1), a row and column tone look-up list is implemented. The DTMF digits are stored in the four LSBs of the 16-bit data word. All DTMF digits are mapped to their hexadecimal numerical equivalent. The * digit is assigned to the hexadecimal number H#E, and the # digit is assigned to the hexadecimal number H#F. Table 14.5 on the next page shows this mapping. The DTMF digits are stored in such a way that you can see the DTMF sequence being dialed directly out of data memory using the ADSP-2100 Simulator or Emulator in the hexadecimal data memory display mode.

When a DTMF digit is read from data memory, the twelve upper MSBs are first masked out. Then the numerical value of the DTMF digit (0, 1,..., 15 decimal) is multiplied by 2 (yielding 0, 2,..., 30 decimal) because each entry within the look-up table is two 16-bit words long. One word holds the row frequency, and the other word holds the column frequency. This offset value is then added to the base address of the beginning of the look-up table. The resultant address is used to read the row frequency and postincremented by one. The incremented address is used to read the column frequency.

DTMF Digit	*Base Address Offset Value*
0	0
1	2
2	4
3	6
4	8
5	10
6	12
7	14
8	16
9	18
A	20
B	22
C	24
D	26
* (E)	28
# (F)	30

Table 14.5 Look-Up Table Offset Values

For example, referring to Listing 14.1 at the label *nextdigit* and Figure 14.4, the DTMF digit is read out of data memory and stored in register AX0 by the instruction AX0=DM(I0,M0). For this example, assume the value in AX0 is H#0004. Control flow is passed to the instruction labeled *newdigit*. The twelve MSBs are set to zero, and the result is multiplied by 2 in the barrel shifter yielding H#0008. AY0 is set to the base address (^*digits*) of the tone look-up table. The base address is added to the offset and placed in the I1 register. I1 now holds the value ^*digits*+8, M0 was previously set to 1, and L1 to 0. The instruction AX0=DM(I1,M0) reads the row frequency (770Hz) from the look-up table and stores it in the variable *hertz1* used by the sinewave generation code. I1 is automatically postmodified by M0 (1), and the next instruction AX0=DM(I1,M0) reads the column frequency (1209Hz) and stores that in the variable *hertz2* used by the sinewave generation code. When the sinewave generation code is executed (at the label *maketones*), a signal of 770Hz + 1209Hz (DTMF digit 4) is generated.

14.4.2 Dialing Demonstration

The program listed in Listing 14.1 is a DTMF dialing demonstration program. DTMF digits are read sequentially out of a linear buffer. The variable *sign_dura_ms* (signal duration in milliseconds) sets the length of time that a DTMF digit is generated. The variable *interdigit_ms* (interdigit time in milliseconds) sets the length of the silent period following a DTMF digit. When a new digit is started, the variable *time_on* is set to 8 x *sign_dura_ms* and the variable *time_off* is set to 8 x *interdigit_ms*. *Time_on* and *time_off* are counters which the program decrements during interrupts to count the passing of time. Because the ADSP-2100 is interrupted at an 8kHz rate, the amount of time between interrupts is 1/8 millisecond (125μs). By taking the time in milliseconds and multiplying that numerical value by 8, the number of interrupts to count results.

The DTMF digit dialing list stored in data memory starting at location ^*dial_list* can contain values other than DTMF digits. It can also contain control words. Refer to the comment immediately following the variable declarations at the top of the source code in Listing 14.1. If a value is encountered in the dialing list which has any bits set in the four MSBs, the dialing stops. This value is used as a delimiter to terminate the dialing list. If a value is encountered in which the four bits 15-12 are zeros, but any of the bits 11-8 are set, the dialing restarts at the top of the dialing list. If a value is encountered in which all eight bits 15-8 are zeros, but any of the bits 7-4 are set, then a quiet space of length *sign_dura_ms* plus *interdigit_ms* is generated. Finally, if all twelve bits 15-4 are zeros, then the four LSBs represent a valid DTMF digit to generate.

A software state machine has been implemented in the program of Listing 14.1. It is partly controlled by IRQ2, which in this example is wired to a debounced switch. The state machine has three states. Pushing the IRQ2 pushbutton moves the state machine into the next state. The current state of the machine is stored in data memory variable *state*. Figure 14.6, on the next page, shows the demonstration program's state machine. The program starts in state 0. In this state, no digits are generated. Pushing the IRQ2 pushbutton moves the machine into state 1, in which a continuous dial tone (350Hz + 400Hz) is generated. The state machine moves to state 2 when the IRQ2 pushbutton is pushed again. In state 2, the DTMF dialing list is sequentially read and DTMF digits generated. The state machine stays in state 2 until IRQ2 is pushed again or a "stop" control word is read out of the dialing list, in which case the machine jumps back to state 0.

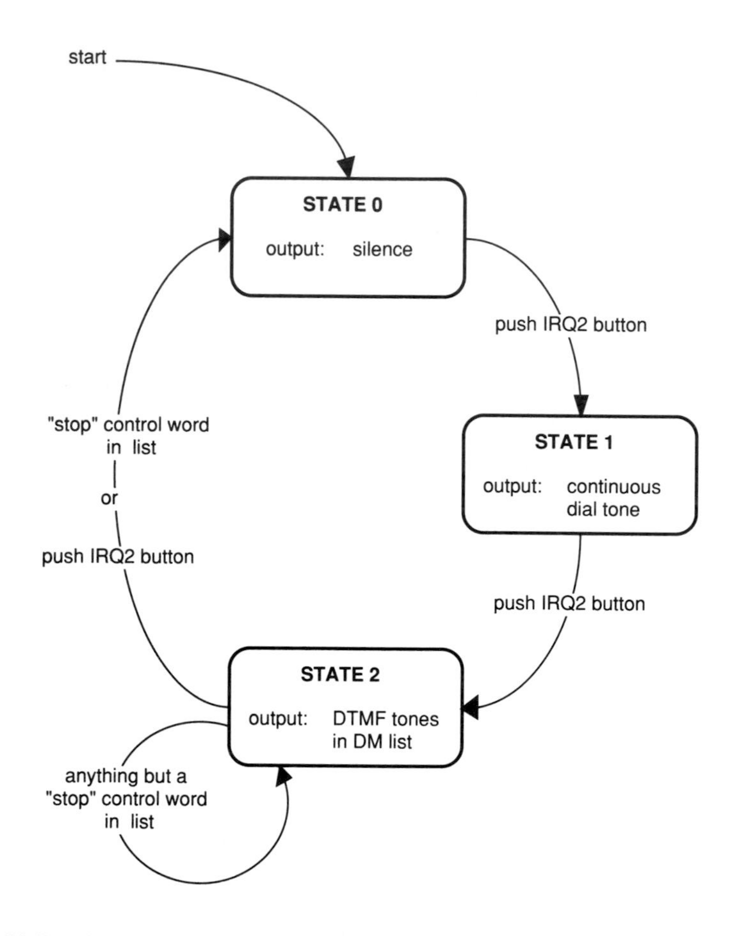

Figure 14.6 Dialing Demonstration Program State Machine

A flowchart of the operation of this demonstration program is shown in Figure 14.7. The data memory variables *row0, ..., row3, col0, ..., col3* in Listing 14.1 are initialized with their appropriate frequencies. These variables are not used by the program at all, but are included as a handy reference so you can look up the frequencies using the data memory display of the ADSP-2100 Simulator or Emulator without having to refer back to any literature.

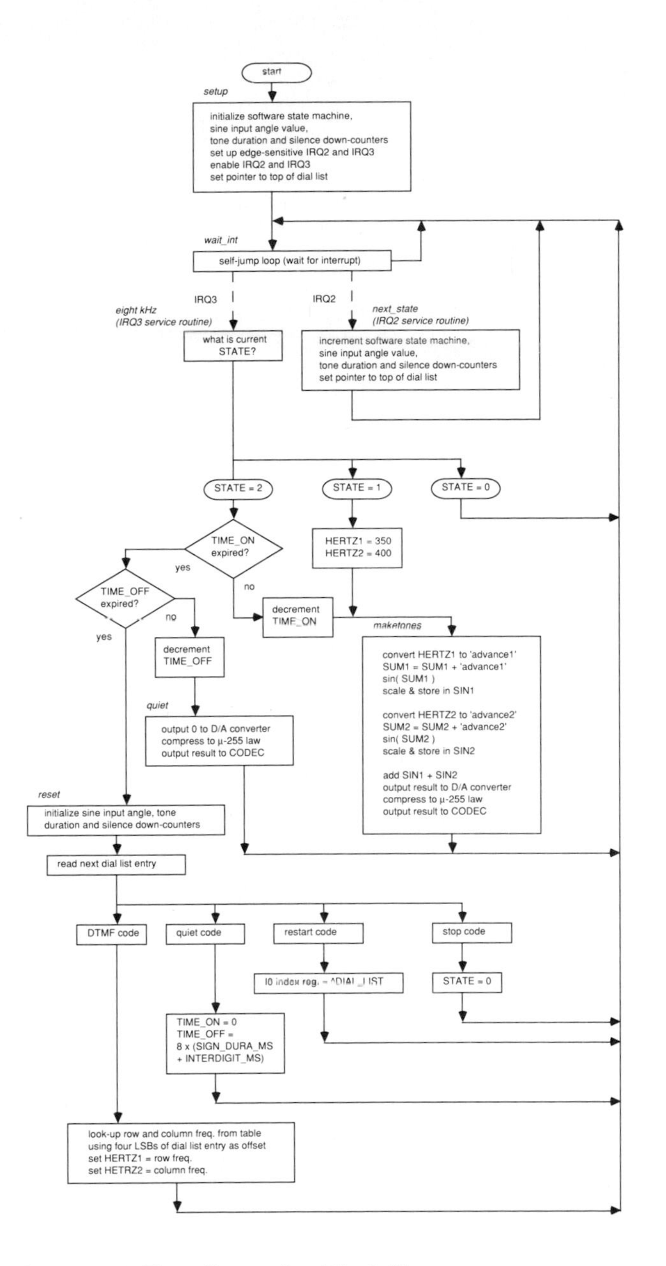

Figure 14.7 Tone Generation Block Diagram

14.4.3 Multi-Channel Generation

A single ADSP-2100 can generate simultaneous DTMF signals for many channels in real time. The multi-channel program uses the same method as the single channel version, but each channel has its own scratchpad variables and memory-mapped I/O ports. The channels are computed sequentially in time every time an interrupt occurs. See Figure 14.8 and Figure 14.9.

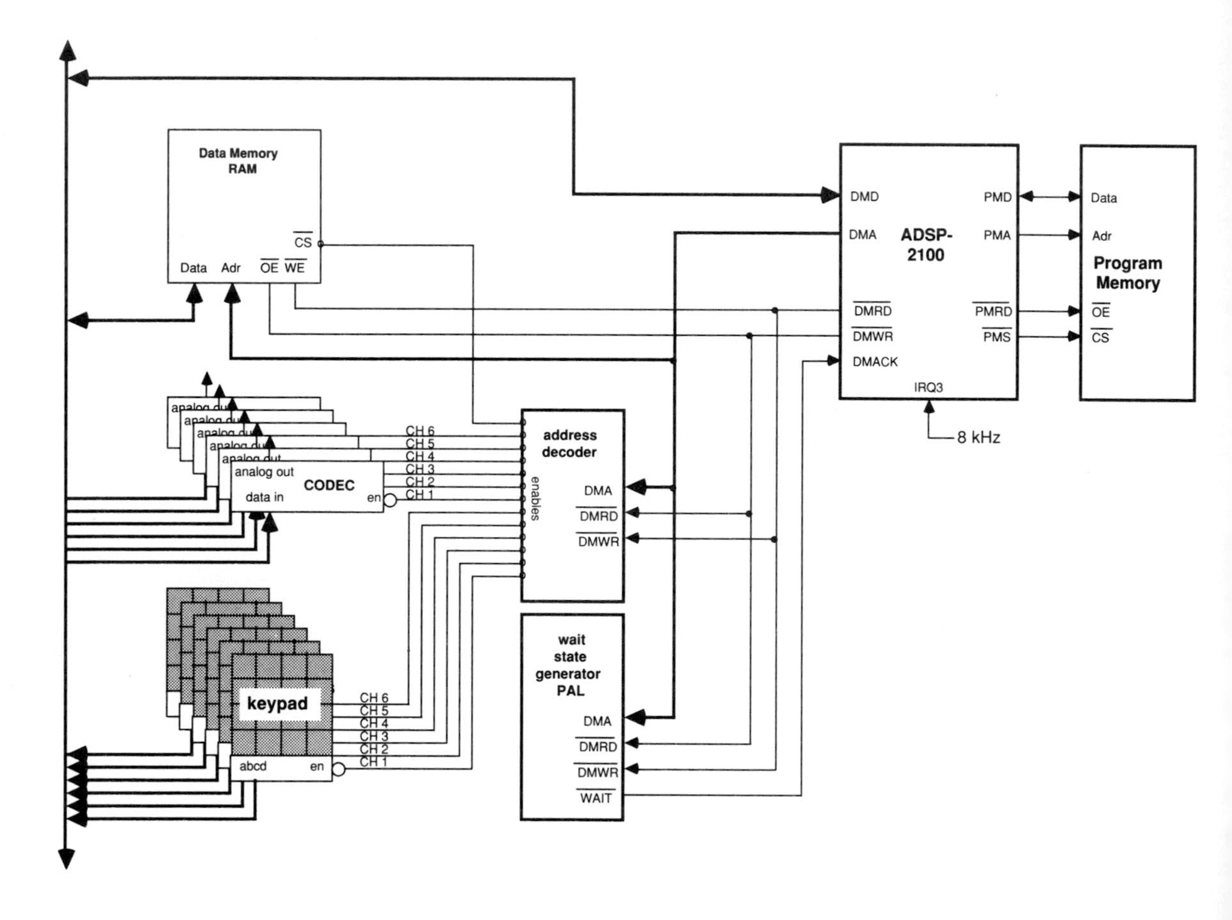

Figure 14.8 Six-Channel Schematic

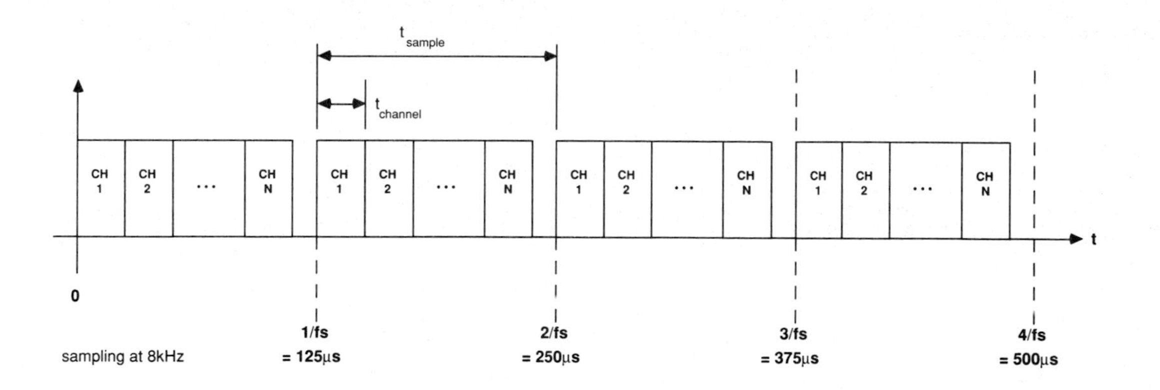

$t_{channel}$ = time to calculate and output a sample within a channel

t_{sample} = time between interrupts (sampling period)

total number of channels possible, N = TRUNCATE ($\frac{t_{sample}}{t_{channel}}$)

Figure 14.9 Maximum Number of Channels (Encoding)

14.5 DECODING DTMF SIGNALS

Decoding a DTMF signal involves extracting the two tones in the signal and determining from their values the intended DTMF digit. Tone detection is often done in analog circuits by detecting and counting zero-crossings of the input signal. In digital circuits, tone detection is easier to accomplish by mathematically transforming the input time-domain signal into its frequency-domain equivalent by means of the Fourier transform.

14.5.1 DFTs and FFTs

The discrete Fourier transform (DFT) or fast Fourier transform (FFT) can be used to transform discrete time-domain signals into their discrete frequency-domain components. The FFT (described in Chapter 7, *Fast Fourier Transforms*) efficiently calculates all possible frequency points in the DFT (e.g., a 256-point FFT computes all 256 frequency points). On the other hand, the DFT can be computed directly to yield only some of the points, for example, only the 20th, 25th, and 30th frequency points out of the possible 256 frequency points. Typically, if more than $\log_2 N$ of N points are desired, it is quicker to compute all the N points using an FFT and discard the unwanted points. If only a few points are needed, the DFT is faster to compute than the FFT. The DFT is faster for finding only eight tones in the full telephone-channel bandwidth.

The definition of an N-point DFT is as follows:

$$X(k) = \sum_{n=0}^{N-1} x(n)W_N^{nk} \qquad \text{where } k = 0, 1, \ldots, N-1 \text{ and } W_N^{nk} = e^{-j(2\pi/N)nk}$$

A single frequency point of the N points is found by computing the DFT for only one k index within the range $0 \le k \le N-1$. For example, if k=15:

$$X(15) = \sum_{n=0}^{N-1} x(n)W_N^{15n} \qquad \text{where } k = 15 \text{ and } W_N^{15n} = e^{-j(2\pi/N)15n}$$

14.5.2 Goertzel Algorithm

The Goertzel algorithm evaluates the DFT with a savings of computation and time. To compute a DFT directly, many complex coefficients are required. For an N-point DFT, N^2 complex coefficients are needed. Even for just a single frequency in a N-point DFT, the DFT must calculate (or look up) N complex coefficients. The Goertzel algorithm needs only two coefficients for every frequency: one real coefficient and one complex coefficient.

The Goertzel algorithm computes a complex, frequency-domain result just as a DFT does, but the Goertzel algorithm can be modified algebraically so that the result is the square of the magnitude of the frequency component (a real value). This modification removes the phase information, which is irrelevant in the tone detection application. The advantage of this modification is that it allows the algorithm to detect a tone using only one, real coefficient.

Not only is the number of coefficients reduced, but the Goertzel algorithm can process each sample as it arrives. There is no need to buffer N samples before computing the N-point DFT, nor do any bit-reversing or windowing. As shown in Figure 14.10, the Goertzel algorithm can be though of as a second-order IIR filter.

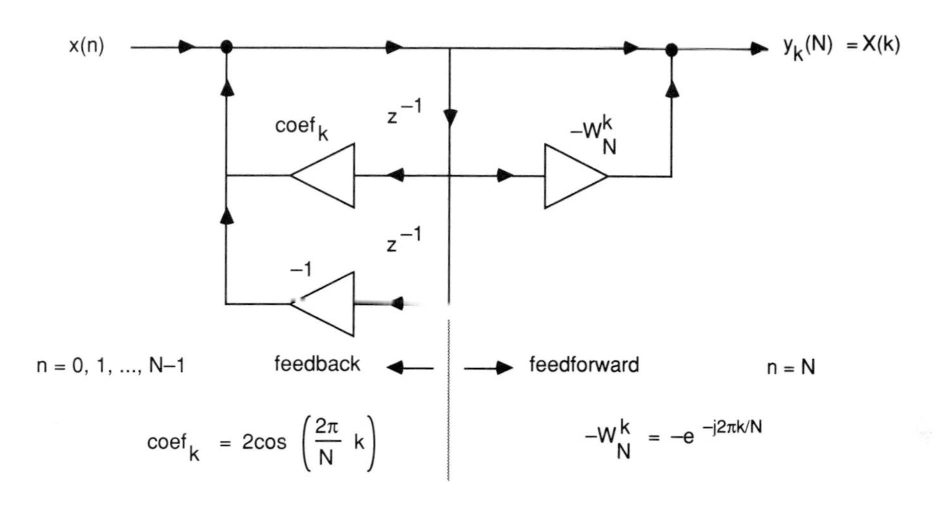

Figure 14.10 Goertzel Algorithm

The Goertzel algorithm can be used to compute a DFT; however, its implementation has much in common with filters. A DFT or FFT computes a buffer of N output data items from a buffer of N input data items. The transform is accomplished by first filling an input buffer with data, then computing the transform of those N samples, yielding N results. By contrast, an IIR or FIR filter computes a new output result with each occurrence of a new input sample. The second-order recursive computation of the DFT by means of the Goertzel algorithm as shown in Figure 14.10 computes a new $y_k(n)$ output for every new input sample x(n). The DFT result, X(k), is equivalent to $y_k(n)$ when n=N, i.e., $X(k)=y_k(N)$. Since any other value of $y_k(n)$, in which $n \neq N$, does not contribute to the end result X(k), there is no need to compute $y_k(n)$ until n=N. This implies that the Goertzel algorithm is functionally equivalent to a second-order IIR filter, except that the one output result of the filter is generated only after N input samples have occurred.

Computation of the Goertzel algorithm can be divided into two phases. The first phase involves computing the feedback legs in Figure 14.10 as depicted in Figure 14.11 on the next page. The second phase evaluates X(k) by computing the feedforward leg in Figure 14.10 as shown in Figure 14.12, also on the next page.

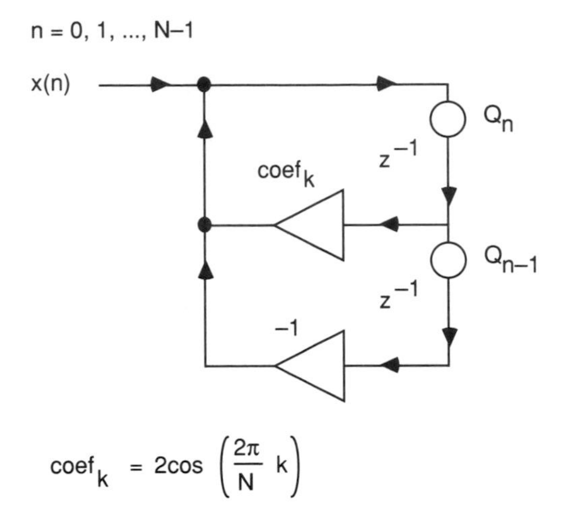

Figure 14.11 Feedback Phase

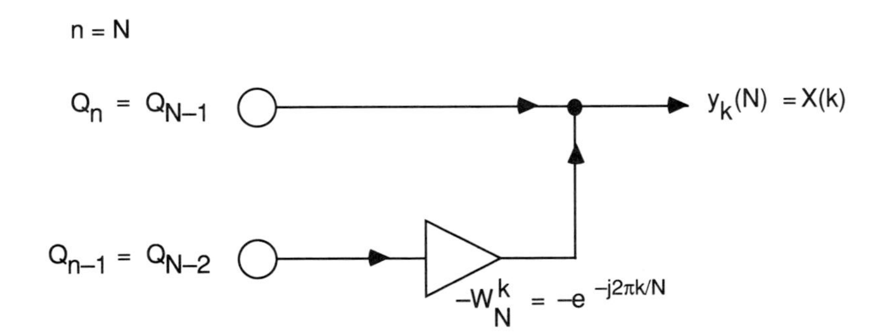

Figure 14.12 Feedforward Phase

14.5.2.1 Feedback Phase

The feedback phase occurs for N input samples (counted as n=0, 1,..., N–1). During this phase, two intermediate values Q(n) and Q(n–1) are stored in data memory. Their values are evaluated as follows:

$$Q_k(n) = coef_k \times Q_k(n-1) - Q_k(n-2) + x(n)$$

where:

> $coef_k = 2\cos(2\pi k/N)$
> $Q_k(-1) = 0$
> $Q_k(-2) = 0$
> $n = 0, 1, 2, ..., N-1$
> $Q_k(n-1)$ and $Q_k(n-2)$ are the two feedback storage elements for frequency point k and x(n) is the current input sample value

Upon each new sample x(n), Q(n–1) and Q(n–2) are read out of data memory and used to evaluate a new Q(n). This new Q(n) is stored where the old Q(n) was. That old Q(n) value is put where the old Q(n–1) was. This action updates Q(n–1) and Q(n–2) for every sample.

14.5.2.2 Feedforward Phase

The feedforward phase occurs once after the feedback phase has been performed for N input samples. The feedforward phase generates an output sample. During computation of the feedforward phase, no new input is used, i.e., new inputs are ignored. As shown in Figure 14.12, the complex value X(k), equivalent to the same X(k) calculated by a DFT, is computed using the two intermediate values Q(n) and Q(n–1) from the feedback phase calculations. At this time, those two intermediate values are Q(N–1) and Q(N–2) since n=N–1. X(k) is calculated as follows:

$$y_k(N) = Q_k(N-1) - W_N^k Q_k(N-2) = X(k)$$

As stated previously, tone detection does not require phase information, and through some algebraic manipulation, the Goertzel algorithm can be modified to output only the squares of the magnitudes of X(k) (magnitudes squared). This implementation not only saves time needed to compute the magnitude squared from a complex result, but also eliminates the need to do any complex arithmetic. The modified Goertzel algorithm is exactly the same as the Goertzel algorithm during the feedback phase, but the feedforward phase is simplified. The magnitude squared of the complex result is expanded in terms of the values available at the end of the feedback iterations. The complex coefficient thereby becomes unnecessary, and the only coefficient needed is conveniently the same as the real coefficient previously used in the feedback phase. The formula for the magnitude squared is derived as follows:

$$
\begin{aligned}
y_k(N) &= Q_k(N-1) - W_N^k Q_k(N-2) \\
&= A - B\,W_N^k \\
&= A - B\,e^{-j\left(\frac{2\pi}{N}k\right)} \\
&= A - B^{-j\,ø} \\
&= A - B\,[\cos ø - j \sin ø] \\
&= A - B\cos ø + j\,B \sin ø
\end{aligned}
$$

$$
\begin{aligned}
\left|y_k(N)\right|^2 &= (\text{Real Part})^2 + (\text{Imag. Part})^2 \\
&= (A - B\cos ø)^2 + (B \sin ø)^2 \\
&= A^2 - 2AB\cos ø + B^2\cos^2 ø + B^2 \sin^2 ø \\
&= A^2 - 2AB\cos ø + B^2(\cos^2 ø + \sin^2 ø) \\
&= A^2 - AB\,(2\cos ø) + B^2
\end{aligned}
$$

$$\left|y_k(N)\right|^2 = A^2 + B^2 - AB\,\text{coef}_k \qquad \text{where } \text{coef}_k = 2\cos\left(\frac{2\pi}{N}k\right)$$

$$\left|y_k(N)\right|^2 = \left|X(k)\right|^2$$

The DTMF decoder calculates magnitudes squared of all eight fundamental tones as well as the magnitude squared of each fundamental's second harmonic. This information is used in one of the validation tests to determine if tones received make up a valid DTMF digit. Specifically, it will be used to give the DTMF decoder the ability to discriminate between pure DTMF sinusoids and speech. Speech waveforms may also contain sinusoids similar to DTMF digits, but speech also has energy in higher-order harmonics, typically the second harmonic. This test is described later.

The modified Goertzel algorithms (one for each k value) have the ability to detect tones using less computation time and fewer stored coefficients than an equivalent DFT would require. For each frequency to detect, the modified Goertzel algorithms need only one real coefficient. This coefficient is used both in the feedback and feedforward phases.

14.5.2.3 Choosing N and k

Determining the coefficient's value for a given tone frequency involves a trade off between accuracy and detection time. These parameters are dependent on the value chosen for N. If N is very large, resolution in the frequency domain is very good, but the length of time between output samples increases, because the feedback phase of Goertzel algorithm is executed N times (once on each input sample) before the feedforward phase is executed once (yielding a single output sample).

If tone detection had been implemented using FFTs, the values of N would have been limited to those that were a power of the radix of the FFT: 16 point, 32 point, 64 point, 128 point, 256 point, etc. for radix-2 (power of 2) FFTs and 16 point, 64 point, 256 point, 1024 point, etc. for radix-4 (power of 4) FFTs. DFTs and Goertzel algorithms, however, are not limited to any radix. These can be computed using any integer value for N.

When an N-point DFT is being evaluated, N input samples (equally spaced in time) are processed to yield N output samples (equally spaced in frequency). The N output samples are:

$$X(k) \quad \text{where } k = 0, 1, 2, \dots, N-1$$

The spacing of the output samples is determined by half the sampling frequency divided by N. If some tone is present in the input signal which does not fall exactly on one of these points in the frequency domain, its frequency component appears partly in the closest frequency point, and partly in the other frequency points. This phenomenon is called leakage. To avoid leakage, it is desirable for all the tones to be detected to be exactly centered on a frequency point. The discrete frequency points are referenced by their k value. The value of k can be any integer within the range 0, 1, 2, ..., N–1. The actual frequency to which k corresponds is dependent on the sampling frequency and N as determined by the following formula:

$$\left(\frac{f_{tone}}{f_{sampling}}\right) = \frac{k}{N}$$

or

$$k = \left(\frac{N}{f_{sampling}}\right) \bullet f_{tone}$$

where f_{tone} is the tone frequency being detected and k is an integer.

Since the sampling frequency is set at 8kHz by the telephone system, and the tones to detect are the DTMF tones, which are also set, the only variable we can modify is N. The numbers k must be integers, so the corresponding frequency points may not be exactly the DTMF frequencies desired. The corresponding absolute error is defined as the difference between what k would be if it could be any real number and the closest integer to that optimal value. For example:

$$\text{absolute k error} = \left|\left(\frac{Nf_{tone}}{f_{sampling}}\right) - \text{CLOSEST INTEGER}\left(\frac{Nf_{tone}}{f_{sampling}}\right)\right|$$

Bell Labs specifically chose the DTMF tones such that they would not be harmonically related. This makes it difficult to choose a value N for which all tones exactly match the DTMF frequency points. A solution could be to perform separate Goertzel algorithms (each with a different value of N) for each tone, but that would involve a lot of non-computational processor overhead. Instead, in this example, values of N were chosen for which the maximum absolute k error of any one of the tones was considered acceptably small. Then, the length of time to detect a tone (which is proportional to the sampling rate multiplied by N) was taken into consideration. The value of N best suited for detecting the eight DTMF fundamental tone frequencies was chosen to be 205. The value of N best suited for detecting the eight second harmonic frequencies was chosen to be 201. See Table 14.6 for the corresponding k values and their respective absolute errors. These values of k allow Goertzel outputs to occur approximately once every 26 milliseconds.

Fundamental Frequency (N=205)	*k value floating-point*	*k value nearest integer*	*Absolute k error*	*coef*$_k$
697.0Hz	17.861	18	0.139	1.703275
770.0Hz	19.731	20	0.269	1.635859
852.0Hz	21.833	22	0.167	1.562297
941.0Hz	24.113	24	0.113	1.482867
1209.0Hz	30.981	31	0.019	1.163138
1336.0Hz	34.235	34	0.235	1.008835
1477.0Hz	37.848	38	0.152	0.790074
1633.0Hz	41.846	42	0.154	0.559454
Second Harmonic (N=201)				
1394.0Hz	35.024	35	0.024	0.917716
1540.0Hz	38.692	39	0.308	0.688934
1704.0Hz	42.813	43	0.187	0.449394
1882.0Hz	47.285	47	0.285	0.202838
2418.0Hz	60.752	61	0.248	–0.659504
2672.0Hz	67.134	67	0.134	–1.000000
2954.0Hz	74.219	74	0.219	–1.352140
3266.0Hz	82.058	82	0.058	–1.674783

sampling frequency = 8kHz

Table 14.6 Values of k and Absolute k Error

14.5.3 DTMF Decoding Program

For DTMF decoding, the ADSP-2100 solves sixteen separate modified Goertzel algorithms, eight of length 205 to detect the DTMF fundamentals, and eight of length 201 to detect the DTMF second harmonics. To implement concurrent Goertzel algorithms of lengths 205 and 201, the feedback phase iterations of all the Goertzel algorithms (fundamentals and second harmonics) are performed for 201 samples (n=0, 1,..., 200). For the next four samples, only the Goertzel algorithms of length 205 (fundamentals) are iterated (n=201, 202, 203, 204). The other Goertzel algorithms of length 201 (second harmonics) ignore the new samples. On the last iteration, when n=N=205, all the feedforward phases are evaluated (both fundamentals and second harmonics), and any new input samples at that time are ignored. In the DTMF decoder application presented here which uses the modified Goertzel algorithm, the magnitude squared calculations are performed for the feedforward phases.

DTMF decoding is done in two major tasks, as shown in the block diagram in Figure 14.13. The first task solves sixteen Goertzel algorithms to calculate the magnitudes squared of tone frequencies present in the input signal, then the second task tests the frequency results to determine if the tones detected constitute a valid DTMF digit. The first task spans N+1 processor interrupts, N for the feedback phases of the Goertzel algorithms, then one more for the feedforward phases. The second task immediately follows completion of all sixteen feedforward phases. The length of time required by the processor for this testing may span the next few interrupts (during which time new input samples are ignored), but since the number of input samples lost is small compared to the number of interrupts serviced during the Goertzel evaluations, the loss is insignificant (see Figure 14.15). The Goertzel algorithms are not sensitive to incoming signal phase, and therefore no phase synchronization is attempted.

14.5.3.1 Input Scaling

It is important to notice that the input sample values are scaled down by eight bits to eliminate the possibility of overflows within 205 iterations of the feedback phase. Scaling by eight bits increases the quantization error of the input samples, but this does not affect the effectiveness of the decoder. Input samples are read from a μ-law-compressed codec. The 8-bit data values are used as offset values to a μ-law-to-linear conversion look-up table. The corresponding linear values are scaled such that the input samples range from H#007F to H#FF80 instead of the normalized equivalent range of H#7FFF to H#8000.

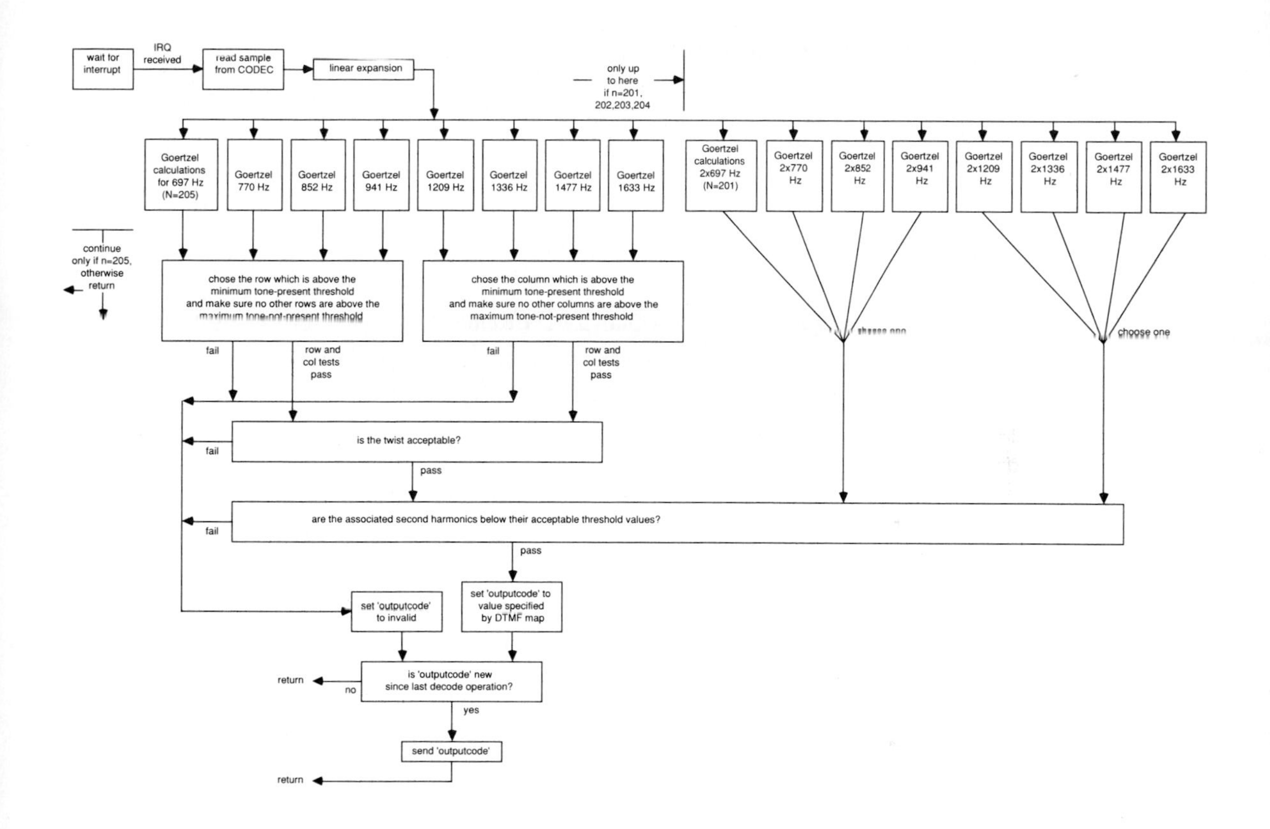

Figure 14.13 Tone Decoder Block Diagram

14.5.3.2 Multi-Channel DTMF Decoder Software

An example of a multi-channel DTMF decoder is given in Listing 14.2. This example is a 6-channel version, although at least twelve channels can be decoded by an ADSP-2100A running at a 12.5MHz instruction rate. The six channels are labeled channel A, channel B, ..., channel F. The following software description gives an overview of the variables and constants used, then outlines the core executive routine and describes the interrupt service routine. The macros and subroutines are explained as needed.

14.5.3.3 Constants, Variables and I/O Ports

The number of channels to decode is specified in the .CONST declaration section. The number of channels must be greater than one and less than or equal to the maximum number of channels allowed, which is dependent on the processor cycle time. The faster the processor cycle time, the more channels can be decoded in real time. The limiting factor is how many of the Goertzel feedback operations can be performed between successive processor interrupts (see Figure 14.14). To decode a single channel only, the source code must be slightly altered. The only necessary alteration involves the circular buffer of length *channels* which stores the input samples; this buffer must be changed to a single data memory variable. Although not necessary, other alterations can be done to optimize the software for a single channel if desired.

The hexadecimal value which the decoder outputs when no DTMF digit is received is defined by the constant called *baddigitcode*. For debugging purposes, a variable called *failurecode* was incorporated into each channel. This variable is assigned a value whether or not a valid DTMF digit was received. If a valid digit was received, *failurecode* is set to zero. A nonzero *failurecode* means that the signals received failed one of the qualifying tests. The *failurecode* value, defined in the constant declaration section, identifies which test failed.

The data variables are separated into two functional groups: housekeeping variables and individual-channel variables. The housekeeping variables are used by the software shared by all sections of the decoder. The individual channel variables are used to keep track of specifics for each channel separately. The variables are explained in detail in Listing 14.2.

The input port for each channel is a codec. The output port for each channel is a D/A converter in this example. In a more realistic application, such as a PBX, switching machine, or electronic voice-mail system, the output would probably be sent to dual-port memory or a mailbox register for use by another processor.

The initialization directives set up some of the static, housekeeping variables. The Goertzel coefficients are initialized as well as the μ-law-to-linear look-up table, thresholds for the received signal levels and twist test limits. (The file containing the μ-law look-up values is not included in the program listings; however, it is included in the diskette that contains the programs in this manual, which is available from your local Analog Devices Sales Office.) These initializations could be adjusted for application-specific requirements, such as meeting the specifications of other administrations or operating as a DTMF signal tester.

To decode more or fewer than six channels, simply add or remove sections identified by the comment "edit this for more channels" in the source code listing.

14.5.3.4 Main Code

The main code section is very simple. The processor is initialized for the decoding operation, then put into an endless loop, waiting for interrupts that occur at the sampling rate (8kHz). After the ADSP-2100 is reset, the first task is to set up the static environment, such as the M and L registers in both data address generators and the ICNTL register. This task is done only once since these initializations are never changed. This set-up is performed by the subroutine called *setup*. Another subroutine called *restart* then initializes other variables which are needed for the decoding operation, but which change and must therefore be reinitialized after each decode operation. The specific tasks here include resetting the I registers of the data address generators to the top of their associated buffers, zeroing out the delay elements for the Goertzel algorithm implementation (Q values), and resetting the counters which keep track of the input sample number (n). The *restart* routine is called after every decode operation, immediately after completion of the digit validation tests, as well as before the very first decode operation.

14.5.3.5 Interrupt Service Routine

The ADSP-2100 is interrupted at an 8kHz sampling rate. The first task done by the processor is to read a new input sample from each codec and store the input samples in a buffer. Next, counters are decremented and tested to determine which sample (n = 0, 1, 2, ..., 205) is currently being processed. If n is between 0 and 200 (inclusive), Goertzel feedback operations are performed on sixteen frequencies per channel, eight fundamental tones and eight second harmonic tones. If n is 201, 202, 203, or 204, then Goertzel feedback operations are performed on the eight fundamental tones for each channel only. The second harmonics are skipped. If n is 205, then magnitude-squared calculations are performed on the eight fundamental tones of each channel, the post testing and digit validation takes place, a new digit is output if necessary, and the *restart* routine is called to reinitialize the processor for the next decoding operation.

Since interrupt nesting mode is not enabled, and since no other interrupts are being used, the processor should never see another interrupt until it has finished executing the interrupt service routine. You need to ensure that the processor does not attempt to decode too many channels, in which case the length of time to perform the sixteen Goertzel operations per channel during samples 0 through 200 would be greater than the sampling period (125µs) (see Figure 14.14). There is one special case in which an interrupt may be overlooked. This is when n=205. In this case, the length of time it takes to perform the magnitude-squared calculations and all the digit validation tests for all channels may exceed the sampling period. Since losing a few input samples out of 205 samples is relatively insignificant, and since incoming signal phase is unimportant, this overlap can be disregarded.

14.5.3.6 Post-Testing and Digit Validation

When calling the macros in the source code listing, the various channels are identified by prefixing each channel's variables with an alphabetic character and underscore. For example, the macro *maxrowcol* is called thus:

```
maxrowcol (^A_mnsqr,A_maxrowval,A_whichrow,A_maxcolval,A_whichcol)
maxrowcol (^B_mnsqr,B_maxrowval,B_whichrow,B_maxcolval,B_whichcol)
```
etc.

Each channel is tested sequentially, identifying each channel's variables by alphabetic prefixes.

Maxrowcol

After completion of the magnitude-squared computations for the eight fundamental tones of each channel, those results reside in each channel's *mnsqr* buffer. The *maxrowcol* macro scans through the results and picks out the largest row result, storing its value in the variable *maxrowval* and its index in the variable *whichrow*. *Whichrow* can be 1, 2, 3, or 4. These values correspond to the frequencies 697Hz, 770Hz, 852Hz, and 941Hz. The column results are likewise scanned, and the largest value and its index assigned to *maxcolval* and *whichcol*. Subsequent testing could set *whichrow* or *whichcol* to zero, indicating that some validation test has failed.

Minsiglevel

The *minsiglevel* macro checks the largest row result and the largest column result chosen to determine whether or not each value exceeds the minimum level necessary for a valid tone. Each tone has its own minimum level threshold in the buffer called *min_tone_level*. Each tone was given its own threshold because of the absolute k error (see previous section on chosing k and N). The DTMF tone frequencies do not correspond exactly to integer multiples of 205/8000; in fact, each tone has a different absolute k error making it necessary to test each magnitude-squared result independently.

The macro *minsiglevel* takes the address of the *min_tone_level* buffer and adds to it the value of *whichrow* minus one. The resulting address is used to look up the minimum signal threshold for that particular row tone. The magnitude squared, stored in *maxrowval*, is compared to the threshold. Failure here sets *whichrow* and *whichcol* each to zero, sets *failurecode* to H#0001 and exits. Otherwise, the row tone passes the test, and the column tone is checked in the same manner as row tone.

No_Other_Peaks

DTMF specifications require that the decoder detect a digit if and only if one row tone is present as well as one and only one column tone is present. The *no_other_peaks* macro makes sure that all the tones other than the maximum row and column tones are below the non-digit threshold.

As in the *minsiglevel* test, this test uses independent thresholds for each tone. The thresholds are stored in the buffer called *max_notone_level*. However, instead of computing each address independently as in the minimum signal level test, this test scans through the whole result (*mnsqr*) buffer and increments a counter (AF register) once for each tone which exceeds the maximum no-tone level. At the end of the scan, the AF register should contain the value H#0002, for one valid row tone and one valid column tone. If any other number is in the AF register, this test fails. If the test fails, *whichrow* and *whichcol* are each set to zero and *failurecode* is set to H#0002.

Twisttests

Twist is the difference, in decibels, between the row tone level and the column tone level. Forward twist (also called standard twist) exists when the column tone level is greater than the row tone level. Reverse twist exists when the column tone is less than the row tone level. DTMF digits are often generated with some forward twist to compensate for greater losses at higher frequencies within a long telephone cable. Different administrations recommend different amounts of allowable twist for DTMF receivers. For example, CEPT recommends not more than 6dB of either twist, Brazil allows 9dB, Australia allows 10dB, Japan allows only 5dB, and AT&T recommends not more than 4dB of forward twist or 8dB of reverse twist.

The twist test macro uses the variables *maxrowcol, maxcolval, whichrow,* and *whichcol* to compute the twist value, setting *twistval* and comparing that value against the predefined twist limits stored in the variables *maxfortwist* and *maxrevtwist*. The macro sets *failurecode* to H#0003 upon failure and sets either the flag *fortwistflag*, or *revtwistflag* as appropriate.

First, the row tone level is compared to the column tone level. If the row tone is greater, program flow jumps to the label *reverse;* otherwise, it continues at *standard*. The standard twist test divides *maxrowval* by *maxcolval* and compares the resultant ratio to *maxfortwist*. (A ratio of powers is equivalent to a difference in decibels.) *Maxfortwist* is the ratio that would result if the greatest allowable twist was encountered. Any ratio which is between that value and unity passes the twist test.

Likewise, if the column tone level was greater, *maxcolval* is divided by *maxrowval,* and the resulting ratio is compared to *maxrevtwist*. If the ratio is greater than *maxrevtwist,* the twist test passes. Of course, a twist value of 0dB would result in a ratio of unity. A very large twist value would result in a very small ratio. The ratio (row or column) is calculated in such a way to ensure that the numerator is always smaller than the denominator. This is done because of how the ALU of the ADSP-2100 performs division.

Check2ndharm

The last item to verify is the level of second harmonic energy present in the detected row and column tones. This test is performed to help the decoder reject speech which might be detected as DTMF tones. This property is referred to as talk-off. DTMF tones should be pure sinusoids, and therefore contain very little second harmonic energy, if any. Speech, on the other hand, contains significant amount of second harmonic energy.

To test the level of second harmonic energy present, the decoder must concurrently evaluate Goertzel algorithms for the second harmonic frequency of all eight DTMF fundamental tones. The second harmonic frequencies (1394Hz, 1540Hz, 1704Hz, 1882Hz, 2418Hz, 2672Hz, 2954Hz, and 3266Hz) can be detected at an 8kHz sampling rate (concurrently with the fundamentals) using Goertzel algorithms of length N=201. This is conveniently close to the length N=205 chosen for the fundamental tones.

During the execution of the Goertzel feedback section of code, both the fundamentals and the second harmonic tones are processed until the counter variable called *count201* expires. For the next four interrupts, another counter called *count4* controls the Goertzel feedback operations. In the latter case, only the eight fundamentals are processed for each channel. The second harmonics are skipped. After the 205th sample has been processed, the *Q1Q2_buff* buffer contains the Goertzel feedback results of the fundamentals at N=205 and of the second harmonics at N=201. When the next interrupt is received, the magnitude-squared computations are carried out for the eight fundamentals of each channel, but no processing is done on the second harmonics yet.

After all the other DTMF digit validation tests are performed, the *check2ndharm* macro carries out the magnitude-squared computations for the second harmonics. But at this time, only two magnitude-squared calculations are performed, one for each detected DTMF tone. This saves the time which would have been wasted if all eight second harmonics had been computed concurrently with the eight fundamentals.

To check the second harmonic level, the address of the channel's Goertzel feedback buffer is passed the *check2ndharm* macro along with the detected tone index variables *whichrow* and *whichcol*. The addresses of the Goertzel feedback values are calculated from the base address plus the index values. The magnitude-squared subroutine is called, and the results are stored in the variables called *rowharm* and *colharm*. Those results are compared to each tone's maximum second harmonic level threshold. The thresholds are stored in the buffer *max_2nd_harm* and, like the fundamental thresholds, are each independently adjustable for each tone.

Outputcode

The last task performed during a decoding sequence is to output a code representing the DTMF digit to the output port. In many applications, this would probably involve writing a hexadecimal code to dual-port RAM or a mailbox for use by the host processor in a PBX, electronic mail system, or digital telephone switch. In this software example, the code is written to a D/A converter. The output of the D/A converter can be used to deflect the vertical trace of an oscilloscope to monitor decoder activity. The extent of the deflection varies according to which DTMF digit is received, if any. If nothing is received, or an invalid digit is received, the code sent to the D/A converter is the constant *baddigitcode*. If a valid digit is received, the index variables *whichrow* and *whichcol* are used to compute the code as follows:

$$\text{16-bit code} = 4096 \times [4(\mathit{whichrow}-1) + (\mathit{whichcol}-1)]$$

In effect, a hexadecimal digit is generated and placed in the most significant hexadecimal digit (4 bits) position of a 4-digit hexadecimal (16 bits) word. The three less significant hexadecimal digits are set to zeros. See Table 14.7 for the one-to-one relationship between received DTMF digits and hexadecimal code output.

DTMF Digit	*Output Code*	*DTMF Digit*	*Output Code*
1	H#0000	7	H#8000
2	H#1000	8	H#9000
3	H#2000	9	H#A000
A	H#3000	C	H#B000
4	H#4000	*	H#C000
5	H#5000	0	H#D000
6	H#6000	#	H#E000
B	H#7000	D	H#F000

Invalid H#FFFF (*baddigitcode*)

Table 14.7 DTMF Tones and Output Codes

The *outputcode* macro not only outputs the appropriate code to the output port, but it also decides whether or not to output anything at all. There must be some distinction made between a long, sustained DTMF signal and several short DTMF signals of the same digit. In other words, it would be undesirable to have the DTMF decoder interrupt a host processor informing it of a stream of new DTMF digits when actually only one DTMF digit was received, but sustained for a long period of time.

For each channel, three decoder-output codes are compared: the current code, the last code, and the next-to-last code. The last and next-to-last codes are stored in the *digit_history* buffers.

The current code is written to the channel's D/A converter if and only if the current code is equal to the last code, but different than the next-to-last code. Whether or not a new digit was detected, the *digit_history* list is updated every time by overwriting the last code with the current code, and overwriting the next-to-last code with the last code. This updates the history list for the next decode operation.

Restart

Before starting the next decode operation, the *restart* subroutine is called. This routine sets all the Goertzel feedback elements to zero, restoring the Goertzel algorithm's initial conditions. The *restart* routine also resets data memory pointers and the two counters (*count201* and *count4*) which keep track of which input sample is being processed.

14.5.3.7 Performance Considerations

The fastest rate of detecting DTMF digits is dictated by how long it takes to perform two successful (all tests passed), sequential decode operations. The amount of time it takes to execute a single decode operation can be calculated as follows:

$$201(T_{sample}) + 4(T_{sample}) + T_{posttest}$$

where

$$T_{sample} = \frac{1}{f_{sample}} = 125\ \mu s$$

and

$$T_{posttest} = \text{number of channels}\ (T_{8\_fund\_mnsqr} + T_{each}) + T_{restart}$$

$$T_{each} = T_{maxrowcol} + T_{minsiglevel} + T_{no\_other\_peaks} + T_{twisttests} + T_{check2ndharm} + T_{outputcode}$$

The exact execution time for $T_{posttest}$ varies with the number of channels being decoded, and also with the actual data processed on those channels. Some of the tests could fail early in their executions, thereby skipping subsequent instructions and testing. Figures 14.14 and Figure 14.15, on the following pages, show processor loading for the 6-channel decoder example in Listing 14.2.

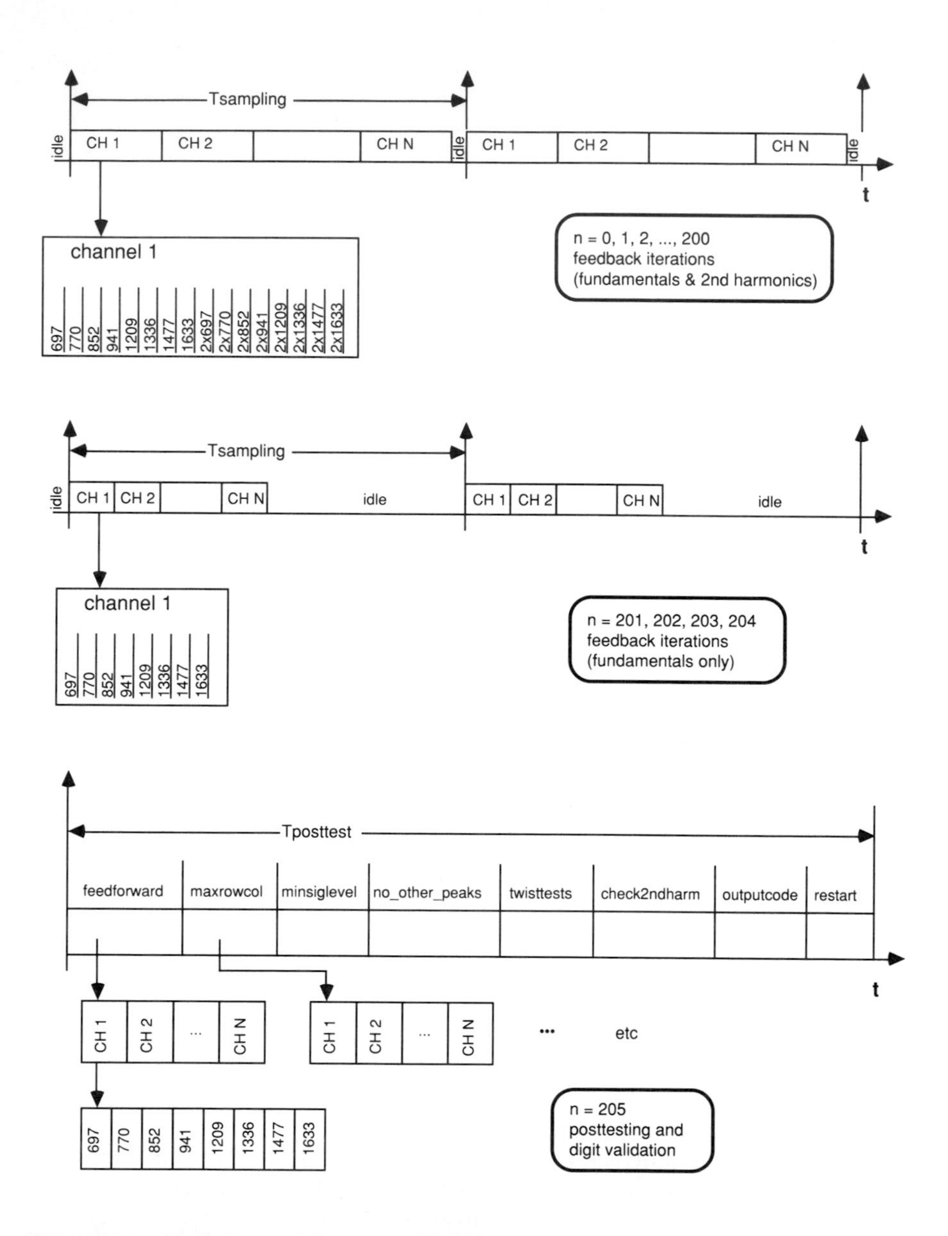

Figure 14.14 Multi-Channel Decoding Timing

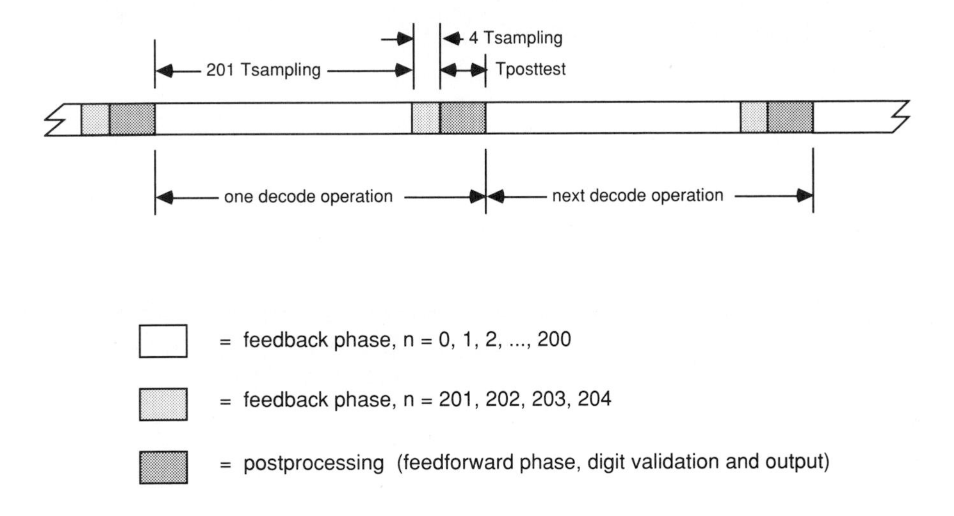

Figure 14.15 Multi-Channel Decoder Output Rate

14.6 REFERENCES

Bellamy, John. 1982. *Digital Telephony*. New York: John Wiley & Sons.

Blahut, Richard E. 1985. *Fast Algorithms for Digital Signal Processing*. Reading, MA: Addison-Wesley.

–. 1982. *Reference Data for Radio Engineers*. New York: Howard Sams.

–. 1986. *IEEE Standard for Functional Requirements for Methods and Equipment for Measuring the Performance of Tone Address Signaling Systems*. (ANSI/IEEE Standard 752-1986). New York: IEEE Press.

Burrus, C.S. and T.W. Parks. 1985. *DFT/FFT and Convolution Algorithms*. New York: John Wiley & Sons.

Oppenheim, Alan V. and Ronald W. Schafer. 1975. *Digital Signal Processing*. New York: Prentice-Hall.

Pearce, J. Gordon. 1981. *Telecommunications Switching*. New York: Plenum Press.

Some Administration Specifications

CCITT:
Recommendations Q.23 and Q.24 in Section 4.3 of the CCITT Red Book, Volume VI, Fascicle VI.1.

AT&T:
Compatibility Bulletin No. 105. *TOUCHTONE Calling - Requirements for the Central Office*. 1975.

CEPT:
Recommendations T/CS 46-02, T/STI 46-04, T/CS 28-01, T/CS 34-08, T/CS 34-09, T/CS 42-02, T/CS 49-04, T/CS 49-07, T/CS 01-02, T/CS 14-01. 1986.

British Post Office:
POR 1151. Section 2. Issue 4. 1979.

14.7 PROGRAM LISTINGS

The complete listings for both the DTMF encoder and 6-channel DTMF decoder are presented in this section.

14.7.1 DTMF Encoder Listing

The code below encodes DTMF digits from a list in data memory. It implements the 3-state software state machine described earlier in this chapter. In state 0, no digits are generated; the ADSP-2100 is idle. In state 1, a continuous dial tone (350Hz + 400Hz) is generated. In state 2, the DTMF dialing list is sequentially read and DTMF digits generated. The state machine stays in each state until the IRQ2 interrupt (connected in the example to a pushbutton) is received. Also, in state 2 when a "stop" control word is read out of the dialing list, the machine jumps back to state 0. Output from the encoder is sent via an I/O port to a D/A converter and is also logarithmically compressed and send to a codec.

Dual-Tone Multi-Frequency 14

```
{  DTMF Signal Generator using ADSP-2100          }

.MODULE/RAM/ABS=0              DTMF_Dialer;

.VAR     row0,row1,row2,row3,col0,col1,col2,col3;
                              {for reference only}
.VAR     hertz1, hertz2;      {scratchpad storage: frequency}
.VAR     sum1, sum2;          {scratchpad storage: phase accum}
.VAR     sin1, sin2;          {scratchpad storage: sine result}
.VAR     scale;               {divisor for scaling sine waves}
.VAR     state;               {current state of state machine}
.VAR     sign_dura_ms;        {signal duration time in milliseconds}
.VAR     interdigit_ms;       {interdigit time in milliseconds}
.VAR     time_on;             {down counter: tones on }
.VAR     time_off;            {down counter: tones off (silence)}
.VAR     digits[32];          {lookup table for row,col freqs}
.VAR     dial_list[100];      {stores sequence to dial}

{  store values in dial_list as follows:
                                                      (E=*, F=#)
   DTMF tone:      h#000y, y=0,1,2,3,4,5,6,7,8,9,A,B,C,D,E,F
   quiet space:    h#00yX, y is non-zero, X is don't care
   redial:         h#0yXX, y is non-zero, X is don't care
   stop:           h#yXXX, y is non-zero, X is don't care
}

.PORT    dac;      {DTMF output to D/A for inspection}
.PORT    codec;    {DTMF output also to µ-law codec}

.INIT    row0:     h#02B9;     { 697 Hz}
.INIT    row1:     h#0302;     { 770 Hz}
.INIT    row2:     h#0354;     { 852 Hz}
.INIT    row3:     h#03AD;     { 941 Hz}
.INIT    col0:     h#04B9;     {1209 Hz}
.INIT    col1:     h#0538;     {1336 Hz}
.INIT    col2:     h#05C5;     {1477 Hz}
.INIT    col3:     h#0661;     {1633 Hz}

.INIT    digits[00]:       h#03AD,h#0538, h#02B9,h#04B9,
                           h#02B9,h#0538, h#02B9,h#05C5;
.INIT    digits[08]:       h#0302,h#04B9, h#0302,h#0538,
                           h#0302,h#05C5, h#0354,h#04B9;
.INIT    digits[16]:       h#0354,h#0538, h#0354,h#05C5,
                           h#02B9,h#0661, h#0302,h#0661;
.INIT    digits[24]:       h#0354,h#0661, h#03AD,h#0661,
                           h#03AD,h#04B9, h#03AD,h#05C5;
```

(listing continues on next page)

```
.INIT   scale:            h#FFFC;
.INIT   sign_dura_ms:     h#0032;    {50 ms}
.INIT   interdigit_ms:    h#0032;    {50 ms}
.INIT   dial_list:        h#00F0, h#0003, h#0000, h#0005,
                          h#0008, h#FFFF;

.EXTERNAL       u_compress;
.EXTERNAL       sin;

IRQ0:           RTI;
IRQ1:           RTI;
IRQ2:           JUMP next_state;
IRQ3:           JUMP eight_khz;

setup:          SI=0;
                DM(state)=SI;
                CALL reset;
                L0=0; L1=0; L2=0; L3=0;
                L4=0; L5=0; L6=0; L7=0;
                I0=^dial_list;
                M0=1;
                M3=1; L3=0;     {used by sine routine}
                ICNTL=b#01111;
                IMASK= b#1100;
wait_int:       JUMP wait_int;

eight_khz:      AY0=2;
                AX0=DM(state);
                AR=AX0-AY0;
                IF EQ JUMP state2;
                AF=AY0-1;
                AR=AX0-AF;
                IF EQ JUMP state1;
state0:         RTI;
state1:         AX0=350;
                DM(hertz1)=AX0;
                AX0=440;
                DM(hertz2)=AX0;
                JUMP maketones;
state2:         AY0=DM(time_on);
                AR=PASS ay0;
                IF EQ JUMP quiet;
                AR=AY0-1;
                DM(time_on)=AR;
```

```
maketones:      SE=DM(scale);

tone1:          AY0=DM(sum1);
                SI=DM(hertz1);
                SR=ASHIFT SI BY 3 (HI);    {mult Hz by 8}
                MY0=h#4189;                {mult by 0.512}
                MR=SR1*MY0(RND);           {mult by 2}
                SR=ASHIFT MR1 BY 1 (HI);   {i.e. Hz * 8.192}
                AR=SR1+AY0;
                DM(sum1)=AR;
                AX0=AR;
                CALL sin;
                SR=ASHIFT AR (HI);         {scale value in SE}
                DM(sin1)=SR1;

tone2:          AY0=DM(sum2);
                SI=DM(hertz2);
                SR=ASHIFT SI BY 3 (HI);    {mult Hz by 8}
                MY0=h#4189;                {mult by 0.512}
                MR=SR1*MY0(RND);           {mult by 2}
                SR=ASHIFT MR1 BY 1 (HI);   {i.e. Hz * 8.192}
                AR=SR1+AY0;
                DM(sum2)=AR;
                AX0=AR;
                CALL sin;
                SR=ASHIFT AR (HI);         {scale value in SE}
                DM(sin2)=SR1;

add_em:         AX0=DM(sin1);
                AY0=DM(sin2);
                AR=AX0+AY0;

sound:          AY0=h#8000;
                AR=AR XOR AY0;
                DM(dac)=AR;
                AR=AR XOR AY0;
                CALL u_compress;
                DM(codec)=AR;
                RTI;

quiet:          AY0=DM(time_off);
                AR=PASS AY0;
                IF EQ JUMP nextdigit;
                AR=AY0-1;
                DM(time_off)=AR;
```

(listing continues on next page)

```
                AY0=h#8000;
                AR=h#7FFF;
                DM(dac)=AR;
                AR=AR XOR AY0;
                CALL u_compress;
                DM(codec)=AR;
                RTI;

nextdigit:      CALL reset;
                AX0=DM(I0,M0);       {read next digit out of list}
                AY0=h#F000;
                AR=AX0 AND AY0;
                IF EQ JUMP notstop;
stop:           AR=0;
                DM(state)=AR;
                RTI;
notstop:        AY0=h#0F00;
                AR=AX0 AND AY0;
                IF EQ JUMP notredial;
redial:         I0=^dial_list;
                RTI;
notredial:      AY0=h#00F0;
                AR=AX0 AND AY0;
                IF EQ JUMP newdigit;
space:          AX0=DM(time_on);
                AY0=DM(time_off);
                AR=AX0+AY0;
                DM(time_off)=AR;
                AR=0;
                DM(time_on)=AR;
                RTI;
newdigit:       AY0=h#000F;
                AR=AX0 AND AY0;
                SR=LSHIFT AR BY 1 (HI);
                AY0=^digits;
                AR=SR1+AY0;
                I1=AR;
                AX0=DM(I1,M0);       {look up row freq}
                DM(hertz1)=AX0;
                AX0=DM(I1,M0);       {look up col freq}
                DM(hertz2)=AX0;
                RTI;
```

```
reset:          SI=0;
                DM(sum1)=SI;
                DM(sum2)=SI;

                SI=DM(sign_dura_ms);
                SR=ASHIFT SI BY 3 (HI);
                AY0=SR1;
                AR=AY0-1;
                DM(time_on)=AR;

                SI=DM(interdigit_ms);
                SR=ASHIFT SI BY 3 (HI);
                AY0=SR1;
                AR=AY0-1;
                DM(time_off)=AR;
                RTS;

next_state:     I0=^dial_list;
                CALL reset;
                AY0=DM(state);
                AR=AY0+1;
                DM(state)=AR;
                AY0=3;
                AR=AR-AY0;              {mod 3, no state 3 exists}
                IF NE RTI;
                AR=0;
                DM(state)=AR;
                RTI;

.ENDMOD;
```

Listing 14.1 DTMF Encoder Program

14.7.2 DTMF Decoder Listing

The code below is the multi-channel DTMF decoder described earlier in this chapter. The six channels are labeled channel A, channel B, ..., channel F. Channels can be added or removed by modifying sections identified by the comment "edit this for more channels."

Input samples are assumed to come from a μ-law codec, and output codes are sent to a D/A converter via an I/O port. DTMF decoding is done in two major tasks. The first task solves sixteen Goertzel algorithms to calculate the magnitudes squared of tone frequencies present in the input signal, then the second task tests the frequency results to determine if the tones detected constitute a valid DTMF digit. The first task spans N+1 processor interrupts. The length of time required by the processor for the second task may span the next few interrupts (during which time input samples are ignored), depending on the tones detected.

```
{                                                                          }
{  Multi Channel DTMF Decoder (six channels shown here)                    }
{                                                                          }
{  INPUT:   one voiceband telephone codec per channel                      }
{  OUTPUT:  DTMF digits are detected within each channel, with a           }
{           corresponding hexadecimal code written to that channel's       }
{           output port (D/A converter) as an activity monitor             }
{                                                                          }

.MODULE/RAM/ABS=0                   Multi_Channel_DTMF_Decoder;

.CONST     channels = 6;                     {must be 2 or more}
           {edit this for more channels}
.CONST     channels_x_32 = 192;
           {edit this for more channels}
.CONST     baddigitcode        = h#FFFF;     {output code for non-digit}
.CONST     pass_posttests      = 0;
.CONST     fail_minsig         = 1;
.CONST     fail_relpeak        = 2;
.CONST     fail_twist          = 3;
.CONST     fail_2ndharm        = 4;
```

```
{ === housekeeping variables === }
{16.0 fixed-point integers}
.VAR       count201;              {counts samples 0 to 200}
.VAR       count4;                {counts samples 201 to 204}

{1.15 fixed-point fractions}
.VAR       min_tone_level[8];     {min "tone-present" mnsqr level}
.VAR       max_notone_level[8];   {max "tone-not-present" mnsqr level}
.VAR       max_2ndharm_level;     {2nd harmonic must be LT this value}
.VAR       maxfortwist;           {quotient row/col must be GT this value}
.VAR       maxrevtwist;           {quotient col/row must be GT this value}
.VAR       mu_lookup_table[256];  {mu-expansion lookup table (scaled 8 bits)}
.VAR/CIRC  in_samples[channels];  {linear input samples (scaled down 8 bits)}

.VAR/CIRC  A_Q1Q2_buff[32],       {Goertzel feedback storage elements}
           B_Q1Q2_buff[32],
           C_Q1Q2_buff[32],
           D_Q1Q2_buff[32],
           E_Q1Q2_buff[32],
           F_Q1Q2_buff[32];
           {edit this for more channels}

.VAR       A_mnsqr[8],            {1.15 Goertzel result values}
           B_mnsqr[8],
           C_mnsqr[8],
           D_mnsqr[8],
           E_mnsqr[8],
           F_mnsqr[8];
           {edit this for more channels}

.VAR/PM
/RAM/CIRC  coefs[16];             {2.14 Goertzel coefs}

{ === individual channel variables === }
.VAR       A_maxrowval;           {1.15 value of max row frequency}
.VAR       A_maxcolval;           {1.15 value of max col frequency}
.VAR       A_whichrow;            {0,1,2,3,4 = invalid, row1, row2, row3, row4}
.VAR       A_whichcol;            {0,1,2,3,4 = invalid, col1, col2, col3, col4}
.VAR       A_fortwistflag;        {1 = forward twist}
.VAR       A_revtwistflag;        {1 = reverse twist}
.VAR       A_twistval;            {1.15 quotient row/col or col/row}
.VAR       A_rowharm;             {1.15 value of row 2nd harmonic}
.VAR       A_colharm;             {1.15 value of col 2nd harmonic}
.VAR       A_digit_history[2];    {stores last 2 output codes}
.VAR       A_failurecode;         {see .CONST definitions above}
```

(listing continues on next page)

```
.VAR        B_maxrowval;
.VAR        B_maxcolval;
.VAR        B_whichrow;
.VAR        B_whichcol;
.VAR        B_fortwistflag;
.VAR        B_revtwistflag;
.VAR        B_twistval;
.VAR        B_rowharm;
.VAR        B_colharm;
.VAR        B_digit_history[2];
.VAR        B_failurecode;

.VAR        C_maxrowval;
.VAR        C_maxcolval;
.VAR        C_whichrow;
.VAR        C_whichcol;
.VAR        C_fortwistflag;
.VAR        C_revtwistflag;
.VAR        C_twistval;
.VAR        C_rowharm;
.VAR        C_colharm;
.VAR        C_digit_history[2];
.VAR        C_failurecode;

.VAR        D_maxrowval;
.VAR        D_maxcolval;
.VAR        D_whichrow;
.VAR        D_whichcol;
.VAR        D_fortwistflag;
.VAR        D_revtwistflag;
.VAR        D_twistval;
.VAR        D_rowharm;
.VAR        D_colharm;
.VAR        D_digit_history[2];
.VAR        D_failurecode;

.VAR        E_maxrowval;
.VAR        E_maxcolval;
.VAR        E_whichrow;
.VAR        E_whichcol;
.VAR        E_fortwistflag;
.VAR        E_revtwistflag;
.VAR        E_twistval;
.VAR        E_rowharm;
```

```
.VAR       E_colharm;
.VAR       E_digit_history[2];
.VAR       E_failurecode;

.VAR       F_maxrowval;
.VAR       F_maxcolval;
.VAR       F_whichrow;
.VAR       F_whichcol;
.VAR       F_fortwistflag;
.VAR       F_revtwistflag;
.VAR       F_twistval;
.VAR       F_rowharm;
.VAR       F_colharm;
.VAR       F_digit_history[2];
.VAR       F_failurecode;
{edit this for more channels}

{ individual channel I/O ports }
.PORT      A_codec;
.PORT      A_dac;

.PORT      B_codec;      {telephone audio 8-bit parallel codec input}
.PORT      B_dac;        {monitor decoder output with voltage level}

.PORT      C_codec;
.PORT      C_dac;

.PORT      D_codec;
.PORT      D_dac;

.PORT      E_codec;
.PORT      E_dac;

.PORT      F_codec;
.PORT      F_dac;
{edit this for more channels}
```

(listing continues on next page)

```
{ variable initializations }
.INIT      coefs[00]: h#6D0200, h#68B200, h#63FD00, h#5EE700;
.INIT      coefs[04]: h#4A7100, h#409100, h#329100, h#23CE00;
.INIT      coefs[08]: h#3ABC00, h#2C1700, h#1CC300, h#0CFB00;
.INIT      coefs[12]: h#D5CB00, h#C00000, h#A97700, h#94D000;

.INIT      mu_lookup_table:        < MU255.Q8 >;
.INIT      min_tone_level:
   h#0003,h#0003,h#0003,h#0003,h#0003,h#0003,h#0003,h#0003;

.INIT      max_notone_level:
   h#0002,h#0002,h#0002,h#0002,h#0002,h#0002,h#0002,h#0002;

.INIT      max_2ndharm_level:      h#0100;
.INIT      maxfortwist:            h#32F5;
.INIT      maxrevtwist:            h#1449;

{%%%%%%%%%%%%%%%%%%%%%%%%%%%%%%%%%%%%%%%%%%%%%%%%%%%%%%%%%%%%%%%%%%%%%%}
{  8-bit mu-law sample read from specified codec, then converted      }
{  to linear (1.15 scaled down 8 bits) via look-up table              }
{                                                                      }
{  INPUT:   channel codec to read                                     }
{  OUTPUT:  scaled linear sample in "in_samples" circular buffer      }
{                                                                      }
.MACRO      get_sample( %0 );    {make sure I3,M0,L3,M4,L6 initialized}

            AY0=DM(%0);          {read codec, mu-law data}
            AF=AX0 AND AY0;      {make sure AX0 initialized to h#00FF}
            AR=AX1+AF;           {make sure AX1 initialized to LUT base}
            I6=AR;
            SI=DM(I6,M4);        {look-up scaled, linear value}
            DM(I3,M0)=SI;        {store input sample}
.ENDMACRO;

{%%%%%%%%%%%%%%%%%%%%%%%%%%%%%%%%%%%%%%%%%%%%%%%%%%%%%%%%%%%%%%%%%%%%%%}
{  pick largest row and col freq values                               }
{                                                                      }
{  INPUT:   pointer to top of channel's mnsqr buffer                  }
{  OUTPUT:  largest row and col values and their indexes              }
{                                                                      }
.MACRO      maxrowcol( %0, %1, %2, %3, %4 );
            {^mnsqr, maxrowval, whichrow, maxcolval, whichcol}
.LOCAL      findmaxrow;
.LOCAL      findmaxcol;
```

```
            I2=%0;

            AX1=0;          {set up variables in case nothing found}
            AY1=0;          {set up variables in case nothing found}
            AY0=0;          {initialize BIGGEST-VALUE-SO-FAR to zero}
            CNTR=4;
            DO findmaxrow UNTIL CE;
                            AX0=DM(I2,M0); {read CURRENT-MNSQR-VALUE}
                            AR=AX0-AY0;    {compare to BIGGEST-VALUE-SO-FAR}
                            IF LE JUMP findmaxrow;
                            AX1=5;
                            AY0=AX0;       {if CURRENT is bigger, store value}
                            AY1=CNTR;      {if CURRENT is bigger, store index}
findmaxrow:                 NOP;

            DM(%1)=AY0;         {store the largest mnsqr value}
            AR=AX1-AY1;
            DM(%2)=AR;          {store index of biggest row (1,2,3,4)}

            AX1=0;
            AY1=0;
            AY0=0;
            CNTR=4;
            DO findmaxcol UNTIL CE;
                            AX0=DM(I2,M0);
                            AR=AX0-AY0;
                            IF LE JUMP findmaxcol;
                            AX1=5;
                            AY0=AX0;
                            AY1=CNTR;
findmaxcol:                 NOP;
            DM(%3)=AY0;
            AR=AX1-AY1;
            DM(%4)=AR;
.ENDMACRO;
```

(listing continues on next page)

```
{%%%%%%%%%%%%%%%%%%%%%%%%%%%%%%%%%%%%%%%%%%%%%%%%%%%%%%%%%%%%%%%%%%%%%%}
{  checks whether selected freqs are GT minimum signal power level     }
{                                                                      }
{  INPUT:      index of detected row and col tones                     }
{              value of detected row and col tones                     }
{  OUTPUT:     failurecode set if test fails                           }
{                                                                      }
.MACRO         minsiglevel( %0, %1, %2, %3, %4 );
               {whichrow, maxrowval, whichcol, maxcolval, failurecode}
.LOCAL         failsiglevel;
.LOCAL         done;

               AX0=^min_tone_level;
               AY0=DM(%0);
               AF=AY0-1;
               AR=AX0+AF;
               I5=AR;       {I5 points to ^min_tone_level+whichrow-1}
               AX1=DM(%1);
               AY1=DM(I5,M4);
               AR=AX1-AY1;
               IF LT JUMP failsiglevel;

               AY0=3;
               AR=AX0+AY0;
               AY0=DM(%2);
               AR=AR+AY0;
               I5=AR;       {I5 points to ^min_tone_level+4+whichcol-1}
               AX1=DM(%3);
               AY1=DM(I5,M4);
               AR=AX1-AY1;
               IF GE JUMP done;

failsiglevel:  AX0=0;
               DM(%0)=AX0;
               DM(%2)=AX0;
               AY0=fail_minsig;
               DM(%4)=AY0;

done:          NOP;
.ENDMACRO;
```

```
{%%%%%%%%%%%%%%%%%%%%%%%%%%%%%%%%%%%%%%%%%%%%%%%%%%%%%%%%%%%%%%%%%%%%%%%%%%}
{  verify that only one valid row freq and col freq are present           }
{                                                                         }
{  INPUT:      pointer to top of channel's mnsqr buffer                   }
{              index of detected row and col tones                        }
{  OUTPUT:     failurecode set if test fails                              }
{                                                                         }
.MACRO         no_other_peaks( %0, %1, %2, %3 );
               {^mnsqr, whichrow, whichcol, failurecode}
.LOCAL         looper;
.LOCAL         failrelpeak;
.LOCAL         done;

               I2=%0;
               I6=^max_notone_level;
               AF=PASS 0;
               CNTR=8;
               DO looper UNTIL CE;
                         AX0=DM(I6,M4);
                         AY0=DM(I2,M0);
                         AR=AX0-AY0;
looper:                  IF LT AF=AF+1;
               AX1=2;                  {see if only 2 tones are over their}
               AR=AX1-AF;              {max notone level thresholds}
               IF EQ JUMP done;
failrelpeak:   AX0=0;                  {clear whichrow,col}
               DM(%1)=AX0;
               DM(%2)=AX0;
               AY0=fail_relpeak;
               DM(%3)=AY0;             {set failurecode}
done:          NOP;
.ENDMACRO;
```

(listing continues on next page)

14 Dual-Tone Multi-Frequency

```
{%%%%%%%%%%%%%%%%%%%%%%%%%%%%%%%%%%%%%%%%%%%%%%%%%%%%%%%%%%%%%%%%%%%%%%%%%%%%%}
{  checks difference between row tone level and col tone level (twist) }
{                                                                       }
{  INPUT:   index of detected row and col tones                         }
{           value of detected row and col tones                         }
{  OUTPUT:  forward twist flag or reverse twist flag set                }
{           twist value(row/col [fwd] or col/row [rev])                 }
{           failurecode set if test fails                               }
{                                                                       }
.MACRO      twisttests( %0, %1, %2, %3, %4, %5, %6, %7 );
            {maxrowval, maxcolval, fortwistflag, revtwistflag,
            whichrow, whichcol, twistval, failurecode}
.LOCAL      standard;
.LOCAL      reverse;
.LOCAL      failtwist;
.LOCAL      done;

            MX0=0;
            MX1=1;
            AX0=DM(%0);
            AY0=DM(%1);
            AR=AX0-AY0;
            IF GT JUMP reverse;
standard:   DM(%2)=MX1;                   {column tone is stronger}
            DM(%3)=MX0;
            AX0=DM(%1);
            AY0=DM(%0);
            AF=PASS AY0;
            AY0=0;
            DIVS AF,AX0;
            DIVQ AX0; DIVQ AX0; DIVQ AX0; DIVQ AX0; DIVQ AX0;
            DIVQ AX0; DIVQ AX0; DIVQ AX0; DIVQ AX0; DIVQ AX0;
            DIVQ AX0; DIVQ AX0; DIVQ AX0; DIVQ AX0; DIVQ AX0;
            DM(%6)=AY0;                   {AY0 = maxrowval / maxcolval}
            AX0=DM(maxfortwist);
            AR=AX0-AY0;
            IF GT JUMP failtwist;
            JUMP done;
reverse:    DM(%2)=MX0;
            DM(%3)=MX1;                   {row tone is stronger}
            AF=PASS AY0;
            AY0=0;
            DIVS AF,AX0;
            DIVQ AX0; DIVQ AX0; DIVQ AX0; DIVQ AX0; DIVQ AX0;
            DIVQ AX0; DIVQ AX0; DIVQ AX0; DIVQ AX0; DIVQ AX0;
            DIVQ AX0; DIVQ AX0; DIVQ AX0; DIVQ AX0; DIVQ AX0;
```

```
            DM(%6)=AY0;                    {AY0 = maxcolval / maxrowval}
            AX0=DM(maxrevtwist);
            AR=AX0-AY0;
            IF GT JUMP failtwist;
            JUMP done;
failtwist:  DM(%4)=MX0;
            DM(%5)=MX0;
            AY0=fail_twist;
            DM(%7)=AY0;
done:       NOP;
.ENDMACRO;

{%%%%%%%%%%%%%%%%%%%%%%%%%%%%%%%%%%%%%%%%%%%%%%%%%%%%%%%%%%%%%%%%%%%%}
{  checks energy levels in second harmonics of detected tones        }
{                                                                    }
{  INPUT:  pointer to top of channel's Goertzel feedback buffer      }
{          index of detected row and col tones                       }
{          pointers to variables holding channel's 2nd harmonic levels }
{  OUTPUT: value of channel's row and col 2nd harmonic levels        }
{          failurecode set if test fails                             }
{                                                                    }
.MACRO         check2ndharm( %0, %1, %2, %3, %4, %5, %6, %7 );
               {^Q1Q2_buff, whichrow, ^rowharm, rowharm,
               whichcol, ^colharm, colharm, failurecode}
.LOCAL         fail2ndharm;
.LOCAL         done;

               AX0=%0;
               AY0=DM(%1);
               AR=AY0-1;                      {range: 1,2,3,4 => 0,1,2,3}
               SR=ASHIFT AR BY 1 (HI);        {range: 0,1,2,3 => 0,2,4,6}
               AY1=16;
               AF=SR1+AY1;
               AR=AX0+AF;
               I0=AR;          {I0 points to ^Q1Q2_buff+16+2*(whichrow-1)}
               AX0=^coefs;
               AF=AX0+AY0;
               AX0=7;
               AR=AX0+AF;
               I4=AR;          {I4 points to ^coefs+8+whichrow-1}
               I2=%2;          {I2 points to ^rowharm}
               CALL mnsqr;
               AX0=DM(%3);
               AY0=DM(max_2ndharm_level);
               AR=AX0-AY0;
               IF GT JUMP fail2ndharm;
```

(listing continues on next page)

```
                AX0=%0;
                AY0=DM(%4);
                AR=AY0-1;                        {range 1,2,3,4 => 0,1,2,3}
                SR=ASHIFT AR BY 1 (HI);          {range 0,1,2,3 => 0,2,4,6}
                AY1=24;
                AF=SR1+AY1;
                AR=AX0+AF;
                I0=AR;         {I0 points to ^Q1Q2_buff+16+8+2*(whichcol-1)}
                AX0=^coefs;
                AF=AX0+AY0;
                AX0=11;
                AR=AX0+AF;
                I4=AR;         {I4 points to ^coefs+8+4+whichcol-1}
                I2=%5;         {I2 points to ^colharm}
                CALL mnsqr;
                AX0=DM(%6);
                AY0=DM(max_2ndharm_level);
                AR=AX0-AY0;
                IF LT JUMP done;
fail2ndharm:    AX0=0;
                DM(%1)=AX0;
                DM(%4)=AX0;
                AY0=fail_2ndharm;
                DM(%7)=AY0;
done:           NOP;
.ENDMACRO;

{%%%%%%%%%%%%%%%%%%%%%%%%%%%%%%%%%%%%%%%%%%%%%%%%%%%%%%%%%%%%%%%%%%%%%%%%%}
{  hexadecimal code for a given DTMF digit is generated  and output if }
{  necessary, digit_history updated, failurecode cleared               }
{                                                                      }
{  INPUT:   index of detected row and col tones                        }
{           failurecode                                                }
{  OUTPUT:  digit_history updated with latest hex output code          }
{           hex output code written to output port if both:            }
{           (1) the current code is the same as the previous code      }
{           (2) but different from the one before that                 }
{           failurecode cleared for next DTMF decode operation         }
{                                                                      }
.MACRO      outputcode( %0, %1, %2, %3, %4 );
            {whichrow, whichcol, digit_history, failurecode, dac}
.LOCAL      checkfailures;
.LOCAL      digitdetected;
.LOCAL      nodigit;
.LOCAL      readlist;
.LOCAL      pushlist;
```

```
checkfailures: AY0=DM(%3);
               AR=PASS AY0;
               IF NE JUMP nodigit;
digitdetected: AY0=DM(%1);
               AR=AY0-1;
               SR=LSHIFT AR BY 12 (HI);
               AY0=DM(%0);
               AR=AY0-1;
               SR=SR OR LSHIFT AR BY 14 (HI);
               JUMP readlist;
nodigit:       SR1=baddigitcode;
readlist:      AY0=DM(%2);
               AY1=DM(%2+1);
               AR=SR1-AY1;
               IF EQ JUMP pushlist;
               AR=SR1-AY0;
               IF NE JUMP pushlist;
               DM(%4)=SR1;
pushlist:      DM(%2+1)=AY0;
               DM(%2)=SR1;
               AY0=pass_posttests;
               DM(%3)=AY0;
.ENDMACRO;

{————————————————————————————————————————————————————————————}
{————————— M A I N   C O D E ————————————————————————————————}
{————————————————————————————————————————————————————————————}

          RTI; RTI; RTI; JUMP sample;

          CALL setup;
          CALL restart;
          IMASK=b#1000;
here:     JUMP here;

{————————————————————————————————————————————————————————————}
{—— I N T E R R U P T   S E R V I C E   R O U T I N E ——}
{————————————————————————————————————————————————————————————}

sample:   get_sample( A_codec );
          get_sample( B_codec );
          get_sample( C_codec );
          get_sample( D_codec );
          get_sample( E_codec );
          get_sample( F_codec );
          {edit this for more channels}
```

(listing continues on next page)

```
dec201: AY0=DM(count201);
        AR=AY0-1;
        DM(count201)=AR;
        IF LT JUMP dec4;

in201:  CNTR=channels;              {number of channels}
        DO chan201 UNTIL;
           AY1=DM(I3,M0);           {get sample for channel, AY1=1.15}
           CNTR=16;                 {8 fundamentals, 8 2nd_harmonics per channel}
           DO freq201 UNTIL CE;
              MX0=DM(I0,M0), MY0=PM(I4,M4);    {get Q1,COEF Q1=1.15,COEF=2.14}
              MR=MX0*MY0(RND), AY0=DM(I0,M1); {mult,get Q2, MR=2.30, Q2=1.15}
              SR=ASHIFT MR1 BY 1 (HI);         {change 2.30 to 1.15}
              AR=SR1-AY0;                      {Q1*COEF - Q2, AR=1.15}
              AR=AR+AY1;                       {Q1*COEF - Q2 + input, AR=1.15}
              DM(I0,M0)=AR;                    {result = new Q1}
freq201:   DM(I0,M0)=MX0;                      {old Q1 = new Q2 }
chan201:NOP;  {do next channel}
        RTI;

dec4:   AY0=DM(count4);
        AR=AY0-1;
        DM(count4)=AR;
        IF LT JUMP last;

in4:    CNTR=channels;              {number of channels}
        DO chan4 UNTIL CE;
           AY1=DM(I3,M0);
           CNTR=8;                  {8 fundamentals only}
           DO freq4 UNTIL CE;
              MX0=DM(I0,M0), MY0=PM(I4,M4);
              MR=MX0*MY0(RND), AY0=DM(I0,M1);
              SR=ASHIFT MR1 BY 1 (HI);
              AR=SR1-AY0;
              AR=AR+AY1;
              DM(I0,M0)=AR;
freq4:        DM(I0,M0)=MX0;
           MODIFY(I0,M2);           {skip 2nd harmonic Q1Q2s}
chan4:     MODIFY(I4,M5);           {skip 2nd harmonic COEFs}
        RTI;
```

```
last:    CNTR=channels;
         DO chanlast UNTIL CE;
            CNTR=8;
            DO freqlast UNTIL CE;
               CALL mnsqr;
freqlast:      NOP;
            MODIFY(I0,M2);        {skip 2nd harmonic Q1Q2s}
chanlast:   MODIFY(I4,M5);        {skip 2nd harmonic COEFs}

maxrowcol(^A_mnsqr,A_maxrowval,A_whichrow,A_maxcolval,A_whichcol);
maxrowcol(^B_mnsqr,B_maxrowval,B_whichrow,B_maxcolval,B_whichcol);
maxrowcol(^C_mnsqr,C_maxrowval,C_whichrow,C_maxcolval,C_whichcol);
maxrowcol(^D_mnsqr,D_maxrowval,D_whichrow,D_maxcolval,D_whichcol);
maxrowcol(^E_mnsqr,E_maxrowval,E_whichrow,E_maxcolval,E_whichcol);
maxrowcol(^F_mnsqr,F_maxrowval,F_whichrow,F_maxcolval,F_whichcol);
{edit this for more channels}

{———————— START OF DIGIT VALIDATION TESTS ————————}

minsiglevel(A_whichrow,A_maxrowval,A_whichcol,A_maxcolval,A_failurecode);
minsiglevel(B_whichrow,B_maxrowval,B_whichcol,B_maxcolval,B_failurecode);
minsiglevel(C_whichrow,C_maxrowval,C_whichcol,C_maxcolval,C_failurecode);
minsiglevel(D_whichrow,D_maxrowval,D_whichcol,D_maxcolval,D_failurecode);
minsiglevel(E_whichrow,E_maxrowval,E_whichcol,E_maxcolval,E_failurecode);
minsiglevel(F_whichrow,F_maxrowval,F_whichcol,F_maxcolval,F_failurecode);
{edit this for more channels}

no_other_peaks(^A_mnsqr,A_whichrow,A_whichcol,A_failurecode);
no_other_peaks(^B_mnsqr,B_whichrow,B_whichcol,B_failurecode);
no_other_peaks(^C_mnsqr,C_whichrow,C_whichcol,C_failurecode);
no_other_peaks(^D_mnsqr,D_whichrow,D_whichcol,D_failurecode);
no_other_peaks(^E_mnsqr,E_whichrow,E_whichcol,E_failurecode);
no_other_peaks(^F_mnsqr,F_whichrow,F_whichcol,F_failurecode);
{edit this for more channels}

twisttests(A_maxrowval,A_maxcolval,A_fortwistflag,A_revtwistflag,
            A_whichrow,A_whichcol,A_twistval,A_failurecode);
twisttests(B_maxrowval,B_maxcolval,B_fortwistflag,B_revtwistflag,
            B_whichrow,B_whichcol,B_twistval,B_failurecode);
twisttests(C_maxrowval,C_maxcolval,C_fortwistflag,C_revtwistflag,
            C_whichrow,C_whichcol,C_twistval,C_failurecode);
twisttests(D_maxrowval,D_maxcolval,D_fortwistflag,D_revtwistflag,
            D_whichrow,D_whichcol,D_twistval,D_failurecode);
twisttests(E_maxrowval,E_maxcolval,E_fortwistflag,E_revtwistflag,
            E_whichrow,E_whichcol,E_twistval,E_failurecode);
twisttests(F_maxrowval,F_maxcolval,F_fortwistflag,F_revtwistflag,
            F_whichrow,F_whichcol,F_twistval,F_failurecode);
```

(listing continues on next page)

```
{edit this for more channels}

check2ndharm(^A_Q1Q2_buff,A_whichrow,^A_rowharm,A_rowharm,
                      A_whichcol,^A_colharm,A_colharm,A_failurecode);
check2ndharm(^B_Q1Q2_buff,B_whichrow,^B_rowharm,B_rowharm,
                      B_whichcol,^B_colharm,B_colharm,B_failurecode);
check2ndharm(^C_Q1Q2_buff,C_whichrow,^C_rowharm,C_rowharm,
                      C_whichcol,^C_colharm,C_colharm,C_failurecode);
check2ndharm(^D_Q1Q2_buff,D_whichrow,^D_rowharm,D_rowharm,
                      D_whichcol,^D_colharm,D_colharm,D_failurecode);
check2ndharm(^E_Q1Q2_buff,E_whichrow,^E_rowharm,E_rowharm,
                      E_whichcol,^E_colharm,E_colharm,E_failurecode);
check2ndharm(^F_Q1Q2_buff,F_whichrow,^F_rowharm,F_rowharm,
                      F_whichcol,^F_colharm,F_colharm,F_failurecode);
{edit this for more channels}

{———————— END OF DIGIT VALIDATION TESTS ————————}

outputcode(A_whichrow,A_whichcol,A_digit_history,A_failurecode,A_dac);
outputcode(B_whichrow,B_whichcol,B_digit_history,B_failurecode,B_dac);
outputcode(C_whichrow,C_whichcol,C_digit_history,C_failurecode,C_dac);
outputcode(D_whichrow,D_whichcol,D_digit_history,D_failurecode,D_dac);
outputcode(E_whichrow,E_whichcol,E_digit_history,E_failurecode,E_dac);
outputcode(F_whichrow,F_whichcol,F_digit_history,F_failurecode,F_dac);
{edit this for more channels}

   CALL restart;
   RTI;

{—————————————————————————————————————————————————————}
{———————— S U B R O U T I N E S ——————————————————}
{—————————————————————————————————————————————————————}

{%%%%%%%%%% O N E   T I M E   O N L Y   S E T U P %%%%%%%%%%%}
{  initializes digit_history lists, M and L registers in        }
{  address generators, and sets ICNTL to edge-sensitive         }

setup:      SI=baddigitcode;
            DM(A_digit_history)=SI;          DM(A_digit_history+1)=SI;
            DM(B_digit_history)=SI;          DM(B_digit_history+1)=SI;
            DM(C_digit_history)=SI;          DM(C_digit_history+1)=SI;
            DM(D_digit_history)=SI;          DM(D_digit_history+1)=SI;
            DM(E_digit_history)=SI;          DM(E_digit_history+1)=SI;
            DM(F_digit_history)=SI;          DM(F_digit_history+1)=SI;
            {edit this for more channels}
```

```
            L0 = channels_x_32;
            L1 =  0;
            L2 =  0;
            L3 = channels;
            L4 = 16;
            L5 =  0;
            L6 =  0;
            L7 =  0;

            M0 =  1;
            M1 = -1;
            M2 = 16;
            M4 =  1;
            M5 =  8;

            ICNTL=b#01111;
            RTS;

{%%%%%%%%%%%% E V E R Y   T I M E   S E T U P %%%%%%%%%%%%%%%%}
{  resets pointers to top of buffers, resets counter values,   }
{  clears Goertzel feedback buffers to zero, etc               }

restart:    I0=^A_Q1Q2_buff;
            CNTR=channels_x_32;
            DO zloop UNTIL CE;
zloop:                  DM(I0,M0)=0;
            I2=^A_mnsqr;
            I3=^in_samples;
            I4=^coefs;
            AX0=201;    DM(count201)=AX0;
            AX0=4;      DM(count4)=AX0;
            AX0=h#00FF;
            AX1=^mu_lookup_table;
            RTS;
```

```
{%%%%%%%%%% S Q U A R E D   M A G N I T U D E   C A L C %%%%%%%%%%%%%%%}
{  calculates squared magnitude (mnsqr) from Goertzel feedback results  }

mnsqr:  MX0=DM(I0,M0);                          {get two copies of Q1, 1.15}
        MY0=MX0;
        MX1=DM(I0,M0);                          {get two copies of Q2, 1.15}
        MY1=MX1;
        AR=PM(I4,M4);                                       {get COEF, 2.14}
        MR=0;
        MF=MX0*MY1(RND); {Q1*Q2, 1.15}
        MR=MR-AR*MF(RND);                               {-Q1*Q2*COEF, 2.14}
        SR=ASHIFT MR1 BY 1 (HI);        {2.14 -> 1.15 format conv., 1.15}
        MR=0;
        MR1=SR1;
        MR=MR+MX0*MY0(SS);                       {Q1*Q1 + -Q1*Q2*COEF, 1.15}
        MR=MR+MX1*MY1(RND);            {Q1*Q1 + Q2*Q2 + -Q1*Q2*COEF, 1.15}
        DM(I2,M0)=MR1;                         {store in mnsqr buffer, 1.15}
        RTS;

.ENDMOD;
```

Listing 14.2 DTMF Decoder Program

Sonar Beamforming 15

15.1 OVERVIEW

This chapter describes a real-time digital beamforming system for passive sonar. The design of this system is based on several ADSP-2100s that independently perform the beamforming calculations under the supervision of an ADSP-2100 master processor. The modular architecture allows you to tailor the size of the system to your performance needs. Code listings for the master and slave processors are included.

A sonar system can use two different methods to analyze and evaluate possible targets in the water. The first is called *active* sonar. This method involves the transmission of a well defined acoustic signal which can reflect from objects in water. This provides the sonar receiver with a basis for detecting and locating the targets of interest. The limitations of this method are mainly due to the loss of the signal strength during propagation through the water and reverberation caused by the signal reflections. Simplistically, active sonar can be thought of as the underwater equivalent of radar.

The second method is called *passive* sonar. This one bases its detection and localization on sounds which are emitted from the target itself (machine noise, flow noise, transmissions of its active sonar). Its limitations are due to the imprecise knowledge of the characteristics of the target sources and to the dispersion of the target signals by the water and objects in the water. A generic passive sonar system is shown in Figure 15.1, which can be found on the next page.

Sonar systems have a wide variety of military and commercial uses. Some of the military applications include detection, localization, classification, tracking, parameter estimation, weapons guidance, countermeasures and communications. Some of the commercial applications include fish location, bottom mapping, navigation aids, seismic prospecting and acoustic oceanography. More detailed information about sonar technology can be found in Winder, 1975 and Baggeroer, 1978.

15 Sonar Beamforming

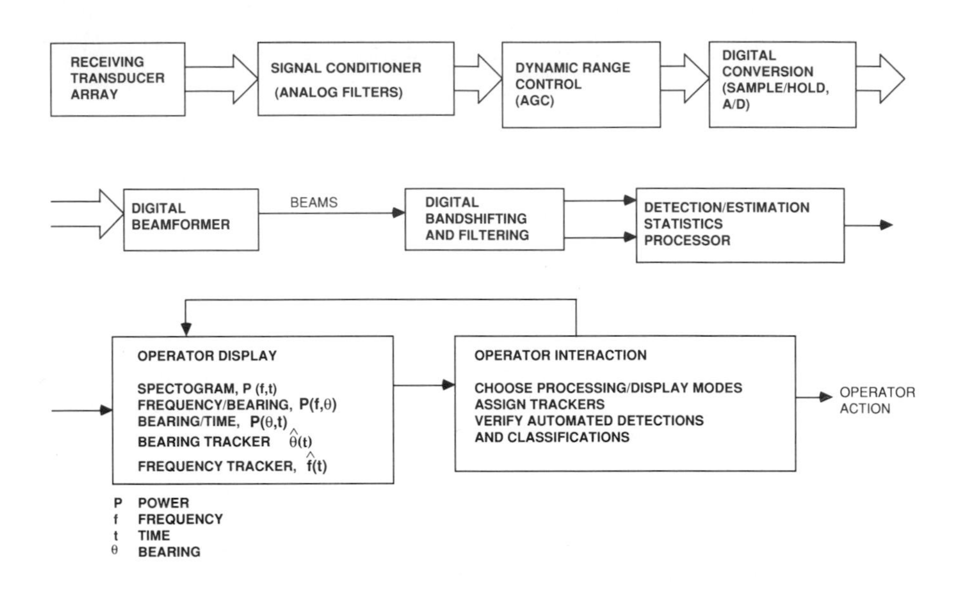

Figure 15.1 Generic Passive Sonar System

15.2 SONAR BEAMFORMING

In its simplest form, sonar beamforming can be defined as "the process of combining the outputs from a number of omnidirectional transducer elements, arranged in an array of arbitrary geometry, so as to enhance signals from some defined spatial location while suppressing those from other sources" (Curtis and Ward, 1980). Thus, a beamformer may be considered to be a spatial filter. It is generally assumed that the waves arriving at the transducers all propagate with the same speed *c*, so that the signals of interest lie on the surface of the cone defined by $\omega = c\,|\,k\,|$ in (k, ω) space. Ideally the passband of the beamformer lies on the intersection of this cone with the plane containing the desired direction vector.

The beamforming operation is accomplished through a series of operations that involve the weighting, delay and summation of the signals received by the spatial elements. The summed output that contains information about a particular direction is called a beam. This output is then sent to a signal processor and/or a display for frequency and temporal discrimination. A beamforming system can employ analog or digital components and techniques; this chapter focuses on a digital beamforming technique.

Sonar Beamforming 15

Beamformers are used both in passive and active sonar systems. In passive sonar, the beamformer acts on the received waveforms. Active sonar also utilizes a conventional beamformer which acts on the waveforms that are reflected from the targets (most active sonars use the same array for receiving and transmitting). There are several well known techniques that can be utilized in forming beams from receiver arrays. The discussion in this chapter focuses on weighted delay-and-sum beamforming technique (also referred as time-delay beamforming) which is very commonly used. Discussion on other techniques, such as FFT beamforming or phase shift beamforming, may be found in Baggeroer, 1978 and Knight, et al., 1981.

15.2.1 Time-Delay Beamforming

In time-delay beamforming, beams are formed by averaging weighted and delayed versions of the receiver signals. Each receiver has a known location and samples the incoming signals spatially. To steer the beams (i.e. to choose beamforming directions), each receiver's output has to be delayed appropriately relative to the other receivers. The time delays compensate for the differential travel time between sensors for a signal from the desired beam direction.

In order to describe this operation mathematically, let us assume that the array of receivers is composed of a three dimensional distribution of equally weighted omnidirectional sensors. Their spatial locations are specified in the Cartesian coordinate system of Figure 15.2. The

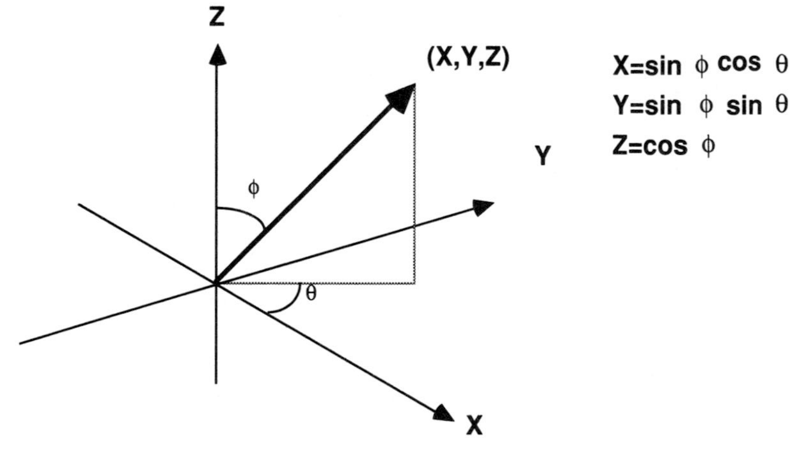

Figure 15.2 Cartesian Coordinate System

15 Sonar Beamforming

beamforming task consists of generating the waveform $b_m(t)$ (or its corresponding sample sequence) for each desired steered beam direction $\mathbf{B}_m$. Each $b_m(t)$ consists of the sum of suitably time delayed replicas of the individual sensor outputs $e_n(t)$.

Let the output of an element located at the origin of coordinates be s(t). Under the assumption of plane wave propagation, a source from direction $\mathbf{S}_l$ (the lth source direction unit vector) produces the following sensor outputs

$$(1) \qquad e_n(t) = s(t + ((\mathbf{E}_n \bullet \mathbf{S}_l)/v))$$

where $\mathbf{E}_n$ is the nth element position vector and v is the speed of propagation for acoustic waves in the ocean ($v \approx 1500$ m/s). Delaying each individual sensor output by an appropriate amount to point a beam in the direction $\mathbf{B}_m$ yields the beamformer output

$$(2) \qquad b_m(t) = \sum_{n=1}^{N} e_n(t - ((\mathbf{E}_n \bullet \mathbf{B}_m)/v))$$

The operation defined in equation (2) is known as *beamforming*. The beamforming operation is computationally demanding because this summation must be calculated in real time for a large number of sensors and a large number of beams.

For simplicity, the rest of this discussion is limited to one-dimensional (line) arrays with regularly spaced hydrophones (underwater omnidirectional acoustic sensors). This discussion can be generalized to multi-dimensional arrays and line arrays with variable spacing.

15.2.2 Digital Beamforming

Assume that the presence of a plane wave signal $s(t - (\mathbf{E} \bullet \mathbf{B})/v)$ needs to be detected. It is propagating with a known direction **B**, and is measured at **E** in a background of spatially white noise (**B** and **E** are vectors). The line array of Figure 15.3 is used. The signal has the same value at each wavefront and the noise is uncorrelated from sensor to sensor. Thus, in order to enhance the signal from the noise, the sensor outputs are delayed and summed. The delays account for the propagation delay of the wavefront to each sensor. This yields

$$(3) \qquad b(n\Delta) = 1/N \sum_{i=0}^{N-1} X_i(n\Delta + ((\mathbf{E}_i \bullet \mathbf{B})/v))$$

where $X_i(n\Delta)$ is the sampled output of the *ith* sensor. Note that for the *ith* sensor the following relationship holds:

$$(4) \qquad (\mathbf{E}_i \bullet \mathbf{B})/v = -i\,(d/v) \sin \theta$$

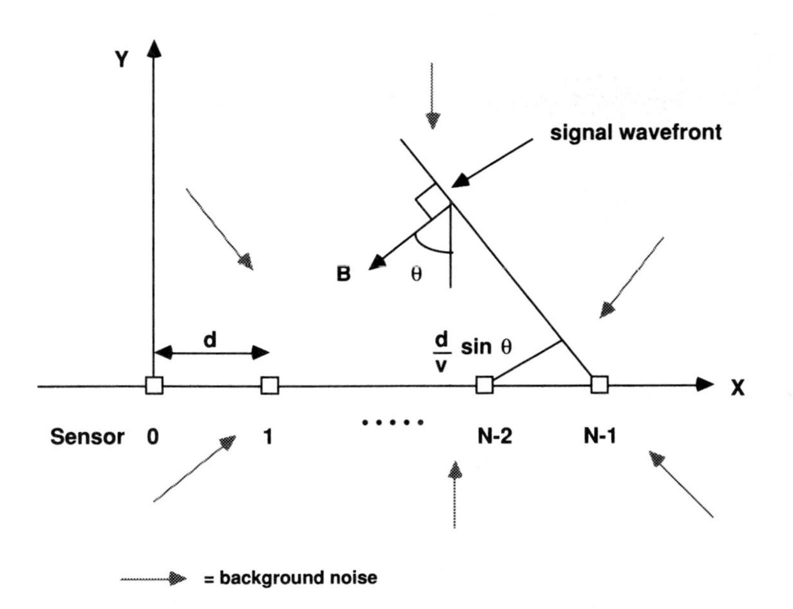

Figure 15.3 Line Array of Equally Spaced Hydrophones

The input to the beamformer is a set of time series. The input is usually one set of time series for each sensor, while the output of the beamformer is another set of time series, generically referred to as beams.

The beamformer is spatially discriminating because for a plane wave with a propagation direction θ, different than θ_0 assumed by the beamformer, the sensor outputs are not coherently combined. This leads to partial cancellation of the incoming signals with $\theta \neq \theta_0$. Thus, for a plane wave signal

$$(5) \qquad X_i(n\Delta) = e^{j(\omega n\Delta + (\omega i d / v \sin\theta))}$$

and

$$(6) \qquad b(n\Delta) = 1/N \sum_{i=0}^{N-1} e^{j\omega i d(\sin\theta - \sin\theta_0)/v} e^{j\omega n\Delta}$$

where d is the spacing between the sensors. Thus the amplitude of the plane wave arriving from a direction θ at the output of a beamformer steered to θ_0 has an attenuation given by

$$(7) \qquad |B(\omega, \theta)| = \frac{\sin[N(\omega/2)\ d(\sin\theta - \sin\theta_0)/v]}{N\sin[(\omega/2)\ d(\sin\theta - \sin\theta_0)/v]}$$

This function is known as the *beam pattern* of the array. An example beam pattern for a line array is shown in Figure 15.4. For a given wave frequency and steering direction θ_0, all plane waves with $\theta \neq \theta_0$ are attenuated, leading to the interpretation of a beamformer as a spatial filter. In most applications, it is desirable to have a beam pattern with a very narrow main lobe and very low level sidelobes for maximum noise rejection.

Increasing ω (the operating frequency) and/or N results in narrower main lobes even though the side lobe level does not change. Increasing the sensor spacing also results in narrower main lobes. But this is limited by the fact that spatial aliasing will occur for $\Delta x > \lambda_{min}/2$, where λ_{min} is the signal wavelength for the highest frequency of interest (Knight, et al., 1981). Spatial aliasing exhibits itself in terms of extra main lobes near the endfire region. In order to reduce the sidelobe levels, the sensor outputs must be weighted. This procedure is known as *shading*. Thus, in equation (6), we replace the 1/N factor by w_k. The corresponding beam pattern is

$$(8) \qquad |B(\omega, \theta)| = \sum_{i=0}^{N-1} w_i\ e^{j\omega i d(\sin\theta - \sin\theta_0)/v}$$

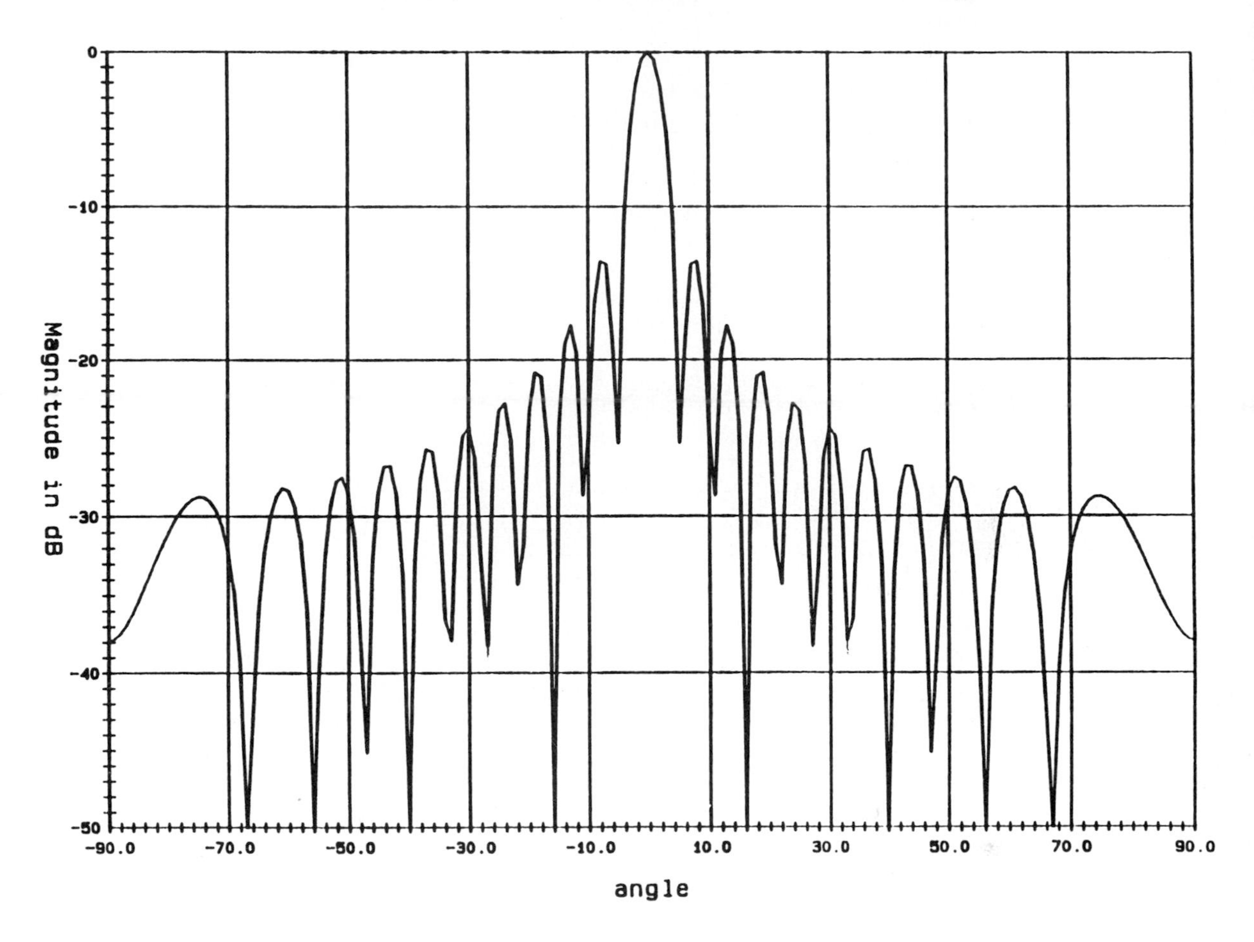

Beam pattern at 0 degrees (perpendicular to line of array) with a sampling rate of 500Hz, for a line of 32 sensors spaced 1m apart

Figure 15.4 Example Beam Pattern

The usual windowing techniques of Fourier transform theory can be used to reduce the sidelobes (Knight, et al., 1981). By employing a Hamming window, for example, the sidelobes may be reduced to –40db at the expense of widening the main lobe.

Another problem that has to be dealt with in a line array with fixed shading is the quantization errors that are introduced from the insertion of the delays. In a digital beamformer, for ideal operation, the beamforming directions are limited such that the delays t are multiples of the sampling interval. Any other choice of directions (and thus delays) introduces errors onto the beam pattern (Gray, 1985). One method that is used to reduce such errors involves the interpolation of the incoming samples. Further discussion on this topic can be found in Pridham and Mucci, 1978.

An overall block diagram for a conventional time delay digital beamformer is shown in Figure 15.5.

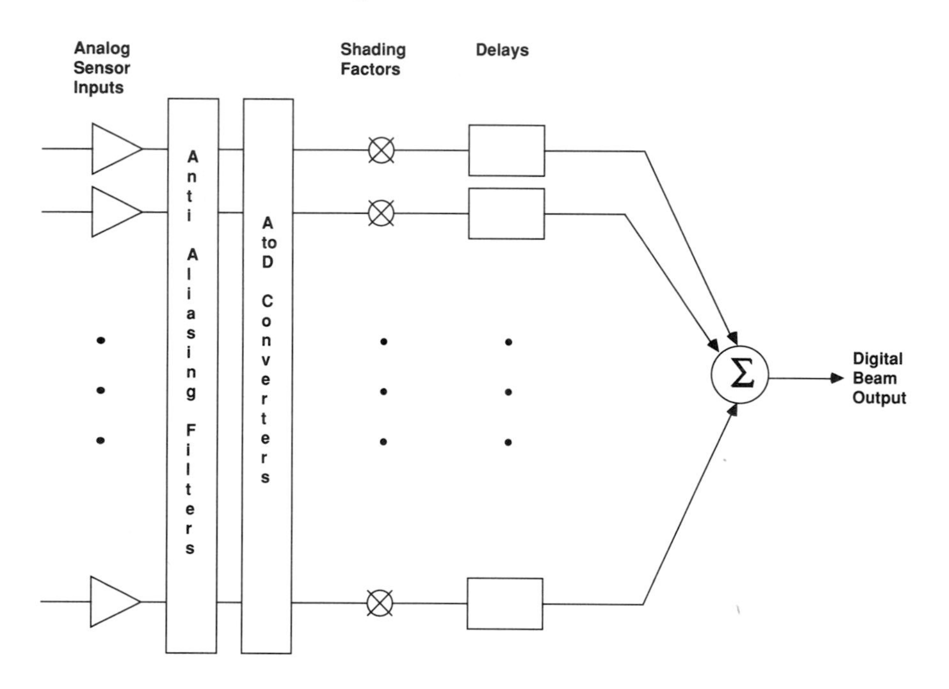

Figure 15.5 Conventional Time-Delay Beamformer

15.3 DIGITAL BEAMFORMER IMPLEMENTATION

The beamforming task can be performed using analog or digital systems. Implementing analog tapped delay lines and forming multiple beams in real time using analog hardware results in big, inflexible and cumbersome systems. A digital beamformer results in a smaller, more accurate, and much more flexible hardware/software unit than an analog technique. In this section, some of the important issues in digital time-delay beamforming systems are discussed.

15.3.1 Computational Power

The computational capacity required by a real-time beamformer can be computed from equation (3) as roughly Nf_s multiplications and additions per second per beam, where N is the number of sensors and f_s is the input sampling frequency. This requirement does not seem by itself very demanding. However, multiple simultaneous beams are usually needed in order to span the space around the sensors and consequently a more realistic requirement is N_bNf_s multiplications and additions per second, where N_b is the number of beams formed. A passive sonar system utilizing ≈30 sensors and requiring ≈30 or more beams at sampling rates of 10kHz or higher needs to perform at least 12 million multiplications and additions per second. Computation demands in the near future will increase rapidly because new generations of quieter sources (e.g., submarines) will require passive sonar to use more sensors (for higher resolution) and more beams.

15.3.2 Memory Usage

Another concern in designing real-time beamformers is the amount of storage that is needed in order to implement the digital delays. For example, in the case of a line array with sensor spacings *d*, the storage necessary to form all the synchronous beams is on the order of N^2f_sd/v. The *synchronous* beams are all the beams that can be formed using delays which are multiples of the input sampling period. In the typical beamforming system that we considered earlier, the size of storage required is on the order of 10000 memory locations. The memory word width chosen in a particular application could be 16 bits or more, which would require at least 20 Kbytes of RAM per beam (unless RAM is shared). The demand for fast accessible storage will be rising in the coming years along with the demand for computation power.

15.3.3 Other Issues

Further discussion on beamforming system issues can be found in Knight, et al., 1981, Janssen, 1987, and Hodgkiss and Anderson, 1981.

15.4 EXAMPLE BEAMFORMER

The example beamformer is able to take inputs from an arbitrary array of up to 32 sensors, form multiple beams in real time, and is user-configurable from an IBM PC personal computer (or compatible). The acoustic frequencies of interest in this example range from 0Hz to 2000Hz. The input sampling rate is 10kHz, which is higher than the Nyquist rate because of the need for a higher resolution in the beamforming directions. The acoustic data is collected by hydrophones and the analog data is digitized to 12 bits. The gain applied to the analog inputs is manually selectable (1, 10, 100 or 1000). The output beams are sent to another system over a specific parallel interface for further processing. The output beams are also available at analog output points, one at a time, for testing, monitoring and some other signal processing tasks.

15.4.1 System Architecture

Several important issues were considered in selecting an architecture for this beamforming system. One consideration is the need for modularity. A modular system gives a user the ability to start small with the option of expanding the system's capabilities later.

Another consideration is the demand for speed. As discussed earlier, in order to form multiple simultaneous beams, a large number of summations have to be computed in real time. This requirement, along with modularity, led to a distributed processing architecture.

One more consideration is flexibility. A user is able to specify any array configuration and form any beam. Ease of use requires a friendly and interactive user interface, through which the users can specify many different system parameters.

15.4.2 Building Blocks

The beamformer consists of the building blocks that are shown in Figure 15.6. There are several parallel buses in the system for data transfers and communication. A wide common bus is used for the subsystems to communicate and exchange data with each other. Another bus facilitates the communications with the IBM PC. This bus is used to download the user system configuration data from the PC into the beamformer. Finally, a bus is used to send the output beams to a sonar signal processing system which performs further processing on the data.

There are three main types of subsystems in this beamformer; each of these subsystems is implemented as a separate board in the example system:

- The *master* module, which is responsible for controlling the data exchange among all internal modules and the data flow over the I/O buses.

- The *slave* module, which is responsible for the actual beamforming task. Each slave can beamform in multiple directions and more slaves may be added in order to form a larger number of beams.

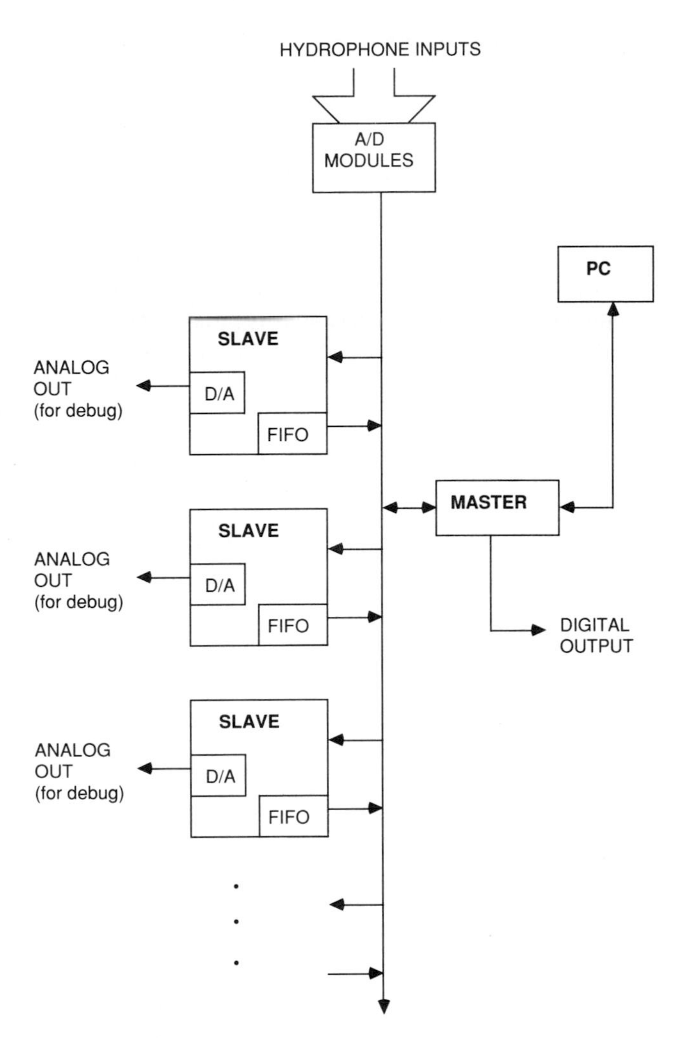

Figure 15.6 Example Beamformer Block Diagram

- The *analog-to-digital conversion* (A/D) module. This module takes the hydrophone outputs as inputs, is responsible for sampling the incoming analog signals and converting them into a digital format. Each A/D module can handle a limited number of hydrophones, but more modules may be attached to the common bus in order to handle a larger number of inputs.

15.4.3 System Operation

The internal operation of the system is controlled by the master. The handling of the incoming samples and the I/O exchanges are also under the master's control. The operation of the system is synchronized to the input sampling clock which has a period of 100 μs (10kHz). This implies that the calculated beam samples have to be sent out every 100μs. Let's call this duration a *system cycle.*

Initially, the master has to accomplish a one-time task of handling of the system configuration data sent from the PC. Thereafter, the master has to go through its duties within one system cycle and be ready to handle the next set of incoming samples. The system events that make up the system configuration (illustrated in Figure 15.7) are as follows:

1. The operator enters system configuration variables (number of sensors, beam directions, shading factors, etc.) through an interactive program on the PC. The same program does some calculations and downloads the data to the master ADSP-2100. Communications between the master and PC are accomplished using a simple protocol over a parallel interface card located in the PC.

2. The master keeps the configuration variables in several of its internal registers. It sends this information to all the slaves so that each of them can identify the beams that they are responsible for.

3. The master waits for a signal from each one of the slaves confirming that they are ready to beamform.

The sampling clock is running during the configuration, but the interrupts initiated by A/D conversions are not recognized until all the slaves are ready. Once each of the slaves has sent the signal that it is ready to beamform, the master starts the cyclic operation of the beamformer.

1. The master responds to the A/D conversion interrupt by reading the results of the A/D conversions, which correspond to a simultaneous snapshot of the incoming waveforms at the hydrophone locations. It reads all the results in sequence and writes them into the memories of all the slaves. Thus, all slaves receive identical copies of the incoming waveform samples.

2. The master initiates an interrupt which orders all the slaves to start beamforming.

MASTER	SLAVE
Power up (or RESET)	Power up (or RESET)
Wait for the user to input the system setup parameters	
PC communications	Idle
Download setup data to slaves	Accept setup data
Idle	Self-prepare using the setup data
Sample all the A/Ds 255 times and send samples to slaves	Receive the first 255 sets of conversion results (fill the sample buffer)
Idle	Beamform
Sample A/Ds, send samples	Receive samples
Read FIFOs, output beams	Beamform
Sample A/Ds, send samples	Receive samples
Read FIFOs, output beams	Beamform
etc.	etc.

Figure 15.7 Sequence of Events

3. Once a slave finalizes a summation (i.e. forms a beam sample), it shifts the result into a FIFO memory for collection by the master. Each slave computes and stores its own beam samples independent of other slaves. Once a slave has completed its assigned set of beams, it waits for the next interrupt (initiated by the master) that orders all the slaves to beamform again.

4. While the slaves are busy forming and storing the current set of beam samples, the master reads the sets of beam samples that were formed during the previous system cycle. After finishing this output duty, the master waits for the next A/D interrupt.

Each slave keeps its input samples in a circular buffer (255 slots) which is located in the slave's data memory space. New incoming samples are put into consecutive locations in this buffer. Once the end of the buffer is reached, the oldest snapshot of samples gets overwritten by the newest samples and this cyclic process goes on.

Each slave performs the beamforming task by reading the appropriate locations in its sample buffer, by multiplying those values with a shading factor and by keeping a running sum of these weighted samples until the summation is finished.

Each slave writes out beam samples to its own dedicated a first-in-first-out memory (FIFO). Only the master can shift the beam samples out of the FIFOs. The master reads and sends each beam sample, one at a time, over its output bus.

The user must realize that the first 255 sets of beam samples produced are invalid. This is due to the fact that the sample buffer is not full until the end of the 255th system cycle. Therefore, during that period, the locations read by the slaves contain meaningless data. Valid system outputs are produced ≈25.5ms after the system starts beamforming.

15.4.4 Timing Issues

There are a number of important operational timing issues due to the length of the system cycles. The number of different beam samples that can be formed by each slave is limited by the system cycle length. This constraint exists because the slaves have to release the control of their individual memory buses in order to allow write operations by the master. Another constraint is that the master needs to read all the incoming samples and also send all the beam samples out within a system cycle. The maximum number of beams that can be formed in this system are directly limited by these timing constraints. The beam allocations per slave must be calculated carefully by the PC during the system configuration phase. Otherwise, incomplete beams and invalid outputs may result because of the master not having enough time to send out all the beam samples or other complications.

15.4.5 Digital Output

The example system provides the ability to send out all the computed beams through a 16-bit parallel output port. This parallel port is located on the master module. It can be used to communicate the beam data to an external signal processor for further processing. The parallel port is comprised of octal latches, D-flops and address decoding circuitry which allow the master to communicate with the outside world using a simple protocol.

15.4.6 Analog Output

The example system provides the ability to observe some beams through analog outputs. A digital-to-analog (D/A) converter is included on each slave board and the desired analog beam output can be selected using a thumbwheel switch. Each slave sends the desired beam sample to its D/A converter before shifting it into the FIFO. The overhead of this analog port is minimal, and the port is a very useful test point for system debugging.

15.4.7 System Configurations

The minimum system configuration consists of a master, a slave and an A/D module. The maximum possible system configuration is limited by the speed of the internal hardware and the maximum data rate capability of the output port. All buses must be present in any system configuration.

15.5 SYSTEM HARDWARE

The system hardware includes ADSP-2100 DSP processors, high speed hybrid A/D converters and very high speed CMOS and bipolar LSI components. Hardware selection, design, operation and interface issues in the system are discussed in the following sections.

15.5.1 Component Selection

The ADSP-2100 fulfills the high computational requirements of the slave modules. It also fulfills the CPU requirements of the master module by enabling high speed input data transfers between the A/D modules and the slaves as well as the output transfers. It provides easy handling of memory mapped peripherals and can handle four external interrupts. Design time is saved by using the same processor for both master and slave.

The 12-bit fast A/D converters, high precision sample and hold circuits, low noise operational amplifiers used in the input gain section and anti-aliasing filters are also critical for a high performance system. Front-end analog signal conditioning and A/D circuit design and production using discrete components is a difficult task in noisy digital environments such as the one assumed in the example system. A hybrid A/D converter with on-board voltage references along with sample and hold circuits can perform the required duties better than any discrete circuit with similar functionalities.

Analog Devices' AD1332 A/D converter is appropriate and convenient for several reasons:

- The AD1332 integrates several of the necessary components inside. Its central element is a 12-bit 5µs AD7672 A/D converter. A voltage reference for the converter is included. The converter is preceded by an on-board AD585 sample and hold circuit which itself is preceded by an optional 4-pole Butterworth low-pass filter (anti-aliasing filter). The converter output is fed into a 12-bit latch with tristate output buffers (an optional integrated FIFO is also available for the temporary storage of the conversion results).

- The AD1332 is easily addressed from a microprocessor, which makes it ideal for this application.

- Because the front end circuitry must be duplicated for all incoming sensor inputs, the use of a hybrid helps reduce design, prototyping and production times. Multiplexing the sample and hold outputs into fewer converters is not desirable because of system performance considerations.

The selection of the rest of the high speed VLSI and LSI components in the system is not as crucial to system performance. Several levels of address decoding and buffering that are present in the system result in high demands on the memory components. Integrated Device Technologies' (IDT) 2Kx8 CMOS static RAMs with 25ns access times are used as the data memory components on the slaves. The program memories for the master and the slaves also need to be very fast. Cypress Semiconductor's 2Kx8 CMOS EPROMs with 35ns access times are used as the program memory components for all ADSP-2100s.

The slave FIFOs are IDT's 72413L35 64x5 CMOS FIFOs. Analog Devices' AD569 16-bit D/A converters provide analog output ports on the slaves. The rest of the LSI components are off-the-shelf Advanced Schottky (Fairchild's FAST and Texas Instruments' 74AS series), Advanced Low Power Schottky (Texas Instruments' 74ALS series) or very high speed CMOS (IDT's 74FCT series) integrated circuits. More detailed information and specifications for these components are available from their manufacturers.

The interface card between the PC and the master should be a parallel I/O card that can easily be plugged into the PC's backplane and addressed from a high level program. There are a large number of such I/O boards available. A short development time discourages spending effort on a complicated protocol, and thus a simple protocol and a very flexible I/O card is preferable. Analog Devices' RTI-817 parallel I/O board contains three 8-bit bidirectional ports that accept user-configured directions. These ports are memory mapped and addressable from high level programs. More detailed information about the card can be found in the *RTI-817 User's Manual* published by Analog Devices.

15.5.2 Master Board Hardware

A high level block diagram for the master board is shown in Figure 6.8. The circuitry on the master board is centered around an ADSP-2100 processor running at 8MHz. This master CPU takes its instructions from three CY7291-35 2Kx8 EPROMs (program memory) mapped to the CPU's program memory address space (see Figure 15.9 on the next page). The master board does not contain any data memory. The ADSP-2100's internal registers are sufficient for most operations except during the configuration phase. The details of the configuration operation are discussed later in the firmware section.

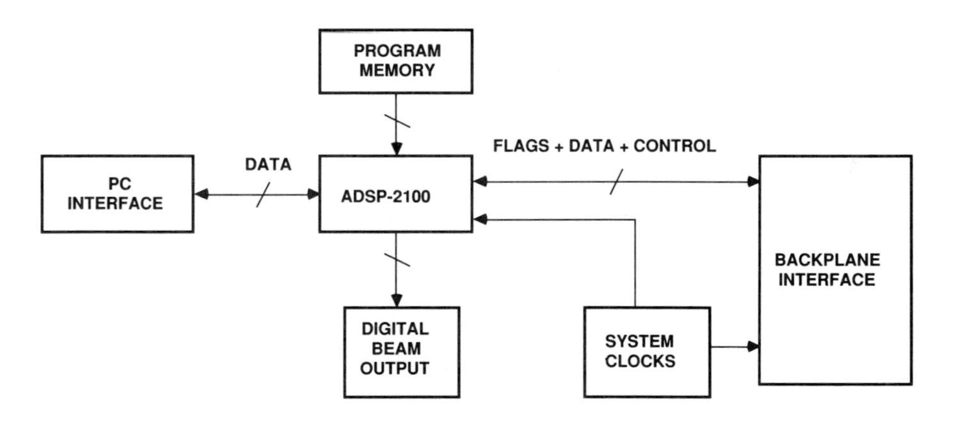

Figure 15.8 Master Module Block Diagram

The master recognizes two interrupts. The first interrupt occurs every 100µs and notifies the master about the availability of new A/D conversion results. This interrupt comes directly from the sampling clock. The second one notifies the master that the external signal processor has received a beam sample and is ready to receive the next one.

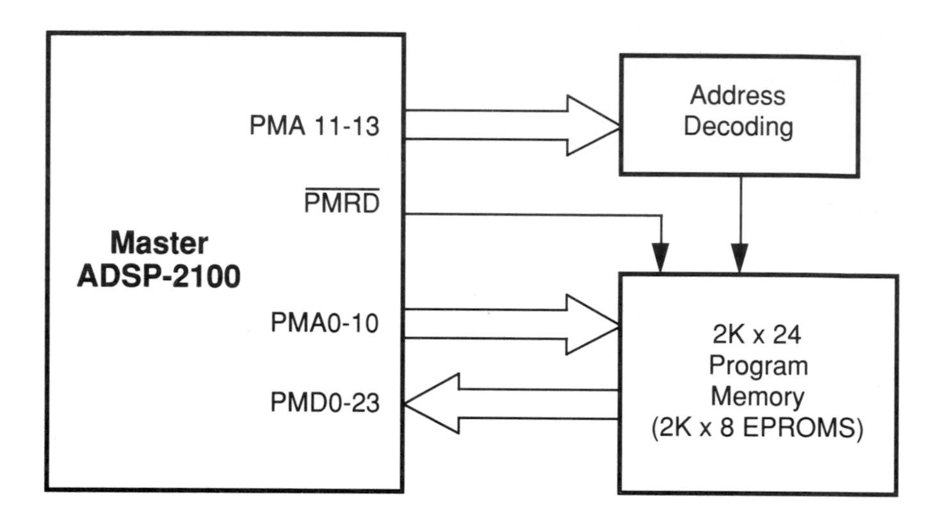

Figure 15.9 Master Module Program Memory Interface

A set of devices generates the sampling clock that is used to sample the analog sensor inputs and to initiate the A/D conversions. These clock signals are sent over the backplane to the A/D boards.

The master board has a number of decoders, latches and buffers that it uses to write to slave data memories, to read slave FIFOs, to send control information to the slaves, to access the PC communication ports and to receive status flags over the backplane bus. See Figure 15.10.

The master board contains a large number of devices dedicated for the CPU's external bus interfaces. These are shown in Figure 15.11. A bank of bus drivers and transceivers provide the necessary buffering for the signals that are traveling over the backplane bus. The communications with the PC are handled through three octal latches that provide a direct interface to the I/O board that is plugged into the PC's backplane. This board also has two inverting octal latches which facilitate the beam sample transfers over the digital output bus.

The interface to the three buses requires four separate connectors: a 96-pin Eurocard connector which is the connection to the backplane, a 50-pin flat cable connector which connects the master and the PC, a 28-pin connector and a 50-pin flat cable connector which are used on the output bus

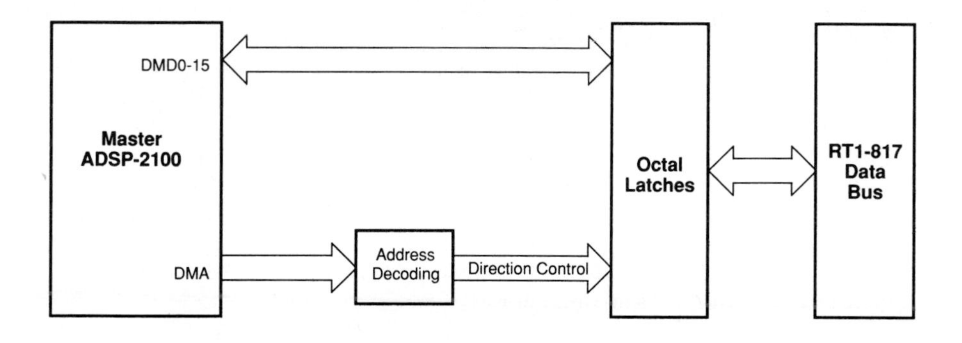

Figure 15.10 Master Module PC Interface

connections between the master and the external signal processor. The master only requires +5V power.

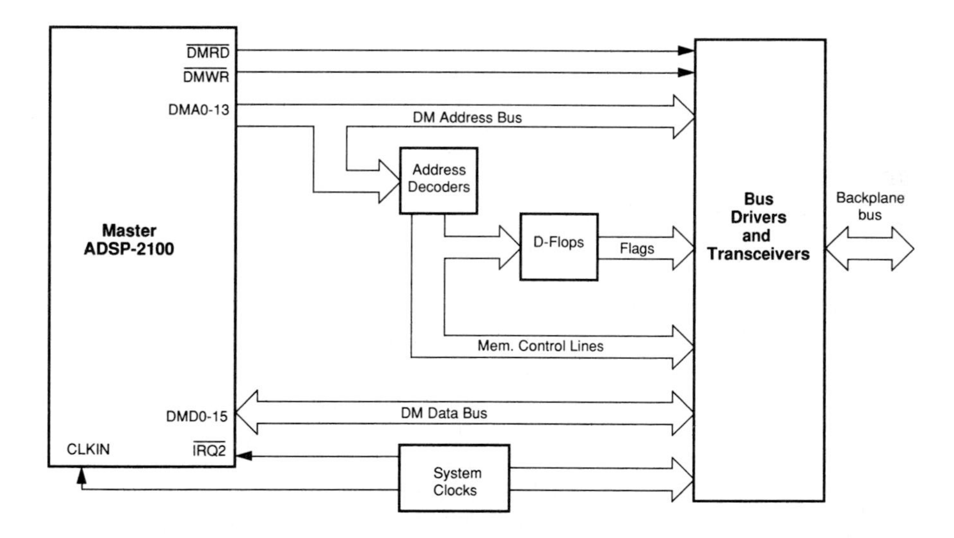

Figure 15.11 Master Module Backplane Interface

15.5.3 Slave Board Hardware

A high level block diagram for the slave board is shown in Figure 15.12. The circuitry on the slave board is centered around an ADSP-2100 processor running at 8MHz. The slave CPU takes its instructions from three CY7291-35 2Kx8 EPROMs (program memory) which are mapped into the program memory address space of the slave CPU (see Figure 15.13). Two 2Kx8, 35ns static RAMs are also mapped into the program memory address space and available for data storage using the upper 16 bits of the ADSP-2100's PMD bus. A decoder provides the necessary address decoding for the PMA lines.

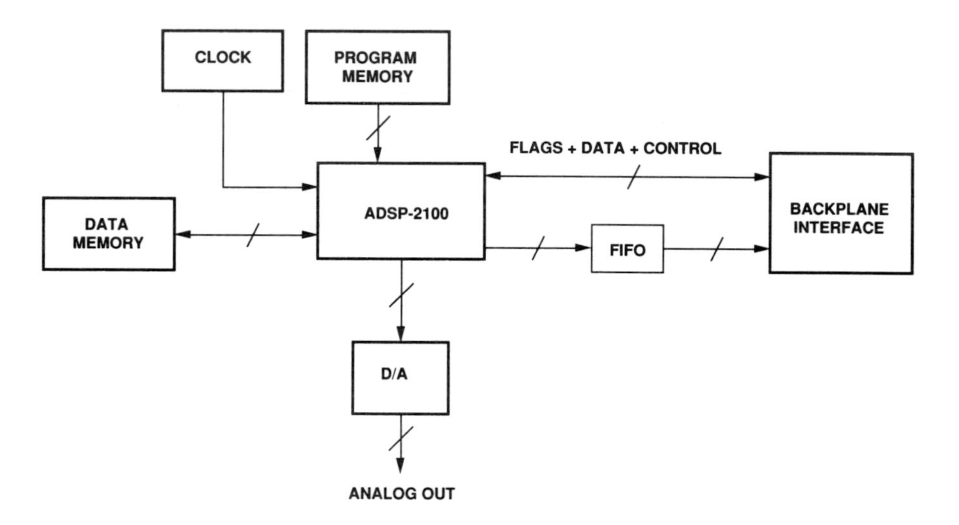

Figure 15.12 Slave Module Block Diagram

There are ten IDT6116LA-25 2Kx8 static RAM chips on the slave board. These devices are mapped onto the data memory address space of the slave CPU (see Figure 15.14). The first 8K locations of this space are dedicated to the circular sample buffer. The remaining 2K locations are used for additional data storage. A decoder provides the necessary address decoding for the static data RAMs as well as the data memory address mapping for some additional devices and flags.

The output FIFO for the slave module consists of four IDT72413L35 64x5 FIFO components. These can be loaded (written to) from the slave DMD bus. The contents of the FIFO can be read from the backplane data bus

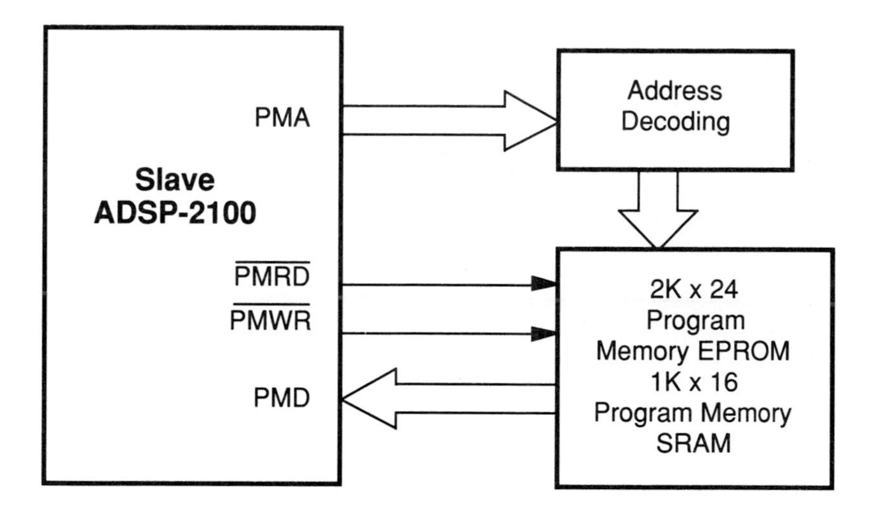

Figure 15.13 Slave Module Program Memory Interface

through buffers (see Figure 15.15). Each FIFO has a unique location in the master's data memory address space; this location is determined by DIP switch settings on the slave board.

The slave CPU only recognizes one external interrupt, the one generated by the master in order to start the beamforming operation after a new sample buffer update. The slave CPU clears this interrupt immediately after it finishes forming its assigned beams.

The slave board, like the master, contains a large number of bus drivers and transceivers for easy interface with the backplane bus. Several lines on the backplane bring control information from the master. Some of these controls cause the slave CPU to halt its operation and surrender the control of its buses to the master. Some status flags are also sent to the master over the backplane.

The AD569 16-bit D/A converter is mapped into the slave data memory address space (see Figure 15.16). An AD588 ±5V voltage reference is used with this D/A converter. Because the D/A converter has a slow access time, the slave write cycle must be extended using the DMACK signal (this is an input to the slave CPU). A small circuit is used to generate DMACK during a write cycle to the D/A converter.

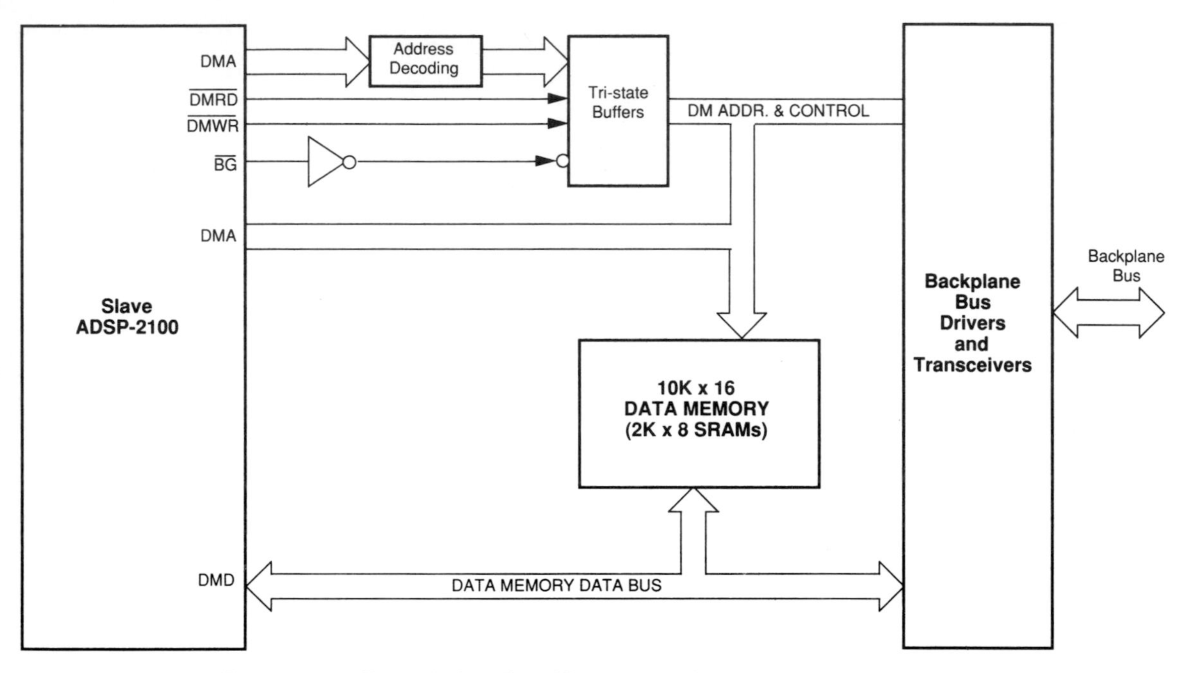

Figure 15.14 Slave Module Data Memory Interface

The slave board has a 16-position thumbwheel switch which selects the beam that is sent out through the D/A converter. A set of four DIP switches give each slave board its own identity. Setting these switches before power-up allows the slave CPU to read them later to determine the beams that are under its responsibility.

There are two connectors on the slave board: a 96-pin Eurocard connector and a male BNC connector. The first is used to interface to the backplane bus, and the second is connected to the analog output of the D/A converter. You can use the BNC connector to send the switch-selected output beam to another device. The slave board requires +5V digital and ±12V analog power supplies.

15.5.4 A/D Board Hardware

A high level block diagram for the A/D board is shown in Figure 15.17. Each A/D board can receive up to four analog inputs ranging between ±5V and can convert them into 16-bit signed fixed-point (1.15 format)

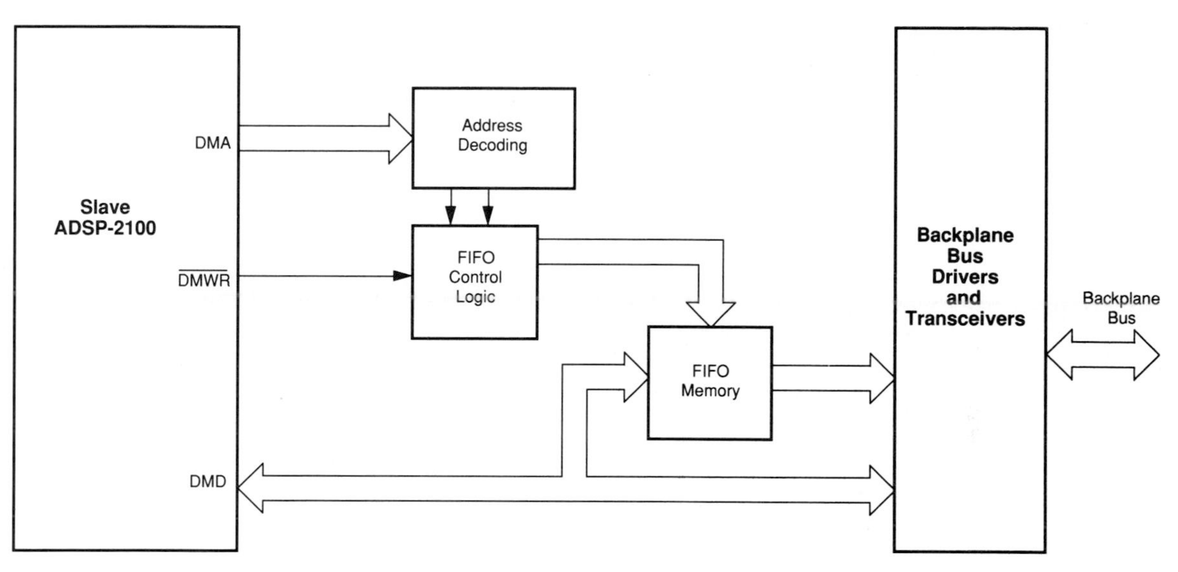

Figure 15.15 Slave Module FIFO Interface

numbers with 12-bit accuracy. The circuitry on the board is designed around four AD1332 12-bit hybrid A/D converters.

For each input, a gain stage which uses ADOP07 operational amplifiers is included on the A/D board to provide the necessary amplification of the

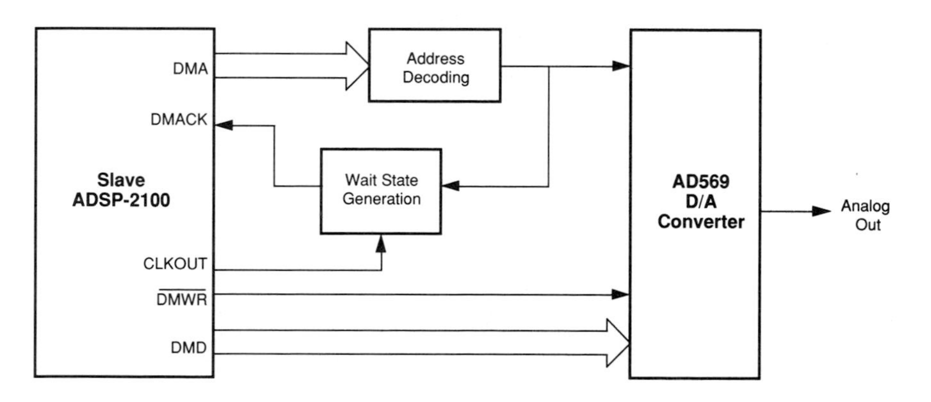

Figure 15.16 Slave Module D/A Converter Interface

hydrophone outputs. The gain is selected (from 1, 10, 100 and 1000) using an external 4-position rotary switch. AD7590 analog CMOS switches select the resistor combination for the amplifier's feedback loop. The CMOS switches are used to allow the future possibility of using CPU-generated signals to make gain selections (i.e., to implement automatic gain control).

The board contains two octal bus drivers to provide adequate buffering for the AD1332 outputs. The outputs of these buffers are tied to the backplane DMD bus (see Figure 15.18). The A/D output, which is in offset binary format, has to have its most significant bit inverted because of the fixed-point format that is used in the system.

A decoder provides a unique location for each AD1332 in the master's data memory address space. The configuration of this decoder must be different in each A/D board for each A/D converter to have a unique location. The A/D board requires +5V digital and ±12V analog power supplies. There are five connectors on the A/D board: a 96-pin male Eurocard connector that is used to interface to the backplane bus and four male BNC connectors that accept the hydrophone outputs.

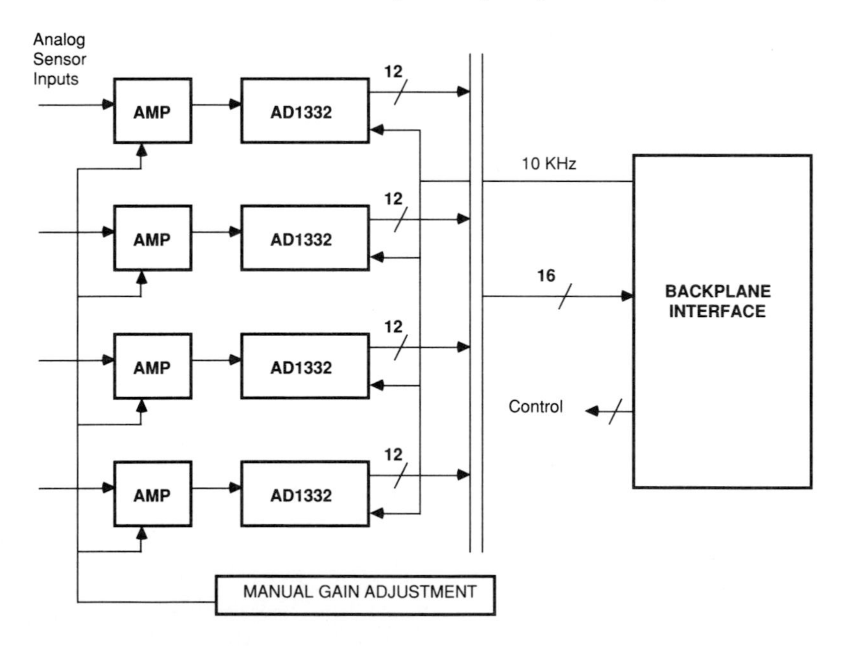

Figure 15.17 A/D Module Block Diagram

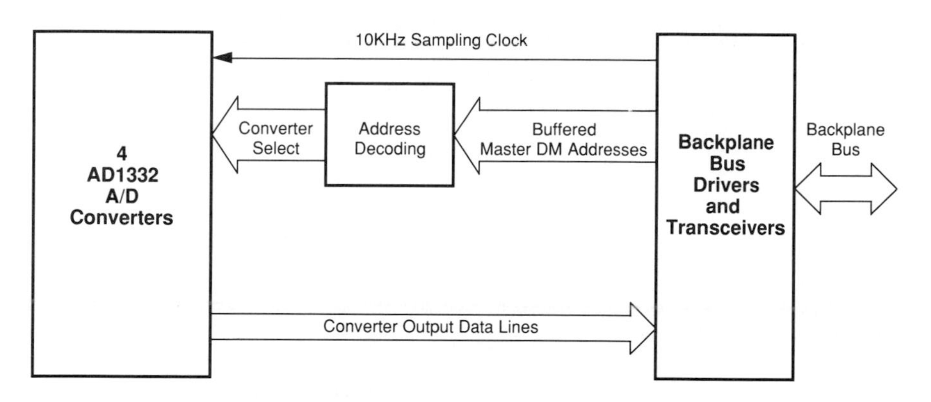

Figure 15.18 A/D Module Backplane Interface

15.6 SYSTEM FIRMWARE

The ADSP-2100 assembly code that is responsible for the master and slave CPUs' operation is discussed in this section.

15.6.1 Master Firmware

The firmware that runs in the master CPU is relatively short. It is assembled using the system specification source file shown in Listing 15.1.

```
.SYSTEM                             master_system;

.SEG/ROM/ABS=0/PM/CODE              rom_program_storage[2048];

.SEG/RAM/ABS=0/DM/DATA              sample_mem[8160];
.SEG/RAM/ABS=8160/DM/DATA           system_info[32];
.SEG/RAM/ABS=8192/DM/DATA           shading_coeff_mem[32];
.SEG/RAM/ABS=8224/DM/DATA           scratch_mem[2016];
.SEG/RAM/ABS=10240/DM/DATA          ad_converters[32];
.SEG/RAM/ABS=14337/DM/DATA          fifos[7];

{The ports declared below are used to set and clear various flags
as well as to communicate with the PC and the external signal
processor}

.PORT/ABS=H#300F                    setbmoutrdy;
.PORT/ABS=H#304F                    clrbmoutrdy;
.PORT/ABS=H#308F                    pcwe;
.PORT/ABS=H#310F                    setsbr;
.PORT/ABS=H#314F                    clrsbr;
```

(listing continues on next page)

```
.PORT/ABS=H#318F                    clrslhalt;
.PORT/ABS=H#31CF                    setslhalt;
.PORT/ABS=H#30C8                    setbpoe;
.PORT/ABS=H#30C9                    clrbpoe;
.PORT/ABS=H#30CA                    clrbmtaken;
.PORT/ABS=H#30CD                    setsmemrd;
.PORT/ABS=H#30CE                    clrsmemrd;
.PORT/ABS=H#30CF                    setslint;
.PORT/ABS=H#3007                    pcrd;
.PORT/ABS=14336                     beamsend;

.ENDSYS;
```

Listing 15.1 System Specification for Master Firmware

The master code, shown in Listing 15.2, only occupies 180 locations in program memory. It can be divided into two sections: the PC communications section and the A/D result handling section.

The PC communications section of the master code starts at the beginning of the file and ends at the *wait* routine. The beginning contains a series of port, variable and interrupt declarations. The program starts by clearing certain flags and then entering the *pc_init_wait* routine which causes the master CPU to wait until the PC is ready to download the system configuration data. Then, the *pc_comm* routine sets some of the internal CPU registers to be used during the communication. At the end of this routine, the PC is notified that the master is ready and the master CPU enters the *pc_wait1* loop.

The first six pieces of data downloaded by the PC are handled differently than the rest. These first six pieces are the following information:

- The number of sensors in the system
- The number of beams to be formed
- The total number of indexes to be used (the indexes are used by the slave CPUs to pick the desired input samples from the circular buffer)
- The number of slaves in the system
- The number of beams assigned to each slave
- The number of beams assigned to the last slave

This data is stored by the *init_sto* routine in the master CPU's data memory (which actually is the same as the slave data memories) in consecutive locations that are declared at the beginning of the program.

Next, the indexes that were calculated by the PC are received and stored in data memory by the *index_store* routine. Then in the *checksum* routine, the master compares the number of pieces of data it received with the number that the PC indicates that it sent (the PC sends a count value to the master in this routine). If these values are equal, that implies that the download was successful.

The *comm_end* routine prepares the master, using the downloaded data, for its next set of tasks (responding to A/Ds, etc.). This routine and the *init_sto* routine are the only times the master CPU reads a value from a slave data memory. This type of read operation is only allowed from a single slave which must be plugged into a designated slot on the backplane.

Another piece of data that is calculated in this routine is the shading coefficient that will be used in the current set of beams. The system currently can only handle a rectangular window with a magnitude of 1. Additional code to handle several different shading windows can be added later.

After configuration, the master CPU issues a signal which commands the slaves to prepare themselves using the recently downloaded data (the slave CPUs were in a TRAP state until now). The master waits for 1000 cycles, more than enough time for the slaves to get prepared. It then enables two nested interrupts and enters the *wait* loop.

The second major section of the code starts at the *wait* loop. The master waits for the "conversion complete" interrupt to occur, then it jumps to the *adcomplete* routine. During this routine, the master reads the output latches (conversion results) of each AD1332 and writes these values into all of the slaves' sample buffers. The master keeps track of a few pointers in order to address the circular buffers and the A/D boards properly.

Next, in the *sendbeam* routine the master reads beam samples from all the slave FIFOs in sequence and writes them to the system output port. Output beam samples are sent one at a time. The handshaking with the external signal processor is executed for each beam sample. The master starts the handshake by asserting the BMOUTRDY signal and then waits for the BMTAKEN interrupt (set by the external processor) that indicates that the external processor is ready to receive the next value. Then the interrupt is cleared by the master and the next beam sample is sent out in the same manner.

For a given number of desired beam directions, it is not always possible to divide the job evenly among the slaves. Thus, the number of beams formed by the last slave may be different than the number formed by each of the rest of the slaves. Consequently, the output FIFO of the last slave is handled by the *oneslave* routine, which serves a similar purpose to the *sendbeam* routine except it only handles the last slave.

Once the beam output tasks are completed, a RTI (return from interrupt) instruction is executed and the program returns to the *wait* loop where the master waits for the next *adcomplete* interrupt.

The master continues its cyclic, double-interrupt-driven operation until the assertion of $\overline{\text{RESET}}$ or a system power-down.

```
.MODULE/ROM/ABS=0     master_code;

{The following are declarations for ports that are used to set and
clear several flags. Data memory mapped D-flipflops are used to
generate flags}

.PORT                 setbmoutrdy;  {Beam output ready flag}
.PORT                 clrbmoutrdy;
.PORT                 pcwe;       {PC output port}
.PORT                 setsbr;     {Slave bus request flag}
.PORT                 clrsbr;
.PORT                 setslhalt;  {Slave halt flag}
.PORT                 clrslhalt;
.PORT                 setbpoe;    {Backplane output enable flag}
.PORT                 clrbpoe;
.PORT                 clrbmtaken;
.PORT                 setsmemrd;  {Slave memory read flag}
.PORT                 clrsmemrd;
.PORT                 setslint;   {Slave interrupt}
.PORT                 pcrd;       {PC input port}
.PORT                 beamsend;   {Interrupt the external processor}

{The following are variables that contain the system configuration
info.}

.VAR/DM/ABS=8160      sensor_num;
.VAR/DM/ABS=8161      beam_num;
.VAR/DM/ABS=8162      index_num;
.VAR/DM/ABS=8163      slave_num;
.VAR/DM/ABS=8164      beams_per_slave;
.VAR/DM/ABS=8165      last_slave_beam_num;
```

```
{The following is the main body of the master program}

{Interrupt vectors occupy the first four PM locations}
                    JUMP adcomplete;  {Vectored address for IRQ0}
                    RTI;
                    JUMP beamtaken;   {for IRQ2}
                    RTI;

{Program execution starts here}
                    IMASK=b#0000;
                    ICNTL=b#00000;
                    AY0=h#0FFF;
                    DM(clrbmoutrdy)=MX0;    {Clear all flags by}
                    DM(clrbmtaken)=MX0; {writing a value into}
                    DM(clrsbr)=MX0;     {their respective DM}
                                        {mapped port locations}
                    DM(clrslhalt)=MX0;
                    DM(clrbpoe)=MX0;
                    DM(clrsmemrd)=MX0;
pc_init_wait:       AX0=DM(pcrd);      {Wait for the ready}
                    AR=AX0-AY0;        {message from the PC}
                    IF EQ JUMP pc_comm;
                    JUMP pc_init_wait;

pc_comm:            MX0=h#FF0F ;       {Initial set up before the}
                    AY0=5;             {PC communications.}
                    AF=AY0+1;   {A message is sent to the PC}
                    AY0=h#1FFF; {at the end of this routine}
                    AX1=h#FFFF; {signaling that the master is}
                    MX1=h#FF0F; {ready}
                    M0=0;
                    I1=8160;
                    L1=6;
                    I2=0;
                    M2=1;
                    L2=8160;
                    M3=-1;
                    I4=8166;
                    M4=1;
                    I5=9000;
                    M5=0;
                    L5=1;
                    I6=0;
                    M6=1;
                    L6=2055;
                    DM(setsbr)=AY0; {Set slave Bus Request flag}
```

(listing continues on next page)

```
                        DM(pcwe)=MX0;    {Send Ready mesage to the PC}
                        JUMP pc_wait1;

pc_comm_end_check:      AY1=I3;          {Counter register to check for}
                        AR=AY1-1;        {the end of the index}
                        IF LT JUMP pc_end;       {downloading}

pc_wait1:               AY1=DM(pcrd);        {Wait for FFFF from the PC}
                        AR=AY1-AX1;
                        IF EQ JUMP pc_read;
                        JUMP pc_wait1;

pc_read:                MODIFY(I6,M6);  {Write the initial 6 pieces}
                        AF=AF-1 ;        {of data and then start}
                        IF LT JUMP index_store; {the index storage}

init_read:              AX0=DM(pcrd);    {Read data sent from the PC}
                        AY1=h#E000;
                        AR=AX0 AND AY1;   {Mask out bottom 13 bits}
                        AY1=h#A000;
                        AR=AR-AY1;
                        IF EQ JUMP init_sto; {Compare to h#A000}
                        JUMP init_read;
init_sto:               AR=AX0 AND AY0;        {Mask out top 3 bits}
                        DM(setbpoe)=MX0;  {Enable backplane drivers}
                        DM(I1,M1)=AR;
                        DM(setsmemrd)=MX0; {Enable reading from}
                                           {slave DM}
                        L4=DM(beam_num);
                        I3=DM(index_num);
                        DM(clrsmemrd)=MX0; {Disable reading from}
                                           {slave DM}
                        L3=I3;
                        DM(clrbpoe)=MX0;  {Disable backplane drivers}
                        JUMP pc_wait1;

index_store:            AX0=DM(pcrd);     {Read the PC output}
                        AY1=h#E000;
                        AR=AX0 AND AY1;   {Mask out bottom 13 bits}
                        AY1=h#A000;
                        AR=AR-AY1;
                        IF EQ JUMP sto;   (Compare to h#A000}
                        JUMP index_store;

sto:                    AR=AX0 and AY0;
                        DM(setbpoe)=MX0;
```

```
                    DM(I2,M2)=AR ;      {Store index}
                    MODIFY(I3,M3);
                    DM(clrbpoe)=MX0;
                    JUMP pc_comm_end_check;
pc_end:             AX1=h#FF0F;
pc_end_loop:        AY1=DM(pcrd);       {Read PC output and decide}
                    AR=AX1-AY1;     {whether communication should}
                    IF EQ JUMP checksum;     {be completed or not}
                    JUMP pc_end_loop;
checksum:           AX1=I6;             {Perform cheksum operation}
checksum_loop:      AY1=DM(pcrd);
                    AR=AX1-AY1;
                    IF EQ JUMP comm_end;
                    JUMP checksum_loop;

comm_end:           MX0=h#F0;           {This routine ends the PC}
                    DM(pcwe)=MX0;   {communications and does some}
                    DM(setbpoe)=MX1; {reads from the slave DM in}
                    AY0=h#0201;         {order to prepare for the}
                    AF=AY0-1 ;          {beamforming tasks}
                    AY0=0;
                    DM(setsmemrd)=MX0;
                    SI=DM(sensor_num); {Read #of sensors from}
                    DM(clrsmemrd)=MX0; {slave DM and do the}
                    AY0=SI;     {necessary division to obtain a}
                    AR=AY0-1;   {scaled magnitude value for the}
                    IF EQ JUMP onesensor;     {rectangular shading}
                    AY0=0;                    {window}
                    SR=LSHIFT SI BY 9 (LO);
                    AX1=SR0;
                    ASTAT=0;
                    DIVQ AX1;
                    DIVQ AX1;DIVQ AX1;DIVQ AX1;
                    DIVQ AX1;DIVQ AX1;DIVQ AX1;
                    DIVQ AX1;DIVQ AX1;DIVQ AX1;
                    DIVQ AX1;DIVQ AX1;DIVQ AX1;
                    DIVQ AX1;DIVQ AX1;DIVQ AX1;
                    DM(8192)=AY0;       {Shading coefficient}
                    JUMP allsensor;
onesensor:          AY0=h#7FFF; {This is shading coefficient}
                    DM(8192)=AY0; {in case of a single sensor}
allsensor:          DM(setsmemrd)=MX1;
                    AY0=DM(slave_num); {Do the rest of the}
                    AX1=DM(beams_per_slave); {necessary reads}
                    MY1=DM(last_slave_beam_num);
                                {before releasing slave BR flag}
```

(listing continues on next page)

```
                MX0=DM(sensor_num);
                DM(clrsmemrd)=MX0; {Clear the slave DM read}
                MY0=255;           {flag and set up some}
                MR=MX0*MY0(UU);{pointers to be used later on}
                SI=MR0 ;           {#of samples to be placed}
                SR=LSHIFT SI BY -1 (HI);
                                   {into the sample buffer}
                L2=SR1;
                AR=AY0-1;
                DM(clrsbr)=MX1;
                AX0=AR;
                I1=10240;
                L1=MX0;
                I2=0;
                M2=1;
                IF EQ JUMP fix_base;
                I3=h#3800;         {Base addr. for the FIFOs}
                AR=AY0+1;          {with multiple slaves}
                L3=AR;
                JUMP normal;
fix_base:       I3=h#3801;         {Base addr. for the FIFO}
                L3=0;                    {with single slave}
normal:         DM(setslhalt)=MX1; {HALT the slaves, this}
                DM(clrslhalt)=MX1; {will cause them to get}
                DM(clrbpoe)=MX1; {out of their TRAP state}
                CNTR=1000;
                DO slave_wait UNTIL CE; {Wait until all}
slave_wait:        NOP;            {slaves are ready to go}
                AR=1;
                AY1=1;
                ICNTL=b#00101;
                IMASK=b#0001;
                           {Enable sampling interrupt IRQ0}

wait:           JUMP wait;

adcomplete:     AR=AR-AY1;               {Ignore the first}
                IF EQ JUMP first_adcomp; {IRQ0 by using this}
                DM(setsbr)=MX1;
                AF=AY0-1;
                DM(setbpoe)=MX1;  {This routine is used}
                CNTR=MX0;         {to fill up the sample}
                DO sample_store UNTIL CE; {memory after each}
                   MX1=DM(I1,M1);        {A/D conversion}
sample_store:      DM(I2,M2)=MX1;
                DM(clrsbr)=MX1;
```

```
                      DM(setslint)=MX1;
                      DM(clrbpoe)=MX1;

sendbeam:             IF EQ JUMP oneslave; {Read out the beams in}
                      MODIFY(I3,M1);       {a sequential manner in}
                      CNTR=AX0;    {each FIFO in order of numbering}
                      DO beamout UNTIL CE; {Send beam outputs via}
                         CNTR=AX1;         {the digital output port}
                         DO fifo_out UNTIL CE;
                            DM(setbpoe)=MX1;
                            MX1=DM(I3,M0);
                            DM(clrbpoe)=MX1;
                            DM(beamsend)=MX1;
                            IMASK=b#0100;{Enable data receive intr}
                            DM(setbmoutrdy)=MX1;    {Set a flag}
                            CNTR=6;  {for the external processor}
                            DO resp_wait UNTIL CE;
resp_wait:                     NOP;
fifo_out:                   NOP;
beamout:                 MODIFY(I3,M1);


oneslave:             CNTR=MY1;
                      DO endfifo UNTIL CE;
                         DM(setbpoe)=MX1; {This routine is for}
                         MX1=DM(I3,M0);   {single slave}
                         DM(clrbpoe)=MX1; {configurations and is}
                                          {also used for handling}
                         DM(beamsend)=MX1; {the last FIFO read out}
                         IMASK=b#0100;         {It handles the}
                         DM(setbmoutrdy)=MX1; {irregularity of the}
                         CNTR=6;               {last set of beams,}
                         DO resp_wt UNTIL CE; {i.e. possibly fewer}
resp_wt:                    NOP;               {beams}
endfifo:                 NOP;
first_adcomp:         AR=0;
                      RTI;

beamtaken:            DM(clrbmoutrdy)=MX1;
                               {Interrupt routine to handle}
                      DM(clrbmtaken)=MX1;
                               {the data receive confirmation}
                      RTI;     {from the external processor}

.ENDMOD;
```

Listing 15.2 Master Firmware

15.6.2 Slave Firmware

The slave board requires less firmware than the master. It is assembled using the system specification source file shown in Listing 15.3.

```
.SYSTEM                             slave_system;

.SEG/ROM/ABS=0/PM/CODE              rom_program_storage[2048];
.SEG/RAM/ABS=2048/PM/DATA           index_mem[2048];

.SEG/RAM/ABS=0/DM/DATA              sample_mem[8160];
.SEG/RAM/ABS=8160/DM/DATA           system_info[32];
.SEG/RAM/ABS=8192/DM/DATA           shading_coeff_mem[32];
.SEG/RAM/ABS=8224/DM/DATA           scratch_mem[2016];

{The ports declared below are used to set and clear some flags as
well as to write to the DAC and to read from some hardware
switches}

.PORT/ABS=H#2800                    beamout;
.PORT/ABS=H#3000                    beamdac;
.PORT/ABS=H#3800                    clrslint;
.PORT/ABS=H#3900                    slave_id;
.PORT/ABS=H#3A00                    dac_beam_sel;

.ENDSYS;
```

Listing 15.3 System Specification for Slave Firmware

The slave firmware code occupies only 100 locations in program memory. The slave firmware, shown in Listing 15.4, can be divided into two sections: the system set-up section and the beamforming section.

The set-up section of the code starts at the beginning of the file and ends at the *wait* routine. The beginning contains a series of port, variable and interrupt declarations. Only one interrupt, SLINT, is recognized by the slave; this is the interrupt initiated by the master to start the beamforming operation.

After system parameters are declared, the slave enters a TRAP state. The slave gets reactivated after the master is finished communicating with the PC. The master asserts the SLHALT signal, which wakes up the slave and causes the program execution to continue from the location following the TRAP instruction.

In the first part of the program, the slave moves the indexes from its data memory into its "index memory" which is located in its program memory space (the indexes are in data memory initially because the master has to store them there temporarily). Then, the slave sets up its address registers using the downloaded beamforming information. Next, the slave enters the *wait* loop to wait for the beamforming interrupts issued by the master.

The slave constantly monitors the D/A beam selection switch while it is in the *wait* loop. Since the slave returns to the *wait* loop every 100µs, it can decide, in real time, which beam to send out through the analog port.

The second major section of the program starts at the *wait* loop. As soon as the SLINT interrupt is received, the slave jumps to the *beam_form* routine. The *beam_form* routine contains very tight loops which allows the slave to form a large number of beams. The routine reads the index memory and picks the indexed samples from the sample buffer. These samples are the delayed samples that are needed for the beam summation.

The frame of the circular buffer is rotated every time a new set of samples comes in. Therefore the indexes that are read must be modified before being used, because they are referenced to the absolute origin of the circular buffer.

The samples that are read are multiplied by the shading factor (which is currently 1) and accumulated in the MR register. The resulting beam sample is written into the FIFO. If the beam sample belongs to the beam that is requested at the analog port, it is then written to the *beamdac* port (the D/A converter). Before returning from the interrupt, the slave clears the SLINT flag. Then the program returns to the *wait* loop to wait for the next SLINT interrupt.

The slave continues its cyclic, single-interrupt-driven operation until the assertion of $\overline{\text{RESET}}$ or a system power-down.

15 Sonar Beamforming

```
.MODULE/ROM/ABS=0     slave_code;

{The following are declarations for data memory mapped ports. One
is used to clear a flag, while others are used to write data to
the DAC, FIFO and read the hardware switches on the slave board}

.PORT              beamout;          {FIFO}
.PORT              beamdac;          {DAC}
.PORT              clrslint;         {Clear the slave interrupt}
.PORT              slave_id;         {Slave identity dipswitch}
.PORT              dac_beam_sel;     {Analog output selection switch}

{The following are variables that contain the system configuration info}

.VAR/DM/ABS=8160      sensor_num;
.VAR/DM/ABS=8161      beam_num;
.VAR/DM/ABS=8162      index_num;
.VAR/DM/ABS=8163      slave_num;
.VAR/DM/ABS=8164      beams_per_slave;
.VAR/DM/ABS=8165      last_slave_beam_num;

{The following is the main body of the slave program}

                      JUMP beam_form;    {Vectored addr. for IRQ0}
                      RTI;
                      RTI;
                      RTI;

   IMASK=b#0000; {Disable interrupts}
   ICNTL=b#00000;
   DM(clrslint)=MX0;
   TRAP;    {TRAP until pc_comm ends}

pc_comm_end:          I1=0;          {This initial routine is used}
                      M1=1;          {to transfer the indexes from}
                      L1=DM(index_num); {sample_mem into the}
                      I4=2048;                  {index_mem in PM}
                      M4=1;
                      L4=L1;
                      CNTR=L1;
                      DO index_store UNTIL CE;
                         MX0=DM(I1,M1);
index_store:             PM(I4,M4)=MX0;

                      MX0=DM(sensor_num);
                      MY0=255;
```

```
                    MR=MX0*MY0(UU);
                    SI=MR0;
                    SR=LSHIFT SI BY -1 (HI);
                                        {SR1 contains the length}
                    I1=MX0;             {of the circular sample}
                    M1=MX0;             {buffer}
                    L1=SR1;
                    SI=DM(slave_id);  {Determine this slave's}
                    SR=LSHIFT SI BY -12 (HI);
                                        {ID# and the starting}
                    AX0=DM(slave_num); {location of the first}
                    AY0=SR1;            {index.Also determine}
                    AF=AX0-AY0;         {the # of indexes}
                    AF=AF-1;            {for this slave}
                    IF EQ JUMP last_slave;
                    SE=DM(beams_per_slave);
                    JUMP all_slave;
last_slave:         SE=DM(last_slave_beam_num);

all_slave:          MX0=DM(sensor_num);
                                        {This routine calculates}
                    MY0=DM(beams_per_slave);
                                        {the starting address of}
                    MR=MX0*MY0(UU);   {this slave's indexes}
                    SI=MR0;
                    SR=LSHIFT SI BY -1 (HI);
                    MX0=SR1;
                    AR=AX0-AY0;
                    AY1=AR;
                    AR=AY1-1;
                    MY0=AR;
                    MR=MX0*MY0(UU);
                    SI=MR0;
                    SR=LSHIFT SI BY -1 (HI);
                    AX1=2048;
                    AY1=SR1;
                    AR=AX1+AY1;
                    I5=AR;              {Starting address of the}
                    M5=1;               {indexes for this slave}
                    M7=-1;
                    MX0=SE;
                    MY0=DM(sensor_num); {# of indexes per beam}
                    MR=MX0*MY0(UU);
                    SI=MR0;
                    SR=LSHIFT SI BY -1 (HI);
                    L5=SR1;             {Total # of indexes to}
                    M3=0;               {be used by this slave}
```

(listing continues on next page)

```
                        L3=L1;
                        I6=8192;
                        M6=0;
                        L6=1;
                        MY1=DM(I6,M6); {Shading coefficient; there}
                        AY1=DM(sensor_num);
                                    {is only one now since it is}
                        AR=AY1-1;   {a rectangular window}
                        AX1=AR;
                        ICNTL=b#00001;
                        IMASK=b#0001; {Enable slave interrupt IRQ0}

wait:                   SI=DM(dac_beam_sel); {Setup down counter to}
                        SR=LSHIFT SI BY -12(HI);
                                          {be used in deciding}
                        AY1=SR1;          {which beam to send out}
                        AF=AY1+1;         {to the DAC}
                        JUMP wait;  {Wait for sample buffer update}

beam_form:              CNTR=SE;
                        DO beam_end UNTIL CE;
                           MR=0; AY0=PM(I5,M5); {Read index}
                           M3=I1;
                           I3=AY0;
                           MODIFY(I3,M3); {Modify index}
                           MX0=DM(I3,M3); {Get first sample}
                           AY0=PM(I5,M5);
                           CNTR=AX1;
                           DO single_beam_sample UNTIL CE;{Beamform}
                              I3=AY0;
                              MODIFY (I3,M3);
single_beam_sample:           MR=MR+MX0*MY1(SS), MX0=DM(I3,M3),AY0=PM(I5,M5);
                           MR=MR+MX0*MY1(RND), AY0=PM(I5,M7);
                           AF=AF-1;              {Check which beam to}
                           IF EQ JUMP dac_write; {send to the DAC}
beam_end:                  DM(beamout)=MR1;      {Write result to FIFO}
                        MODIFY(I1,M1);            {Advance circular sample}
                        DM(clrslint)=MX0;         {buffer pointer}
                        RTI;

dac_write:              DM(beamdac)=MR1;        {Write result to DAC}
                        JUMP beam_end;

.ENDMOD;
```

Listing 15.4 Slave Firmware

15.7 SYSTEM SOFTWARE

The system software consists of the PC program that is responsible for the user interface and the downloading of the system configuration data. The code is written in the C language and compiled on the Microsoft C Compiler.

The declaration section at the beginning of the program includes certain useful libraries, defines a number of variables and declares the 8-bit parallel I/O port addresses. These ports are located on an Analog Devices RTI-817 parallel I/O card which is plugged into the PC's backplane. Following this section there are a series of function definitions and the execution loop of the program.

The program interactively takes in the system variables from the user. Some questions are displayed on the screen which are answered by the user via the keyboard. The values that have to be entered by the user are: the number of sensors, the number of beams to be formed, the number of slaves in the system, the cartesian coordinates for the sensor locations and the spherical coordinates for the desired beams. The program assigns these values to variables and arrays in order to calculate the necessary tap delays for beamforming.

The program converts the spherical coordinate beam directions to cartesian coordinates. Then it calculates the necessary delays using the equation (2). The propagation speed of sound in water is assumed to be 1470 m/s, which is typical for the ocean water. (This value can be changed easily in the code to conform to the application environment.) The program identifies the beams that the system is unable to produce with the given array configuration. This task is accomplished by checking whether any one of the required delays falls outside of the sample buffer length.

Next, the program determines the maximum number of beams that can be formed with the given system configuration. It also calculates the number of beams to be assigned to each slave and the number of beams to be assigned to the last slave. These values are stored as variables to be downloaded to the master.

The program downloads the system configuration information to the master, beginning with six pieces of system set-up data, as explained earlier in the master firmware section. Then the program sends all the calculated indexes to the master, followed by a checksum, which, as explained earlier, corresponds to the number of pieces of data (number of indexes + 6) just downloaded. If the master acknowledges that the download was successful, the downloading operation is completed by sending a confirmation message to the PC screen. If the master indicates an unsuccessful download, the downloading operation is terminated by sending a failure message to the the screen. In this case, the user is given the choice of aborting or retrying the download.

The download is easily executable by the user. Once the system parameters are entered, it takes at most a few seconds for the PC to download all the information to the master.

15.8 ENHANCEMENTS

There are several ways to improve the performance, functionality and the user interface of the example beamformer. Possible additional features as well as some architectural and circuit level enhancements are briefly discussed in this section.

15.8.1 Additional Features

A large number of features can be added to this system without great difficulty. One feature is the choice of frequencies for the input sampling clock. It is possible to route an external clock to the A/D boards by incorporating additional hardware on the master board. The external sampling clock option would an additional piece of data collected by comm.*c* during the system parameter configuration. The program would download the information to the master, which would activate the necessary signals for clock selection. The program would also have to modify the beam assignments, because the system cycle may be shorter or longer depending on the choice of sampling frequency.

A software enhancement is the ability to save the current system parameters into a file. This allows the system to be restarted by instructing *comm.c* to boot up the system using the saved parameters instead of getting them from the user. A user could edit this file to restart the system with a new set of configuration parameters in a very short time. This feature would make *comm.c* even more user-friendly.

Another feature is the addition of a filtering and smoothing circuit for the output of the D/A converter. Smoothing the output of the D/A converters would make it possible to feed the analog output beams into a spectrum analyzer or a general purpose data acquisition system for further signal analysis.

The availability of various shading windows for the inputs is another useful enhancement. The modifications would have to be done in *comm.c* and also the system firmware programs. In *comm.c*, the shading window option would be gotten from to the user and the shading coefficients would be calculated on the fly and downloaded to the master. The master would send these coefficients to the slaves instead of the unit rectangular window.

The shading factors would reside in the program memory space of each slave and could ultimately be used by the slaves during beamforming. The downloading overhead would be minimal. The additional program memory accesses during the beamforming loop should not result in a performance degradation since they can be performed in parallel with the data accesses. Careful calculations are needed to determine the exact performance consequences of such a system modification.

15.8.2 Performance Improvements

There are several ways to improve the performance of the beamformer described in this chapter by making relatively minor modifications to the hardware and software. Some modifications, with ascending levels of complexity, are discussed in this section.

An important performance issue for a real-time beamformer is the beam throughput. The main goal of such a system is to form as many simultaneous beams as possible using the existing technology. There are several ways to improve the beam throughput within the existing distributed processing architecture. The most obvious and relatively easy way to achieve this goal is to replace the ADSP-2100s with ADSP-2100As. The ADSP-2100A has an 80ns instruction cycle time as opposed to 125ns cycle time of the ADSP-2100. This upgrade would result in $\approx$50% increase in system beam throughput.

The ADSP-2100A is pin and source code compatible with the ADSP-2100. This allows the easy upgrade of the system with no firmware changes. The modifications that are needed are mostly in hardware. The timing requirements during the ADSP-2100A's data and program memory access operations must be carefully analyzed and faster devices should be placed on the critical data paths. It is possible to upgrade only the existing memory components to compensate for the new shorter data and program memory access cycles.

A modification to *comm.c* would also be necessary because it would have to be able to assign a larger number of beams per slave. The input bandwidth of the external signal processor would have to be carefully evaluated, because it is likely that the output bandwidth of the upgraded system, in maximum configuration, would be higher than the input capacity of the external processor. If such an incompatibility resulted, you could use fewer slaves to match the output bandwidth requirements. The overall consequence would be a cost reduction for the less demanding users and higher performance for the more demanding users.

Another improvement is increasing the maximum number of sensor inputs. It is possible to add more input channels to the beamformer. However, each added A/D converter would have to be placed in a unique location in the master CPU's data memory address space, requiring some additional address decoding circuitry on the A/D boards. There is enough room for more digital components on these boards and the changes in the wiring would be relatively simple.

One important effect of adding channels is a reduction of the maximum number of beams that can be formed simultaneously, because the master will have to read more inputs within one system cycle and consequently will have less time to read the results from the slave FIFOs. It is possible to keep the beam throughput at the current level by upgrading the processors while increasing the number of sensor inputs. These performance tradeoffs should be considered carefully before expanding the A/D capabilities of the system.

A major improvement is to redesign the A/D boards with multi-channel A/D converter hybrids replacing the single-channel AD1332s. Analog Devices' AD1334 would be the optimal choice. The system redesign effort that is necessary to implement the substitution of AD1334s is of moderate difficulty. The savings in the number of A/D cards would prove this redesign effort to be very valuable, especially if the need for input channels is expected to rise.

The AD1334 contains four sample and hold circuits, a 4-to-1 analog multiplexer, an AD7672 12-bit, 5μs A/D converter and an output FIFO. The sample and hold circuit (AD585) and the A/D converter are the same as the ones used in the AD1332. The output FIFO is 12 bits wide and 64 locations deep. It is possible to use this hybrid in a mode where all of the sample and hold circuits sample the inputs simultaneously.

Some overhead analog circuitry must be added externally because of the lack of on board low-pass filters in the AD1334. The AD1334 has the same package as the AD1332, so it would be possible to fit as many as three

A/D hybrid packages on the same board even with the additional digital and analog overhead circuitry that is needed. Such a construction strategy would allow each A/D board to handle up to 12 sensor inputs.

Some minor modifications in the master firmware would also be needed in order to properly address the AD1334s.

15.9 REFERENCES

Baggeroer, A. B. 1978. "Sonar Signal Processing." In *Applications of Digital Signal Processing*, A. V. Oppenheim, Ed. Englewood Cliffs, NJ: Prentice-Hall, Inc.

Curtis, T. E. and R. J. Ward. 1980. "Digital Beam Forming for Sonar Systems." *IEEE Proceedings*, Vol. 127, Pt. F, No. 4.

Gray, D. A. 1985. "Effect of Time-Delay Errors on the Beam Pattern of a Linear Array." *IEEE Journal of Oceanic Engineering*, Vol. OE-10, No. 3.

Hodgkiss, W. S. and V. C. Anderson. 1981. "Hardware Dynamic Beamforming." *Jour. Acoust. Soc. Am.* 69(4).

Janssen, R. J. 1987. "Sonar Beamforming and Signal Processing." *Electronic Progress*, Vol. 28, No. 1, Raytheon Co.

Karagozyan, K. 1988. "A Multi-Processor Based Digital Beamforming System." Master's thesis, Massachusetts Institute of Technology, Cambridge, Mass.

Knight, W. C., R. Pridham and S. M. Kay. 1981. "Digital Signal Processing For Sonar." *IEEE Proceedings*, Vol. 69, No. 11.

Pridham, R. G. and R. A. Mucci. 1978. "A Novel Approach to Digital Beamforming." *Jour. Acoust. Soc. Am.* 63(2).

Winder, A. A. 1975. "Sonar System Technology," *IEEE Transactions on Sonics and Ultrasonics*, Vol. SU-22, No. 5.

Memory Interface 16

16.1 OVERVIEW

This chapter presents some examples that illustrate basic considerations for interfacing memory to ADSP-2100 Family processors. An example of a multiple paging scheme for data memory is included. Memory-mapped I/O is also demonstrated.

16.2 PROGRAM MEMORY

ADSP-2100 Family processors have a 14-bit program memory address (PMA) bus and a 24-bit program memory data (PMD) bus.

The ADSP-2100A has an additional address pin, the PMDA pin. (See the *ADSP-2100 User's Manual* for details on PMDA). The PMDA signal can be used as a fifteenth PMA bit. The lower half (16K) of program memory stores the instructions, which are 24 bits wide. The upper half (16K) of program memory stores program memory data that is 16 bits wide and left-justified (occupies data bits 23-8). Generally, the program memory data space is used to store constants, such as filter coefficients.

The following program memory example consists of 6 Cypress 8K x 8 static RAMs (CY7C185). They are configured into two banks of 8K x 24-bit wide memory. One bank is shown in Figure 16.1, on the next page. In each bank of RAM, PMD0-7 from the ADSP-2100 Family processor is connected to the D0-D7 pins of one RAM, PMD8-15 to the D0-D7 pins of the second RAM and PMD16-23 to the D0-D7 pins of the third RAM. PMA0-12 from the ADSP-2100 Family processor are connected to the A0-A12 pins of each of the RAMs.

The $\overline{RD}$ and $\overline{WR}$ lines are connected directly to the memories while the $\overline{PMS}$ line is used by the decoder.

16 Memory Interface

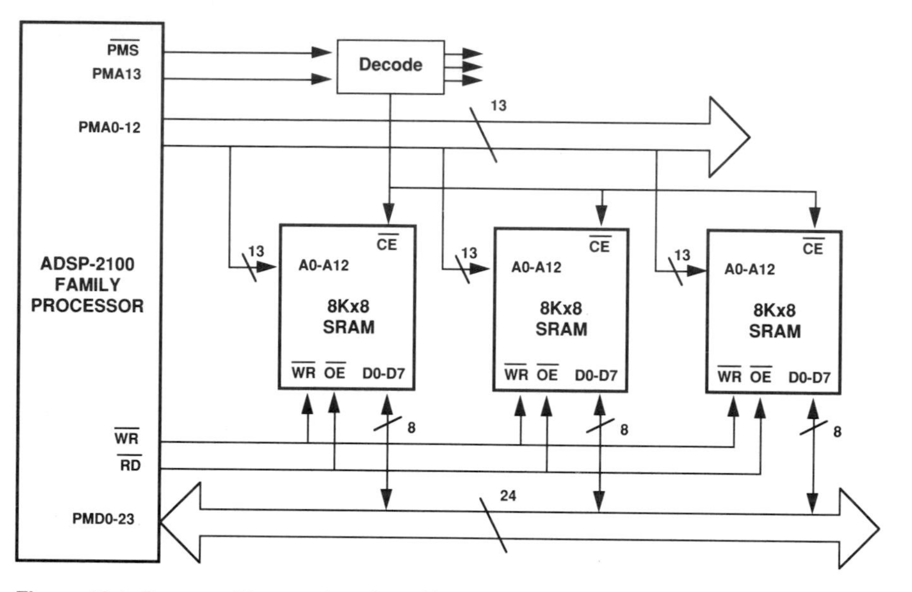

Figure 16.1 Program Memory Interface (One Bank)

16.2.1 Program Memory Bank Enables

Only one bank of program memory RAM is enabled at a time. Bank enables are generated by decoding PMA13 and $\overline{\text{PMS}}$.

On the ADSP-2100A only, the PMDA signal acts as a fifteenth address bit. When PMDA is low, it selects the lower half of program memory (H#0000-3FFF, instruction space); when it is high, it selects the upper half of program memory (H#4000-7FFF, data space). Each half of program memory consists of two banks. PMA13 enables either the upper or lower half of memory.

16.3 DATA MEMORY

Data memory is generally used to store acquired data. The ADSP-2100 Family processors have a 16-bit wide data memory data (DMD) bus and 14-bit wide data memory address (DMA) bus. The 16-bit wide data bus allows direct interface to 16-bit A/D and D/A converters. The 14-bit wide address bus permits addressing of up to 16K of data memory.

Memory Interface 16

Data memory paging can be used to allow ADSP-2100 Family processors to access several 16K x 16 data memory pages. The pages are in parallel, that is, they each use the same address space. One data memory page is enabled at a time via a data memory page select register, which is mapped into program memory data space.

Each page of data memory in the example uses four 8K x 8 Cypress static RAMs (CY7C185), configured in two 8K x 16 banks of RAM as shown in Figure 16.2. In each bank, DMA0-12 from the ADSP-2100 Family processor are connected to the A0-A12 pins on both chips. DMD0-7 are connected to the D0-D7 pins on one RAM chip, and DMD8-15 are connected to the D0-D7 pins of the second RAM chip.

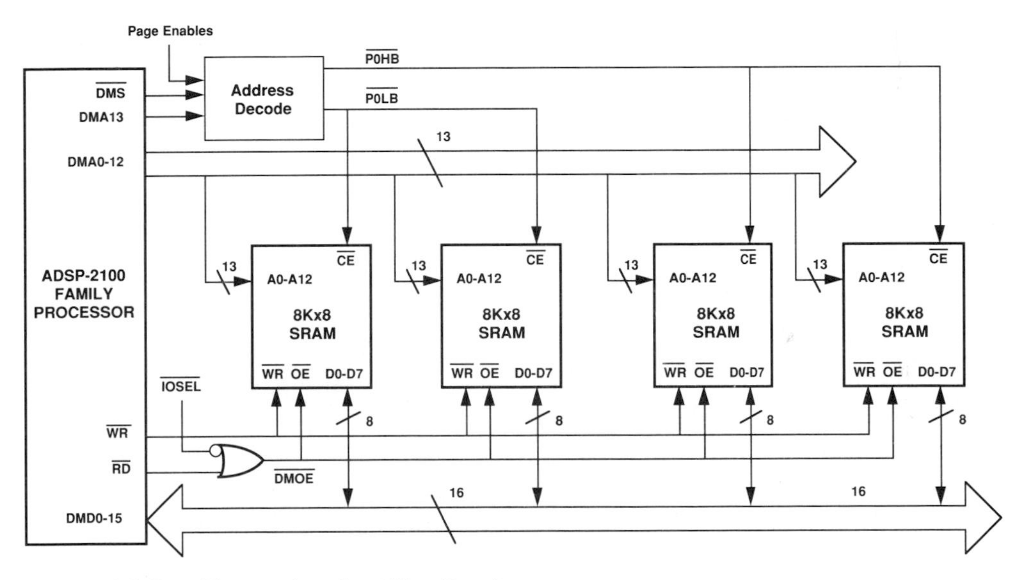

Figure 16.2 Data Memory Interface (One Page)

A bank enable signal, as described in "Data Memory Bank Enables," is connected to the chip enable (CE) pin of each RAM. The $\overline{\text{WR}}$ signal is connected to the $\overline{\text{WR}}$ pins of all RAM chips. The Data Memory Out Enable ($\overline{\text{DMOE}}$) signal is connected to the $\overline{\text{OE}}$ pin on all RAMs. $\overline{\text{DMOE}}$ is generated by ORing $\overline{\text{RD}}$ and the inverse of the I/O Select signal ($\overline{\text{IOSEL}}$), to prevent bus contention during I/O accesses. (See "I/O Memory Mapping" for more information on $\overline{\text{IOSEL}}$.) When a bank enable signal goes low, both RAM chips in the corresponding bank are enabled.

16.3.1 Data Memory Page Enables

The data memory page is selected through a reserved location in program memory. For example, the software can select a data memory page by writing the page number in binary to program memory location H#3FFE. This value appears on PMD8-23, and the LSBs (bits 8, 9 and 10) are decoded in hardware to generate the data memory page enable (see Table 16.2 for values).

	Data Memory Page Enable (Active Low)							
PMD10-8	DMPG0	DMPG1	DMPG2	DMPG3	DMPG4	DMPG5	DMPG6	DMPG7
000	0	1	1	1	1	1	1	1
001	1	0	1	1	1	1	1	1
010	1	1	0	1	1	1	1	1
011	1	1	1	0	1	1	1	1
100	1	1	1	1	0	1	1	1
101	1	1	1	1	1	0	1	1
110	1	1	1	1	1	1	0	1
111	1	1	1	1	1	1	1	0

Table 16.2 Data Memory Page Enables

PMD8-10 from the ADSP-2100 Family processor are input to a 3-to-8 decoder with latchable outputs, as shown in Figure 16.3. The 3-to-8 decoder is always enabled; its latched outputs are only updated on the rising edge of its Latch Enable (LE) input. The outputs of the 3-to-8 decoder are the data memory page enables ($\overline{\text{DMPGx}}$, where x is the page number).

The LE input for the 3-to-8 decoder is generated as follows. Location H#3FFE is decoded using two 8-bit bus comparators (74AHCT521). The Q inputs to the bus comparators are hardwired to H#3FFE. The P inputs are connected to the program memory address bus. The outputs of the comparators, which go low when the Q and P inputs are the same, are ORed to create the data memory page select enable signal ($\overline{\text{DMPSE}}$). $\overline{\text{DMPSE}}$ is NORed with the $\overline{\text{PMWR}}$ signal to generate the LE input to the 3-to-8 decoder.

When $\overline{\text{PMWR}}$ goes high, the decoded value of PMD8-10 is latched into the output of the 3-to-8 decoder, selecting the data memory page. A data memory page remains selected until another is selected or the system is reset. At power-up and after reset, the data memory page defaults to page zero.

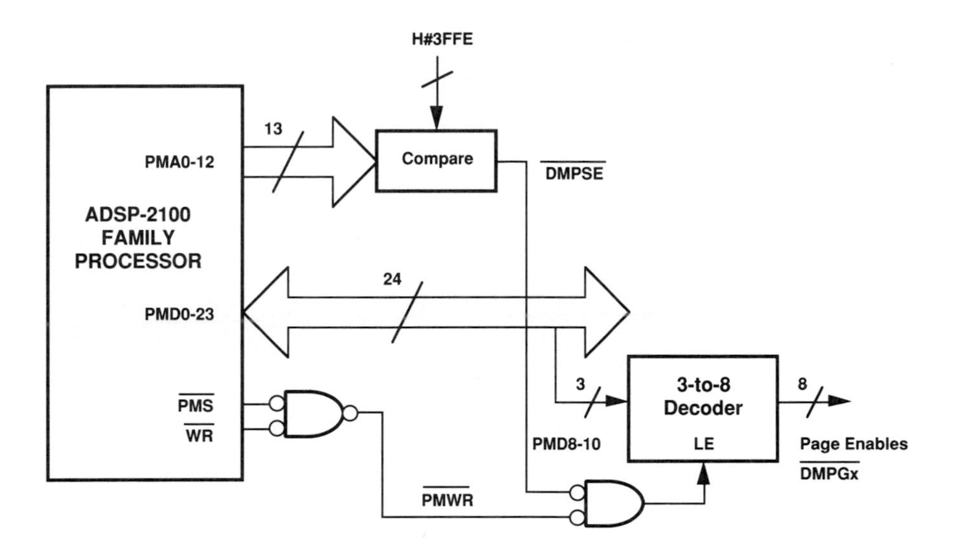

Figure 16.3 Data Memory Page Enables

16.3.2 Data Memory Bank Enables

The bank enables are generated from the most significant DMA bit, DMA13, and the page enable signals, DMPG0 and DMPG1, as shown in Table 16.3. DMA13 and an inverted version of DMA13 are each ORed with the data memory page enable signal to select one of the two banks in the enabled page. If a data memory page is not enabled, no RAM banks can be enabled.

			Data Memory Bank Enables			
DMPG0	DMPG1	DMA13	$\overline{P0LB}$	$\overline{P0HB}$	$\overline{P1LB}$	$\overline{P1HB}$
1	0	0	0	1	1	1
1	0	1	1	0	1	1
0	1	0	1	1	0	1
0	1	1	1	1	1	0

Table 16.3 Data Memory Bank Enable Decoding

16.4 I/O CONFIGURATION

This example maps I/O accesses into data memory; I/O devices can occupy certain memory space locations in place of memory. The I/O devices are written to and read from as if they were memory. The memory locations that map to I/O are called I/O locations.

I/O devices can be memory-mapped into program memory data space as well as data memory. A key difference to consider is:

- The program memory data bus is 24 bits wide, to accommodate instructions. In program memory data accesses, PMD7-0 are read from or written to the PX register.

See the *ADSP-2100 User's Manual* for further information.

16.4.1 I/O Memory Mapping

This example maps 16 I/O locations into the last 16 locations of each data memory page, H#3FE0-3FFF. Either I/O or memory can be used at these locations; a hardware switch selects I/O or memory. When I/O is enabled and a data memory access to an I/O location is executed, the $\overline{IOSEL}$ signal is generated. The $\overline{IOSEL}$ signal enables the I/O device to drive the data memory data bus and is also used to disable memory, preventing an I/O device and memory from driving the data bus at the same time. $\overline{IOSEL}$ should be qualified with a data memory read or data memory write to ensure a proper I/O access. When I/O is not enabled, data memory accesses to the I/O locations are executed as memory accesses and no $\overline{IOSEL}$ is generated ($\overline{IOSEL}$ remains high).

There are eight I/O switches, one for each of the eight pages of data memory that the example system can use. The I/O switch selects either memory or I/O for the last 16 locations of its data memory page. In this way, I/O can be enabled for any combination of data memory pages.

The $\overline{IOSEL}$ signal is generated when DMA5-13 are high (for addresses H#3FE0-3FFF). DMA5-13 are connected to a 12-input NAND gate (74S134) as shown in Figure 16.4. (Unused inputs are pulled high.) The output of the NAND gate is the $\overline{IOSEL}$ signal. Any address equal to or greater than H#3FE0 generates the $\overline{IOSEL}$ signal.

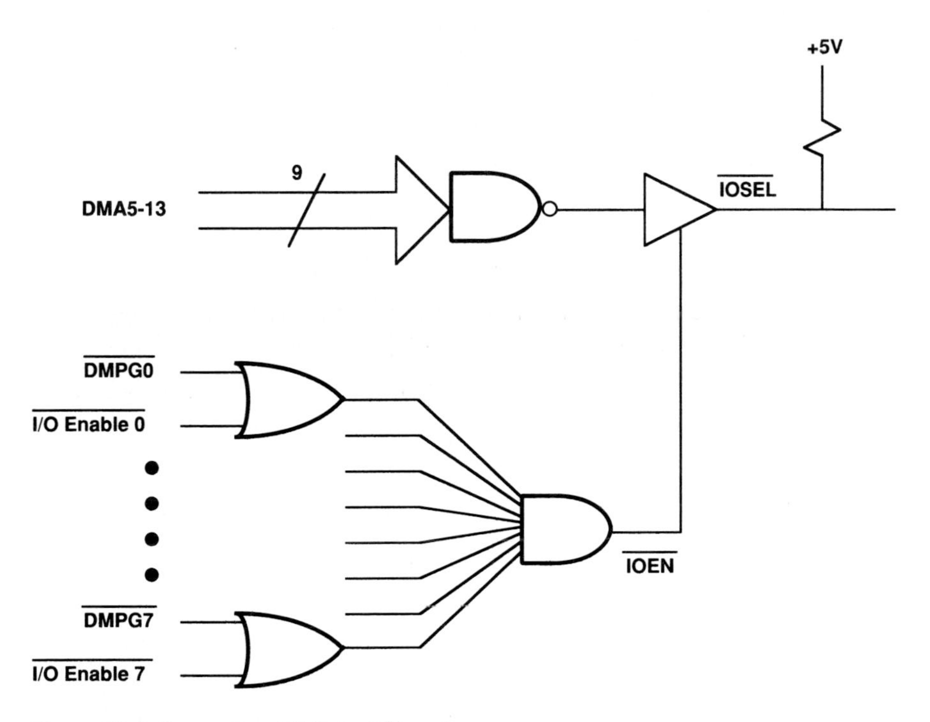

Figure 16.4 Generating I/O Select Signal

The 74S134 that generates $\overline{\text{IOSEL}}$ has a tristate output which is controlled by the $\overline{\text{IOEN}}$ signal. Each of the eight (active low) data memory page enable signals is ORed with its data memory page I/O enable switch (active low). The eight signals are connected to an 8-input NAND gate. The inverted output of the 8-input NAND gate ($\overline{\text{IOEN}}$) is active only if a data memory page enable and the corresponding I/O enable switch are both active. Because only one data memory page is enabled at a time, only one input to the 8-input NAND gate can go low at a time. This allows you to set the I/O switches in any combination; $\overline{\text{IOSEL}}$ is not active until a page with I/O enabled is selected.

16.4.2 I/O Switches

I/O location H#3FFF is used by the example system to read the settings of four SPDT switches. The four general-purpose switches set the values of bits 0-3 at that location. An executing program can read the bits to determine the switch settings and then branch accordingly.

The values of the SPDT switches are output through a tristatable buffer (74F244A) to the 4 LSBs of the DMD bus. The buffer is enabled by decoding address H#3FFF as shown in Figure 16.5. The $\overline{\text{IOSEL}}$ signal is generated when DMA5-13 are high. Location H#3FFF (DMA0-13 high) can therefore be decoded from an inverted $\overline{\text{IOSEL}}$ signal and DMA0-4. These signals are input to an 8-input NAND gate (74ALS30). (The unused inputs are pulled high.) When the data memory address is to H#3FFF and I/O is enabled, the output of the NAND gate, $\overline{\text{EN}}$, is asserted low.

The $\overline{\text{EN}}$ signal and the $\overline{\text{DMRD}}$ signal are ORed and input to the enable pin on the 74F244A. When a read to I/O location H#3FFF is executed, first $\overline{\text{EN}}$ goes low, then $\overline{\text{DMRD}}$ goes low, enabling the 74F244A. On the rising edge of $\overline{\text{DMRD}}$, the output of the 74F244A is clocked into the ADSP-2100 Family processor via the DMD bus.

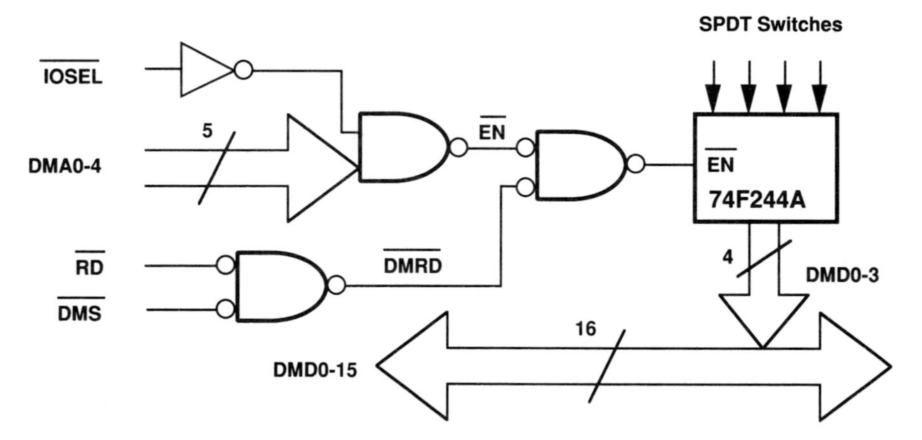

Figure 16.5 Decoding for I/O Switches

Multiprocessing 17

17.1 OVERVIEW

Complex signal processing applications may demand higher performance than a single DSP processor can provide. When a single processor falls short, a multiprocessor architecture may boost throughput. However, the law of diminishing returns applies. As more processors are added, additional computation time is spent in interprocessor communication, degrading the overall performance for each processor. Four processors, for example, cannot deliver four times the computation power of one processor. A processor pair almost doubles the speed of a single processor while keeping the architecture and interprocessor coordination as simple as possible. This chapter develops a processor-pair architecture, based on a dual-port RAM. The design is easy to implement and provides a significant computational boost over a single processor.

17.2 SOFTWARE ARCHITECTURE

To complement the hardware design, a hypothetical application is presented. Data is input and low-pass filtered by one processor, then the second processor determines the peak location within a filtered window. Although the software implementation is simplistic, it shows a technique for programming in a multiprocessing environment: alternating buffers and flags.

The alternating buffers in this application are two identical buffers located in dual-port RAM so both processors can access them. The first processor fills buffer 1 with information, while the second processes the information in buffer 2. Each buffer has a flag that indicates completion of operations on that buffer. When processor 1 has finished its operations on the buffer data, it sets the flag, signaling processor 2 to begin operations on that buffer. The sequence of operations is shown in Figure 17.1, on the next page.

The alternating buffer scheme is easier to implement than a single buffer scheme. If only one buffer were used, careful timing analysis or extensive handshaking would be required to ensure that the processors did not use old or invalid data.

Processor 1 (Filter)	**Processor 2 (Peak Locator)**
Initialize flags, coefficients delay line, pointers	Initialize pointers
Perform low pass filter operation on data in buffer 1	Check flag 1; wait if not set
Set flag 1	
	Check flag 1; if set, perform peak locating operation on data in buffer 1
Perform low pass filter operation on data in buffer 2	
	Clear flag 1
Set flag 2	
	Check flag 2; if set, perform peak locating operation on data in buffer 2
Perform low pass filter operation on data in buffer 1	
	Clear flag 2
Set flag 1; etc.	
	Check flag 1; etc.

Figure 17.1 Alternating Buffers and Flags

17.3 HARDWARE ARCHITECTURE

This system includes two ADSP-2100s, each with its own private memories. Private memories are accessible to one processor only. Common memory is accessed by both. Figure 17.2 shows a block diagram of the system.

Each processor has a private memory of 32K of 24-bit program memory and 14K of 16-bit data memory. In addition, 2K of 16-bit dual-port RAM is shared by both processors. This area of memory allows inter-processor communication and data transfers.

17.3.1 Using Dual-Port Memory

The 2K x 8-bit dual-port RAMs used in this design are the IDT7132 and the IDT7142 produced by Integrated Device Technology. A useful feature of the IDT7132 is its on-chip arbitration support. The IDT7142 acts as a

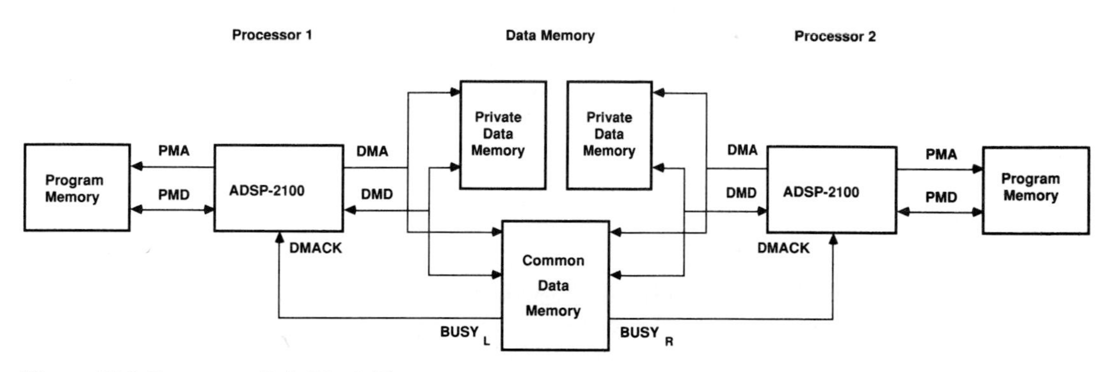

Figure 17.2 Processor Pair Block Diagram

slave chip to the IDT7132 and does not require arbitration circuitry. Most memory accesses can be completed without arbitration, but contention situations require arbitration. Contention occurs if both processors are writing the memory at the same time or if one processor is writing while the other is reading. A simultaneous read by both processors does not cause contention.

Without arbitration, simultaneous writes to a location result in an indeterminate value. The actual value stored is dependent on component timing and other variables.

A simultaneous read/write might occur when one processor is updating variables that the other processor is using in its computations. When one processor is reading a location the other is writing, the result without arbitration depends on the timing of the individual components. The result of the read might be the old value, the new value, or something in between.

The on-chip arbitration of the IDT7132 prohibits simultaneous access of a single memory location. The arbiter circuit consists of two address comparators and two BUSY output signals, one signal for each side of the dual-port memory. If both addresses are equal (and $\overline{\text{CE}}$ for that memory is active) the processor whose address arrived last is held with the $\overline{\text{BUSY}}$ signal. With the $\overline{\text{BUSY}}$ output of the RAM connected to the DMACK input of the corresponding ADSP-2100, the ADSP-2100 will insert wait states while the other ADSP-2100 completes a memory access. In this way, the access for one side is delayed.

When programming in this environment, it is important to be aware of the delay that can occur from contention. Wherever possible, contention should be avoided. One way to avoid contention is to synchronize program flow using the flags in software, so both processors do not access the same buffer concurrently. In the example software shown in this chapter, the alternating buffer scheme prevents the processors from trying to access the same data. Contention can occur only if both processors try to access the same flag.

17.3.2 Dual-Port Memory Interface

Two 2K x 8-bit dual-port RAM chips are shared between the two ADSP-2100s. This memory is used to transfer information between the two processors. The lowest 2K of data memory space on each processor (locations 0000-2048) is dual-ported. Additional dedicated memory can be added to each processor as needed; decoding is provided for a full 16K of data memory. If your application requires less memory, only include the amount you require. Additional dual-port memory can be used in place of private memory as needed.

The eleven low address bits (DMA0-DMA10) of each processor are connected to either the left-side or right-side address bits of the dual-port memory. The upper three address lines (DMA11-DMA14) are connected to a 1-of-8 decoder to determine which memory bank is enabled. Each BUSY output from the dual-port memory is connected directly to the DMACK input of the corresponding ADSP-2100. When contention occurs, the memory's arbitration circuitry pulls one of the $\overline{\text{BUSY}}$ lines low.

17.3.3 Decoder Timing

Because the memory consists of multiple 2K blocks of memory, a 1-of-8 decoder is necessary to produce the appropriate chip enable ($\overline{\text{CE}}$) signal for each memory chip. The delay incurred by decoding the address bits is important because the $\overline{\text{BUSY}}$ output delay from the IDT7132 is relative to $\overline{\text{CE}}$, and $\overline{\text{BUSY}}$ must be returned to the DMACK input of the ADSP-2100 well before the end of the access cycle.

The $\overline{\text{CE}}$-to-$\overline{\text{BUSY}}$ delay of the IDT7132 is 30ns (for a 45ns part). The ADSP-2100 has three timing requirements for DMACK. DMACK must be returned a specified time after DMA becomes stable (#75 timing parameter from the *ADSP-2100 Data Sheet*). During a read cycle, DMACK must be returned a specified time after $\overline{\text{DMRD}}$ goes low (#74). During a write cycle, DMACK must be returned a specified time after $\overline{\text{DMWR}}$ goes low (#99). Calculations for each of these three requirements determine the maximum permissable decoder delay, as shown in Figure 17.3. The decoder must be faster than the minimum value of all three.

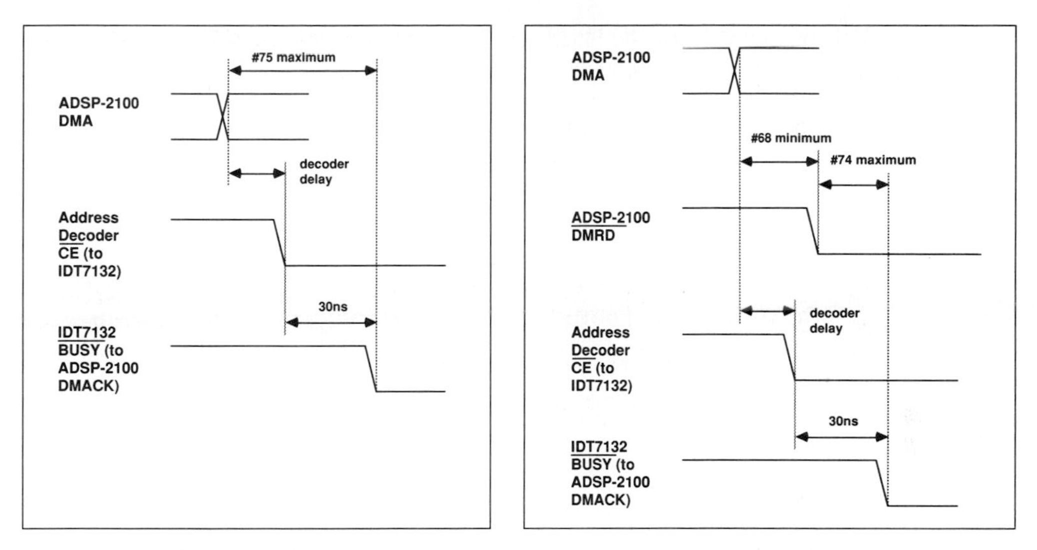

Requirement 1

Requirement 2

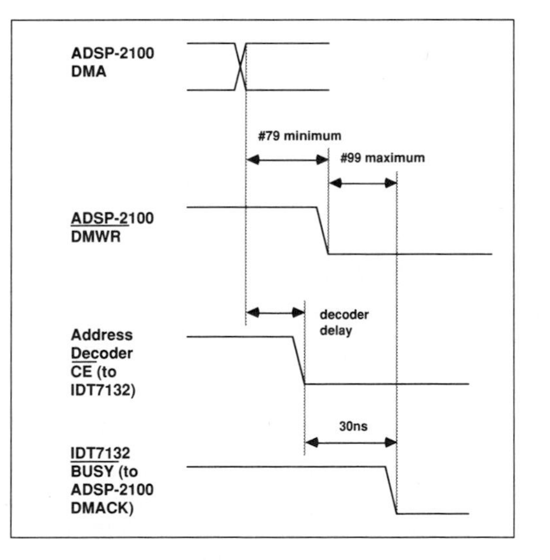

Requirement 3

Figure 17.3 Calculating Decoder Delay Requirement

Requirement 1, for the DMA to DMACK specification (#75), determines the maximum decoder delay as follows:

(DMACK to DMA Valid) – $\overline{\text{BUSY}}$ Delay

or

#75 – 30ns

Requirement 2, for the $\overline{\text{DMRD}}$ to DMACK specification (#74), must include the DMA to $\overline{\text{DMRD}}$ delay (#68). The maximum decoder delay is determined as follows:

(DMA to $\overline{\text{DMRD}}$) + ($\overline{\text{DMRD}}$ to DMACK) – $\overline{\text{BUSY}}$ delay

or

#68 + #74 – 30ns

Likewise, Requirement 3 for the $\overline{\text{DMWR}}$ to DMACK specification (#99), must include the DMA to $\overline{\text{DMWR}}$ delay (#79).

(DMA to $\overline{\text{DMWR}}$) + ($\overline{\text{DMWR}}$ to DMACK) – $\overline{\text{BUSY}}$ delay

or

#79 + #99 – 30ns

For an 8MHz ADSP-2100, the minimum value of all three calculations is 7ns. The 74FCT138A 1-of-8 CMOS decoder has a 6ns maximum delay, providing the necessary speed to use the 45ns dual-port memory.

17.4 SYNCHRONIZING MULTIPLE ADSP-2100S

Although processor clock synchronization is not absolutely necessary in a multiprocessor environment, it is advisable. Synchronization ensures that both processors are executing the same internal phase at the same time.

In a system which contains more than one ADSP-2100, synchronization is guaranteed if the processors share a common clock and a common $\overline{\text{RESET}}$ signal. When the $\overline{\text{RESET}}$ line is asserted, both ADSP-2100s become synchronized so that their internal clock phases are the same.

17.5 DEVELOPMENT TOOLS

The ADSP-2100 Development Tools can be used with multiprocessing architectures. Following a few guidelines in writing the software ensures proper operation. Each part of the Development Tools that requires special consideration is described below.

17.5.1 System Builder

The System Architecture file for the system presented in this chapter is shown in Listing 17.1. The program memory space is allocated with 16K of memory for instructions and 16K for data. The data memory space is also divided into two blocks. The lowest 2K is the dual-port memory that shared by both processors. Each processor has an additional 14K of private data memory.

Both processors use the same architecture file in this example. In other applications, each processor might have different memory requirements. The dual-port memory could be mapped into different locations on each processor. For example, one processor could map the dual-port memory in the 0 to 2K range, while the other maps the common memory in the 14K to 16K range.

```
.SYSTEM                         multiprocessor;

{Program Memory Section}
.SEG/ROM/ABS=0/PM/CODE           code_area[H#4000];
.SEG/RAM/ABS=H#4000/PM/DATA      data_area[H#4000];


{Data Memory Section}
.SEG/RAM/ABS=0/DM/DATA           common_memory[H#0800];
.SEG/RAM/ABS=H#800/DM/DATA       private_memory[H#3800];

.ENDSYS;
```

Listing 17.1 System Architecture File

17.5.2 Assembler

Two common arrays and associated flags store the filtered data. They are declared in each processor's main routine (shown in Listings 17.2 and 17.3).

The absolute (ABS) directive causes the variable to be stored in the absolute location specified in the code. Without the ABS directive, the

Linker is free to place data anywhere in available memory. The Linker places variables in contiguous areas of memory if their declarations are all on the same line.

Because the dual-port memory is defined in both processors' data memory space, the processors must not have conflicting allocations in shared memory. The best way to avoid this is to dedicate sections of the dual-port memory not needed for communication to one processor only. This can be done by allocating a dummy array in one processor's code over the range of memory dedicated to the other processor.

Listing 17.2 shows the variable declarations for the filter processing module. This code executes an FIR filter on the data. The output of the filter is stored in one of the two data arrays. When the array is full, the appropriate full_flag is set, informing the peak processor to start its operations. The filter processor can then filter another window of data into the alternate data array.

The Linker is free to place additional variables anywhere in the unallocated areas. No contention can occur within the common memory, because it is entirely allocated to one processor or the other.

```
.MODULE      data_filter;

.INCLUDE     <const.h>;

{Dual Port Memory Declarations}
.VAR/DM/RAM/ABS=0              full_flag_1, data_1[256];
.VAR/DM/RAM/ABS=256            reserved_for_peak_module[767];

.VAR/DM/RAM/ABS=1024           full_flag_2, data_2[256];

{Private Memory Declarations}
.VAR/PM/RAM/CIRC               coefficient[taps];
.VAR/DM/RAM/CIRC               delay[taps];

.INIT        coefficient : <coeff.dat>;
.INIT        full_flag_1 : 0;
.INIT        full_flag_2 : 0;
```

```
            I0=^delay;          L0=%delay;
            I4=^coefficient;    L4=%coefficient;
            I1=^data_1;         L1=0;
            M0=0;               M1=1;               M4=1;

main:       I1=^data_1;
            CALL zero_delay;
            CALL do_filter;
            AX0=H#FFFF;
            DM(full_flag_1)=AX0;
            I1=^data_2;
            CALL zero_delay;
            CALL do_filter;
            AX0=H#FFFF;
            DM(full_flag_2)=AX0;
            JUMP main;                              {continue loop}

zero_delay: CNTR=%delay;
            AX0=0;
            DO zero_it UNTIL CE:
zero_it:       DM(I0,M1)=AX0;
            RTS;

do_filter:  CNTR=%data_1;
            DO filter_data UNTIL CE;
               CNTR=taps-1;
               SI=DM(I1,M0);
               DM(I0,M1)=SI;
               MR=0, MX0=DM(I0,M1), MY0=PM(I4,M4);
               DO tap_loop UNTIL CE;
tap_loop:         MR=MX0*MY0 (SS), MX0=DM(I0,M1), MY0=PM(I4,M4);
               MR=MX0*MY0 (SS);
               IF MV SAT MR;
filter_data:   DM(I1,M1)=MR1;

            RTS;

.ENDMOD;
```

Listing 17.2 Source Code for Filter Processor

The array called *reserved_for_peak_module* is an array of common memory that used only by the peak processor to avoid conflicting memory allocations. Listing 17.3 shows the variable declarations for the peak processor. The two ranges of memory shared by the processors have identical declarations. The only difference is that the peak processor has the reserved space in the upper 1K of the dual-port area for the filter processor.

```
.MODULE/ABS=0                 peak_processor;

{Dual Port Memory Declarations}
.VAR/DM/RAM/ABS=0             full_flag_1,data_1[256];

.VAR/DM/RAM/ABS=1024          full_flag_2,data_2[256];
.VAR/DM/RAM/ABS=1281          reserved_for_filter_module[767];

          RTI;
          RTI;
          RTI;
          RTI;

          L0=0;
          M0=1;

main:     AX0=DM(full_flag_1);
          AR=PASS AX0;
          IF EQ JUMP main;
          I0=^data_1;
          CALL peak;
                              {Do something with the peak value}
          AX0=0;
          DM(full_flag_1)=AX0;
check_2:  AX0=DM(full_flag_2);
          AR=PASS AX0;
          IF EQ JUMP check_2;
          I0=^data_2;
          CALL peak;          {Do something with the peak value}
          AX0=0;
          DM(full_flag_2)=AX0;
          JUMP main;
```

```
peak:        CNTR=%data_1-1;
             AY0=DM(I0,M0);
             AR=PASS AY0;
             DO find_peak UNTIL CE;
                AF=AR-AY0, AY0=DM(I0,M0);
find_peak:      IF LT AR=PASS AY0;
             AF=AR-AY0;
             IF LT AR=PASS AY0;
             RTS;

.ENDMOD;
```

Listing 17.3 Source Code for Peak Processor

17.5.3 Simulation

The multiprocessing environment can be tested using the Simulator. When simulated, the filter program produces output data and stores it in the common data memory. You can then use the Simulator command to dump from data memory to store the dual-port data memory image on disk. Restart the Simulator, loading the peak processor program, and execute the Simulator command to reload the image of the common memory. Then simulate the peak processor program, which operates on the data generated by the filter program. A batch file that automatically executes this sequence of commands (using ADSP-2100 Cross-Software version 1.5x commands) is shown in Figure 17.4.

Command	*Comment*
`load filter`	Load filter program
`readimage test.dat`	Load memory image of input data
`run`	Execute filter program
`dumpdm full_flag_1 257 hold.dat i`	Write the file *hold.dat* with the flag and buffer data
`load peak`	Load peak program
`readimage hold.dat`	Load memory image of flag and data buffer from *hold.dat*
`run`	Execute peak program

Figure 17.4 Batch File Example

Host Interface ■ 18

18.1 OVERVIEW

This chapter describes host interfacing techniques in general using the example of the Motorola 680x0 family of processors. Today's computer CPUs are very powerful. They have large, versatile instruction sets and addressing modes. They can handle complicated sequencing and stacking, address and manage large data spaces, even while keeping track of virtual memory, user and supervisor modes, etc. However, these CPUs do not perform fast numerical operations. The ADSP-2100 can act as a fast numerical coprocessor to a host CPU. The host CPU takes care of administrative duties while the ADSP-2100 processes numerical data concurrently. This coprocessor configuration is useful in computationally intensive applications, such as graphics, spectral analysis, data compression, linear algebra, vector estimation, encryption, error coding, image processing, and speech recognition.

18.2 INTERFACE CONFIGURATIONS

One of the most common host/coprocessor architectures is the master-slave arrangement. The host CPU acts as the master in a system, passing data and/or instructions to the slave coprocessor. Communication from the slave coprocessor back to the host is usually very minimal, consisting of interrupt signals sent to the host or flags set for the host to poll.

Host interfacing is also an issue for interprocessor communications in distributed processing architectures. Hardware and software interface issues are similar to those of the master-slave configuration. Although only the master-slave(s) architecture is discussed in this chapter, distributed processing intercommunications can be developed from the principles outlined here.

There are four general methods of communication between processors:

- Memory bus sharing/arbitration
- Hardware communication ports separate from the memory interface
- Concurrent memory sharing using dual-ported memory
- Memory swapping

18.2.1 Bus Sharing

The ADSP-2100 has bus request ($\overline{BR}$) and bus grant ($\overline{BG}$) pins for memory bus sharing and arbitration protocol (see Figure 18.1). A host CPU asserts the $\overline{BR}$ input of the ADSP-2100 to request access to the ADSP-2100's local program or data memory. The ADSP-2100 responds by asserting its $\overline{BG}$ output and releasing control of its memory interface. The host CPU can then drive the ADSP-2100's memory interface with its own signals. Data stored in data memory or program memory can be read, written or modified, and program instructions in program memory can be initialized or inspected. When the host CPU is done driving the ADSP-2100's memory interface, it deasserts the $\overline{BR}$ input. The ADSP-2100 responds by re-establishing its memory interface, deasserting $\overline{BG}$ and continuing program execution.

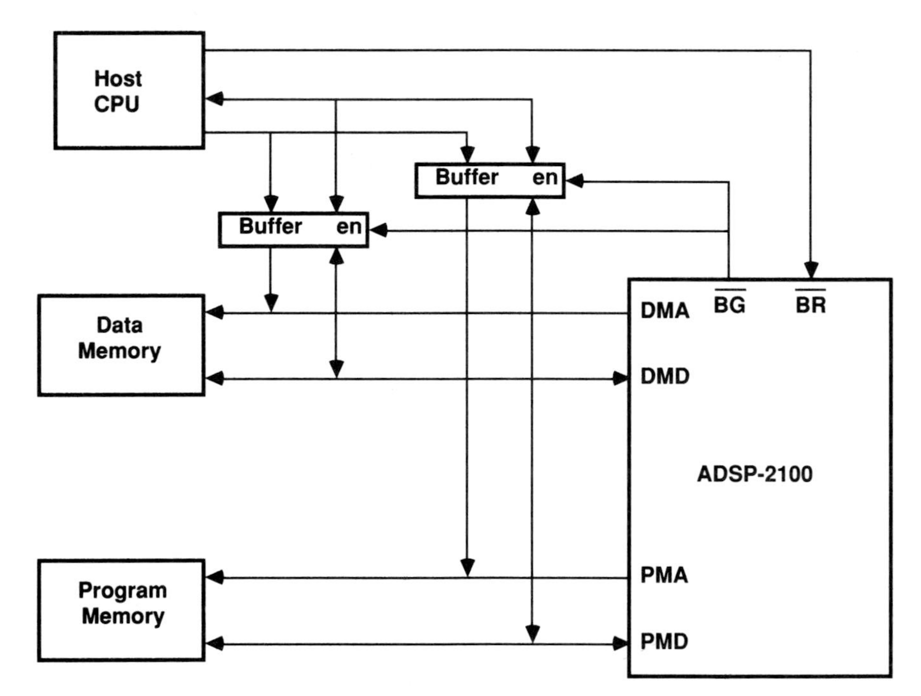

Figure 18.1 Bus Sharing Interface

18.2.2 Communication Ports

Hardware communication ports on chip provide a communication channel separate from the memory interface. Whether serial or parallel, these ports usually operate at a rate considerably slower than direct memory access through bus sharing. The ADSP-2100 does not provide any hardware communication ports. However, the ADSP-2101/2 microcomputer has two serial communication ports (see Figure 18.2). Use of the serial ports is explained in the *ADSP-2101/2 User's Manual* and the application note "Loading an ADSP-2101 Program via the Serial Port."

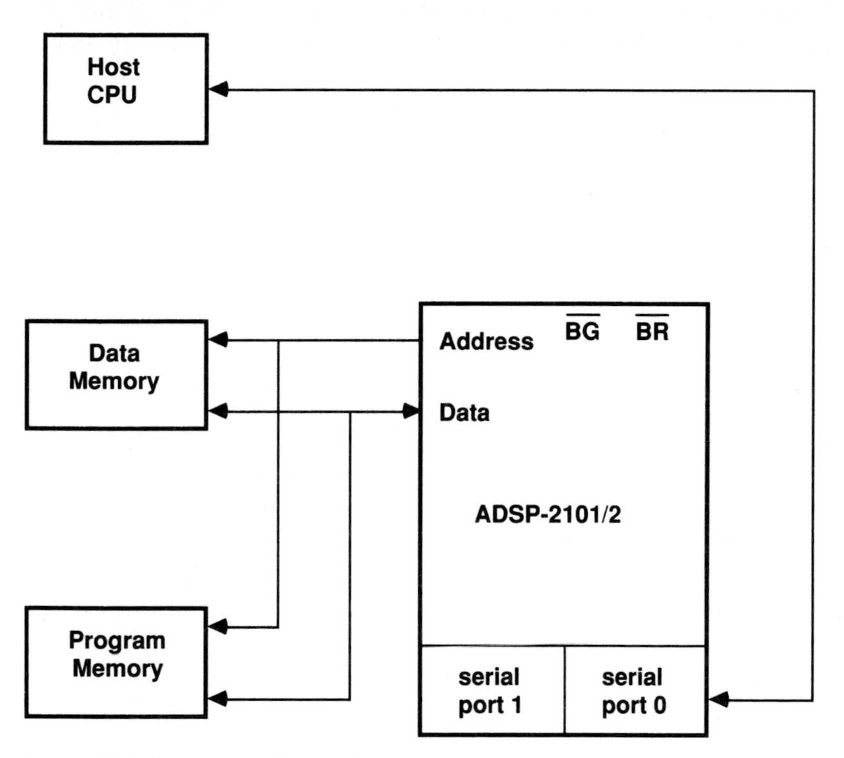

Figure 18.2 Communication Port Interface

18.2.3 Dual-Port Memory

Dual-port memories are high-speed SRAM memories which have two sets of address, data and read/write control signals (see Figure 18.3). Each set of memory controls can independently and simultaneously access any word in the memory; both sides of the dual-port RAM can access the same memory location at the same time. For host interfacing, you connect one side of the dual-port memory to the main CPU and the other side to the ADSP-2100. Because the two processors can write to the same location in memory at the same time, or one can read a location while the other is changing the same location, there must be some arbitration to decide which device has precedence, and there must be a way to hold off the processor of lower priority so it can extend its bus cycle until the memory is accessible.

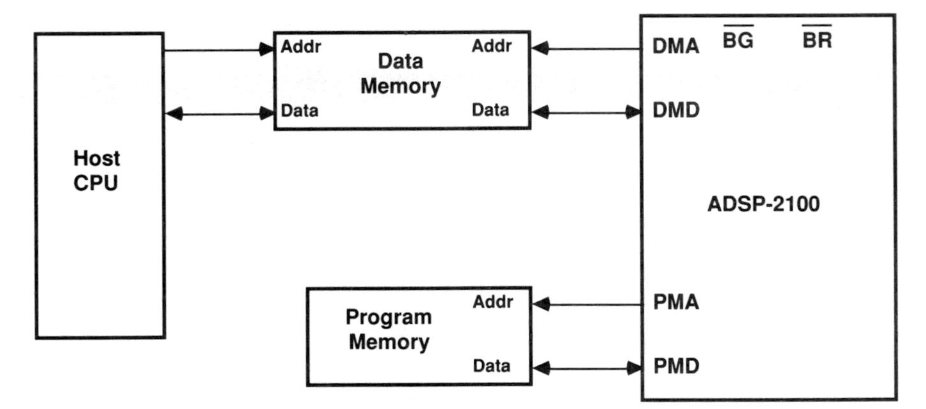

Figure 18.3 Dual-Port Memory Interface

Dual-port memories offer four types of arbitration:

- *Hardware busy logic* generates a busy signal to one port when the other port is writing the same location. The busy signal holds off Port B access to a location to which Port A is writing, and vice versa.

- *Semaphore logic* has on-chip semaphore latches that can be set and polled; each port passes a flag, or token, to the other to indicate that a shared resource is in use. Semaphore logic allows one processor to lock out the other when accessing a particular block of data.

- *Interrupt logic* generates an interrupt at Port B when Port A writes to a special location. Port B clears the interrupt by reading that location. When Port B writes to another special location, a similar interrupt is generated at Port A.

- *No arbitration logic* at all is used on some dual-port RAMs. If these devices are used, the system must have been designed so that contention never occurs.

See Chapter 17, *Multiprocessing*, for an example of the use of dual-port memory.

18.2.4 Memory Swapping

Memory swapping uses several hardware copies of the same memory space; only one copy (called a frame) is enabled at a time per processor (see Figure 18.4, on the following page). A different frame is enabled for each processor. The host processor can load one frame with data while the slave processor is processing data in another frame. When both processors are done, the enable signals to the memories are switched, and it appears to each processor as if all the data in its frame has changed instantaneously. This scheme is often used in systems in which multiple processors compute assigned sections of a larger algorithm. When two frames are used, or when two buffers within a frame are exchanged in software, this is called ping-ponging.

18.3 ADSP-2100 INTERFACE CONSIDERATIONS

This section describes in detail the considerations for implementing a bus sharing interface to the ADSP-2100.

18.3.1 Bus Request

The normal synchronous mode of granting a bus request proceeds as follows. The external device (host CPU) requests the buses by asserting the $\overline{BR}$ input of the ADSP-2100. This input is recognized by the ADSP-2100 at the end of the next internal clock state three, and the ADSP-2100 halts in state eight of the same instruction cycle. The ADSP-2100 handshakes with the host CPU by asserting $\overline{BG}$ at the end of state three of what would have been the next instruction cycle (four CLKIN cycles after the bus request is recognized). Typically, the $\overline{BG}$ output of the ADSP-2100 is used to enable bus transceiver chips (74F244 or 74F245), establishing a connection between the host CPU's bus signals and the ADSP-2100's memory.

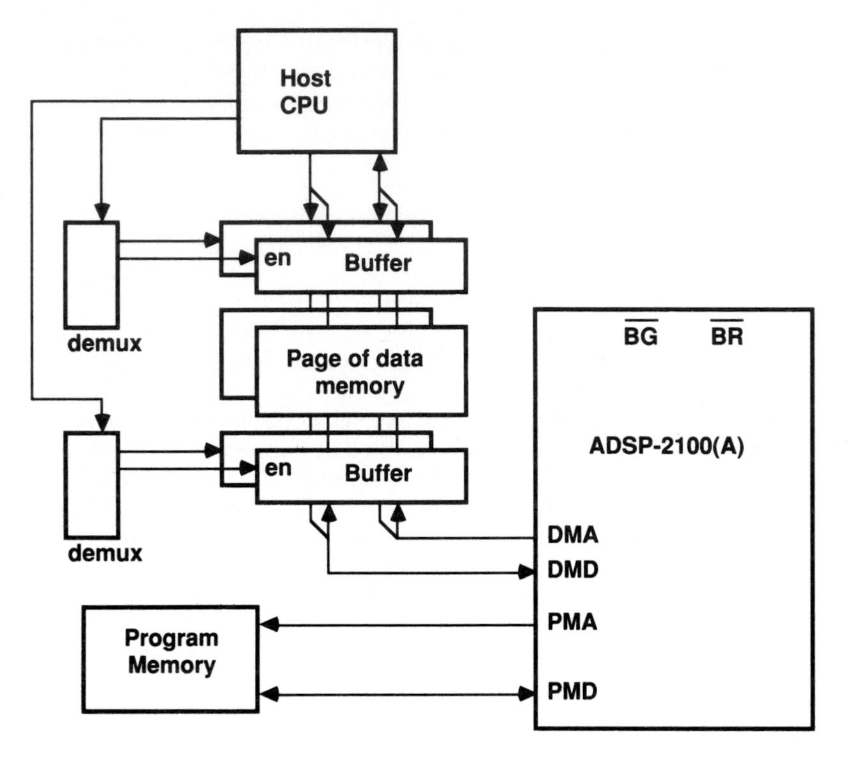

Interface to only data memory shown

The host must ensure that it and the ADSP-2100 use different memory pages

Figure 18.4 Memory Swapping Interface

The ADSP-2100 then tristates all outputs, including PMA, PMD, PMDA, $\overline{\text{PMWR}}$, $\overline{\text{PMRD}}$, and $\overline{\text{PMS}}$ for program memory, and DMA, DMD, $\overline{\text{DMWR}}$, $\overline{\text{DMRD}}$, $\overline{\text{DMS}}$ for data memory. The PMD, $\overline{\text{PMS}}$, $\overline{\text{PMWR}}$, $\overline{\text{PMRD}}$, DMD, $\overline{\text{DMS}}$, $\overline{\text{DMWR}}$ and $\overline{\text{DMRD}}$ signals are internally pulled up by 50kΩ while $\overline{\text{BG}}$ is active. The PMA, PMDA, and DMA signals are not internally pulled up.

When the host CPU is finished accessing the ADSP-2100's memory, it deasserts $\overline{\text{BR}}$. The ADSP-2100 deasserts $\overline{\text{BG}}$ four CLKIN cycles (internal clock states) later and re-establishes control over its program and data memory signals in state one of the next instruction cycle, resuming operation as it left off.

A bus grant can occur even in the double instruction cycle execution of a program data memory fetch when the cache is not valid. It is possible to interrupt that instruction's execution with $\overline{\text{BR}}$ in between the two cycles.

The ADSP-2100's internal state is not affected by the bus granting operation. The only restriction is that $\overline{\text{RESET}}$ should not be changed during a bus grant. Activity on the $\overline{\text{RESET}}$ input while the ADSP-2100 is asserting $\overline{\text{BG}}$ causes indeterminate operation. When downloading a program to program memory from an external device (on power-up for example), you should proceed in the following order. This order ensures that program execution starts at PC=h#0004 with the ADSP-2100 in a known state.

1. Assert the $\overline{\text{RESET}}$ signal for the standard full instruction cycle
2. Assert the $\overline{\text{BR}}$ signal
3. Download the information
4. Deassert the $\overline{\text{BR}}$ signal
5. Deassert the $\overline{\text{RESET}}$ signal

During $\overline{\text{RESET}}$, the timing of $\overline{\text{BR}}$ and $\overline{\text{BG}}$ is different from their timing during normal operation; the operation is asynchronous during $\overline{\text{RESET}}$. See the *ADSP-2100/2100A Data Sheet* or the *ADSP-2100 User's Manual* for details.

18.3.2 Software Handshake

Handshaking information is needed to tell the slave coprocessor when new data is available and the host when the slave coprocessor is done processing its data. A simple method uses two locations in ADSP-2100 memory reserved as flags: one, which the ADSP-2100 polls, to indicate the presence of new data, and the other, which the host polls, to indicate the completion of the data processing task. The ADSP-2100 can poll the new-data flag until data is available, process the data, set the data-done flag, and go back to polling for new data. An example of the ADSP-2100 code to execute this loop is shown below:

```
.VAR/ABS=h#3FFE   new_data;
.VAR/ABS=h#3FFD   data_done;
.INIT             new_data:    0;         {0=FALSE, else =TRUE}
.INIT             data_done:   0;         {0=FALSE, else =TRUE}

idlestart:        DO idleloop UNTIL NE;   {wait for new data}
                     AR=DM(new_data);
idleloop:            AR=PASS AR;
                  AR=PASS 0;
                  DM(new_data)=AR;        {clear the flag}
                  CALL process_it;        {process the data}
                  AX0=h#FFFF;
                  DM(data_done)=AX0;   {indicate processing done}
                  JUMP idlestart;
```

18.3.3 Hardware Handshake Using Interrupts

The ADSP-2100 and host CPU can also handshake by sending hardware interrupts to indicate, for example, that new data is available or that data processing is complete. The simplest configuration connects the ADSP-2100's TRAP output directly to the host CPU's interrupt input. The routine that the ADSP-2100 executes is terminated by a TRAP instruction. This halts the ADSP-2100 and asserts the TRAP output high. The host's interrupt service routine can use $\overline{\text{BR}}$ and $\overline{\text{BG}}$ to download new data to the ADSP-2100's memories and then assert and release the ADSP-2100's $\overline{\text{HALT}}$ line, causing the ADSP-2100 to continue processing on the next instruction. The next instruction can jump back to the beginning of the same routine.

Another way to handshake with interrupts is to have address decoding logic generate an interrupt to one processor when the other processor accesses a particular memory address. The interrupt signal can be tailored for an active low, active high, or edge-triggered interrupt, depending on the input of the particular processor.

18.3.4 Software and Hardware Handshake Comparison

The advantage of a hardware handshake over a software flag handshake is speed. The receiving processor responds to the interrupt as soon as it becomes active. Code is reduced as well because there is no need to perform the polling functions using $\overline{BR}$ and $\overline{BG}$.

The advantage of software flag polling over the hardware interrupts is that the whole system remains in a static state on a cycle-by-cycle basis. The handshaking status can be inspected using the CPU's software development tools, making the system easier to debug. In addition, there is no need to redesign interrupt logic if the interrupts are used for other purposes. The example described in this chapter performs software flag handshaking.

18.4 68000 INTERFACE CONSIDERATIONS

This section describes the characteristics of the 68000 microprocessor that are relevant to the ADSP-2100 interface.

18.4.1 68000 Addressing

The 68000 processor has a 16-bit data bus and a 24-bit address bus. Its 16 general purpose internal registers are each 32 bits wide. The 68000 recognizes five data types:

- Bit
- BCD (4-bit)
- Byte
- Word
- Long word (32-bit)

The 24-bit address bus allows the 68000 to access 16 megabytes of external memory; however, only 23 address bits actually are externally available as address pins. The least significant address bit (A_0) is not externally available. Therefore, the 16 megabytes of memory are actually organized as 8 megawords of 16-bit words located at even-numbered addresses (LSB = 0).

Byte addressing allows the 68000 to access 8-bit byte data on both even and odd address boundaries in memory. The $\overline{UDS}$ (upper data strobe) and $\overline{LDS}$ (lower data strobe) outputs of the 68000 are used for byte-oriented addressing.

18.4.2 68000 Bus Signals

The block diagram of the 68000 in Figure 18.5 shows the control signals to the 68000. The interface with the ADSP-2100 described in this chapter involves only the address bus, data bus, and asynchronous bus control signals. All other signals are left to the 68000 system designer's implementation because they do not affect the ADSP-2100 interface directly.

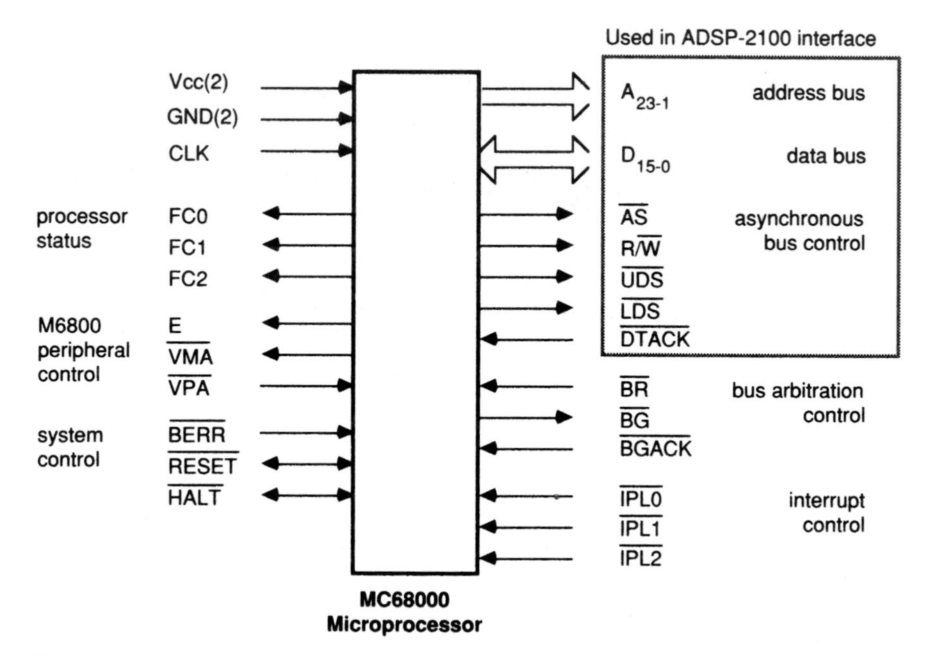

Figure 18.5 68000 Block Diagram

The control of the 68000's bus is asynchronous. Once a bus cycle is initiated, it is not completed until a handshaking signal ($\overline{DTACK}$) is asserted by external circuitry. The signals that control address and data transfers are:

- Address strobe ($\overline{AS}$)
- Read/write (R/$\overline{W}$)
- Upper data strobe ($\overline{UDS}$)
- Lower data strobe ($\overline{LDS}$)
- Data transfer acknowledge ($\overline{DTACK}$)

The 68000 asserts $\overline{AS}$ when an address is available and asserts R/$\overline{W}$ high or low to indicate whether a read or a write is to take place over the bus. Because the bus cycle is asynchronous, external circuitry must signal the 68000 when the bus cycle can be completed by asserting the $\overline{DTACK}$ input to the 68000. During a read cycle, $\overline{DTACK}$ low tells the 68000 that valid data exists on the data bus. In response, the 68000 latches in the data into the chip and terminates the bus cycle. Similarly, on a write cycle, $\overline{DTACK}$ low informs the 68000 that the data has been successfully written to memory or a peripheral device, and the 68000 responds by ending the bus cycle.

On the ADSP-2100, DMACK can be tied high if all memory accesses can complete in a single instruction cycle; in the absence of a DMACK low, the ADSP-2100 data memory read or write cycle ends in one cycle. Similarly, $\overline{DTACK}$ on the 68000 can be hardwired low if the design of the system is such that the 68000 bus cycle can always be completed without any wait states.

The $\overline{UDS}$ and $\overline{LDS}$ signals act as an extension of the address bus, replacing the address LSB A_0. In the case of a byte transfer, they indicate whether the data is on the upper data lines (D_{15-8}) or the lower data lines (D_{7-0}) as shown below:

R/$\overline{W}$	*$\overline{UDS}$*	*$\overline{LDS}$*	*Operation*
0	0	0	word —> memory or peripheral
0	0	1	high byte —> memory or peripheral
0	1	0	low byte —> memory or peripheral
0	1	1	(invalid data)
1	0	0	word —> 68000
1	0	1	high byte —> 68000
1	1	0	low byte —> 68000
1	1	1	(invalid data)

18.5 68000-TO-ADSP-2100 BUS SHARING INTERFACE

The example in this chapter shows how a 68000 CPU can control DMA to and from an ADSP-2100 processor using bus sharing and arbitration. Other CPUs in the 680x0 family have similar interfaces. A schematic for the interface is shown in Figure 18.11 later in this chapter.

18.5.1 Memory Mapping

Data transfers between the 68000 and the ADSP-2100 occur in a 16-bit word format because that is the data type expected by the ADSP-2100.

Note that 16-bit words are stored in ADSP-2100 memory on every address boundary, whereas the 68000 expects 16-bit words only on even address boundaries. In this example, the ADSP-2100 data memory space overlays the 68000 memory in the address range $02xxxx. The interface circuit maps the ADSP-2100 data memory addresses h#0000 to h#3FFF into 68000 memory addresses $020000 to $027FFE, as shown in Figure 18.6.

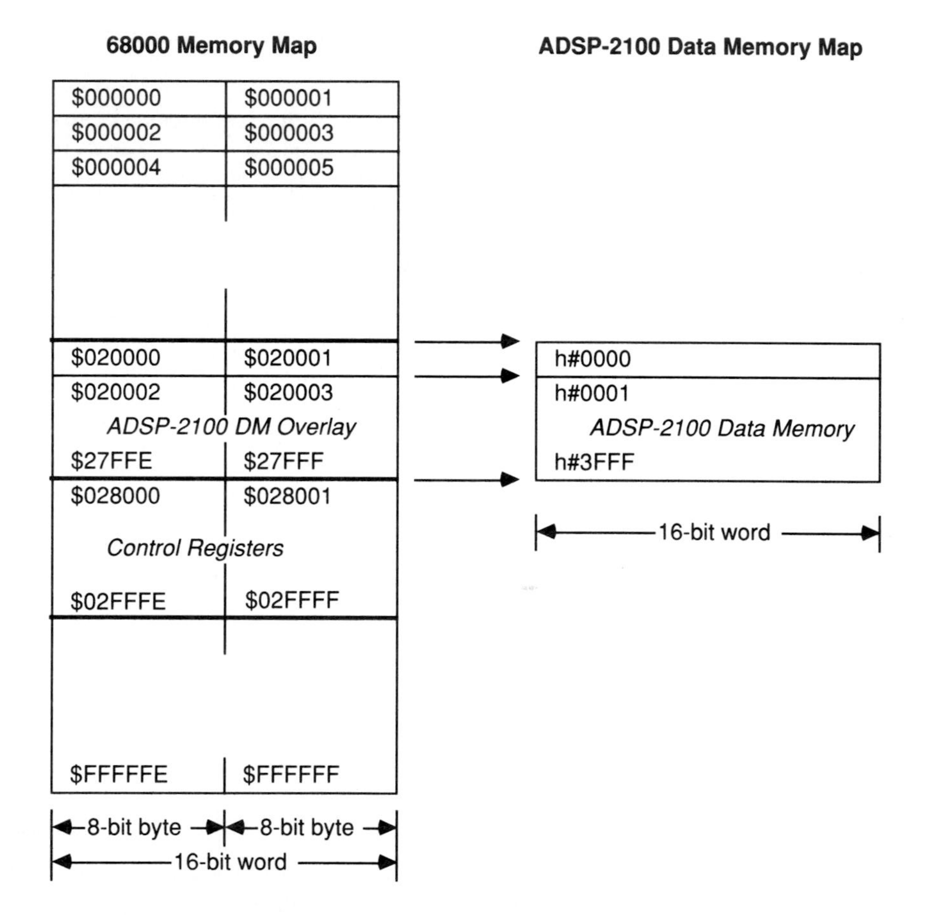

Figure 18.6 Memory Map Overlap

Software handshaking is implemented by memory-mapping two hardware control registers in the 68000's address space. Figure 18.7 shows the 68000 decoding logic to select between ADSP-2100 memory and the transmit and receive control registers. The control registers are used to pass hardware signals between the 68000 and the ADSP-2100, under software control. These registers are described in the next section. The 68000 address bit A_{15} differentiates between accesses to the "68000 memory" that is actually the ADSP-2100 data memory space and accesses to the control registers. The control registers are memory-mapped in the 68000 memory space at address $028000. However, because only $02xxxx and A_{15} are decoded, the registers can be accessed at any address from $028000 to $02FFFE. The 68000 $R/\overline{W}$ output differentiates between the transmit and receive control registers.

From 68000
(E1 from address
decoder)

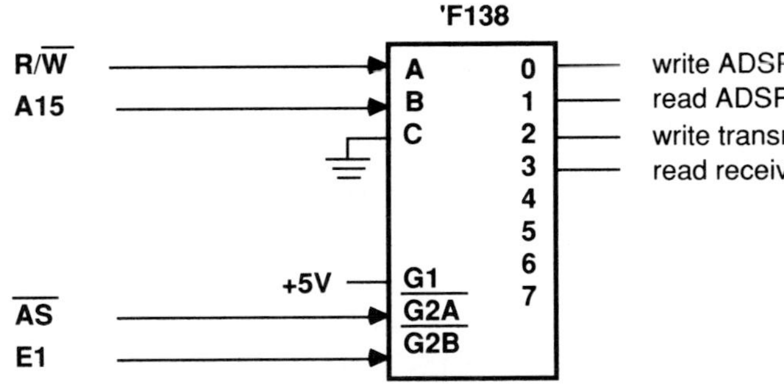

A15	$R/\overline{W}$	$\overline{AS}$	E1	
0	0	0	0	write ADSP-2100 data memory
0	1	0	0	read ADSP-2100 data memory
1	0	0	0	write transmit control register
1	1	0	0	read receive control register
x	x	x	1	none
x	x	1	x	none

Figure 18.7 Decoder for Data Memory and Control Registers

18.5.2 Control Registers

The 68000 requests access to the ADSP-2100 memory by writing to the transmit control register. The 68000 reads back the status of $\overline{\text{BG}}$ from the ADSP-2100 by reading the receive control register. An example circuit for the transmit control register is shown in Figure 18.8. The register in this example consists of 74F74 D flip-flops. One of the 74F74 Q output pins is tied directly to the $\overline{\text{BR}}$ input of the ADSP-2100. To assert or deassert $\overline{\text{BR}}$, software writes a logic 0 or 1 to the $\overline{\text{BR}}$ bit position.

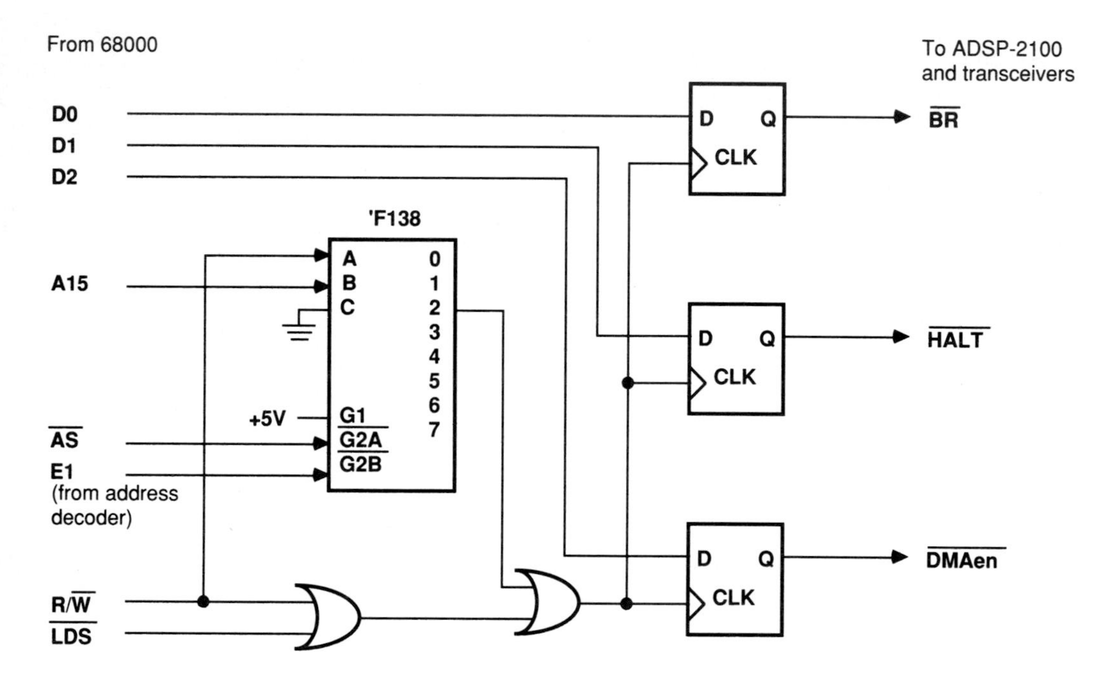

Figure 18.8 Transmit Control Register

An example receive control register circuit is shown in Figure 18.9. The receive control register is implemented by simply connecting the $\overline{BG}$ output of the ADSP-2100 to an input of a 74F244 bus transceiver and connecting the output of the transceiver to the 68000's data bus. The 74F244 bus transceiver is normally deselected, but when the 68000 outputs the address for the control register, the transceiver is enabled, and $\overline{BG}$ is then electrically connected to the 68000's data bus. The same transceiver can be used to inspect seven other hardware logic levels or to provide a read-back function to the transmit control register or can even be expanded in parallel to monitor more signals.

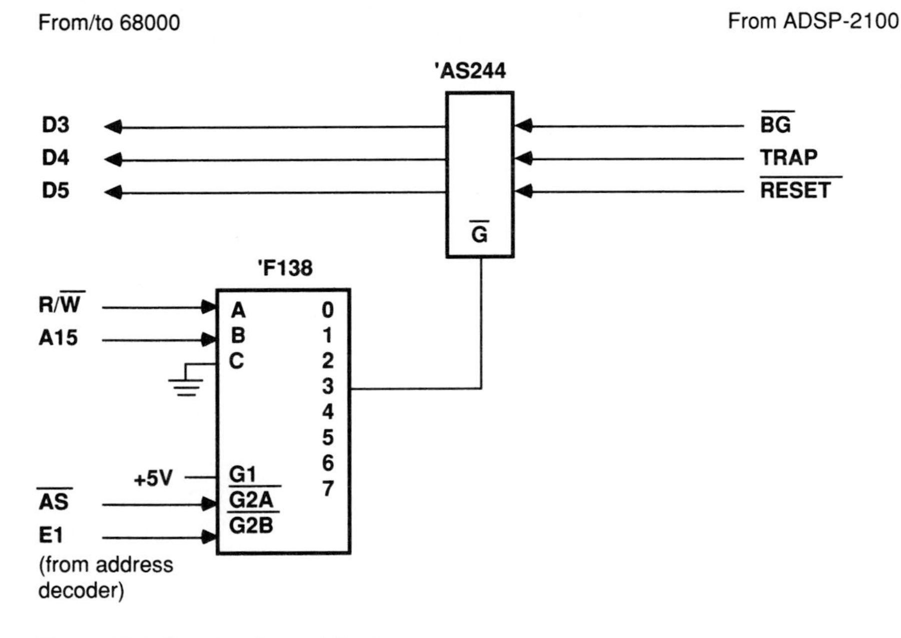

Figure 18.9 Receive Control Register

Examples of the 68000 address decoding logic used to select the hardware control registers are shown in Figure 18.10. Each register's address is decoded with a 74F138 and/or 74F521 and is qualified by $\overline{AS}$. The output of the decoder clocks the data from the 68000 data bus into the 74F74 flip-flops of the transmit control register or enables the transceiver of the receive control register.

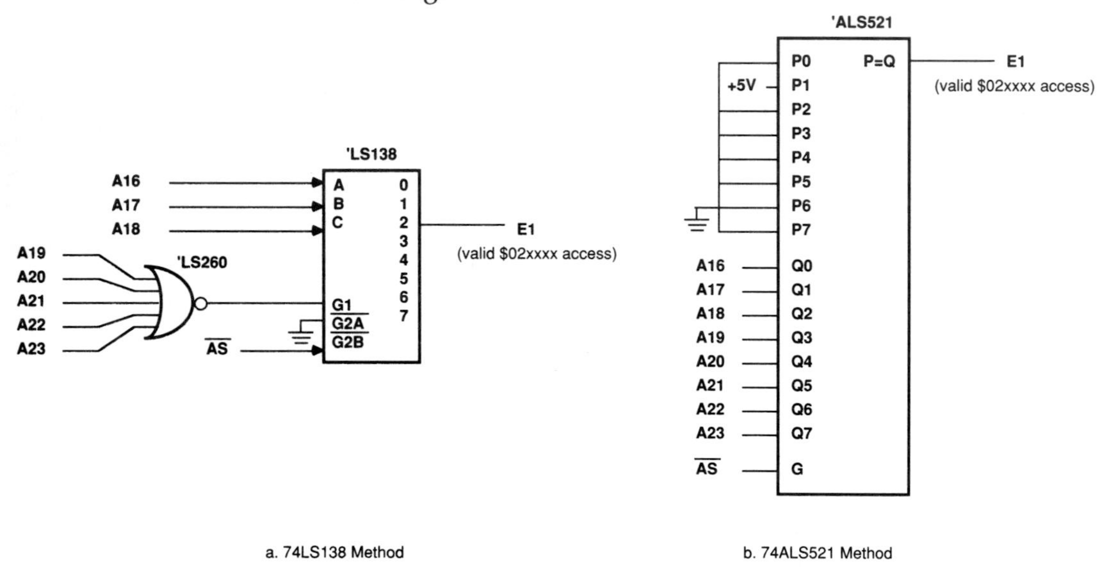

Figure 18.10 Two Methods for Decoding $02xxxx Accesses

When the 68000 writes to the transmit control register, $\overline{DTACK}$ is returned to the 68000 by a 74F244 buffer that is enabled whenever a valid access to address $02xxxx memory space (the interface circuitry) is detected. Because the 68000 in this example runs so much slower than the ADSP-2100, there is no need to extend the 68000 bus cycle. Therefore, the $\overline{DTACK}$ input to the 74F244 buffer is simply tied to ground. In a different application requiring wait states in the 68000 bus cycle, the grounded input of the 74F244 would be replaced by the output of a wait state generator.

18.5.3 Interprocessor Data Transfers

Once control of the ADSP-2100's buses has been granted to the 68000, the 68000 must be able to send and receive data and memory control signals. Another bit in the transmit control register enables a pair of 74F244 bus buffers which provide a path for 68000-generated addresses to the ADSP-2100's data memory address bus. Because the two 74F244 buffers can pass 16 signals, but the ADSP-2100 requires only 14 address signals, the two extra lines can be used for memory control signals. They can be used to send $\overline{\text{DMRD}}$ and $\overline{\text{DMWR}}$, while the existing ADSP-2100 address decoding logic generates $\overline{\text{DMS}}$. In this example, the ADSP-2100 data memory has all $\overline{\text{DMRD}}$ inputs grounded, and $\overline{\text{DMS}}$ is used as an enable to the decoding logic (74F138), so the two lines are used to send $\overline{\text{DMS}}$ and $\overline{\text{DMWR}}$. A 4.7kΩ resistor pulls up the $\overline{\text{DMWR}}$ signal so that when the 74F244 buffers are deselected (outputs are tristated, but $\overline{\text{BG}}$ is still asserted), the $\overline{\text{DMWR}}$ signal is forced high.

Two 74F245 bus transceivers connect the 68000's data bus to the ADSP-2100's data bus. In this case, all 16 bits are used for data. The transceivers' enables (G) are connected to address decoding logic. The direction controls (DIR) are tied directly to the 68000's R/$\overline{\text{W}}$ output.

The direct memory access operation between the 68000 and the ADSP-2100 proceeds as follows:

1. The 68000 and ADSP-2100 are processing separate tasks
2. The 68000 writes a 1 to the $\overline{\text{BR}}$ bit in the transmit control register
3. The 68000 reads the receive control register, and continues to read it until the $\overline{\text{BG}}$ bit is low
4. The 68000 writes a 1 to the bit in the transmit control register which enables the address buffers to the ADSP-2100 data memory (make sure to keep the $\overline{\text{BR}}$ bit set)
5. The 68000 reads or writes ADSP-2100 memory by accessing addresses $020000 to $027FFE in its own memory space
6. The 68000 writes a 0 to the bit in the transmit control register to deselect the address buffers
7. The 68000 writes a 0 to the $\overline{\text{BR}}$ bit in the transmit control register to deassert $\overline{\text{BR}}$
8. The 68000 reads the receive control register, and continues to read it until the $\overline{\text{BG}}$ bit is high
9. The 68000 and ADSP-2100 continue processing separate tasks

Important: If the 68000 is running much slower than the ADSP-2100, it is possible to simplify the 68000 software by not checking for the deassertion of $\overline{BG}$ in step 8, and steps 6 and 7 can be combined into one data write. Step 3 should not be skipped, however, because the ADSP-2100 might be waiting for DMACK to be asserted, in which case the ADSP-2100 would not immediately assert $\overline{BG}$. When the ADSP-2100 is halted or in a TRAP state, the bus request is latched by the ADSP-2100 and serviced after the normal synchronization delay; in this case, the ADSP-2100 remains halted, and tristates its buses. See the *ADSP-2100 User's Manual* for details.

18.6 USING PALS

The ADSP-2100 interface to a host processor using bus arbitration involves some logic functions which can be performed by a PAL (programmable array logic; a registered trademark of Monolithic Memories, Inc.) device rather than discrete logic devices. The savings in chip count is not significant because most of the chips needed for the interface are bus buffers and transceivers. It may be desirable, however, to integrate discrete gate functions into a PAL to allow more flexibility; the PAL can be reprogrammed, if necessary, to debug the circuit. A GAL (generic array logic; a registered trademark of Lattice Semiconductors, Inc.) device can emulate several types of PALs, providing even more flexibility.

To program a PAL or a GAL, you determine Boolean expressions for all the output signals as functions of the input signals. PAL programming software generates a file from these expressions which is used to burn (program) the PAL.

The following example of generating the $\overline{BR}$ signal shows how to derive the Boolean equations for PAL programming. $\overline{BR}$ is one bit of the transmit control register and is shown in the schematic of Figure 18.11 as the output of a D-latch or flip-flop. A registered PAL has internal D-latches (registers); one of the D-latches on a registered PAL can be used instead of discrete logic to generate $\overline{BR}$. The three signals needed for the D-latch are D (input), Q (output), and CLK. The input is the LSB (D_0) of the 68000 data bus, and the output is $\overline{BR}$. CLK must be generated internally by the logical combination of other input signals. The transmit control register is clocked by a combination of R/$\overline{W}$ low, $\overline{LDS}$ low, A_{15} high, $\overline{AS}$ low, and E1 ($02xxxx access) low (see the schematic in Figure 18.11). These signals are inputs to the PAL and are logically combined to yield an output signal which is tied directly to the PAL's clock input (pin 1 on a 16R4A).

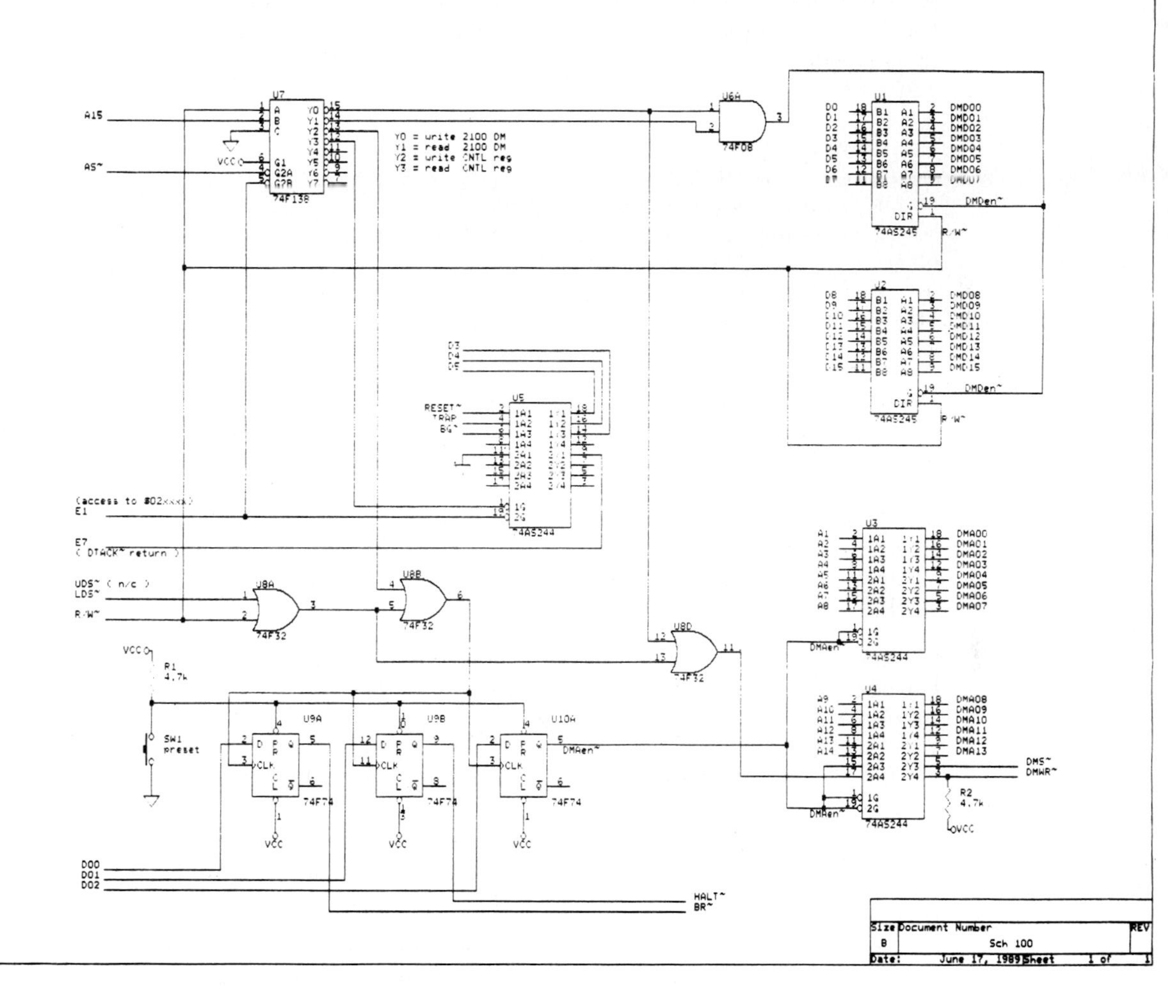

Figure 18.11 68000 Interface Schematic

18.7 680X0 FAMILY OF MICROPROCESSORS

The example in this chapter shows the ADSP-2100 interface with the Motorola 68000 processor, a typical host CPU. Any other CPU requires a similar interface, including the other processors in Motorola's 68000 family. Some of the differences are described in this section. See Johnson in *References* for more information about the differences between the 680x0 processors.

The other processors in the 68000 family are:

68008	8-bit byte version of the 68000
68010	same as 68000, but supports virtual memory
68012	extended physical address space as well as virtual memory support
68020	complete 32-bit virtual memory processor, coprocessor interface, instruction cache, dynamic bus sizing, more instructions, more addressing modes
68030	similar to the 68020, but more of everything, including data cache

Because all processors in this family are object-code upward compatible, the interface software is easily adapted from one family member to another. And because the 68000's asynchronous bus signals are practically the same throughout the 68000 family, the hardware is likewise easily adapted from one family member to another.

The 68010 hardware interface is exactly the same as the 68000 interface because the two processors are pin compatible. The 68010 supports virtual memory; if the ADSP-2100 has a full complement of memory physically present, there is no need to be concerned with page faults (access to virtual memory outside the currently active physical memory).

The 68012 is similar to the 68010 in its virtual memory support, but it has 30 external address bits (A_{29-1}, A_{31}) instead of 23 (A_{23-1}). Its asynchronous bus signals are the same as those of the 68010 and 68000, so the 68012 interface is the same as the 68000 interface except that it must decode more address bits. The 68012 has an additional asynchronous bus signal, $\overline{RMC}$ (Read-Modify Cycle), which indicates that an indivisible read-modify-write cycle is being executed. Because read-modify-write operations are not necessary for the ADSP-2100 interface, this signal can be ignored.

The 68008 is similar to the 68000, but it has only an 8-bit wide external data bus. Its asynchronous bus signals are the same as the 68000's, except that $\overline{\text{UDS}}$ and $\overline{\text{LDS}}$ are consolidated into $\overline{\text{DS}}$. If ADSP-2100 uses only 8 bits of its data bus, the 68008 interface is the same as the 68000 interface; however, if 16-bit data is needed, you must multiplex the 68008 data bus into high-byte and low-byte operations or, better yet, use a 68000.

The 68020 is a complete 32-bit processor with virtual memory support. All 32 address bits (A_{31-0}) and all 32 data bits (D_{31-0}) are available externally. The 68020 supports more instructions, more addressing modes, and more supervisor functionality than the 68000, plus a 256-byte instruction cache. The 68020 also features a coprocessor interface, intended for instruction-mapped coprocessors, such as the MC68882 floating-point coprocessor, not a memory-mapped coprocessor, such as the ADSP-2100 in this example.

An important additional feature of the 68020 is dynamic bus and data sizing. Because the 68020 can operate on many data types (bit, byte, word, long-word), the $\overline{\text{UDS}}$ and $\overline{\text{LDS}}$ signals of the 68000 are replaced by a data strobe ($\overline{\text{DS}}$) used in conjunction with two data-size output encoding signals (SIZ1, SIZ0). The SIZ1 and SIZ0 outputs indicate the number of bytes of an operand remaining to be transferred during a given bus cycle. Because all transfers between the 68020 and the ADSP-2100 are assumed to be 16-bit words, the SIZ1 and SIZ0 signals can be ignored.

$\overline{\text{DTACK}}$ is replaced by two signals, $\overline{\text{DSACK1}}$ and $\overline{\text{DSACK0}}$, on the 68020. The $\overline{\text{DSACKx}}$ pins perform the same asynchronous bus transfer acknowledge function, but with greater functionality. The $\overline{\text{DSACKx}}$ pins are defined as follows:

$\overline{\text{DSACK1}}$	*$\overline{\text{DSACK0}}$*	*Result*
H	H	insert wait states in current bus cycle
H	L	cycle complete - data bus port size is 8 bits wide
L	H	cycle complete - data bus port size is 16 bits wide
L	L	cycle complete - data bus port size is 32 bits wide

For example, if the processor is executing an instruction that requires a read of a long-word operand, it attempts to read 32 bits in the first bus cycle. If the port is 32 bits wide, the 68020 latches all 32 bits of data and continues with the next operation. If the port is 16 bits wide, the 68020 latches the 16 bits of data and runs another cycle to obtain the remaining 16 bits. An 8-bit port is handled similarly, requiring four read cycles.

In the ADSP-2100 interface, the 68020 operates on 16-bit word data (specified in 68020 assembly language by the .W qualifier). Therefore, the ADSP-2100 interface keeps $\overline{\text{DSACK0}}$ always high, and returns $\overline{\text{DSACK1}}$

high to insert wait states and low to complete the cycle. Notice that the 16 bits of the ADSP-2100 data must be tied to the 16 MSBs (D_{31-16}) of the 32-bit 68020 data bus. Additional asynchronous bus signals ($\overline{ECS}$, $\overline{OCS}$, $\overline{RMC}$ and $\overline{DBEN}$) can be ignored for the ADSP-2100 interface.

The 68030 has all the features of the 68020, plus a data cache with several external cache control pins. Its asynchronous bus signals are identical to those of the 68020, and thus its ADSP-2100 interface is the same as that of the 68020.

18.8 SUMMARY

There are four general methods of interfacing a host processor to the ADSP-2100. The same methods can be used to interface multiple ADSP-2100s with a host processor or with each other. The method of bus arbitration using the $\overline{BR}$ and $\overline{BG}$ pins of the ADSP-2100 to provide an interface with a Motorola 68000 host CPU has been presented in this chapter. The same interface can be easily adapted to other members of the 68000 family.

18.9 REFERENCES

Johnson, Thomas L. 1986. "A Comparison of MC68000 Family Processors." *BYTE Magazine.* Volume 11, Number 9. New York: McGraw-Hill, Inc.

Motorola, Inc. 1986. *M68000 8-/16-/32-Bit Microprocessors Programmer's Reference Manual, fifth edition.* New Jersey: Prentice-Hall.

Motorola, Inc. 1985. *MC68020 32-Bit Microprocessor User's Manual, second edition.* New Jersey: Prentice-Hall.

Motorola, Inc. 1982. *MC68000 Educational Computer Board User's Manual, second edition.* Motorola, Inc. MEX68KECB/D2.

Motorola, Inc. *32-Bit Solution Information.* Motorola, Inc. Literature Package M32BITPAK.

Motorola, Inc. *M68000 Family Information.* Motorola, Inc. Literature Package M68KPAK/D.

Triebel, Walter A. and Avtar Singh. 1986. *The 68000 Microprocessor.* New Jersey: Prentice-Hall.

Index

Index

D

E

F

Index

Index

U-V

W